T5-AGA-404

AUTOMATION AND WORK DESIGN

AUTOMATION AND WORK DESIGN

A Study prepared by the International Labour Office

Edited by

Federico BUTERA
Istituto di Ricerca Intervento sui Sistemi Organizzativi (RSO)
Milan
Italy

and

Joseph E. THURMAN
International Labour Office
Geneva
Switzerland

1984

NORTH-HOLLAND
AMSTERDAM • NEW YORK • OXFORD

ISBN: 0 444 87538 7

Publishers:
ELSEVIER SCIENCE PUBLISHERS B.V.
P.O. Box 1991
1000 BZ Amsterdam
The Netherlands

Sole distributors for the U.S.A. and Canada:
ELSEVIER SCIENCE PUBLISHING COMPANY, INC.
52 Vanderbilt Avenue
New York, N.Y. 10017
U.S.A.

The responsibility for opinions expressed in studies and other contributions rests solely with their authors, and publication does not constitute an endorsement by the International Labour Office of the opinions expressed in them.

The designations employed and the presentation of material do not imply the expression of any opinion whatsoever on the part of the International Labour Office concerning the legal status of any country or territory or of its authorities, or concerning the delimitation of frontiers.

Library of Congress Cataloging in Publication Data

Main entry under title:

Automation and work design.

Bibliography: p.
Includes index.
1. Work design. 2. Automation. I. Butera, Federico, 1940- . II. Thurman, Joseph E. III. International Labour Office.
T60.8.A8 1984 658.5'14 84-8169
ISBN 0-444-87538-7 (U.S.)

PRINTED IN THE NETHERLANDS

PREFACE

The modern workplace is confronted by accelerating changes in both work and workers. Technological and economic developments - especially microprocessor-based automation - are rapidly altering the knowledge, skills and responsibilities required of workers. At the same time, a broad range of social developments have changed the nature of the modern workforce. Workers' educational attainments, community and family roles, degrees of participation, attitudes and expectations are in sharp contrast with the realities of work many of them experience.

Can the design of work, aided by advances in automation, help to bridge the gap between the nature of work and the legitimate expectations of workers? This question is of obvious significance to governments, employers and workers throughout the world.

This book, the result of a project of the International Labour Office, addresses itself to the issues posed at national and enterprise level by automation and work design. Its scope is broad, because differences in national situations and institutions, in types of automation, in the particularities of enterprises and in the characteristics and preferences of individuals and categories of workers are all critical to a full understanding of the experiences which are reported and the analyses which are made.

In addition to ILO staff, experts from 12 countries participated in the project. The results are necessarily heterogeneous, but always interesting. Special thanks are due to the outside experts, for it was their knowledge of national and enterprise-level experience which formed the basis of this book. The quality of their efforts justifies a careful reading of each chapter.

The trends, controversies, policies and practices analysed in this book are likely to become even more relevant as automation directly affects an increasing number of workers. The ILO will continue to take an active part in ensuring that developments of interest to governments, employers and workers receive careful analysis and appropriate international attention.

George Spyropoulos,
Chief,
Working Conditions and
Environment Department,
International Labour Office.

CONTENTS

* * *

Figures

* * *

Tables

ABOUT THE AUTHORS

The authors of Part I, and general editors

FEDERICO BUTERA is the founder and director of RSO, the Institute for Action Research on Organisation, Milan, Italy, where he has been involved in various large industry-based projects on joint design or redesign of technology, organisation and the social system, using detailed and participative methods of analysis. Before founding RSO in 1974, he spent 11 years with Olivetti SPA, where he was director of the Centre for Sociological Research and Organisational Studies. He has written several books, as well as numerous articles, and has actively promoted the improvement of quality of working life (QWL) in Italy and internationally through teaching, consultancy and other activities. He was associated from its inception with the project which resulted in this book, and joined the ILO in late 1981 as a short-term official to participate in its completion.

JOSEPH E. THURMAN joined the ILO in 1973 after attending Stanford University and the Stanford University Graduate School of Business. He is presently Senior Research Officer in the Conditions of Work and Welfare Facilities Branch, where his responsibilities include projects on work organisation and job content. Within the framework of the International Programme for the Improvement of Working Conditions and Environment (PIACT), he is frequently involved in practical activities to improve working conditions in developing countries. His working papers, articles and contributions to ILO reports cover working conditions and related topics. Current research includes practical means for improving work content during the introduction of new technologies, work organisation and new office technology, and actual hours of work in developing countries.

Authors and contributors to Part II

Federal Republic of Germany

WOLFGANG H. STAEHLE has been Professor of Business Administration at the Free University of Berlin since 1979. From 1973 to 1979 he held the same position at the University of Darmstadt. He received the degree of Dr. oec. publ. from the University of Munich and Dr. habil from the University of Augsburg. He has published a number of works on job design and automation.

France

YVES CHAIGNEAU was director of the National Agency for the Improvement of Working Conditions (ANACT), Paris, from 1976. He holds a degree in law from the Institut d'Etudes Politiques de Paris. He has published several books in the field of work and work organisation.

FRANÇOISE PIOTET has also been employed at ANACT. She holds a doctorate in sociology and a degree from the Institut d'Etudes Politiques de Paris, where she is a lecturer. She is co-author of a monograph with Yves Chaigneau and author of a number of reports for European institutions.

Hungary

CSABA MAKO is a graduate of the Karl Marx University of Economics in Budapest and holds a Ph.D. in sociology from the Hungarian Academy of Sciences. He heads a group at the Institute of Sociology in Budapest, his main research topic being the study of socio-organisational changes within the socialist enterprise. In 1981-82 he was visiting professor at the Universities of Montreal and Quebec. He has published books and articles - the majority of them together with Lajos Héthy - in Hungary and abroad.

Italy

FEDERICO BUTERA is described above.

Japan

HAJIME SAITO has been director of the Institute for Science of Labour since 1962. He graduated from the Faculty of Medicine of the University of Tokyo and has specialised in work physiology. At present, he is also an associate senior researcher of the National Diet Library, a permanent member of the Board of Directors of the Tokyo Foundation of Promotion of Aged People's Work and Scientific Adviser to the Ministry of Labour. He has published a number of works on the subject, among the most recent being monographs on shift work and on monotonous work.

Netherlands

C.L. EKKERS, who holds a doctorate in psychology, is a staff member of the Netherlands Institute for Preventive Health Care, Leiden. At present he is leader of a Dutch project, sponsored by the Socio-economic Council, aimed at assessing the effects of automation on the quality of work, and is the author of publications in this field.

Norway

MAX ELDEN became the Director of the Institute for Social Research in Industry (IFIM) at the Norwegian Technical University in Trondheim in 1977. He has a master's degree in political science from California State University, San Diego, and a master's degree in public administration and a Ph.D. in politics and organisation from the University of California at Los Angeles (UCLA). He moved to Norway in 1972 to accept a Royal Norwegian Council for Scientific and Technical Research Post-

Doctoral Fellowship at the Work Research Institute in Oslo. He has published in professional journals on the topics of QWL, industrial democracy, participative design and action research.

VIDAR HAVN has a master's degree in engineering from the Norwegian Technical Institute (NTH). He is a research assistant at IFIM and a teaching assistant at NTH while completing an advanced graduate degree in sociology, with a thesis concerning work-related consequences of new technology.

MORTEN LEVIN has a master's degree in engineering (operations research) from NTH and an advanced graduate degree in sociology. He has been a full-time researcher at IFIM since 1973 and recently became an assistant professor in work sociology at NTH. His main research interest is worker participation in the design of new technology, with a special focus on the role of trade unions. He also collaborates with unions in participatory research and worker education.

TORE NILSSEN is employed as a research assistant at IFIM while completing his advanced degree in sociology. His thesis assesses the relation between new technology and worker solidarity.

BENTE RASMUSSEN has an advanced degree in social sciences from the University of Amsterdam (1978). She has been a full-time researcher at IFIM where she has been responsible for research projects on the work-related consequences of automation in the chemical process industry, in the engineering industry and, more recently, in the area of women's working situation in offices.

KNUT VEIUM has an advanced degree in political science from the University of Oslo (1978). After acting as a research assistant at the Work Research Institute in Oslo, he joined IFIM in 1980. He has studied QWL in the public sector and has contributed to an international study of industrial democracy in Europe (IDE). More recently, he has been responsible for a research project on the QWL consequences of new technology in the paper-making industry.

Poland

WŁODZIMIERZ PAŃKÓW, KRZYSZTOF MREŁA and MARIAN J. KOSTECKI are senior research fellows at the Institute of Philosophy and Sociology, Polish Academy of Sciences. They obtained their masters' degrees in sociology from Warsaw University and their Ph.Ds from the Polish Academy of Sciences. Their research interests include continuity and change of organisations within the social structure, organisational control, and power and organisation. Pańków, Mreła and Kostecki are authors of numerous articles and of several books.

Union of Soviet Socialist Republics

ARKADI I. PRIGOZHIN is head of section of the Institute for Systems Studies, Moscow. He holds the degree of Candidate of Science and is a sociologist engaged in applied research and consultancy. He has numerous scientific publications in the sphere of organisational behaviour, processes of social change and socio-psychological aspects of management.

ALEXANDER E. LOUZINE is an industrial engineer and Candidate of Science (Moscow State University). He carried out consultancy and training in work organisation in the USSR before joining the Conditions of Work and Life Branch of the ILO in 1978, and has published a number of papers in this field. His current activities include action study training to improve QWL in developing countries.

United Kingdom

DAVID W. BIRCHALL is currently Director of Graduate Studies at the Management College, Henley. He was previously Senior Research Fellow in the Henley Work Research Group and has been involved in action research in the United Kingdom mass-production industry since 1974. He holds a Ph.D., and is author of a number of publications in the field of work organisation and technological change.

ANNETTE DAVIES has, for several years, been undertaking research at the Management College, Henley, into personnel policies and technological change in the brewing industry in the United Kingdom.

DON HELSBY is a senior industrial engineer at Cadbury Typhoo Ltd., where he is responsible for implementing technological change. He has been involved in collaborative projects with the Henley Work Research Group, where the focus has been on changes in work organisation to improve job satisfaction and performance.

CHRISTIAN SCHUMACKER was, for some time, responsible for advising on work structuring at British Steel. He then became Sear Fellow at the London School of Economics. More recently, he has been involved as a consultant in the field of work structuring in many large organisations and runs his own consultancy.

United States

JOEL FADEM has been Senior Research Fellow, Center for Quality of Working Life, Institute of Industrial Relations, UCLA since 1975. He holds post-graduate degrees in administrative sciences and industrial relations from Yale and Oxford Universities respectively, and was a Fulbright Scholar in the Economics Department at Melbourne University. He has served as a consultant to industrial organisations, trade unions and government agencies in the United States, United Kingdom and Canada, and has published several articles on QWL and the effects of technological change on labour and management.

PAUL GUSTAVSON is Manager of Organisational Behaviour and Employee Relations, Zilog Division of Exxon Computer Systems. He holds an M.A. degree in organisational behaviour from Brigham Young University. He has been invited to present his work on job and organisational design at several recent conferences.

JAMES C. TAYLOR is President, Socio-technical Design Consultants, Inc., and Adjunct Research Fellow at the Center for Quality of Working Life, Institute of Industrial Relations, UCLA. He holds a Ph.D. in organisational psychology from the University of Michigan. He is author of several books, and is currently carrying out research into reinvention in office automation and into participative organisational research.

Yugoslavia

VALENTIN JEŽ is Senior Research Fellow at the Centre for Research on Self-Management, Association of Slovene Trade Unions, and Lecturer on Organisational Theories at the High School for Social Work, University of Ljubljana. He studies industrial psychology and sociology of organisations at the University of Ljubljana. He is author of numerous articles, primarily concerning technology and its social consequences, organisation and various aspects of self-management.

PART 1

COMPARATIVE ANALYSIS

by

Federico Butera and Joseph E. Thurman

CHAPTER I

AN APPROACH TO AUTOMATION AND WORK DESIGN

1. Introduction

Twenty years ago, automation was a subject of major international concern. Large productivity gains from various techniques of process control, materials handling and data processing - all of which tended to be called automation - led to fears of mass unemployment, or at least enforced leisure. Research and development programmes were instituted, some with the purpose of advancing the technical applications of automation and others which concentrated on possible social effects.

The economic and engineering advantages of automation have continued to spur its increasing use. However, empirical findings had shown by the mid-sixties that the worst fears of unemployment were unfounded.[1] The level of academic and popular concern fell sharply, and more subtle findings that automation affected the quality rather than the quantity of work received little attention. The impact on the distribution of skill requirements, organisational and managerial structures and practices, and social and psychological demands of work was largely ignored until the acceleration of automation due to the advent of the microprocessor.

Perhaps because of the earlier research it was quickly realised that microprocessor-based automation was likely to have profound effects on the nature or content of work. Thus, international tripartite interest led to an ILO project to examine the impact of automation on working life.[2] Given the complexity of both the technical and social factors examined, the ILO project was designed with a broad geographical base but with specific and intensive theoretical and practical focuses. Essentially restricted to industrial work,[3] it covered the characteristics of automated technologies and their implications for work organisation at shop-floor level and for the quality of working life (QWL) of the workers involved. An attempt was made to identify strategies and practical methods for the design of automated technologies and the concurrent design of tasks, jobs, roles and organisation in appropriate ways. This volume reports the results of the project.

The project began in 1980 with the preparation of a statement of its objectives, technical model and main concepts, together with practical information about procedures and outputs. This "research design"[4] proved to be very useful, not least because it was possible at a later stage to use it as a basis for discussion. Several factors which contributed to the design tended to shape its structure and content: the desire to have a wide geographic distribution of both national information and case studies; the need to provide theoretical guidance without artificially restricting the possible approaches; and the desire to emphasise practical experience (through case studies) as well as theories and national situations.

The second stage of the project involved associating experts from 12 countries (Federal Republic of Germany, France, Hungary, Italy, Japan, Netherlands, Norway, Poland, Union of Soviet Socialist Republics, United Kingdom, United States and Yugoslavia) who were in a position, through previous practical experience, to describe national situations and/or case studies in a way which met the criteria set out in the research design. The contributions of these experts went through several stages of revision. Most of the experts were able to come to Geneva in May 1981 to discuss the research design and the first drafts of their contributions.

The national backgrounds and case studies have been placed in the national chapters of Part II. Information is provided on theoretical developments, national programmes, positive and negative experience at shop-floor level, and much more - all of which take place in a variety of economic, social and technological circumstances influenced by differing managerial practices, industrial relations systems, historical contexts, consultant interventions and ultimately by the different individual workers involved. Each of the national chapters necessarily contains information which will be of varying interest to public authorities, employers, managers, trade unionists, the academic community and others who may be concerned with policies and practices concerning automation and work design.

While there was no easy way to squeeze this wealth of information into a common mould, the third stage of the project resulted in the comparative and concluding material of Chapters 1-5. These five chapters constitute Part I of this volume.

2. Technical model and hypotheses

Much of the debate about automation and work is deeply embedded in what is called "technological determinism". This implies two assumptions -

(1) Technical developments have an intrinsic logic based upon technical and economic requirements which determines their speed and direction of growth. Social considerations like the occupational structure, the social system within a factory, the needs and behaviour of the "automated" workers are not and cannot be taken into account in the design phase of new technologies, at least not in a way as binding as other economic and technical variables. Both technocrats and social reformers often share the same assumption that technology is too serious a business to be oriented by social parameters.

(2) A given set of machines, tools, transport systems, information systems, technical procedures and knowledge (a given technology) allows basically only one optimum pattern of organisation, social system functioning, working conditions, etc. Any significant

difference in organisation and job design could result in economic ineffectiveness or in internal incoherence.

The project which led to this volume started from a different set of assumptions concerning technology, which take into account the more advanced research in this field.[5] These are presented in summary form below.

Technology design has always encompassed definite models of people and society. This will also be true of future designs with different models and values. In the past this happened not as a result of explicit moral, cultural or ideological preferences of the technology designers, but through complex social and economic processes possessing strong differences according to various technological levels and stages of implementation. For the future, putting "new models" and "new values" into technology means a complex intervention in socio-technical systems whose properties and dynamics have to be subjected to both scrutiny and experiment. Could some properties of technical systems be redesigned using social considerations?

The most famous model for interpreting the interdependence between the technical and social systems in a concrete situation is the socio-technical approach developed by the Tavistock Institute.[6] For the Tavistock researchers, technical systems throughout the history of industrialism were designed in close relation to social systems. However, the traditional social parameters of such joint designs are no longer acceptable to workers and to society. In addition to this, in the last 40 years the tendency to optimise the technical system alone has increased and the already poor equilibrium between the technical and social systems has worsened.

For the Tavistock school of thought, improving one of the two systems (social or technical) implies a joint reconsideration of the other. In order to make a significant improvement in QWL, technical systems must be adapted to social parameters, taking advantage of the different opportunities for technological choices within each industry.

Each technology allows a larger or smaller set of alternative designs of organisation and jobs. Recent studies of the technological constraints on organisational and job design are very important in that they differentiate between actual technological imperatives and the other organisational design characteristics which are derived from mere social choices, even though the latter have often been justified as technological consequences. The boundaries of organisational design may be fixed by technology and other variables, but concrete alternatives can and must be chosen.

Davis and Taylor[7] clarify how technological alternatives affect social systems. If the latter are taken into consideration, choices can be made in terms of organisational and job design as well as in terms of more appropriate design of technology. Otherwise, technology design brings with it an embedded and unperceived set of social choices.

We assume in fact that within the same "basic technology", an investigation will meet with differences in -

- "operational technologies";
- task structures;
- individual jobs and formal organisation structures;
- patterns of individual roles and co-operation (different social sub-systems); and
- work environments.

These will have different consequences on the various dimensions of QWL. These concepts are explained in greater detail below.

3. Automation

There are many definitions of automation: each of them stresses one aspect or another of this complex phenomenon. A definition may emphasise, for example, the degree to which an operation is performed by a technical device; automation as a step towards a cybernetic system; the use of feedback systems; or the substitution of operations performed by machine (electronic or mechanical) for human tasks.[8] Some definitions suggest that automation is a gradual process, while others stress a discontinuity between automation and mechanisation.

Two elements of these definitions should be emphasised: significant passage of conversion tasks into machine operations; and relevant use of machine-managed feedback systems. These essential elements of automation refer to replacement of human work by machines which, in turn, can profoundly affect the remaining human work.

Concerning the replacement of conversion tasks, Wild writes that -

> It seems reasonable to view both mechanisation and automation as trends rather than states, in both cases the trends being associated with the replacement of human activities by the activities of inanimate objects. Mechanisation, for our purposes, is therefore seen as an aspect of, or component part of, automation, mechanisation being concerned with activities, whilst automation of such activities implies the use of control procedures which are also largely independent of human involvement ...[9]

In other words, automation subsumes mechanisation. This definition evokes the problems of the elimination of skilled work (Touraine's phrase A)[10] and of designing the remaining work in a way which has some meaning for the workers and respects their needs.

Concerning the use of machine-managed feedback systems, Crossman[11] defines automation as "replacement of human information processes by mechanical ones" (where mechanical is used in a sense including electro-mechanical, electronic and so on). This definition evokes the problems of change in the control of work and co-operation patterns, and of the need to redesign it in such a way that there is a meaningful interconnection between each worker, other workers and production processes.

4. Quality of working life

QWL has come to be popular as a general expression for various job-related factors in the overall quality of life. The term encompasses parameters of jobs, subjective and objective reactions of workers, procedures for improvement and other areas depending on the point of view of the person employing it. It is related to such terms as humanisation of work, industrial democracy, conditions of work, scientific work organisation, etc.

Walton has proposed the following eight categories as a framework for analysis of QWL:[12]

- adequate and fair compensation;
- safe and healthy working conditions;
- immediate opportunity to use and develop human capacities;
- opportunity for continued growth and security;
- social integration in the work organisation (freedom from prejudice, egalitarianism, mobility, supportive primary groups, community, interpersonal openness);
- constitutionalism in the work organisation;
- work and the total life space;
- the social relevance of work life.

The following set of "psychological job demands" proposed in Norway[13] is narrower:

- variation and meaning on the job;
- continuous learning on the job;
- participation in decision-making;
- mutual help and support from fellow workers;

- meaningful relationship between job and social life outside;

- a desirable future in the job - not only through promotion.

Certain dimensions of QWL are likely to be particularly useful in analysing the impact of automation. One such concept is that of the control exerted by workers in the conduct, co-ordination, maintenance and design of production processes. This seems to be an aspect of the value and meaning of work that automation might reduce or destroy. Lack of complexity, variety and mental challenge are, in many cases, a result of lost positive influence of the workers in production processes.

Another concept which is extremely important but which is often ignored in QWL cases is that of the intensity or stressful nature either of the work or of the change process. For some workers (or, more accurately, some worker-job combinations), more variety, complexity and mental challenge could be harmful rather than helpful. It is therefore necessary to examine the possibility that "more is not always better", even for such attractive-sounding job dimensions as mental challenge and social interaction. Workers can be overloaded on these dimensions as well as deprived.

As far as dimensions of QWL which go beyond an objective description of the work itself are concerned, the integrity of social roles seems fundamental. A major problem for workers in some automated systems is that they may spend so much of their working lives merely waiting or watching that they cease to have any sense of being a worker with a useful social function. A basic dimension of QWL seems to be, in this case, the workers' need to preserve their role of "producer" and their growth in terms of technical knowledge and skills. In other words, there is a fear that automation leads to alienation.

In order to help clarify the sometimes vague or subjective formulations which have been used to describe the content, nature or quality of work, the concept of task should be emphasised. This allows concrete descriptions of those activities which are carried out by machines or automated equipment and of those which are carried out by workers. The set of tasks which makes up each individual worker's job can then be analysed to show the skills and qualifications required, the level of responsibility and autonomy, the physical and psychological demands, etc. Additional information comes from considering the roles or functions of workers and their patterns of co-operation, together with organisational and occupational structures.

5. Technology and organisation

We may define technology in Woodward's terms as "the collection of plant, machines, tools and recipes available at a given time for the execution of the production tasks and the rationale underlying their utilisation";[14] or with Meissner, "we would say that the technology of the workplace consists of pieces of steel, stone, wood in certain shapes. But this is not what is meant ... [that] ... These material things ... are the product of designs, the manifestations of ideas of those who planned a process and the means of facilitating it".[15]

Hickson and colleagues[16] distinguish three correlated dimensions of technology (operations technology, materials technology and knowledge technology), where the first is central because of its role in controlling the workflow.

Whatever definition is given (or implied) for technology, it is very important to make clear the rationale underlying the technology examined: the goals, reference problems and measures which were taken into account by the person using the definition.

Organisation could be defined as "the structure of main relations formally identified and codified within a firm ...".[17] Others identify the content and character of organisational structure as "span of control, ratio supervisors/staff, formalisation of communications, specialisation, etc."[18] or dimensions of organisation like "specialisation, standardisation, formalisation, centralisation, configuration".[19]

Formal organisation is only one layer of the organisation: the perceived organisation, the informal organisation and what may be called the real organisation are other coexisting layers.[20] They have a basic role when diagnosing tensions, conflicts, development trends and redesign criteria.

A basic aspect of formal organisation is the job: the sum of elementary tasks of which an individual is in charge. Dimensions such as prescribed actions, rules, duties, control systems, authority structures, interaction systems, etc., are intrinsic to the job concept: the job is an elementary block of formal organisation-building. One could say also that the job is the boundary within which a worker experiences physical or psychological fatigue or stress, autonomy or its absence, mental or professional challenge or their absence, control over the work environment or its absence. For the purpose of this project, however, job boundaries are not the most basic element for exploring the problems of QWL.

Technology and formal organisation (or organisational structure) both seem to have the same function: to define a situation in which a product can be produced and in which work is structured and controlled. Technology and organisation have the joint function of defining quantity, quality and co-ordination of the work to be done by workers in a particular setting. If we are to study the possible future of human work, we need to consider the interdependence of technology and organisation in coping with this joint function.

An analytical tool for examining in empirical terms this "joint teleology" of technology and organisation is the notion of task as an event to which a person or machine operation is purposefully devoted (human tasks, machine cycles). Herbst illustrates how human tasks and machine cycles in industry are two faces of the same coin: it is the whole coin which defines what people have to do and how they live at a workplace. He writes -

> An event can be analysed with respect to its role within the technological frame of reference and with respect to its role within the social behaviour frame of reference and, by implication, every technical event has a corresponding behavioural representation, and every behavioural event has a corresponding representation within the technological frame of reference ... If operations, independently of who carries them out, are looked at in terms of their sequential and mutual dependence relationships, we arrive at a representation of the production task structure. If activities are analysed with respect to the sequence of an inter-relation of material changes, we arrive at a representation of the material production processes.[21]

The work relationship structure is a substantive concept which can help to identify the content of human tasks in production processes (in terms of physical and psychological load, skill content, importance, etc.) and the relationships of interdependence among tasks. It is largely designed by activities usually classified as technological design (worker/machine allocation, layout, transport, instrumentation, physical communication control, feedback systems, etc.). Its design takes place before the design of jobs, system of job allocation, organisational unit boundary definitions, co-ordination systems or interaction patterns.

The concept of work relationship structure clarifies the reason why one basic technology (material technology and knowledge technology) does not determine one organisational system. Some good "social choices" concerning the structuring of social organisation are made too late, when some particular work relationship structure has already been designed. But if a specific work relationship structure makes certain job and organisation designs economically and socially unfeasible, it very rarely determines just one solution: organisational choice, within larger or smaller systems of constraints, is always possible.

Some of the main elements of work organisation which help to elucidate the relationship between automation and QWL and the different strategies for improving the latter are described below.

5.1 Human tasks

These remain after automation. Particularly important is the content of such tasks. The technical content gives us important information. If feeding or transport tasks are largely present, we perhaps have "transitional tasks" characteristic of partial automation; if human control and regulation tasks are prevalent, automation is a support to the decision-making of the operator; if both data input tasks and control tasks are present, we have an interactive system between worker and machine; if only monitoring tasks are present, the human role is occasional intervention. Thus tasks identify the pattern of work which was designed. The relationships or interdependence among tasks are also important because they clarify the basic unit of organisation (individual jobs or groups).

5.2 Jobs

Before studying the influence of automation on jobs, it is necessary to note whether the tasks can meaningfully be grouped into individual jobs. Often this is not the case. When manipulative tasks disappear and detection, transmission, comparison or regulation tasks are jointly performed by several people, we tend to speak of a "set of jobs", a "group of performed activities", etc.

5.3 Roles and co-operation patterns

Two related concepts help to capture the nature of the variable "work": the concept of role (what an individual "does in his relations with others seen in the context of its functional significance for the social system")[22] and "patterns of co-operation" ("technical co-ordination among people", "quality of a productive system where functions are interconnected").[23] In the case of advanced automation, the notion of "individual jobs" becomes useless both for describing the organisation and for diagnosing QWL. Instead, there are several workers and the work is a "group of performed activities". Perhaps for this reason the majority of studies on the social impact of automation are about "automation and control systems", "automation and supervision" or "automation and the structure and behaviour of work groups".

Each worker has different roles with respect to the different aspects of the socio-technical system. Meissner identifies three basic "role expectations" stemming from technology: behaviour permitted by technology, behaviour required by technology and indifferent behaviour.[24] Davis classifies the main roles concerning a technology as energy supplier, guider of tools, controller and regulator of a working situation or of a system, and diagnoser or adjuster of difficulties.[25] Susman identifies three types of roles: transformation roles of physical or informative input into required output; co-ordination roles; and maintenance and innovation roles.[26] Butera and D'Andrea distinguish extrinsic co-operation (as imposed from above) from intrinsic or self-regulated co-operation (where workers control and implement the rules of the co-operation). Concerning the latter, he identifies four aspects found in empirical research: operational co-operation; co-operation in control and regulation of productive and social processes; co-operation in task and role allocation; and co-operation in system maintenance and innovation.[27]

Roles and co-operation patterns are not merely the result of techno-organisational design. They are determined also by the larger social system and, in some cases, by people themselves. They are analytical tools for interpreting and designing the mutual relationships between the technical and social systems.

6. The national context

Societal variables in each national context may explain differences in similar technological cases within different countries or similarities in different technological cases within the same country. Variables to be kept in mind include -

- level of industrialisation of the country;
- technological level of the country;
- level of employment;
- structure of the labour market;
- migration;
- the industrial relations system;
- the administrative system of work regulation; and
- the educational system.

In interpreting automation and work design at shop-floor level, it is necessary to examine the national situation regarding such key areas as -

- major developments concerning automation in the country; the main sources (industry, universities, research centres, etc.) and the chief areas of application;
- the main approaches to design of work organisation during automation and the principal institutions involved;
- the attitudes and actions of social partners (government, employers, trade unions, etc.); and
- policies, programmes or trends concerning humanisation of work, humanisation of technology, etc.

7. Work design - a positive approach

The impact of technology on the content of jobs has long been considered inevitable. The design of technological processes and related machinery should be, according to this view, given over to engineers working on the basis of scientific principles. Jobs emerge rigorously from the tasks required to operate and maintain previously designed equipment. Corrective action can be taken if severe problems such as unsafe conditions appear, but such action is also the domain of the engineer.

Such an approach to technology and work organisation gives little hope for the creation of jobs with desirable content. On the other hand, there is some evidence that another approach to the design of technology is possible, one which takes at least partly into consideration the social as well as the technical side of production. This latter view, if valid, means that it is useful to explore the concrete potential for various technologies and production systems to operate efficiently and competitively while offering a desirable working environment and jobs with positive elements.

The technical model and hypotheses adopted in the preparation of this book, together with the approach to the key variables involved, imply that there is scope for positive intervention in the design process. This has guided the analysis of the material gathered from various countries, and especially in the preparation of conclusions. However, this view of work design was not expected to lead to standardised solutions applicable everywhere. Variables such as the age, education, sex and skill distributions of the working population; economic and technological changes in the industrial sector; the economic situation of different groups of workers; the operation of the industrial relations system; the changing attitudes of individuals to work and so on - all form the context in which automation and work design take place, condition the results which are obtained and are sometimes affected in their turn. At enterprise level, the local demographic, economic, social and cultural environment can be even more important, though it may be quite different in some respects from the aggregate national environment. Most important, the effects which occur inside the enterprises can be appropriate or inappropriate in terms of individuals, groups, the enterprise as a whole, the locality or the entire society. Work design thus calls for planning and various forms of effective workers' participation so that these factors can be taken into account in each instance.

Notes

[1] These fears have been revived recently. See, for example, J. Rada: The impact of micro-electronics: A tentative appraisal of information technology (ILO, Geneva, 1980).

[2] This volume results from a project which responds to the following ILO programme objective: to secure widespread adoption of technologies and forms of work organisation which lead to greater challenge, learning and self-expression on the job; increased opportunities for desirable occupational futures; greater possibilities for communication and co-operation among workers; greater worker control over the pace and timing of tasks; increased worker influence in the planning and organisation of work; and reduced fatigue and stress. Widespread tripartite agreement with the objectives of this project was recently expressed in a resolution concerning the improvement of working conditions and the working environment in Europe adopted unanimously by the Third European Regional Conference of the International Labour Organisation, Geneva, 1979. The Conference underlined "the key role of work organisation in all areas of improving working

conditions and the working environment and its additional importance in promoting the worker's well-being and fulfilment and the efficiency of production systems". The resolution also refers to "the possible negative effects of automation and computerisation as well as their positive potential".

[3] A complementary ILO project concentrates on office work.

[4] ILO: Research design for an ILO international comparative study on job design and automation (Geneva, doc. CONDI/T/1980/6, 1980; mimeographed).

[5] L.E. Davis and J.C. Taylor: "Technology, organisation and job structure", in R. Dubin (ed.): Handbook of work, organisation and society (Chicago, Rand McNally, 1976).

[6] F. Emery and E. Trist: "The characteristics of socio-technical systems", in F. Emery (ed.): Systems thinking (London, Penguin, 1969).

[7] Davis and Taylor, op. cit.

[8] L. Gallino: Dizionario di Sociologia (Turin, Utet, 1979), pp. 55-56.

[9] R. Wild: Work organisation: A study of manual work and mass production (New York, John Wiley, 1975), pp. 31-32.

[10] A. Touraine: Sociologie de l'action (Paris, Editions du Seuil, 1965).

[11] E.R.F.W. Crossman: "European experience with the changing nature of jobs due to automation", in OECD: The requirements of automated jobs, North American Joint Conference, Paris, Dec. 1964 (1965; mimeographed).

[12] Richard E. Walton: "Criteria for quality of working life", in Louis E. Davis and Albert Cherns (eds.): The quality of working life, Vol. I (New York, Free Press, 1975), pp. 91-104.

[13] ILO: New forms of work organisation, Vol. I (Geneva, 1979), p. 23.

[14] J. Woodward: Industrial organisation: Behaviour and control (London, Oxford University Press, 1979), p. 73.

[15] M. Meissner: Technology and the worker: Technical demands and social processes in industry (San Francisco, Chandler Publishing Co., 1969), p. 14.

[16] D.J. Hickson et al.: "Operations technology and organisation structure: A reappraisal", in Administrative Science Quarterly, Vol. 14, No. 4, pp. 378-397.

[17] Gallino, op. cit., p. 489.

[18] Woodward, op. cit., pp. 51-60.

[19] D. Pugh et al.: "Dimensions of organisation structure", in Administrative Science Quarterly, Vol. 13, No. 1, pp. 65-105.

[20] F. Butera: La divisione del lavoro in fabbrica (Venice, Marsilio, 1977), Ch. 7.

[21] P. Herbst: Socio-technical system design (London, Tavistock, 1974), pp. 114-115.

[22] T. Parsons: The social system (New York, Free Press, 1951), p. 25.

[23] T. Caplow: Sociology of work (New York, McGraw Hill, 1955), pp. 27-28.

[24] Meissner, op. cit., p. 25.

[25] Davis and Taylor, op. cit.

[26] G. Susman: "Task and technological prerequisite for delegations of decision-making to work groups", in Davis and Cherns: The quality of working life, op. cit.

[27] F. Butera and R. D'Andrea: Lavoro umano e prodotto tecnico: Una ricerca sulle acciaierie di Terni (Turin, Giulio Einaudi, 1979), pp. 275-276.

CHAPTER II

AUTOMATION, WORK DESIGN AND SOCIAL POLICY: A REVIEW OF NATIONAL CONTEXTS AND INSTITUTIONAL ACTION

1. Introduction

Concrete applications of automation in workplaces have their roots in wider technological, social and economic developments and decisions. In some cases, the underlying forces seem quite abstract: scientific progress, cultural and social change, economic growth, supply and demand on the labour market. In other cases, specific programmes, practices, institutions, disputes or agreements are more clearly related to particular intentions, problems or results. Based on the national background information gathered on 12 countries and contained in Part II of this volume, this chapter is intended to identify those patterns and trends which suggest the nature of emergent social policy and institutional action as they specifically relate to both automation and QWL.

The 12 countries examined are the Federal Republic of Germany, France, Hungary, Italy, Japan, Netherlands, Norway, Poland, Union of Soviet Socialist Republics, United Kingdom, United States of America and Yugoslavia.

No attempt has been made to supplement the information provided in the individual chapters on each country. Where information from a national chapter is directly referred to below, the name of the country (and, if appropriate, section number) is indicated in the text or in parenthesis.[1] In a few cases, the name of the author of a national chapter is given rather than the name of the country concerned in order to clearly indicate that a personal opinion or theory is being discussed. However, this practice has been kept to a minimum: in general a reference to a country merely means that illustrative information is available in the chapter on that country, rather than any claim that the country itself exemplifies or supports the statement made.

In considering the material presented below, it is useful to keep in mind that it is based on the limited material available at a specific time. Many situations, policies and institutions are very rapidly evolving, for reasons which will become clearer as the analysis proceeds. Comparative judgements in such dynamic conditions are liable to become obsolete very rapidly, even if they were valid when made. Such judgements are therefore avoided. Even the identification of patterns and trends, limited to connections and common points rather than comparisons among countries, is necessarily an inductive and tentative process. The countries covered have very different levels of economic and technological development, as well as differing political and social orientations, and they are in themselves complex, heterogeneous and dynamic.

This chapter is divided into three sections. First is a description of the extent to which automation has been introduced in various countries and situations, accompanied by a brief indication of the policies and motivations behind increasing automation. A second section covers in slightly greater detail the issues concerning the impact on QWL which have accompanied or been raised by automation. The third section describes the policies, programmes and institutions which are concerned with these issues.

2. Automation: motives and scope of introduction

2.1 Motivations for the introduction of automation

National policies towards automation are characterised by an energetic ambivalence. Increased productivity, economic growth and international competitiveness seem to require promotion of the most advanced technologies. Yet the rapid introduction of automation is feared because of its social effects, particularly those concerning employment. In spite of the two-edged effects of automation, it is generally agreed that to discourage its adoption or even to neglect it is to risk being left out of the race towards the future. The result tends to be large, technology-oriented programmes for the promotion of automation and a more or less developed set of institutional reactions to any problems which have emerged.

Economic motivation is the most common reason cited for government involvement in the promotion of the development of automation. This is expressed in different ways: better international competitiveness (France, 1.1; United Kingdom, 1.9); acceleration of the rate of industrialisation or economic growth (nearly all countries); reduction of international economic and technological dependence (France 1.3.2(4); Poland, 1.6). This same economic motivation predominates in decisions of enterprises and is expressed frequently in the promotion of research on the development and application of new techniques (United Kingdom, 1.4; Soviet Union, 1.3; France 1.4.1) or the training of technical personnel (United Kingdom, 1.6.3; France 1.4.1).

It is of interest, however, that purely economic considerations do not fully explain the direction of government policy and action. Motivations related to overcoming possible shortages of skilled workers (United States, 1.8), skilled managers (Soviet Union, 1.2) or even unskilled workers (Soviet Union, 1.2) can direct interest towards a specific type of automation. Industries with strategic implications are often favoured (United States, 1.8). At the same time, governments may be wary of promoting developments which threaten employment (France, 1.4.1); United Kingdom, 1.6.2) or other social goals.

The overall promotion of automation may be facilitated by general government or popular views that such changes are rational, desirable and inevitable or hindered by a more suspicious view of technological change as potentially dangerous. In those countries where large elements of the population are convinced that the unregulated advance of technology may easily result in more harm than good, there are calls for reflection before action (France 1.2.2, 1.4.1) or for an effort to optimise both social and purely technical effects (Federal Republic of Germany, 2.3).

At the level of industries and enterprises, it is the nature of the basic technology which seems to have the most influence on the use of automated technologies. Integrated forms of automation are found mainly in continuous and semi-continuous process industries (chemicals, petroleum, paper, iron and steel, etc.) where they are needed to ensure that the process keeps flowing. The economics of enterprises in these industries are characterised by high capital-labour ratios, pronounced economics of scale and, consequently, large opportunity costs of any interruptions in production. In other industries, automation more often consists of limited solutions in problem areas (automatic welding and painting in motor vehicle manufacture, machinery of parts requiring extremely close tolerances), where human control is difficult, dangerous, unhealthy or inadequate. Activities less directly connected with production itself, such as materials handling and packaging, have in general only recently benefited from mechanised aids, although packaging of continuous process products (some foods, drinks, chemicals, etc.) has been affected by automation for some time. The penetration of various forms of automation into various industries and parts of enterprises is discussed in particular detail concerning two countries (France, 1.3.2; United States, passim), while examples and statistics can be found in all the country reports.

2.2 The scope of introduction of various forms of automation

The descriptions of automation in the sections on national background do not permit a complete statistical statement of the extent of automation in various countries. This would be far beyond the scope of the present project. There are data which suggest that automation in various forms is quite widespread and that its application is accelerating, as has been almost universally reported elsewhere. What is more important in the context of work design, which is principally focused on the shop-floor, is the considerable information which shows that automation is not being adopted in a smooth flow, but is instead extremely heterogeneous in its forms and in the rates at which it is being applied. This idea, that automation is differentiated in many ways and that its concrete applications cannot be logically predicted is essential to an understanding of the diverse ways automation is related to the qualitative aspects of work.

The most common way used to describe the forms of automation is by reference to specific examples which can be easily pictured or which have an accepted definition. This is a particularly useful way of showing the extent to which automation exists in particular industries or types of production process. One classification scheme of this type (France, 1.3.2, table 23) includes computer-assisted design (CAD), numerically controlled (NC) machines, automats and robots, and integrated systems.

Other schemes include distinctions within these categories (e.g. United States, 1.8-1.12) by referring to computer-based process control (see especially Norway, 1.2), etc. An extremely broad form of integration is reported concerning the Soviet Union (1.2) in the form of computer-based management systems (CBMS).

These examples of automation are found to differing extents in different production processes or in different stages of the same process. For the more sophisticated examples, a relatively low level of adoption is currently found. NC machines comprised 2 per cent of the machine tools in the United States in 1978 (United States, 1.8). Similarly, low levels are found for various specific forms of automation in different countries. On the other hand, it would not be appropriate to say there is no automation until one of these specific types is found. Wherever new machines, new inventory and materials handling systems, new monitoring equipment, etc., are introduced, the forms and extent of human control and intervention in work are changing. The cumulative effects of this automation by small, partial, discrete steps (Italy, 3) is reflected in the broad measure of automation found in the French analysis (1.3.2(1), table 24) which, applied to various industries, shows a consistent shift towards increased automation and a preponderance of moderately low levels. In short, many forms of automation are found, but applications of the most advanced types are not very common. A wealth of information is provided on the factors which have promoted or constrained the application of automation. An underlying factor in all the discussions is the development of inexpensive, reliable, versatile microprocessor technology (or for some large integrated systems, computer technology) which provides a hardware basis for modern automated systems. While mechanised, hydraulic and pneumatic systems exist which have the theoretical characteristics of automation, these are rarely rapid enough, complex enough or inexpensive enough to have other than specialised applications.

3. Issues concerning automation and quality of working life

The extent to which a machine or production system is self-operating obviously has implications for the kinds of work which are to be carried out by people. After all, any tasks which are not carried out by machines, computers or other hardware are inevitably carried out by people if they are to be done at all. As production systems become more and more mechanised and automated, the nature of human work is increasingly either whatever is "left over", or whatever is required in terms of operation, maintenance or monitoring of automatic equipment. Thus a new nature of work - and a new QWL for the workers concerned - may be said to have been emerging during the introduction of automation in recent decades.

Much has been learned concerning the impact of automation on the nature of work. Many of the national chapters therefore describe the "history of ideas" which have arisen through experience and research. These ideas are rarely absolute conclusions. Instead, they are issues debated among experts and more recently among policy-makers at national level. The issues are often critical, concerning fundamental questions and choices in a broad spectrum of social policy areas. Before turning to the issues, however, it is necessary to report a number of clarifications made in the national chapters about the nature of automation itself.

3.1 What is automation?

Confronted by the complexity of technological change, the authors of the national chapters use various definitions and classifications to help explain the phenomenon of automation. It is implicit in these discussions that automation is not just another form of technological change. The word automation (autom = self [oper]ation) refers not to the sophistication or recency of a technological process, but to the way it is controlled.

The general question of control is discussed in various ways in the national chapters. The division of labour between worker and machine is used (Federal Republic of Germany, 1) to show stages of the mechanisation and automation process, in which first the direct use of tools and secondly the control, monitoring and maintenance of machines disappear from human work. Another way of seeing this is in terms of management (Soviet Union, 1.1). Production systems are managed through the use of information, communication, regulation, etc., and automation means the carrying out of these functions by machines rather than people. The most advanced states of automation are discussed in terms of systems with emphasis on feedback loops, the position of human operators in such loops and the scope of human versus machine control (especially the Netherlands case study).

The reason that so much care is taken with the abstract concept of control is that a common element is sought which can be used to understand the very different concrete applications of automation in the real world. After all, the mechanical, electrical and information technologies which underlie modern automation (France 1.3.2) are not easy to define and describe separately. Moreover, the real world of the shop-floor often combines different stages of production with very different degrees of automation (e.g. the United Kingdom case study on biscuit manufacturing) or even parallel use of both mechanical and automatic equipment to accomplish the same work, as when NC machines operate next to manually controlled ones (Poland, case-study; United States 1.8.1). While some of the country studies describe very careful systems to measure or classify the degree of automation (Netherlands, case study), these classifications are not used by statistical reporting systems.

3.2 Automation and skills

The relationship of automation to skills, or, more precisely, the likely impact of continuing automation on the distribution of skills required of the labour force, is the most frequently discussed issue concerning automation and QWL. The significance of this question at national level is very pronounced.

The debate about skills takes place in changing demographic contexts. Several of the monographs stress the educational structure of the economically active population concluding, for example, that many workers are already over-educated for their jobs (Poland, 1.5) or that divergences between the expectations and needs of increasingly educated workers and the requirements of their work are likely to grow (Hungary, 1). In addition to the overall effect of educational systems, other factors are shown to raise the educational level of the working population, such as increased labour force participation among educated women, reduced immigration and a lower labour force participation by older workers (France, 1.2.1). On the other hand, concern is expressed about shortage of skilled workers (United States, 1.8) and managerial staff (Soviet Union, 1.2).

It is not usually assumed that automation will affect all occupations or all groups of workers equally. The "polarisation hypothesis", which claims that "as automation advances it produces a small number of highly trained automation workers juxtaposed with a large number of unskilled (or de-skilled) people responsible for the remaining work, which is simple in nature" (Federal Republic of Germany, 1), represents one kind of concern expressed in several countries, though usually indirectly. Such a broad hypothesis, however, seems to pay little attention either to the differing forms of automation or to the social choices made at enterprise level. Three authors (Elden, Fadem, Staehle) suggest that it is important to express the polarisation hypothesis in a way which takes into account the possibility that de-skilling occurs as a result of management decisions, a possibility which is supported by statements in the case studies (thus Hungary, 5: "The impoverishment of a number of machine operators' jobs was by no means a consequence of technology but rather a result of the ignorance and inexperience of a managerial decision ..."). The possibility of using choices inherent in the technology to upgrade skills can lead to a reversal of the polarisation hypothesis. One author, after noting that the polarisation hypothesis is sometimes confirmed and after analysing the possible social choice process, states: "On the whole, it seems that automation requires an increased number of skilled workers even though it does, in fact, tend to decrease or even eliminate a large number of unskilled jobs. Moreover, in all cases it transforms the trades that workers have learnt" (Chaigneau, France, 1.3.3(2)).

Before returning to the important concept of the transformation of occupations, it should be noted that in the case studies the phenomenon of de-skilling is preponderant (see Chapter 3). At times, however, this de-skilling is intentional, as in the cases of shortages of skills mentioned above.

3.3 Automation and the transformation of work content and occupational patterns

The debate about skills focuses on the question of the developments taking place in skills levels. The overall level of skill is an abstract and composite concept which may not be the most revealing way of conceptualising what is happening to skills. When automation is introduced into a workplace, some skills become obsolete, others require adaptation and still other, new skills are required. These new or adapted skills may well fit into a new career structure or occupational hierarchy. Thus two kinds of transformation take place: in work content and in occupational categories. While it is arguable whether new jobs found in automated work are higher or lower in skill than traditional jobs, it is clear that the skills required are very different.

In an example from the United States, the change between the trade of machinist to the occupation of NC machine operator is analysed in Blauner's (1964) terms as a change from "able workman" to "reliable employee". In the latter case, skill level as reflected by pay was sometimes seen as higher, but it was clear that in terms of traditional trades there were neither the same level of required qualifications and experience nor the usual career steps: "... positions offered no prospects for future job progression and were filled by younger, unskilled workers ..." (United States, 1.8).

From the point of view of the worker - especially a worker who has spent many years gradually learning the skills required by a traditional trade - the transformation of work content and occupations is a very threatening phenomenon. There is no guarantee that the worker will be able to adjust to the new tasks required or that these tasks will offer the same status or satisfaction as the previous ones.

The tasks to be carried out by workers in automated production relate to the underlying technology and to management decisions as well as to automation, but some generalisations can be made. One idea already mentioned is that human work is used to enhance the reliability of the work system rather than to carry out operations. Further understanding of this question comes from the national background to France, which notes that there is a change from heuristic to algorithmic work (1.3.3(1)), i.e. a change from gradually learning more and more about a production process itself and the skills required to carry out each of its steps to learning the branchings and loops of a computer model. Almost all the case-studies dealing with advanced forms of automation touch directly or indirectly on these points as well.

A more obvious point about the transformation of occupations is that some old skills will be eliminated and that new ones must be learned. Even where the newer skills are at a "higher" level they are of a different nature, and a considerable personal investment in developing certain skills, especially traditional trade skills, may be lost to the worker. Thus there is considerable interest in training programmes at both national (United Kingdom, pp. 14-15; France 1.4.1) and enterprise level. At the same time, it is not clear that the workers who are displaced by atuomatic equipment will be the same ones who operate this equipment. This can lead either to the technological unemployment of skilled workers or to the "marginalisation" of such workers into occupations of a transitional and unimportant character (France, 1.3.3(2)).

While the transformation of occupations necessarily implies a loss of one set of skills, it is not always viewed negatively. The Polish case-study reports that though NC machine operators lose their manual skills, "they have a sense of influence over the conduct of the work through the setting of the sophisticated machine, as well as the possibility of introducing corrections to the program", a possibility also mentioned elsewhere.

3.4 Automation, communication and co-operation

Few national chapters mention problems in this area, perhaps because design opportunities often exist which can allow various options such as group work. Isolation of workers is mentioned in the chapter on France (1.3.3(1)) as contributing to overall stress placed on the worker. The Yugoslav case-study notes that a very large number of workers have become isolated in the new steel mill in spite of attempts to correct the situation. The facilitation of group work is a goal in several countries (Soviet Union, 1.5.1) and the general evaluations of the possibility of accomplishing this goal under conditions of automation are usually very positive.

The introduction of automation has often been combined with group work arrangements, especially when entire new plants have been designed. In some cases, such group work has been introduced as part of an eclectic process of innovative job design (United States, 1.12) and in others it is encouraged by public policies and institutions (e.g. Norway, 1.3). This type of work obviously facilitates the possibilities for workers to create a desirable social climate and to share responsibilities and problems. On the other hand, certain forms of automation have been related to social isolation.

The size and structure of work groups is one of the principle differences between the experience of market and centrally planned economies. The work groups called "brigades" (Soviet Union, 1.5.1) are considerably larger and more differentiated than their counterparts in market economy countries.

3.5 Automation, responsibility and supervision

Decentralisation of responsibility is mentioned as an accompaniment of automation in several chapters (e.g. Norway, 5), or at least as a subject of debate (Italy, 3). The emergence of the "reliable employee" (United States, 1.8) leads to the reduction of supervisory roles centred on merely watching over the workers rather than accomplishing a technical function (France, 1.3.3(2)). On the other hand, control functions embodied in computer software can "mould" the actions of workers, making them agents of the computer rather than truly responsible (see the section above on transformation of occupations). The role of the foreman is placed in question and the interface between management and worker becomes more complex during automation, but general statements about the level of responsibility of workers after automation are lacking. It seems clear, however, that the roles of immediate supervisors are strongly affected (Norway, 5; United States, 1.11).

The question of responsibility is of course related to the general question of worker participation in decisions at shop-floor level. Various forms of participation and of its encouragement are found in particular countries. Automation can be a focus for such participation, and it can help or hinder the formation of continuing responsibilities for workers. While there is considerable evidence of the former idea, the national chapters do not provide general statements about the impact of automation on the level of responsibility or discretion of workers. Implicitly, perhaps, discussions of level of skills are meant to reflect responsibility as well as skill, and one could speculate that the discussion of transformation of occupations derives partly from the development of skills which are inherently more responsible (monitoring a process is more responsible than operating a machine; theoretical skills are more responsible than empirically learned skills), at least in terms of traditional job evaluation criteria. None the less, discussions of participation concentrate on participation during the introduction of automation, not on the level of responsibility of workers afterwards.

3.6 Automation and the working environment

Nearly all the national chapters note that automation is accompanied by improvements in the working environment and a reduction in the physical load placed on workers. In many cases this may even be a major motivation in taking the decision about which part of a production process to automate or in which industries automation should be promoted (Federal Republic of Germany, Hungary, Poland, Soviet Union, Yugoslavia). Reductions in noise, vibration, exposure to chemicals and dust, extremes of temperature and humidity, etc., are all mentioned as common results. There may also be a decrease in accident risks (France, 1.3.3(1)). The reduction of physical load on workers and the improvement of other ergonomic factors are particularly often mentioned.

On the other hand, a certain number of problems which relate to mental workload, monotony and stress are noted, particularly in the national chapters on France, Japan and the Soviet Union. While the measurement of stress and its causes presents many difficulties, there would seem to be relatively clear indications of the following problems:

- stress due to a combination of monotonous work which none the less requires vigilant attention (France, 1.3.3(1); Japan, 1.2; Soviet Union, 1.4);

- increased shift and night work with its medical and social consequences (France, 1.3.3(1); Japan, 1.2);

- consequences of the use of video display units (VDUs) (France, 1.3.3(1));

- increased subjective complaints (France, 1.3.3(1)) and psychosomatic illness (Japan, 1.5).

4. Policies, programmes and institutions dealing with the social impact of automation

4.1 National policies and legislation

Several countries have adopted national policies, often accompanied by legislation, directed at the improvement of work. In the Federal Republic of Germany (2.1) there is particular emphasis on scientifically developed principles of a more humanised work design. Thus national policy stresses the development and application of such principles, includes legislation requiring their application and gives works councils, along with government agencies, an important role in ensuring that this is done. In Norway (1.3), national policy, together with legislation for its implementation, plays an equally central role. Scientific research has certainly been very influential in this country, but the thrust of policy and legislation is more towards participative or negotiated solutions within the framework of a broad definition of an appropriate working environment. Centralised policies and legislation are also characteristic of countries with centrally planned economies. In these countries, there is a tendency for the overall national policy towards automation and new technology to include detailed statements of principles concerning the improvement of work content (see especially Soviet Union, 1.5.3).

Other countries are marked by the absence of articulated policies at national level and by the absence of legislation which prescribes either appropriate work content or any required way of promoting improvements. In such countries, it is left to the techniques of managers and engineers, the mechanisms of collective bargaining (or, in the case of Yugoslavia, the mechanism of the worker self-management system) and the possible interventions of public or private institutions, consultants or academics to determine the design of work and the content of jobs.

Where there are national-level policies concerning work content, specific reference is sometimes made to automation. In the Soviet Union, the use of mechanisation and automation is a major component of the national policy on work content improvement (1.4). National policies on automation, conversely, sometimes make mention of improvements in work. This is the case in the Federal Republic of Germany (2.1) and in France (1.4.1).

4.2 Institutions

The chapter on France identifies three main priorities which have served to orient public institutions concerned with job design and automation:

> Briefly, the policy of public authorities has been characterised by three priority approaches, all of which are interlinked and aimed at both supply and demand for computerisation: the first is aimed at expanding the use of computerised and automated systems as a means of improving the performance of France's production capacity; the second

concerns the development of a national computer and automation industry; and the third has as its objective the integral consideration of conditions of work as early as possible (France, 1.4.1).

It should be noted, however, that these priorities did not emerge at the same time and that the programmes and institutions established in response to them are not of the same size and scope. Regarding the first priority, all the countries in our sample have policies and related institutions designed to encourage the effective application of automation and other new technologies. In the case of the Soviet Union, for example, the development and application of new technologies for workplaces is seen as a central concern of the entire national research establishment (1.3). The same approach is taken in the chapter on the United Kingdom (1.4), which notes that approximately half of national research expenditure was financed by government.

The second priority, that of the development of a national computer industry, is also widespread. Nearly all country reports refer to government policies and programmes directed at the development of an independent national industry producing the computer and microprocessor hardware found in modern automation.

If the national budgets for research and development in the field of automation are large, the same is not necessarily true for those institutions whose role is the protection of workers during the automation process. It is difficult to be precise about the relative size of programmes and institutions in these fields. On the one hand, institutions concerned with research and development are only concentrating a portion of their efforts on the automation of workplaces. At the same time, national institutions concerned with improvements in working conditions, occupational safety and health, work content and related subjects normally devote only small proportions of their resources to conditions under automation. An idea of the relative scope of such programmes, however, can be derived from the experience of the Federal Republic of Germany, where the action programme on humanisation of work receives 2 per cent of the federal research budget (2.1).

One of the main roles of public institutions whose programmes relate to the quality of work is the promotion of research at enterprise level. Funding bodies such as the Social Science Research Council in the United Kingdom, the Fund for the Improvement of Working Conditions (FACT) in France and the Action Programme for Humanisation of Work in the Federal Republic of Germany have supported research based on the idea that the findings might be applicable to other establishments. At the same time, many countries have created public or semi-public institutions. The following two descriptions suggest the scope of activities of this type of institution:

> The National Agency for the Improvement of Working Conditions (ANACT), a public institution with a tripartite board of directors which includes representatives of employers' organisations, trade unions and various ministerial departments and experts, is under the supervision of the Ministry of Labour. In addition to the advice provided by its

officials concerning all applications to ... FACT and to the Fund for Training Assistance (FAAFE), ANACT has as its mission: to collect and analyse information concerning all activities for the improvement of working conditions; to contribute to the development and promotion of research, experimental work or concrete projects concerning working conditions; to encourage manufacturers to design ergonomic machines and industrial buildings; to support training activities; and to develop methodologies for action. Within this framework, a number of projects deal with computerisation and automation, and these are among the priority themes of the current programme.

Together with the Ministry of Industry (Information Technology Commission) and the Ministry of Labour, ANACT has been involved in the preparation of an "Automation and working conditions plan" (PICT). ANACT is one of the principle instigators of this plan (France 1.4.1).

With reference to the United Kingdom -

... a paper in 1973 on QWL ... recommended amongst other things the setting up of a body to co-ordinate and work out needs for a programme of development in industrial/commercial settings. From this developed the idea of the Department of Employment Work Research Unit which was set up in December 1974 under a joint tripartite steering committee comprising the Trades Union Congress, the Confederation of British Industry and Department of Employment representatives ... the objectives of the Department of Employment Work Research Unit [are] as follows:

- promotion of activities in the field of job satisfaction and restructuring;

- provision of consultancy and advice to companies;

- writing of reports, papers and notes to disseminate the information on the subject;

- compilation of information on the activities of similar bodies in other countries;

- provision of training courses and appreciation courses for individuals and organisations; and

- development, sponsoring and control of research, particularly action research to provide demonstration projects through university and other research groups (1.8).

Other institutions with relatively central national roles concerning quality of work include the Work Research Institutes and the Institute for Social Research in Industry (IFIM) in Norway. In countries where there is no comparable central body, an important role is often played by a variety of semi-public organisations, which generally combine a consultancy function with research and information activities. This situation may be found, for example, in Italy and the United States.

In countries with centrally planned economies, design institutions are set up on an industry-by-industry basis. These institutions are responsible, among other things, for the preparation of model schemes for work organisation used in a standardised way in plants throughout a country. The industrial institutions are in turn co-ordinated by a central body (see especially Soviet Union, 1.3).

4.3 Employers and managers

The national chapters do not dwell on the role of employers and managers in introducing automation or changes in work organisation. One reason for this may be that in most countries the exclusive role and decision making power of employers and managers in this domain has gone unchallenged. The trade union positions described in the next section can be seen as very recent challenges. Management, of course, has considerable power to introduce change. In some cases, these changes come about through a series of "small" management decisions (Italy, 3). The implications of such decisions only emerge gradually and may not be part of any official position taken at national level. Thus, in France -

> Their [the employers] attitudes are expressed more in relation to the conditions of operation, survival and development of industry and national production capacity, especially in the face of competition, and usually in actions rather than words. They aim to exploit technical progress to improve company performance (1.4.2).

It is extremely difficult to separate the attitudes and positions of managers from those of engineers, programmers and other design specialists. The national chapters reflect a large apparatus for the development and application of technological improvements, and of course many enterprises specialise in such work. Within an individual enterprise, however, engineers are seen as part of management or at least as the agents of management. When a new machine or process is introduced, even if its conception and development took place outside the enterprise itself, the decisions required for its introduction are management decisions in principle, though they may in fact be decisions of engineers or engineer-managers.

Where employers have taken official positions, a certain optimism may be noted. In the Federal Republic of Germany -

> Unlike the trade unions, the employers take a predominantly optimistic view of automation. The relief of the worker from physically demanding and even degrading activities is stressed, and faster economic growth (new products and types of production) and increasing property ownership are usually adduced as the consequences of "technological change" (2.3).

4.4 Trade unions

Trade union views have varied widely from country to country, depending upon the historical antecedents, labour market position, and social and political roles of the trade unions in different countries. Their traditional emphasis has been on topics like wages, employment security and occupational safety and health, and the mechanisms which they have developed are designed to be particularly effective in these fields. Since, as noted above, changes in technology and work organisation have often been considered exclusively managerial prerogatives, trade unions in recent years have been engaged in a process of definition or redefinition of the powers they demand in this field and, more important, of the mechanism whereby they expect to exercise their influence.

National variations in the trade union role concerning automation and work design can be analysed according to three important dimensions. First, there is the dimension of trade union power and influence at both national and enterprise level. Second, there is the degree to which the trade union has developed centralised expertise concerning automation and work design. Third, there is the question of the length of time that either automation or work design issues have been important to the trade union, and consequently the degree to which it has developed its experience concerning various means of work improvement.

4.4.1 Power at national and enterprise level

In several countries, trade unions in recent years have enjoyed considerable political power at national level. Perhaps the clearest example is the case of Norway, in which -

> The close connection between the labour movement and the Labour Party has been of major significance in the period since 1935, when the Norwegian Labour Party, with only small interruptions has controlled the Government. This has given the Norwegian trade union movement a unique chance to realise its interests through the apparatus of the State (1.3).

The ability of a trade union to influence work through centralised national means leads naturally to the idea of a legislative approach. The 1977 Norwegian legislation on the working environment is an example of a national-level approach to problems that in some other countries would be treated on a more decentralised basis: "A significant characteristic of this law is the attempt to regulate socio-psychological aspects of working life" (1.3). In the Federal Republic of Germany as well, it is noted that legislation requires that "accepted work science principles about humanised job design must be observed" (2.1).

In other countries, trade union power tends to be concentrated at enterprise level: "The most characteristic feature of industrial relations In Japan consists of intra-enterprise organisation of trade unions, which is linked with the lifetime employment system and seniority-based wages"

(Japan, 1.3). This has resulted in the concentration of automation in new plants or new jobs and thus it has "applied mainly to young workers, female workers or unorganised labour, or ... gradually penetrated ... such mass production systems as conveyor flow systems" (1.3).

In the United Kingdom the post-war period has been marked by "a growth of plant bargaining with its increasing emphasis on the role of the shop steward" (1.7). Emphasis on the plant level seems to have reinforced the traditional concerns of trade union with such issues as job security and seniority rights. Where collective bargaining is highly decentralised, there has been a certain delay between plant-level experimentation and national-level trade union policies. This is particularly true where experimentation has been initiated by management.

4.4.2 Centralised expertise

The formulation of central trade union policies with sufficiently detailed content to be effective has also depended upon the availability within the trade union organisations of expertise capable of analysing the complicated questions of automation and work design from a worker point of view. It is noted that in Norway, "unions still have only limited experience and resources in dealing with such relatively non-conventional and complex issues as socio-psychological health factors and sophisticated computer technology". In spite of increasing trade union interest, this has resulted so far in "only limited progress in practical implementation" (1.4).

The significance of centralised expertise is shown by the role of research-based reports in the development of trade union policies and programmes. In the Federal Republic of Germany, the increasing trade union interest in "qualitative objectives concerned with the actual content of work" was "demonstrated by the paper published by the DGB (German trade union confederation), under the title Humane job design" (2.2). The report to the Swedish Trade Union Confederation on "Solidarity and co-determination" proved very influential in establishing the action programme of that organisation. The chapter on France refers to important articles, books and white papers which have formed the basis of statements of trade union policy. Such publications as Les dégâts du progres[2] would not have been possible without considerable expertise on the part of internal staff (1.4.3).

The various trade union positions in France, which emanate from trade union confederations with varying ideological and practical orientations, illustrate by their complexity the need for analytical sophistication. The General Confederation of Labour (CGT), for example, identifies, among others, the following objectives for automation: reduction in repetitive tasks; lightening of workloads; raising of qualification levels through the full use of acquired skills; development of possibilities for initiative; increased training; reorganisation of work on the basis of co-operation; and increased means of worker participation. The French Democratic Confederation of Labour (CFDT) takes the point of view that "technology itself is not neutral" and emphasises the negative effects on the various categories of work improvement listed by the CGT as well as the positive potential of automation. The Force Ouvriére (FO) adds the problem of ergonomic design, the French Confederation of Christian Workers (CFTC) notes the dangers of forgetting that technology is not an end in itself, while the French Confederation of Executive Staffs (CGC) emphasises the dangers of polarisation, the need to consider careers, the desirability of decentralisation, the dangers of excessive mental load, monotony and depersonalisation of work, and the need to increase initiative while avoiding coercive control. It should be apparent that the consideration of these many different facets during a process of rapid and complicated technological change is a subject in which the assistance of specialists is required. This is especially so since change must take account of circumstances noted in trade union positions but not listed above, such as the overall impact of computerisation on the quality of life and the impact of automation on the quantity of employment.

4.4.3 Duration of concern with automation and work design

It emerges from the various national chapters that trade unions have been concerned with the problems of automation and work design for differing periods of time. Only where trade union concern has been relatively extensive for a considerable period have trade unions been able to concretise their demands in collective agreements and enterprise practices. Those trade unions which have had a continuing programme on work design over a period of years have been able to use the more recent concern with automation as a means to start implementing the programmes they have been developing.

4.5 Collective bargaining

Collective bargaining concerning automation and work design tends to be organised around two separate but related issues. The first of these issues is the question of collective bargaining and the content of work. Such bargaining can take place with or without technological changes being introduced, and tends to be associated with the broad demands for co-determination, industrial democracy, humanisation of work and QWL discussed above. The second central collective bargaining issue is that of the introduction of new technologies. This type of bargaining is usually more concerned with employment and related effects than with work content, and does not usually focus in detail on the various forms of automation or technological change.

An early example of the development of work organisation as a collective bargaining issue is found in Italy (2). It is reported that the period 1969-74 was characterised by negotiations on such "dimension of work organisation as piece-work, production rhythms, qualification and hierarchy". While work organisation "became a social issue and a subject of industrial relations", a limit was reached: "Starting in 1974 the union interest in work reform seemed to become more ritual and verbal: no procedure was generally agreed upon for implementing the changes".

A similar process of interest in the early seventies led to the signing of a landmark collective agreement in the Federal Republic of Germany in 1974. Thus -

> ... claims connected with the actual content of work became the central issue in an industrial dispute for the first time. The reduction of extreme forms of division of labour, increased intellectual involvement through greater freedom of action, improvement of opportunities for social contact and reduction of monotonous tasks are today regarded as valid trade union aspirations (2.2).

Norway has developed a comprehensive system of laws, regulations and agreements, notably an Act relating to the quality of the working environment, which aims to increase the positive factors related to QWL. In the United States, progress has been more tentative. "Sheltering agreements" have been established under which "experiments" have been launched concerning work organisation. However, the organisation of work and the assignment of workers to jobs is normally explicitly reserved to management in collective agreements.

A separate stream of agreements relates to the various implications of technology. In the United Kingdom -

> ... over the past few years many trade unions have spelt out their policy regarding new technology and many have also negotiated new technology agreements. In formulating these technology agreements, trade unions seek agreement in areas such as procedural aspects, job security, training, distribution of benefits and ongoing monitoring (1.9).

Broad agreement has been reached between the Confederation of British Industry and the Trades Union Congress on the need for a joint approach to new technologies with emphasis on "the level of the enterprise and individual plants within it" (1.9). In the Federal Republic of Germany (2.2), collective agreements signed in March 1978 on new technology in the printing and metal industries included provisions on structuring of work and workplaces and on safeguards against loss of pay in the case of reclassification, in addition to job security guarantees. In the United States, technological change provisions were found in 17 per cent of a sample of 400 recent collective agreements. However, these agreements normally seem designed to respond to issues of job security rather than work content. The most common provision is a requirement for prior notification and consultation (United States, 1.4). Recent trade union demands include the prevention of the erosion of bargaining unit skills (1.4) and "full participation in the decisions that govern the design, deployment and use of new technology" (1.4). In France, as well, trade union demands have recently emphasised such questions as a lighter workload, raised level of qualifications, increased responsibility and initiative, etc. (1.4.3). However, this does not yet seem to have resulted in specific collective agreement provisions on work content.

A particularly interesting innovation in collective agreements is the "data agreements" found in Norway at both national and local level. These agreements create the possibility of "data shop stewards" who participate in projects for the introduction of computerised technologies. Some 700 data shop stewards have been elected and partially trained but it would seem too early as yet to evaluate their impact at enterprise level (1.4).

4.6 Final remarks

All of the 12 countries covered in this book are encouraging the rapid development and application of automated technologies. All are, to a lesser extent, encouraging improvements in the QWL. In both cases, there remains much to be done. Advanced forms of automation are common in only a few industries: much more common is the piecemeal introduction of equipment or systems which is only partially integrated into production systems. Improvements in QWL, or more specifically in the organisation of work and job content, likewise show only limited progress, though the amount of experimentation and the development of specialised institutions, expertise and even of legislation has grown in the last decade.

There is widespread recognition at national level that automation is a complex phenomenon with broad but differentiated effects on the nature of work. The relationships between automation and skills, occupational patterns, communication and co-operation of workers, responsibilities, supervision and the working environment are the subject of national concern, debate and programmes. While many of the issues involved remain controversial, there seems to be fairly general agreement that is necessary to consider the way automation is applied - i.e., the work design of which automation is a part - in order to analyse its impact. Hence the increasing interest of trade unions, managers, national institutions and researchers in the subject of design, and especially in finding ways to participate in design without being overwhelmed by the rapid advances in engineering and systems development.

The evidence at national level suggests that a process of dynamic adjustment is under way. The negative results of the failure to consider the social side of work design are increasingly recognised and evoke an increasingly effective response. None the less, it would be unwise to assess current progress with excessive optimism. While new policies and recent developments capture attention, they are far from representative of typical situations. The national chapters deliberately emphasise those developments which suggest a new approach to work design, but they do not fail to report that the resources devoted to technology development are vastly greater than those devoted to design; that there are major difficulties in moving from general principles to shop-floor practices; and that appropriate design practices presently touch only a very small minority of workers. The major positive development seems to be an emerging technical consensus on desirable design characteristics and procedures. There is a movement from problem identification towards programme development.

While a wide variety of individuals and institutions have a stake in the design process, it is not easy to envisage a global mechanism or set of mechanisms which would permit the effective participation of all those concerned. It may be observed, however, that such possibilities are being explored, if only in a tentative and incomplete way. At enterprise level, a practice which has been used to associate the various stakeholders in the design process is the design team. Though this is an effective innovation, the national chapters in this study suggest various constraints on purely enterprise-level activities and a number of possible participants in a wide process of design.

One constraint on enterprise-level teams is that the specific interests of the various parties concerned may be expected to diverge quite widely in concrete design situations, especially in the short run. A second constraint is that all the participants in design - who include, for example, vendors of advanced equipment and systems - may not be present during actual design applications. This is especially true in the case of small enterprises.

Various programmes and institutions at national level can help to overcome these constraints, but many countries have found that without legal obligations it can be very difficult to ensure that anything is done in concrete situations. The legal obligations can relate to the content of work or the structure and procedures for improvement (e.g. social planning, workers' participation, approval or certification of designs).

In view of the complexity and difficulty inherent in the design process, often augmented by the rapidity of technological change, another task will be to develop and apply policies and practices which reflect and cope with these problems. Slogans and broad principles are far from sufficient; it will thus be necessary to emphasise the scientific and technical basis of design. This requires better integration of the relevant academic and pratical disciplines through more joint work at enterprise level.

The potential participants in design are very numerous. Governments as well as employers and workers and their organisations obviously have a primordial role. Design professionals of various sorts, including engineers, systems analysts, ergonomists, occupational psychologists and personnel specialists will need to bring various competences and disciplines into play. Institutional frameworks are needed to encourage research and experimentation, provide pratical advice, promote the exchange of information and spearhead the penetration of work design improvements in those enterprises, industries, sectors or localities where support is most needed. There is a special need to ensure that the massive national expenditure of resources in such fields as technology development and training are influenced by these different participants.

All of the above hints at the possibilities for wide participation in design without really showing how this can be organised and applied. Of course, the differing economic, technological and industrial relations contexts in various countries may require entirely different mechanisms to face similar problems of design. There does, however, seem to be a need for a wider community of involvement than can be found at shop-floor level. Because the focus of interest is work design, perhaps a successful attempt to involve the various interested parties could be called a design community.

At an international level, it is easier to evaluate the need for design communities than the success which has been achieved in various countries and localities. If the interpretations found above are correct, the most recent developments at national level are in fact a reflection of the needs which have been identified. The fuller presentations of these developments in the national chapters are a rich source of ideas for possible adaptation and adoption in differing national situations.

Notes

[1] For convenience, the Union of Soviet Socialist Republics and the United States of America are referred to as the Soviet Union and the United States respectively.

[2] Conféderation Française Démocratique du Travail: Dégats du progrès: Les travailleurs face au changement technique (Paris, Editions du Seuil, 1977).

CHAPTER III

DESIGNING WORK IN AUTOMATED SYSTEMS: A REVIEW OF THE CASE STUDIES

1. Introduction

Case studies are in-depth examinations of the structure and functioning of a unique organisation or organisational unit seen in its own particular context and dynamics. In a wider sense, they may concentrate on certain features of particular organisations or groups of people.[1] A careful examination is necessary to a full understanding of the relationships and the dynamics described.

The case studies in this volume have varying purposes. Some contain factual information unavailable elsewhere or develop new insights, which may lead to hypotheses concerning concepts or relationships between the phenomena being studied; others also suggest solutions to particular problems and models for future research and training. However, because they are illustrative and not representative, their individual aims and characteristics prevent any analysis of statistical similarities or quantitative trends. Some generalisations are possible, on the other hand, as to whether the human activities or organisation elements under examination are meaningfully related to the observed technological changes.[2]

The research design left the authors free to give their own accounts and analyses of the units chosen. Only a few constraints were set regarding the issues to be discussed, the information to be provided and the hypotheses to be explored.[3]

The purpose of this chapter is not direct comparison. Instead, we present information and findings from the case studies which seem most important. From these, conclusions are drawn concerning specific topics such as automation, organisation, work design and QWL. Relationships between the type of automation used, the content of tasks and organisational design are considered. Finally, we discuss the impact of automation on QWL and the implications for design and design processes.

2. Content and context of the case studies

This section presents a synthesis of the diverse characteristics of the case studies. In addition to summarising certain main features relating to content, the material below is intended to help put the cases in context, that is, to show how the constituent elements of the organisation relate to one another and how they are influenced by external factors.

2.1 Format

Most of the case studies are in-depth examinations of the impact of automation on the structure and functioning of work in a specific production unit, sometimes in comparison with other units. The 14 cases of this type are from the Federal Republic of Germany, Hungary, Italy, Norway (3), Poland, the United Kingdom (3), the Soviet Union (2), the United States and Yugoslavia. Other types of cases include one from France, which summarises experience from several plants belonging to a single large enterprise; three from Japan, which gather information on specific enterprise-level working conditions and environmental problems; and one from the Netherlands, which is an exploration of the relation of control tasks to automation and to QWL in different situations.

In addition, material relevant to analysis of the case studies is contained at various points in the national backgrounds, as in the "pre-case" from Norway and the "mini-cases" from the United States.

Table 1 gives the national source, author and content of each case together with a code used throughout this chapter for reference purposes.

Table 1: General summary of case studies

Code	Country	Author	Content
F	France	F. Piotet	Automation and working conditions in the production units of a cement enterprise
FRG	Federal Republic of Germany	W.H. Staehle	Job design and automation in production: printed circuit board assembly
H	Hungary	C. Makó	Automation in a motor vehicle plant: workers' needs and work content
I	Italy	F. Butera	Joint design of technology and organisation in a new automated rolling mill
J 1	Japan	H. Saito	Changes of workload in an automated power plant
J 2	Japan	H. Saito	Work incorporated into automatic production of fluorescent lamps

Table 1 (cont.)

Code	Country	Author	Content
J 3	Japan	H. Saito	Inspection work in soft drink bottling plants
NL	Netherlands	C. Ekkers	Job design and automation: results of a Dutch project on control tasks
N 1	Norway	T. Nilsson V. Havn	Two cellulose factories with different levels of automation
N 2	Norway	K. Veium	Computer-based process control in a cellulose factory
N 3	Norway	B. Rasmussen M. Levin	Technological and work design change over 26 years in an aluminium plant
P	Poland	M. Kostecki K. Mreła W. Pańków	Introduction of numerically controlled milling machines
UK 1	United Kingdom	D.W. Birchall D. Helsby	Automation in biscuit manufacturing
UK 2	United Kingdom	P.C. Schumacker H. Hunter P. Gearing	New ore terminal
UK 3	United Kingdom	A. Davies	Technological changes in two breweries
USA	United States	P. Gustavson J. Taylor	Socio-technical design and new forms of work organisation: integrated circuit fabrication
USSR 1	Soviet Union	A.I. Prigozhin	Raw material kilning
USSR 2	Soviet Union	A.I. Prigozhin	Clinker milling
Y	Yugoslavia	V. Jež	Introduction of new technology in the steel mill of the Ravne steel works

2.2 Methodologies and interpretative approaches

All the case studies deal with the impact of advanced technologies on the content of work, in accordance with the original research design. Each, however, has a different methodology, a different focus of analysis, and different concerns and recommendations. These vary according to many factors, including the professional orientation of the author, the nature of the case and the role played by the research in the change process.

Table 2 gives a synopsis of similarities and differences in methodology found in the case studies. The main methodological difference is that some cases fully describe a unit and the related changes, while others focus on specific phenomena and provide empirical evidence and examples from various cases.

As far as the type of research design is concerned, most of the monographs are ex-post facto. In a few cases (I, USSR 1), the researcher himself played a substantial role in the change. The constraints on and opportunities for such action research are illustrated by cases based on research design made retrospectively, as discussed below.

Descriptive analyses prevail: most of the conclusions drawn by the authors are based on their previous concerns and knowledge, using the descriptive material as a starting point. This does not reduce the value of these conclusions, but it is useful to be aware that scientific evidence is not common in this field of study. The prevailing use of unstructured interviews and close or participant observation should not be considered merely a methodological predilection of the authors. Most of the research in this field is still of a "problem definition" type. Such research does not allow strictly formalised techniques and quantitative elaboration.

Most of the research explores the impact of automation on work at the level of groups or production units. Changes in individual jobs are often examined as results of the changed system of work or of work-group design. Nevertheless, there are three cases where the level of analysis is the individual job, and one case concerning the work system in several different plants in the same industry. Almost all the cases, regardless of the main level of analysis chosen, also give information about other levels of the work situation (i.e. tasks, jobs, organisational structure in production or other units, organisation of the enterprise and of the industry).

As expected, the main focus of analysis varies. Many cases concentrate on the more common social consequences of automation, such as skills (two cases), qualifications and composition of the workforce (two), physical and psychological working conditions (three), social interaction (one) or changing attitudes of workers (one). Some discuss principally the consequences of automation for such organisational criteria as the changing nature of tasks (two), changes in job content (two) or work groups (one). One case discusses a model of the relation between technology and organisation, while two concentrate on economic aspects of automation. The process of change and alternative principles for design are each the main focuses of one other case study.

Table 2: Methodology of case studies

Case study code	Type of research design	Main orientation of the research report[1]	Research techniques	Main focus of the research report	Main level of analysis	Main areas of recommendation and concern
Key	(1) Ex-post facto analysis of single case or of a pair of similar cases (2) History of change in an industry (3) Action research (4) Action-oriented study (5) Participant observation (6) Design (7) Analysis of a phenomenon	(1) Descriptive (2) Explanatory (3) Evaluative (4) Normative			(1) Individual jobs (2) The unit (group, department, etc.) (3) Enterprise (4) Industry (5) Society	(A) Reaction to social consequences of automation (0) Different design process (1) Personnel management policies (2) Jobs and organisation design (3) Different technological design (4) New goals and parameters (5) Actions at societal level
FRG	(1) Ex-post facto analysis of a change at shop-floor level	(1)	Interviews and document analysis	Economic and social consequences of automation	(1)	(A)+(0) - socio-technical situational analysis - participation
F	(2) History of changes in a whole industry	(2)	Document analysis	Qualifications, workforce composition, working conditions	(4)	(A)+(1)+(2)+(5) - personnel policy - job and organisation design - labour market
H	(7) Comparison of workers' attitudes in two technological situations	(3)	Questionnaires and following tabulation	Attitudes and needs of workers	(1)	(A)+(1) - understanding workers' needs
I	(3)+(1) Report of action research and explanatory analysis of a planned change	(2)	Previous socio-technical analysis, following interview and document analysis	Relationships among technology, organisation and design process	(1/2/3)	(A)+(0)+(2)+(3) - planned interdisciplinary participation process - different secondary technology design at early stage
J (1, 2 & 3)	(7) Analysis of psycho-physiological effects of various types of automation	(3)	Clinical observation	Physical and psychological working conditions	(1)	(1)+(2)
NL	(7) Analysis of task structure in different automated system	(2)		Tasks in automation	(1)	Not applicable

Table 2 (cont.)

Case study code	Type of research design	Main orientation of the research report[1]	Research techniques	Main focus of the research report	Main level of analysis	Main areas of recommendation and concern
N 1	(1) Ex-post facto comparison of two plants with same products and different automation and organisation	(1)	Interviews and document analysis	Work group versus job	(1/2)	(A)+(2)+(0) - organisation principles and philosophies - design process
N 2	(1) Ex-post facto analysis at shop-floor level	(1)	Interviews and document analysis	Skills and learning processes	(1/2)	(A)+(0)+(2) - organisation principles and philosophies - design process
N 3	(1) Ex-post facto analysis conducted with trade union	(1)	Interviews and document analysis	Working conditions and QWL of portions of workforce adversely affected by automation	(1/2)	(A)+(0)+(2) - organisation principles and philosophies - design process
P	(1) Ex-post facto analysis at shop-floor level	(1)	Interviews	Restraining and facilitating factors in the industrialisation process	(1/5)	(A)+(5)
UK 1	(1) Ex-post facto analysis of a change	(1)	Analysis of data obtained from consultancy	Change in jobs	(1)	(A)
UK 2	(6) Report of a design project	(4)	Design procedure discussed after the event	Design process	(1/2/3)	(A)+(0)+(3) - involvement of top management, union and engineers in organisation design at very early stage of new system's design
UK 3	(1) Ex-post facto comparison of two plants	(1)	Observation	Recruitment and compensation policies	(3/4)	(A)+(1) - management of technological change - pay, grading and fringe benefits
USA	(4)+(5) Action study of a new plant design	(1)	Interviews, document analysis and participant observation	New design compared with conventional one. Economic and social results	(1/2/3)	(0)+(1)+(2)+(4)

Table 2 (cont.)

Case study code	Type of research design	Main orientation of the research report[1]	Research techniques	Main focus of the research report	Main level of analysis	Main areas of recommendation and concern
USSR (1 & 2)	(4)+(1) Report of social planning after the technological change	(1)	Close observation, time study, training	Social system design during automation	(1/2)	(A)+(0)+(2)+(4) - innovation planning - social planning - new models (autopilot)
Y	(1)+(7) Evaluation of conditions affected by a technological change	(3)	Analysis of existing records in the plant	Operation of system of workers' self-management during technological change	(1/3)	(A)

[1] The column Main orientation of the research report shows the prevailing (never exclusive) strategy of collecting and presenting data: (1) Descriptive: the research design is designed and reported in order to give the best understanding of factual data: evaluation and explanation are given afterwards, in the researcher's conclusion; (2) Explanatory: the report is designed to use the data for illuminating a theoretical model or for testing a hypothesis; (3) Evaluative: the report is oriented mainly to measure a particular phenomenon; (4) Normative: the report is built up so as to provide or to discuss recommendations in the context of a carefully described real situation.

Analysing the recommendations of the case studies is an additional way of indicating the differences and similarities between the authors' approaches. Two cases principally advocate proper planning of the design process, and six mention this as one of several important recommendations. A better understanding of the workers' needs and attitudes and improved personnel administration are the main recommendations of one case, while two cases propose the setting up of social goals at the beginning of the change process as the major recommendation. New and more appropriate models and principles of work organisation to meet both technological and social requirements are advocated by all, and are the main recommendation in four cases. An analysis of social and technical aspects in an interdisciplinary, participative design exercise carried out at a very early stage of technology and plant design is given as the essential recommendation in two cases. In one case, changes in the educational system and the labour market structure are indicated as priorities.

Irrespective of their focus on one or more main recommendations, most of the authors indicate many generally desirable courses of action when automated processes are designed and implemented, due account being taken of national differences.

2.3 Nature of the technical systems observed

The cases differ greatly in terms of type of industry studied, size of unit changed, type of production process, new organisation and type of automation adopted. This is illustrated in table 3.

The case material covers a large spectrum of industries. The steel industry (three cases), cement industry (three), paper production (two) and the electronic appliance industry (three) are each studied in more than one case. The power, motor vehicle, aluminium, brewing, soft drink bottling, semiconductor manufacturing and machine tool industries are each the subject of one study. In addition, enterprise-level information is provided concerning a number of other industries in the NL case and in the United States national background, in particular the motor vehicle, nuclear power and aerospace industries. Indeed, almost all industries where rapid development of automation has been found in recent years are covered. The most noticeable absences in this casual sample are railways and chemical industries, where automation has an older tradition of implantation and of relevant social studies.

Recent data show that the intense development of automation is no longer restricted only to very large enterprises. While five case studies refer to enterprises with more than 10,000 employees, seven apply to enterprises with between 1,000 and 2,000 employees, and seven to enterprises with fewer than 1,000 employees.

The units affected by the technological change are in most cases the core units of the production process, where the highest material output is generated and/or where the final quality of product is largely dictated.

Table 3: Case studies: nature of technical systems

Case study code	Type of industry and production process	Unit changed	Workforce characteristics	Previous technology	Automation adopted	Previous organisation	New organisation	Work content
F	Cement industry; continuous process	Whole cycle of a group of cement industries (redesign)	Poorly educated workers with a weak labour market position	Machines in sequence with traditional operation plus remote decentralised computer control	Centralised automatic EDP system of process control	Stage 1: worker/machine individual relation; Stage 2: the same through a console; only the foreman has a picture of the whole process	Polyvalence: rotation between maintenance and production; same achievable grade for all; elimination of foremen; grade and career independent of posts	Monitoring and maintenance; composite tasks
FRG	Electronic industry; small batch assembly	Printed circuit board assembly department (redesign)	4 male (supervisor and senior workers) and 24 unskilled female workers	Manual assembly	Semi-automatic assembly; light ray indicates where the component is to be placed	1 supervisor, 2 senior workers, 24 assemblers and servicing structure	Same as before (1 senior worker programs the computer)	Senior worker programs; assembler carries out orders of the light ray every 1.3 seconds by transferring components
H	Heavy and light vehicle industry; mass production process	Engine plant, rear bridge plant, tool plant	Large majority of male workers (over half skilled)	Traditional individual machines	Flow mass production with automated machines (transfer, etc.)		Workers work in the same line on different machines	Workers on automated, semi-automated or traditional machines on same transfer line
I	Steel industry; continuous process passing through discrete stages and equipment	Seamless pipe mill plant (new design)	Few skilled workers; many unskilled peasant workers	Discontinuous process, manually operated and controlled	Highly automated new seamless pipe mill	In the old existing mill: hierarchical structure; strong division between production and services; tight job definition on work station basis; few craftsmen and many unskilled workers	No change in the managerial structure; foremen as technical facilitators; technical office as a "two-way memory"; work groups	Role design based on control and regulation of variances; polyvalence and equal grades; overskilling through training concerning process control; control room instrumentation and joint responsibility create group behaviour
J 1	Power plant; automated power production process	Entire plant	-	Operators carried out a variety of physical tasks	Automatic process control with monitoring	-	-	Operators basically involved in inspecting and vigilance of operations from inside control room

Table 3 (cont.)

Case study code	Type of industry and production process	Unit changed	Workforce characteristics	Previous technology	Automation adopted	Previous organisation	New organisation	Work content
J 2	Fluorescent lamp production; assembly-line production process	Assembly line	All workers young females; duration of employment generally short	-	Operation of semi-automatic machines	-	-	Simple, repetitive tasks of feeding or inspecting; work strictly paced by speed of machines or conveyors
J 3	Soft drink bottling plant; continuous flow of discrete items	Bottling line	-	-	Automatic washing, transport and filling machines; manual inspection	-	-	-
NL	Various	Various	---- Comparative analysis of control tasks in various forms of automation ----					
N 1	Paper production industry; continuous process in discrete stages	Bleaching departments; Plant 1: redesign; Plant 2: new design	-	Manually controlled process in the old plants	Plant 1: local computer-based process control; Plant 2: central computer-based process control	Plant 1: hierarchical organisation; 21 job titles different in content and grade; Plant 2: new design	Plant 1: no change; Plant 2: autonomous work groups of former multi-skilled workers; no foremen	Plant 1: Computer-controlled dosage of chemicals; workers do the rest; Plant 2: computer controls 99 per cent of process; workers monitor and run the process in case of a breakdown
N 2	Paper production industry; continuous process in discrete stages	New cellulose production plant (new design)	Older workers had problems with the new technology and had to change jobs	Manually controlled process in the old plants; small-scale experiment with computer control	Fully computer-based process control without human back-up; 5 computers and VDUs	Operators assigned to specific phases of process	2 shifts of 6 process technicians led by 1 foreman; 2 sub-teams in charge of fibre line and reclaiming area; some multi-skilling	Displays provide all information on the process. Operators regulate through a terminal keyboard; they sit with their backs to the window showing the factory floor

Table 3 (cont.)

Case study code	Type of industry and production process	Unit changed	Workforce characteristics	Previous technology	Automation adopted	Previous organisation	New organisation	Work content
N 3	Aluminium industry; continuous process in discrete stages	Aluminium production plant	-	Little mechanisation; heavy equipment	Automatic process control (two-way terminals)	Three stages: manual work (co-operative work groups for process control and specialisation, authoritarian foreman); mechanisation (less rigid style); automation (current stage)	Work groups controlling process in daytime; other operators in shift	Daytime operator: control and maintenance; shift operator: watchman with routine tasks
P	Machine tool industry; small batches and production by units	Production of conventional and NC machine plant (redesign)	Qualified workers (most manual workers or craftsmen)	Conventional machine tools	New NC milling machines	Traditional workshop with industrial craftsmen	Organisation unchanged; no special department for new machines	Workers given responsibility for running and adjusting programs
UK 1	Food industry; semi-continuous process	Biscuit manufacturing plant (redesign)	-	Manual machine operations; manually operated baking processes	Mechanisation of wrapping/packing system; automatic process control in many phases	Hierarchical/ functional with jobs corresponding to work stations	Increased supervisory rotation but elimination of traditional supervisory duties	Increased complexity of maintenance and supervisory tasks; decrease in transformation tasks' complexity; merging of some jobs
UK 2	Steel industry; semi-continuous transportation process	Ore terminal unit (new design)	-	Not applicable	Sophisticated mechanical handling and computer-based control system	None. Principles of work structuring were set up	Alternatives set up; no ways found for an ideal match among group size, geography and technology	Work groups; dual roles for supervisors; relative specialisation; clear-cut communications

Table 3 (cont.)

Case study code	Type of industry and production process	Unit changed	Workforce characteristics	Previous technology	Automation adopted	Previous organisation	New organisation	Work content
UK 3	Brewing industry; continuous process	Plant 1: brewery (new design); Plant 2: brewery (new design)	Plant 1: partly permanent and partly temporary workers	Traditional labour-intensive brewing technology	Plant 1: limited automation of process; Plant 2: extensive automation of process	Plant 1: rigid 9-grade structure with traditional demarcations; Plant 2: a number of semi-skilled and unskilled brewing and processing grades	Plant 1: reduction to three supervisory levels; full flexibility of operators and elimination of traditional demarcations; Plant 2: reduction in number of semi-skilled and unskilled brewing and processing grades and increase in skilled operators	Preservation of decision-making and manual tasks of operators
USA	Semiconductor industry; small batch fabrication process	Circuit fabrication plant (new design)	4 team managers; 8 manufacturing teams with 100 technicians responsible for 18-hour production	Same semiconductor technology	Automatic loaders for diffusion furnaces (minor automation)	Conventional comparison plant has shift supervisors for each technical (similar equipment) unit, plus separate engineering assistants	8 semi-autonomous teams are responsible for 8 state changes over 18-hour production day; much smaller maintenance, engineering and supervisory staff	Technicians (workers) in teams are responsible for all manufacturing equipment for their teams' state changes
USSR 1	Cement industry; continuous process with high uncertainty	Raw material kilning section	Older peasant workers, 60 per cent with secondary or higher education	Manually operated equipment absorbing continuous disturbances (imperfect control)	Decentralised computer-based process control (8 boards)	4 operators and 4 assistant operators (usually sons) in pairs having total control of production processes	Phase 1: board operators overlap with kilning operators; Phase 2: division of functions of the two	Phase 1: board operators control process but switch to manual when trouble occurs; kilning operators control through instrumentation; Phase 2: remote control with different functions and autopilot for kilning operators

Table 3 (cont.)

Case study code	Type of industry and production process	Unit changed	Workforce characteristics	Previous technology	Automation adopted	Previous organisation	New organisation	Work content
USSR 2	Cement industry; continuous process with high uncertainty	Clinker milling section	Older peasant workers, 60 per cent with secondary or higher education	Manually operated equipment absorbing disturbances; more complete control possible	Centralised computer-based process control	4 operators and 4 assistant operators (usually sons) in pairs having total control of production process	Phase 1: two parallel control boards and operators; Phase 2: operator = equipment overseer	
Y	Steel industry; semi-continuous process passing through discrete stages and equipment	Rolling mill	Mostly skilled or semi-skilled workers	Discontinuous manually controlled process	"Detroit" automation (mechanisation of transport and limited use of computer)	Traditional hierarchical organisation with occupations based on crafts	No change in organisation	De-skilling of many workers; specialisation

- = data not included in the case study.

In a few cases the entire production plant is automated, i.e. both machines and flow are computer controlled. This is the case of the new seamless pipe mill (I) and the most advanced cement plants reported in F. The majority of departments described, however, fall into the category of automation of a continuous process which is part of a larger discontinuous production process: two bleaching departments in cellulose processing (N 1, N 2), a kilning and a clinker section in a cement plant (USSR 1, USSR 2) and an aluminium smelting unit (N 3). Many units analysed are workshops where conventional machines and automatic control equipment co-exist: a biscuit manufacturing department (in UK 1), departments with both conventional and NC machines (P), an engine manufacturing line (in H). A different case is printed circuit board assembly (FRG) where automation is developed almost to the very end of production: only the final stage of placing components on boards is still performed by workers. An ore terminal (UK 2) is another special case in which computer control has been introduced for some specific functions.

2.4 Case studies of redesign and new design

Another element which makes for enormous differences in the change processes adopted, as well as in their consequences, is the distinction between redesign and new design (see under "Unit changed", table 3). The demarcation line between the two could be placed differently, but we will consider cases of new design and implementation of a new technical, organisational and social system which has some degree of autonomy in terms of results, structure and resources. This is never the case when the new unit is just a machine or a set of machines fitted into an existing technical system (e.g. a line) or in a previous organisation (e.g. a workshop), irrespective of whether there are more profound effects than in cases of new design.

Cases of new design include the new rolling mill of I where - in a new location - new machines, new computer control procedures and new organisational units were set up. The terminal of UK 2, the bleaching department of N 1, the cellulose plant of N 2, the brewery of UK 3, and the microprocessor plant of USA are similarly new designs.

In one redesign case, new independent machines are put beside older ones (P) while in two others new, more advanced machines or equipment are inserted in an existing flow line (H, UK 1). Another form of redesign is a new computer-based process control system running existing equipment, as in the clinker and kilning section of a cement factory (USSR 1, USSR 2), the bleaching department in N 1 and the two last stages of cement manufacturing systems in F. A case by itself is automated assembly (FRG): within the same organisational unit, with the same product, manufacturing procedures and workers, part of the work performed by people is transferred to the computer. In the old system the workers chose the components from a series of boxes and found the right location on the board to fit the components; in the automated system the computer places one box at a time in front of the workers and commands a light ray indicating the exact place for the component on the board.

2.5 Environmental (or contextual) variables

It is well known that the choice of adopting a particular technology depends heavily upon factors at societal level, such as the economy, general technological developments, the general social system, and the political and administrative structure. More precisely, these external factors (or "environment")[4] relevant to the structure and activity of the organisational unit being considered play an important role in technological choice. The mutual and often unpredictable influence patterns of these external factors, together with the discretionary nature of managerial choices,[5] nevertheless make it difficult to draw any precise cause-and-effect pattern between environment and technology.

A similar but more complicated net of influences connects environmental factors with work organisation and QWL. Environmental factors can have a direct impact on types of organisation (for example, industrial democracy projects at national level have had an impact on the development of semi-autonomous work groups in Norway) or on QWL (economic expansion has had a direct impact on programmes for humanisation of work in enterprises in the Federal Republic of Germany). Such factors can also have an indirect impact on organisation and QWL via the technology introduced (for example, shortage of labour leads to automation of management functions which leads to changes in skills and in social adjustments in the USSR). A further complication results from the fact that external influences have differing effects upon the unique nature of specific organisations, each with its own particular "biology". A cement plant in France is not the same as a cement plant in the USSR, at least in terms of work authority and interconnection of professional and social roles of manual workers (see F, USSR 1, USSR 2).

Tables 4-7 are intended both to provide a guide and to illustrate the possible causal relationships within each case. Understanding them has practical importance: whatever technological or organisational characteristics are considered desirable for improving QWL, each policy-maker needs to be aware of environmental conditions which facilitate or impede their success. Failure to do so is likely to lead to adoption of techniques or models which are not applicable to the specific conditions of the new situation, a practice which has been particularly prevalent in attempts to spread new forms of work organisation.

The tables summarise the available information on each case concerning the relevant environment (table 4) and in relation to the strategic and structural features of the enterprise (table 5). Enterprise-level effects such as employment, economic results and industrial relations are given in table 6, and effects on QWL of workers affected, and the impact on others, in table 7. While we recognise that the social effects of automation are also extremely important, we have not included them in the tables since few case studies refer explicitly to this area.

All this should provide a better picture of the complexity, differentiation and variety of the case studies collected in this volume. Readers may perhaps find this initial mapping useful as a starting point for examination according to their own specific questions, hypotheses and operational needs.

Table 4: Case studies: environmental (or contextual) conditions[1]

Case study code	Economic situation of the product	Technology	Labour market	Social conditions	Political aspects
F	Continuing trends of increased productivity and decreasing costs; competition based upon decreasing prices	3 stages of technical development: traditional, mechanical and automated centralised control	Labour market segment from which workers were recruited was traditionally weak and underprivileged	-	Enterprise is the cement industry "laboratory" of French employers
FRG	Demand for expanding the range of printed circuit boards produced	High technological level of the product	Abundant labour supply (unskilled women)	Immigrants, unskilled and weak in labour market	-
H	High demand for product in the COMECON countries	High level of technological development; first automated transfer line in Hungary's motor industry	Increase in number of industrial workers; general and professional education of workers has also greatly increased	-	-
I	Highly competitive regarding both cost and quality; need to expand market share	Technologically complex equipment producing higher yield on a continuous basis has been available for 10 years; but only recently has the computer made possible its control	Increase in wages; more educated labour force	Labour force more demanding regarding work environment and qualification; powerful trade unions	State-owned industry; assignment from the Government to be the "cutting edge" of steel industry modernisation
N 1	Need to improve quality and resulting total yield	50 per cent of wood plants are computer controlled; 100 per cent of bleaching departments automated	Located in most industrialised part of Norway	-	Pulp and paper mills were main site for the first experiments in industrial democracy in 1960s; data agreements made in 1972
N2	Previous uncertainties about what to with cellulose production	50 per cent of wood plants are computer controlled	Same as for N 1	-	Same as for N 1
N 3	Traditionally cheap foreign raw material (bauxite) and cheap power; foreign competition and restrictions on power utilisation	-	Company located in small town (agriculture and fishing); 12 per cent of the inhabitants work in the company	-	75 per cent state-owned
P	Overcoming international dependence	Need for a nationally based technology	-	-	Investment decision made at central level; plant has a limited role
UK 1	Fluctuations of demand; highly competitive	New mechanised wrapping and packaging technology and cheap automatic control of baking process available	-	Strong community backing for the factory; firm has socially progressive image	-

Table 4 (cont.)

Case study code	Economic situation of the product	Technology	Labour market	Social conditions	Political aspects
UK 3	Competition based on cost control	Micro-electronics allow small breweries to automate and increase use of equipment	-	-	-
USA	US market threatened by Japanese competition	Increasing complexity and accelerating cost of equipment	Very tight in "silicon valley"; labour readily available in Idaho	Silicon valley: high turnover, experienced labour force. Idaho: little relevant experience, well-educated local labour	Some pollution control legislation a factor
USSR 1 and USSR 2	Demand of central planning to improve quantity, quality and time of delivery	Development of knowledge of process and of automatic control techniques makes possible total or partial automatic control of process, using existing heavy equipment	Single-industry town; hereditary occupations, peasant workers	Traditionalism and strong kinship values; professional roles in factory overlap with socio-administrative roles in the community	Increasing demands from central planning to factory management
Y	Competitive	Evolving technology	-	-	Self-management

- = data not included in the case study.

[1] Due to the unavailability of data, the following case studies have been excluded from this table: J 1, J 2, J 3, NL and UK 1.

Table 5: Case studies: the enterprise (or larger unit) [1]

Case study code	Type of industry	Characteristics of production system (incl. type of process)	Organisational structure and size of plant (no. of employees)	Economic situation	Industrial and social relations	Products	Strategy	Process of change (and people involved)
F	Cement industry	Small added value per kilo; high product mix; variability of material input (chemical-mechanical continuous process)	-	Under challenge	High trade unionisation; active negotiation on advanced ideas; innovative contracts; "laboratory" of French employers	Cement	Reduce number of workers required and increase productivity	Major technological changes (top management, engineers, unions and various expert researchers involved)
FRG	Electronic industry	Large number of batches of limited size (20-100 boards per batch); high value of components; high added value (assembly work)	(700 employees in plant)	Good; need for expansion to make full use of capital	10 per cent of workforce trade unionised	Continuously changing products	Quality, flexibility, reduction in absenteeism and in required training	Top-dowm management approach (top management, then engineers)
H	Heavy and light vehicle industry	Mass production (integrated mechanical workshop)	10,000 blue-collar workers	Satisfactory economic situation	-	Major change in products	Vertical development of the production system	Top-down approach
I	Steel industry	High product mix; small added value per tonne (mechanical continuous flow within discrete stages)	Hierarchical/functional; programmes of organisational development (13,000 employees)	Critical	Conflictual but correct and constructive	Very high mix of products; need to expand the production of a particular range	Increase quantity and quality of production at decreasing cost per unit	Planned change with interdisciplinary and participative approaches (at the same time and at an early stage: managers, engineers, social science specialists, unions and workers)
N 1	Paper production industry	Plant 1: bleaching department is part of sequence; Plant 2: bleaching departments are independent units (chemical-mechanical continuous process)	Large corporation with 2,800 employees (Plant 1: 100 workers; Plant 2: 65 workers)	Plant 1: critical; Plant 2: good	Advanced industrial relations system (technology agreements)	Large number of grades of product in Plant 1; only 2 grades in Plant 2	Cope with market variability for Plant 1; increase productivity for Plant 2	Classical top-down process, but following negotiation and training (top and middle management, then engineers, then union informed, lastly QWL experts consulted and trained)

Table 5 (cont.)

Case study code	Type of industry	Characteristics of production system (incl. type of process)	Organisational structure and size of plant (no. of employees)	Economic situation	Industrial and social relations	Products	Strategy	Process of change (and people involved)
N 2	Paper production industry	Traditional conversion technology in five phases (chemical-mechanical continuous process)	Large corporation of 1,500 employees with 3 factories (660 employees in plant)	Good	Advanced industrial relations system (technology agreements)	Conventional cellulose products for paper industry	Close old plant to to open a new automatic factory	Top-level decision to have highly automated plant (top management and engineers; negotiations with unions)
N 3	Aluminium industry	Physical-chemical continuous process in three stages	Part of the largest European corporation producing aluminium (1,200 employees)	Under challenge	Advanced industrial relations system (technology agreements)	-	Reduce costs of energy and raw materials; produce by-products	Top-down process (recent involvement of unions)
P	Machine tool industry	Manufacturing (NC machines)	1,500 employees	-	Most employees manual workers or craftsmen	New products	Obtain resources for technical renewal and advancement	Technocratic top-down approach; adaptive adjustment (administration and engineers)
UK 1	Food industry	Mixing of continuous mechanical and packaging processes on discrete items	Participation programmes (2,000 employees)	Good	Collaborative and far-sighted	High product mix with variable demand; new products	Optimum utilisation of resources; cost reduction	Classical process of introducing automatic process control; diffusion of information (top and local management, engineers; union was informed)
UK 2	Steel industry	Mechanical continuous and semi-continuous processing of heavy items)	Special department for experiments in organisation and job design (200 employees)	-	-	-	Optimum utilisation of manpower and facilities	Organisational design after technological decisions (engineers, ergonomist, behavioural scientist; steering group set up)

Table 5 (cont.)

Case study code	Type of industry	Characteristics of production system (incl. type of process)	Organisational structure and size of plant (no. of employees)	Economic situation	Industrial and social relations	Products	Strategy	Process of change (and people involved)
USA	Semi-conductor industry	Some vertical integration; small batch (20 wafers = 3,000 chips); high value added (automatic machines with manual feeding and assistance)	Conventional organisation structured on technology and time; new (STS) design structured on product state changes (fewer than 2,000 employees)	Markets depressed	Workers in this industry and company not unionised	New products introduced at a rate of 3-4 per year	Manufacture quality circuits at minimum economic and social cost	Management design team used STS before plant start-up; design team plus workers chose redesign after start-up (original design: all managers; redesign: all management and employees affected)
USSR 1 and USSR 2	Cement industry	Small added value per kilo; high product mix; variability of material input (chemical-mechanical continuous process)	Both plants have 1,600 employees each	Improved economic situation	Low conflict; clannish attachment	Cement	Improve quality; ensure timely deliveries of finished products	After introduction of automation, social planning intervention: analysis of technology and organisation; social recommendations accepted (managers, engineers, then sociologists)
Y	Steel industry	Mechanical discontinuous flow within discrete stages	State-owned	Difficult economic situation	Good	High product mix	Modernisation	Change came about through technological innovation; workers' councils were consulted during process (workers' councils, managers, engineers; afterwards when problems arose, foremen and workers)

- = data not included in the case study.

[1] Due to the unavailability of data, the following case studies have been excluded from this table: J 1, J 2, J 3 and NL.

Table 6: Case studies: enterprise-level effects[1]

Case study code	Economic results	Employment	Workforce composition	Industrial relations	Technical-scientific advancement in industry
F	Increased production (from 0.15 tons per worker/day before 1930 to 36 in 1980)	Dramatic decrease	Fading away of traditional manual worker	Improved	Major change in the entire industry
FRG	Fixed cost increased by 25 per cent; variable cost decreased by 50 per cent; expected output could fail to be realised	No change	In the department concerned the entire labour force is female, with an average length of service of 10 years	Good employer/worker relations; only 10 per cent of workforce unionised	A step towards total automation
H	Improved economic results; company pays highest wages in Hungary	No change	Same workforce composition	-	First motor factory to set up automated transfer line
I	Positive economic results	Absolute employment increased (employment per unit of output decreased)	Exclusion of old and "practical" workers; increase in educated "intellectual" manual workers; fading of differences between manual and technical workers	Most advanced union/management agreement on work organisation in Italy	Developed technology sold outside as a product
N 1	Plant 1: flexibility; Plant 2: low unit cost	Plant 1: no decrease; Plant 2: decreased work hours per product unit	Plant 2: change towards younger and more educated workers	Workers' participation in minor implementations; unions insufficiently involved	-
N 2	Improved economic results	Number of existing jobs guaranteed	Only young and educated workers appointed	Unions positive but only slightly involved	Pilot advancement for whole industry
N 3	Fivefold increase in production	Reduction in total number of employees	Increased division between day shift workers with task variety and night shift workers (essentially watchmen)	Increased interest from unions	-
P	Doubts about economic results	No change	Total workforce of 1,500, of which 660 are blue-collar workers; average length of service 16 years	-	Step towards modernisation of industry
UK 1	Lower costs per unit; reduced flexibility	Lower manning but no change in overall employment (part time)	Increased ratio of women workers	Good employer/union relations	Advanced packaging system
UK 3	Improved economic results	Plant 1: reduction of 30 per cent; Plant 2: reduction from 83 employees to 36	Plant 1: older workers either transferred or accepted redundancy; Plant 2: decrease in unskilled workers with corresponding increase in skilled workers	Plant 1: very good union/management relations	Plant 2: most modern technology industry utilised

Table 6 (cont)

Case study code	Economic results	Employment	Workforce composition	Industrial relations	Technical-scientific advancement in industry
USA	Start-up costs improved; product costs reduced	Fewer maintenance, engineers and supervisors; no increase in technicians	Demographic characteristics of workers similar to previous composition	Good employer/worker relations	Improved competitive position regarding Japan
USSR 1 and USSR 2	Improvements in quality, quantity and timely deliveries	No change	Increased division between control room and plant operators	-	Pilot experiment for the industry
Y	Improved economic results due to an important increase in production	Improved employment situation	Some workers became more skilled, but many more de-skilled	-	First to introduce the technology in Yugoslavia

- = data not included in the case study.

[1] Due to the unavailability of data, the following case studies have been excluded from this table: J 1, J 2, J 3, NL and UK 2.

Table 7: Case studies: effects on quality of working life (QWL)[1]

Case study code	Physical working conditions (temperature, noise, dust, fatigue)	Psychological conditions (mental load, monotony stress)	Administrative conditions (job security, compensation)	Work itself (well-being, complexity, autonomy and co-operation, control)	Occupational and social roles; QWL of others
F	Improved physical working conditions in the central part of the cycle; decrease in accidents	Emerging negative effects: isolation, long distances, mental loads	Better qualifications and pay for remaining workers	New type of know-how and skills; better careers; extended control	Segmentation of workforce; problems for older workers; formal qualifications attached to individual growth, not to posts; subcontracting of most arduous tasks to external enterprises
FRG	Muscular tension	Monotony and mental stress	-	Elimination of any skill and training; no complexity; no positive control: only disruptive action possible	No occupational identity
H	Acceptable physical working conditions; however, workers complained of air pollution and lack of cleanliness of workshops	No major difference in workers' attitudes and assessment	Higher pay for working on automated machines	No process control and no discretionary intervention in automated machines	Increased social contacts
I	Better physical working conditions	No stress and monotony	Upgrading	Preservation of traditional skills and new control skills; group working; good level of process control; general satisfaction	New career patterns; quasi-engineering roles; exclusion of "traditional" workers
J 1	Improved physical working conditions; muscular load reduced considerably	Increase in mental load and irritability	-	Physical well-being enhanced; complexity of tasks reduced	-
J 2	Unsatisfactory physical working conditions (cramped space, high noise levels, fatigue)	Considerably monotony due to simple, repetitive tasks	-	No complexity or control	-
J 3	Fatigue caused due to high speed of conveyor belt	Considerably feelings of monotony due to simple, repetitive tasks	-	No complexity or control	-
N 1	Plant 1: still problems of physical working conditions; Plant 2: control room eliminates problems of physical working conditions	Routine and monotonous work in Plant 2	Higher pay in Plant 2	Plant 2: work consists primarily of monitoring: monotonous, physically less complex, mentally more demanding, formal autonomy, less ongoing control; more training and responsibility for controlling events	Plant 2: disappearance of traditional jobs; towards a role of process operator in a work group; social structure more egalitarian and more integrated

Table 7 (cont.)

Case study code	Physical working conditions (temperature, noise, dust, fatigue)	Psychological conditions (mental load, monotony stress)	Administrative conditions (job security, compensation)	Work itself (well-being, complexity, autonomy and co-operation, control)	Occupational and social roles; QWL of others
N 2	Good physical working environment	Lack of feeling of what is really going on "out there"; younger workers accept remoteness of control room, but some older workers also need to go and see the plant because they do not trust the instruments	-	Operators only know and control the process as reflected in the computer algorithm; loss of physical knowledge of the production process	Fading away of old trades; new process technician occupation created
N 3	Less physically heavy work	Isolation increased for shiftworkers	-	Work content better for day workers, worse for night workers	Polarisation between control room and plant operators
P	Improved physical working conditions	Isolation; idle time	Lower pay	Improved work content	Sense of improvement
UK 1	Considerable noise levels	Isolation and monotony	Part-time and shift work increased; technological renewals improved sense of job security	Improvement for low-skilled operators' jobs; worsening for skilled operators	Increased need for reliable workers
UK 3	Improved physical working conditions	-	Initial fears of job security	-	Flexible operators
USA	No change (clean room environment)	Increased challenge and variety for all employees; employees more satisfied; lower turnover	Slightly higher pay; began with "pay for skill" plan, currently undergoing change	Increased discretional intervention by workers	Strong product identity with new design; increased social contact within plant; improved relations; more company and product identity and less trade/craft identity
USSR 1	Exposure to noise, dust and heat decreased; substantial improvement	Initially loss of status and work authority for the operators	No change in pay	Different complexity and modes of control; idle time	Operators became transitional; new occupations of trained EDP operators; expert workers often elected in community administration
USSR 2	As for USSR 1	As for USSR 1	As for USSR 1	Loss of process control, then equipment control	As for USSR 1
Y	Improvement in working conditions	Feelings of deskilling and meaninglessness	Wage improvements for bonuses because of innovations suggested	Isolation; specialisation; worsening of communications	-

- = data not included in the case study.

[1] Due to the unavailability of data the following case studies have been excluded from this table: NL and UK 2.

3. Economic results of the introduction of automation

Almost all case studies report the following positive economic results of automation from the firm's standpoint:

- an increased rate of production (quantity);
- an improved quality of output (fall in quality-related costs); and
- reduced numbers of workers, both manual and supervisory.

Labour savings are explicitly reported in UK 3 (30 per cent reduction in manpower), and over 50 per cent reduction in USSR 1. Where no reduction in the number of jobs or workers is mentioned, it is usually because -

- the right to work is guaranteed, as in P;
- collective labour agreements have guaranteed that employment would not be reduced, as in I; or
- the company belongs to a growth industry so that workers can be employed elsewhere in the company, as in FRG.

In contrast to these cost savings, there is in some cases an increase in costs (especially wages) for the remaining personnel (FRG, H, UK 1). Two reasons can be identified -

- higher work demands and higher qualifications justify higher pay (as in UK 1);
- compensation is provided for the impoverishment of the remaining work (as in FRG and H).

4. The process of designing technology and work organisation: from top-down technocratic to multi-purpose planned approaches

The case studies describe various processes for introducing automation at plant level. All the authors share the concern that an early and careful consideration of the social effects of automation in the design is important. Moreover, this is frequently used as a criterion for distinguishing between good and bad design.

A few of the cases report anticipated consideration of social effects[6] in technical and organisational design (F, I, N 2, UK 2). However, the majority of cases report the lack of initial structured consideration of human consequences of automation. Afterwards, either successive corrective actions were taken when problems arose (P, N 3, UK 3, USSR 2) or there was no remedial action at all, resulting in unsolved problems (FRG, H, N 2).

There is some evidence from the cases that the early structured consideration of some human aspects of automation may result in consideration of only a relatively narrow and carefully selected portion of the labour force. Where the design calls for young, well-educated workers (I, N 1) or subcontracts disagreeable duties (F), some workers may be deliberately left out of the design.

As pointed out by Davis,[7] the issue is more complex than early consideration of human factors: it is a question of planned design. This point is strongly developed in I, where a description is given of a planned joint design of technology and organisation. This involved an initial formal exercise of goal and problem setting, a set of interconnected analytical/design steps covering the major elements to be designed including the social system, the adoption of interdisciplinary approaches and socio-technical methodology in most of these steps, and, last but not least, the participation of people with different occupational backgrounds and interests.

Five basic recommendations for purposely planned design are proposed by Butera (I case study) -

- goal and problem setting in an analytical, explicit, concrete way, including social goals and problems as well as technical ones;

- formal steps for finding design alternatives for the organisation and social system. It is proposed that this be done when secondary technology is designed. A well thought-out interaction procedure with the process of technical design is also recommended;

- training of engineers as well as social specialists to see both the human and the technical sides of the socio-technical systems and possibly to use socio-technical methodologies of analysis and design;

- responsibility for design assigned to a temporary unit including managers and technicians from different functional areas, as well as knowledgeable manual and technical workers who are to be involved in operating the new system. In addition, appropriate ways need to be found of informing the trade union and, in some cases, assuring its participation in the various steps; and

- sanction from top management and the design of an interactive procedure for decision-making.

Ekkers (in NL) provides a detailed flow chart for design of new equipment and work organisation, and makes suggestions about taking alternatives into account in the early stages of technical design. Staehle (in FRG) advocates a socio-technical approach to analysis and design, with the participation of the people concerned. A contingent view is suggested, meaning that specific consideration has to be paid to the concrete problems and particularities of the unit as well as to the performance expected.

The definition of social goals at the very highest level is stressed by Prigozhin (Soviet Union, 1.5.2). At plant level an approach called "social planning" is described which encompasses early and specific goal setting, careful analysis of the social system, and description of the technology with respect to both social goals and social system characteristics.

In the UK 2 case study, the design procedure starts from the definition of an ideal organisation according to previous experience. Some difficulties in implementation lead the author to suggest initiating such exercises at an early stage when the technology and the plant layout are chosen, with the participation of engineers and negotiation with trade unions.

Most of the above recommendations, however, are not reflected in the most common form of design reported, which is the technocratic engineering top-down approach. This approach lacks any explicit, early consideration of social aspects.

More general problems of design are noted in the USSR case study. The contradiction between social values and engineering rationality led to social research lagging behind the technical system design. In USSR 2, it was stated that social solutions had to be looked for at the stage when contradictions had been already detected.

FRG is a typical case in which the new technology is introduced from the top down on the exclusive basis of economic considerations concerning "theoretical output". This approach measures only the variation in cost of production factors such as increased fixed costs of new equipment and decreased variable costs of wages. The author's criticism is addressed both towards the lack of consideration of the substantial worsening of work content and towards the doubtful possibility of achieving an actual output near the "theoretical" one. This is because the necessary "co-operation of workers on the basis of joint responsibility" was not designed into the system.

UK 1 is a slightly different case: a traditional approach influenced by new ideas. The procedure adopted is a classical one: a strategic decision to introduce the new technology is made by top management. Subsequent feasibility studies involve headquarters and local management, while a detailed process design is carried out by project engineers. There is step-by-step informing (but not involvement) of the trade union. Implementation results in final union/management bargaining on work allocation, manning and payment. The author indicates that some social considerations were taken into account during the process because of the existence in the firm of "participation programmes". These did not, however, result in any specific procedure during the actual change.

In some cases a procedure was set up after the change process had already gone through several stages. These can be called cases of late organisation design. In N 1, there is a striking contrast between a cultural and institutional milieu nurtured with new ideas, and a very conventional design procedure. The design of the new bleaching unit followed the top-down engineering approach until the phase where the technical and management system as well as the consequent human tasks had been carefully designed. But at this point, external social scientists were called in to help managers make decisions on some organisational alternatives like group work versus individual jobs, task rotation versus rigid task allocation, etc. Unions were not involved at this late stage either, although they had the right to be, nor did they set up goals during any phase of the process.

Some cases could be labelled "remedial redesign". The actions described in USSR 1 and USSR 2 came at a late stage (when the new equipment was already running) but with a rather sophisticated and analytical procedure. The disturbance of the social system by the new technology led to work study and in-depth analysis of the social system together with the new technical system. As a result, a specific diagnosis was made both of the concurrent alterations of social and professional roles and of the alternatives remaining under the technology for different designs of jobs, co-operation and communication patterns, skills and careers. Redesign of socio-organisational elements was proposed and implemented.

The flexible manufacturing example in the United States national background starts from the same point. Dissatisfaction arose about a new machine tools section designed on the principle of "button pushing". The corrective procedure adopted was a two-step programme of redesign. First individual jobs were enriched, then work groups were set up. Organisational principles and goals were redefined on the basis of a discussion on the causes of problems.

No previous planning is reported in the Polish case, and "spontaneous adaptation" was the result. The traditional occupational/social influence of the workers involved (craftsmen, lathe operators) led to the adoption for new NC machine jobs of similar solutions to those in the planned process of the United States flexible manufacturing example.

It may thus be seen that different types of approaches to design are possible. These are classified in table 8 according to a scheme based on how early or late social considerations are taken into account in the design process.

5. Automation and work organisation

5.1 Tasks: the core of work

The first organisational elements affected by automation are human tasks (I, N), which can be considered as the building blocks of organisation. In this subsection we assess the effects of elimination of tasks, or of certain components, the creation of new tasks and the transformation of tasks, as well as the tasks of designing technology and organisation, and the structure of task relationships.

Table 8: Types of design approaches

	Type of approach	Possible component elements of the approaches						
		Ex-post facto evaluation	Post-design negotiation	Anticipatory sensitivity to social aspects	Anticipatory formal setting of goals and parameters	Socio-technical training	Inter-functional integration and participation	Stepwise procedures for design of technology and organisation
Technocratic engineering top-down approach	1. Purely technocratic (workers considered as mere production factor)	X	X	-	-	-	-	-
	2. Traditional approach influenced by new ideas (psycho-social needs assumed to be known by engineers)	X	X	X	-	-	-	-
	3. Remedial organisational design (redesign when trouble arose)[1]	X	X	X -	X (ex-post facto)	-	- X (ex-post facto)	-
Planned socio-technical design	4. Late organisational design with social considerations (separate design of technology - with economic considerations - and organisation - with social ones)	X	X	X	X	X -	X -	-
	5. Joint design of technology and organisation at early stage	X	X	X	X	X	X	X

X = presence; - = absence.

[1] Type 3 can be sequential to types 1 and 2, and also a new cycle of types 4 and 5.

5.1.1 Elimination of tasks

(1) Positive effects. Some automated systems eliminate or improve heavy, hazardous or monotonous work. Examples include robots which perform welding, painting or casting tasks. However, this is mentioned rather as a side-effect than as the main purpose of these devices (J, USA and USSR). Automatic packaging eliminates tedious wrapping tasks in UK 1, but serious problems of employment result as manning is reduced from 20 workers to one.

A different example is automation of direct inspection close to an unhealthy production process. USSR 1 and F report the elimination of the kilning operator's frequent checking through a peep-hole which exposed him to dust, heat and fumes. Computer control of the work flow, together with highly mechanised cranes with magnetic lifters, abolished the dangerous human tasks of transfer and hanging of ingots and other heavy material in I.

(2) Ambiguous effects. Contradictory effects result when automatic regulation of the production process eliminates the tasks of inspecting local gauges or operating valves and wheels in a highly polluted environment, as in steel mills (I), aluminium plants (N) or cement factories (USSR 1, USSR 2, F). On the one hand, this disassociates work in these industries from occupational diseases and hazards; but on the other hand, loss of status (USSR 1), of skills (N) and of control of the production process (F) are reported when a new, meaningful job of checking and regulating the process is not designed.

(3) Negative effects. Automation abolishes certain manual tasks that encompass a substantial degree of process control and that require knowledge of tools and processes as well as intellectual and manual ability. The manual tasks of running a machine tool (H, P, the United States flexible manufacturing example) or of driving a white billet into the jaws of a rolling mill (I, Y) are examples of such tasks that disappear. A model of human work in which manual and intellectual control of transformation processes are intimately interconnected is being lost.

5.1.2 Elimination of the manual or intellectual components of traditional tasks

The effects of this type of change are extremely negative. In process industries, the basic conversion process has long been performed by equipment. The remaining manual tasks, such as activating devices like valves, wheels, gates, levers or lifting devices, have been reduced or eliminated by automation. In our sample, such tasks were mainly abolished in aluminium smelting (N 3), cellulose bleaching (N 2) and cement production (F). This has the positive side-effect of making work less strenuous (see above) but it also takes the only remaining part of the control/regulation cycle away from the workers, leading to problems of boredom, idle time and the lack of a concrete perception of the actual production process (N 1), as will be seen later on. Manual tasks can be intentionally preserved, as in I. The need for maintaining a balance of manual and intellectual tasks in a job is cited by Prigozhin (Soviet Union national background).

There are instances where the control rather than the manual component is subtracted from human tasks. In the assembly of electronic components (FRG), the computer takes over the human task of searching for the correct part and its position on the board. What remains for the worker is only the grasping and positioning of a component every 1.3 seconds, and the positioning of boards. In this type of case, human tasks are mere physical movements of the hands, pure "therbligs", as Gilbreth has named elementary movements. The human task no longer encompasses any control over the production process, but is only a manual task controlled by the computer, where discretion is no longer required. Such tasks are totally prescribed in real time, a feature new in the history of the treatment of people as machines. The binding rules of assembly not only fully prescribe what has to be done, but are provided in the form of fragmented and unequivocal instructions about what to do in the next 1.3 seconds. Skills disappear because the human task does not require knowledge of any kind: the algorithm of the system and the procedure for getting the work done exist only in the computer's memory. The manifestations of this procedure are instructions distributed in small portions to the workers, as for example a "finger of light" shows the exact place in which to insert the component. The human task is reduced to the supply of energy for physically transferring parts. This suggests a new low point on the scale of mechanisation.

5.1.3 New tasks: negative effects from "transitional tasks"

Automation also creates new human tasks. Some of these are ancillary tasks to the new machines in which the worker performs some physical functions that the machine cannot yet do alone, such as line feeding, transferring, making simple inspections, flow facilitation, monitoring, etc. Such tasks, originating at the periphery of or between machines (J), are a well-known result of advanced mechanisation rather than a new consequence of automation. Similar tasks can be found in automatic packaging (UK 1), in automatic bottling (UK 3, J), in lamp manufacturing (J) and in many other cases (soap making, drilling, casting, etc.). This transformation of human tasks into ancillary services to machines may worsen in computer-controlled systems, for two reasons: (a) the higher speed of production; and (b) the widening contrast between the complexity of functions performed by technical equipment and the simplicity of human tasks. This point is made concerning robots in J.

These simple tasks are commonly called "transitional tasks", which suggests that they are the temporary outcome of an as-yet imperfect automated technology. It is assumed that time will make them disappear, solving all related problems. This view is like the misguided belief - common in the sixties - that fragmented work was a temporary cost of building a bridge to total automation, itself a promise of freedom from all toil. The assumption underlying the term "transitional" is very dangerous, because transitional tasks are a widespread current problem (J, NL, UK 1) with serious consequences for QWL and the composition of the workforce. They will not be eliminated by the passage of time but through proper design of technology. Their negative impact can also be reduced by including them in a job with other tasks.

5.1.4 New complex tasks

Some new tasks, on the contrary, are complex and responsible, such as the ongoing control of production processes with the assistance of automated systems.

A classification of control tasks provided by Ekkers (NL) distinguishes between -

1. manual control;

2. automatic control, divided into -

   (a) the automatic processor prepares information for final decision of the operator; or

   (b) the processor actually controls the process while the operator only sets up initial values or intervenes in case of disturbances;

3. supervisory control: the automatic processor has full control of the process including the setting of values, and the operator merely monitors.

Each of these categories is associated with a different composition of various basic activities: monitoring the system; control based on adjustments of the production process; communication; and activities aimed at the future effectiveness of the system. Figure 1 illustrates in functional terms the different type of control (Netherlands, 5.2.2).

Control tasks in 2(a), 2(b) and 3 are new tasks resulting from automation. They depend upon the nature of design of the automated control system (computer hardware and software systems).

In the I case study, options are described relating to joint design of the automatic information system and the task structure, as well as management of planned change. The following classification of human control tasks connected with automated systems is proposed:

(a) definition of process parameters;

(b) equipment pre-setting;

(c) monitoring for unusual events;

(d) data checking;

Figure 1: Types of control function

Level 1: manual control

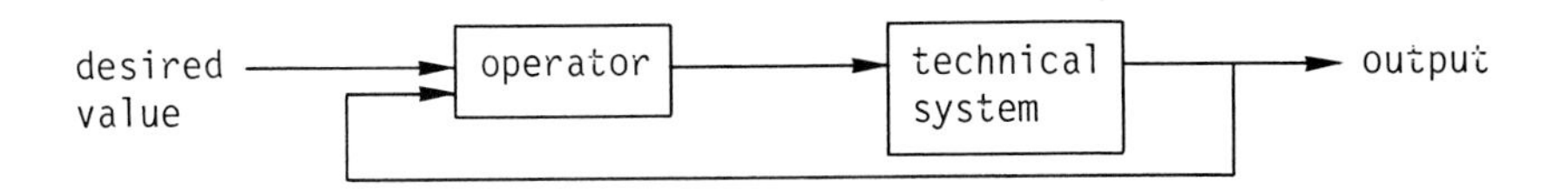

Level 2: automatic control

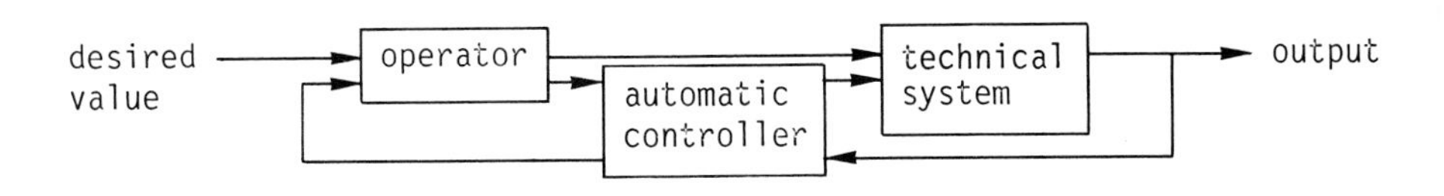

Level 3: supervisory control

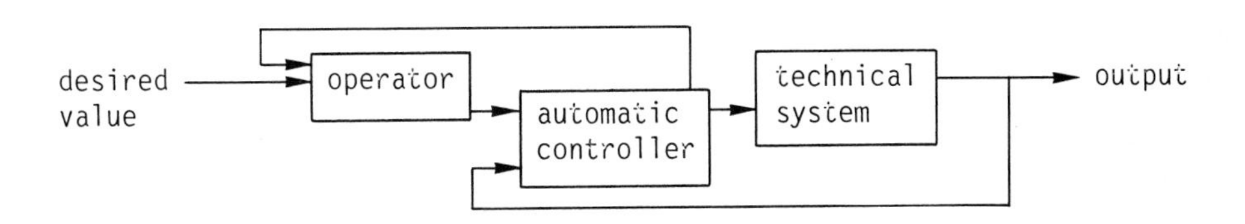

(e) data comparison;

(f) ongoing regulations or adjustments, including -

(i) start-up;

(ii) transmission of information to regulating places;

(iii) ongoing adjustment of parameters;

(iv) process stoppage;

(v) manual plant operation.

The control system was finally designed in a way which benefited from the possibilities of automatic control. However, automatic control was subordinated to human control, in other words, the human control tasks mentioned above were emphasised. As a result, an overall control system was designed with closed-loop feedback to the worker; thus, the worker "has the last word". The control system also has a partly fixed and partly variable computer memory which means that part of the instructions stored in the computer memory can be changed easily by the operators if needed. These tasks were then distributed to both shop-floor operators and engineers, and were combined in group work including manual control and manual duties.

5.1.5 Transformed tasks and worker/computer allocation

(1) Operators. The modification of existing control tasks is frequently reported. One major modification is the shift from direct process control (the operator inspects the physical equipment and the process itself and directly regulates the equipment or the material flow) to control via computer.

This shift is visualised in F as two stages of technical development in the French cement industry: (a) a dialogue between the worker, electrical equipment and control system, and machine; and (b) a dialogue between the work group, integrated control system and machine via local electrical equipment (France, 2.3, figures 16 and 17). The same situation is described in the two bleaching departments of N 1 and in the new steel rolling mill of I.

The worker/computer allocation of control tasks offers a range of options, from leaving workers only pure monitoring control tasks (N 1, N 2, F) to preservation of a full range of control tasks as in I and Prigozhin's recommendations (Soviet Union, 1.4). Prigozhin suggests that workers have to overcome the limitations of formalisation due to automation. This results in a computer-based information system (CBIM) instead of a computerised system. Functional needs of the production process (for example, the likelihood of problems) and characteristics of the workforce (education, qualifications, attitudes) can influence this kind of worker/computer allocation.

However, in all cases, the following radical changes in the nature of human control of production processes take place:

- a pattern of mixing control and actions gives way to purely control tasks with little or no physical intervention by the worker in the production process. A more abstract pattern of work's "building blocks" is emerging;

- a heuristic operating mode (based upon direct, sensorial contact with the material) is replaced by an operating mode based upon algorithms (understanding and interpreting an abstract logical system) (F). This means a more abstract conception of the real production process (N 1);

- human control tasks in automated systems can allow the operators to make choices, but these are now of different types. As Prigozhin says, they are choices among alternatives displayed by an interface device. Human control tasks seem to be moving away from the physical world, or at least from direct perceptual knowledge;

- control tasks usually have a wider scope than a single machine or operation: a segment of production, an integrated planning-manufacturing procedure, a long-range maintenance procedure;

- control (in particular monitoring) of sophisticated equipment usually encompasses higher accountability or responsibility for breakdowns;

- control tasks usually relate not to continuing production processes but to disturbances (N 1, N 2), problems (N) or variances (I). They relate to "stochastic events" (USA), i.e. events which occur from time to time but without any certainty as to when or how (Davis).[8]

Of course, there is much idle time if a job is composed only of this last category of tasks. This situation is vividly illustrated in N 1 which compares two bleaching departments of the same firm at different levels of automation. The less automated department had frequent disturbances and changeovers and the workers spent most of their time with inspections and manual adjustments. The more automated department needed two reliable production workers to monitor for serious but rare disturbances. They say that "the company really makes money when we don't do anything". The same happened in USSR 1 and 2. Formerly, two workers spent 70 per cent of their time controlling the standard content of calcium in an unstable process where objective information was scarce: the need for intuition and individual skill and the right to mistakes came with this task. In the new automated procedure, this work was absorbed by the computer and the problem of filling the jobs with different tasks arose.

(2) Supervisors. Radical changes in supervisory tasks are reported, as most of the hierarchical co-ordination duties were no longer technically required. On the other hand, control tasks, maintenance tasks and tasks related to integration of information became more complex. Of course, major problems arose concerning task allocation between supervisors, workers and technicians of various kinds.

From N 1 we learn that the foreman became superfluous as the production operators were given responsibilities of process control and became more competent in managing the production process. In UK 1, the traditional supervisor was eliminated while in UK 2, many supervisory grades were abolished and the hierarchical levels reduced.

This reduction in personal control through supervisors seems to be compensated by more impersonal control devices. The new data-processing technology can be used to increase control over the worker (time at workplace, quantity and quality of work). Evidence for this trend may be found in USSR 1, where the kiln operators experienced increased outside supervision of their work through the computer.

(3) Process planning. The quantity and complexity of programming tasks increased enormously (e.g. FRG, I, N 1, UK 1, USSR 1). The use of algorithms and knowledge of the rules of operation of the system characterise these tasks; moreover, they encompass the memory of the system (I). Options are available for allocation of these tasks to technicians only or to technicians and production workers (see further I, P, FRG, USSR) and for integration of planning with management of current operations.

(4) Maintenance. No uniform pattern appears concerning maintenance tasks: in general they become more complex, but automation can also imply simplification and differentiation of maintenance operation.

5.1.6 Tasks of designing technology and organisation

It seems paradoxical that almost no case deals much with the completely new tasks related to automation: the tasks of inventing, designing and engineering automation itself. However, some discussion of technological design tasks is found in F and I, and more information, at a higher level of aggregation, is provided in the national background chapters when scientific and technological research is mentioned. It is true that due to the nature of science-based technology, automation developments are often initiated outside the boundaries of an individual enterprise. However, problems of integration between designer and user and between design and implementation are crucial. These are mentioned in unsuccessful experiences (Poland, 1.7; Italy, 3) and in successful ones (F).

Another reason for not including a discussion of design as part of the description of work organisation is that most authors conceive these tasks as belonging to specialists of various disciplines and developed within a process of change. Most of the information on this subject in the case studies is contained in the sections where the process of technological change is described.

5.1.7 The structure of task relationships

The relationships between tasks are as important as the tasks themselves. These influence the pattern of co-operation and communication and ultimately the organisational structure (I).

The following points on task relationships emerge from the case studies:

- the operational tasks relating to isolated automatic machines or equipment are usually themselves isolated (FRG, H, P). However, the structural basis for social isolation which appears in these cases can be overcome by other means;

- tasks related to maintenance, integration, co-ordination and innovation are highly interdependent;

- tasks of monitoring automated process control have low interdependence (J);

- other control tasks for the same automated processes have a high level of interdependence as far as goals, information and disturbances are concerned (I, N, F); and

- the structured operational and time relationships among tasks found in low technology tend to disappear in automation, as does the time dependency common in batch/mass production systems of production (tasks in sequence).

5.1.8 Conclusions

The design of automated control systems substantially determines the content and inter-relationships of human tasks. This does not mean that the nature of human work is technologically determined. On the contrary, technology is always designed with implicit ideas about human work. The case studies have provided evidence that it is possible to design alternative technical systems according to different ideas of human work. Moreover, technical designs that create tasks demanding larger discretion and skills also give better technical and economic results, as illustrated in I and the United States national background, and discussed in FRG.

Critical human performances are connected with anticipation or recovery from breakdowns. Unless appropriate information-processing capacity and decision-making authority are built up in the tasks themselves, it will be very difficult to obtain reliable behaviour, skilful data processing and necessary co-operation from workers. Designs of technical systems intended to be "people proof", as in the nuclear power plant of Three Mile Island, result in human errors. These errors concern seemingly insignificant events which can lead to a chain of events people are not trained to understand.

5.2 Control room design and plant layout

In automated processes human control tasks are usually performed far away from the physical transformation process. The nature and consequences of such remote control are often commented upon in the case studies. In F, N 1 and 2 and I, the technical features of remote control are described. Direct regulation of production processes is taken over by electronic devices (actuators), while the computerised control system commands many such devices and consequently many machines. Computerised control of inputs, production planning, production flow, quality control, etc., can be located in decentralised hardware/software systems or integrated with process control.

The design of control rooms has important implications for the work of people, since it is here that there is an interface between computerised and human control systems. Different strategies of design are described in the case studies.

5.2.1 Centralisation of computer control

In the USSR case, the control room of the cement plant is centralised: all the means for checking how the automatic control is running are located in a single room. Eight computer operators work there with new and totally different roles from the remaining plant operators. The latter remain in the plant where they use four display screens to report data on the status of production in the four kilns.

5.2.2 Centralisation of all control

Centralisation is also present in N 1, but there is no distinction between computer operators and plant operators. All "operators" are stationed in the control room and go out to the plant when necessary. Moreover, the computer controls everything, but if it breaks down the system reverts to use of analogue instruments such as controllers in the instrument panel, or to direct manual control of regulating valves directly in the plant.

5.2.3 Decentralised control room

The same idea of the control room as a place to gather people supervising both computer control procedures and physical processes is adopted in the I case, but the control rooms are decentralised. Four control rooms, located in the major production segments, are intended to house all the instruments and the human resources necessary for controlling the production process.

5.2.4 Control room segregated from the production process

Another alternative in designing control rooms is to decide whether operators are to use only computer input/output devices or whether they are also to follow the physical process by sight and hearing.

Some cases report the first solution. In P, the control room is "shut off from the production room" and is equipped with tables and chairs where workers spend most of the time reading papers, studying textbooks or preparing notes for community work they do in their free time. In N 2, the control room is a large area where the entire process is controlled and monitored, but which also functions as a rest room. It is possible to see the factory from the control room, but the operators sit with their backs to the process in order to avoid glare on VDUs from the windows.

5.2.5 Control room allowing perception of physical processes

In I, a deliberately different design is reported. Control rooms are elevated from the floor and have windows all around. Almost all the process is visible, and closed-circuit television sets give a view of areas not seen from the windows. All operators sit in front of windows, where consoles and overhead indicators are located. No tables for social gathering are provided.

5.2.6 Instrumentation

In automated systems the work of people is strongly influenced by the design of instrumentation provided to operators. The existence of readable instrumentation which allows the operator to take part in process control is stressed in I. There, a case of ergonomic design is described -

> Changes were made to the layout originally presented by the manufacturer of the consoles, which optimised electrical cable routes and connections but made the perceptional load too high. The consoles are now located under the windows and exhibit well-separated areas of emergency control, plant setting ... , manual control ... and digital indicators. Closed-circuit television sets are mounted on the consoles ... together with VDUs showing sequence values. Minicomputers complete with printers and monitors for appraisal of automatic process control are located just behind the console. An intercom system and telephones connect various control rooms with each other and with the rest of the factory (Italy, 4.8).

A totally different instrumentation strategy is to eliminate as much as possible and simplify the remainder. This is criticised in J. In an automated power plant -

> the pressure from the vigilance period increased while the frequency of manipulation became far less than before ... this new situation seemed to cause particular nervous strain among the operators ... The study indicated that further automation would not free the operator from vigilance tasks and accompanying nervous strain. It was suspected that the simplification of the operator's role would aggravate rather than improve such problems of nervous strain (Japan, 2.3).

5.2.7 Remoteness

The remoteness of the operators from the process is reported as a major aspect of automation. On the positive side, this implies that operators working in the control rooms can be separated most of the time from a dangerous or unhealthy working environment, that they can be oriented towards larger segments of the production process and not towards an individual machine and that they can better interact with each other for work and social purposes. Of course, the improved control rooms are well lit, air conditioned, noise protected and well equipped. The larger they are, the better the physical working conditions and social interactions will be. Efforts towards an ergonomic design of control rooms are reported in I and the three N cases. On the negative side, the physical disconnection with the production process is reported as a problem, as has been noted previously.

These mixed consequences are reported in N 2 -

> The new control room offered a good physical work environment, comfortable chairs, silence and possibilities for social contacts with colleagues. But what was lacking was a feeling of what was going on "out there". This explanation is exemplified by older workers who, having, for example, started a motor from the panel, actually went out into the factory to see if the motor was really running ... [on the contrary] newcomers often tend to make quick changes, resulting in very high pressure on equipment ... [the operators] do not learn the process as it is but a picture, or a theoretical model, of it ... this might increase the possibilities for errors and wrong actions (Netherlands, 3.6).

Psychological and occupational remoteness from the production process can transform it - in the operators' eyes - into an abstraction. This is reported by many cases as a major problem in automation, but evidence is also given that it can be overcome. Two avoidable components of role remoteness have been illustrated: the nature of tasks and the design of the control room. Job and work group design, time arrangements, job allocation and careers can also reduce the feeling of remoteness, as discussed in the next section.

5.2.8 Conclusions

Design of control rooms depends upon three sets of alternatives, as follows:

- exclusively computer control rooms versus general control rooms;
- centralisation versus decentralisation of control rooms; and
- a perceptual segregation versus perceptibility of the production process from the control room (visibility, hearing, smelling, etc.).

From these independent dimensions, eight categories result as shown in figure 2. These range from type 1, which is a real "command deck" for the operator, to type 8, where different centralised control rooms for computer and instrumental control are both perceptually segregated from the physical process. In the latter case, the layout does not permit the operator any meaningful control but requires only the accomplishment of a procedure.

5.3 Design of jobs and work groups: the basis of occupations

5.3.1 Definition of the problem

A job can be defined as a patterned series of tasks within a stable set of duties. According to the bureaucratic-Taylorist tradition, the job

Figure 2: Alternative designs of control rooms in automation

Centralisation	Perceptual segregation from physical process	Specialisation: Different control room for EDP and process control	Specialisation: Single control rooms for overall process control - EDP and manual
Centralised	Physical process visible from control room	6	2
Centralised	Physical process invisible from control room	8	4
Decentralised	Physical process visible from control room	5	1
Decentralised	Physical process invisible from control room	7	3

specifications of rank-and file-workers are based upon the detailed description of concrete actions (to assemble, to operate, to disconnect, to fill, etc.). In some cases (e.g. assembly-line work), the job description is replaced by a "work cycle", a detailed definition of physical movements. Sometimes responsibility for an expected result is mentioned, but in most cases this does not include more than accountability for errors. People are usually trained, paid and grouped on the basis of various dimensions of their jobs (job evaluation). Tasks requiring different skills and levels of responsibility are usually grouped into differently paid occupations.

Recent developments oriented to improve QWL tend to diverge from the prevailing pattern of job design described above. These include -

- supplementing more fragmented jobs with some tasks allowing a degree of variety, challenge, learning, etc. (job enrichment tendencies);

- a reduced level of detailed prescription. Instead, the worker's action is oriented towards control of disturbances, variations and results (minimal critical specification, variance control, jobs defined by results, socio-technical tendencies);

- reduced direct correspondence between a job and a worker. Enough tasks of transformation, co-ordination and management are given to a "working unit" in order to assure human control over a meaningful result. This usually requires the contribution of several people. In a work group or a collective job, people have a certain degree of autonomy for allocating various tasks among themselves and for some task definition as well. In semi-autonomous work groups, there often is no permanent occupational differentiation among the workers;

- occupations defined as roles with explicit reference to the relationships with other elements of the system (people, know-how, machines, procedures) and to expected results. Tasks, knowledge, interactions and results are identified on the basis of what actually occurs and not on prescriptions or goals.

5.3.2 Alternative designs of jobs with transformation tasks

New fragmented tasks are reported as the only components of fragmented jobs with low skill requirements in FRG, H and J. In FRG, some tasks are described for which substantial preparation is required, but these have been incorporated in a separate job attributed to the only male worker of the team. The allocation of these tasks to female assembly workers would have been technically feasible, but both a strictly Taylorist management philosophy and the weak labour market position of migrant women militated against the idea.

Similar starting points but different developments are reported in the numerical control example in the United States national background. An aerospace firm initially designed purely push-button jobs for NC lathe operators, with resultant low grading. After production problems, resentment and grievances, a programme was set up for redesigning the work. New tasks were added to the operators' jobs: preventive care of equipment, minor machine adjustments and repairs, design of new tapes, tools and fixtures, etc. Important tasks were not included: programming, because workers dropped out of the necessary training course; and major repairs, because of the trade union's job demarcation restriction. Additional benefits were introduced, including a 10 per cent bonus, flexible working hours and lunch breaks. These changes in jobs and benefits, however, did not have a substantial impact either on elimination of technical disturbances or on objective correlates of workers' morale such as absenteeism.

Then the programme was reoriented from individual jobs to collective group work, and attention was devoted to making the group a viable self-maintaining unit. Tasks formerly performed by supervisors or specialists were assigned to the group when necessary to the attainment of goals. These included tasks of internal co-ordination or of liaison with production control, planning, quality control, materials management, maintenance engineers and payroll officers. Housekeeping and safety functions, training of new operators and record-keeping were also assigned to the group. Because co-ordination tasks related to results were included, people had some control over organisation or "organisational competence". Technical tasks related to the sophisticated transformation processes and to maintenance were progressively taken over by the group whenever operators felt ready to improve their skills and recognised the functional need to perform these tasks within the group. These tasks included micro-planning of tool utilisation, diagnosis of machine failures, programming, tool placement and design, and development of new methods. No limitations were set up by management to such progressive incorporation of tasks. It is reported that this second phase of the project was a success.

Thus a new way of working was experimentally implemented. However, it included a different set of production roles with strong implications for the occupational system. This led to problems: how were manual workers to be graded and paid in the same way as supervisory and technical personnel? Both management and unions seemed incapable of providing answers which did not disturb the existing personnel rules and practices within the firm and the industry, so the experiment was discontinued.

5.3.3 Alternative designs of jobs with control tasks

Control tasks vary from passive monitoring to complex and active supervision of the automated control system. However, they all possess certain properties influencing the kind of jobs possible in automation. The following characteristics of control tasks partly overlap with those mentioned in section 5.1.5:

(a) they absorb only a limited amount of workers' time (N 1, F, I, USSR 1);

(b) they require response to stochastic events, i.e. events necessitating rapid action at unpredictable times;

(c) they imply a high level of functional interdependence and communication among the workers;

(d) they demand patterns of activity which vary over time according to productive and organisational events and performance;

(e) they require the operators to have knowledge which is not restricted to the most frequent activities performed but which extends to understanding large portions of the production process.

The more uncertain the state of the production system and the less formalised the algorithm of the automatic control system, the more marked are characteristics (b) to (e).

These properties each require substantially different job designs. It is no longer appropriate to rely on traditionally prescribed job specifications, full saturation of work time with measurable activity, clear-cut boundaries among jobs, information and co-ordination concentrated in supervisory roles, segmentation of separate activities into distinct jobs held by differently trained and paid people, and vocational training limited to the definable actions permanently performed in these jobs.

However, management and trade union organisational philosophies and strategies often have a stronger influence upon the actual solutions adopted than the functional requirements of the production process. This is explicitly stated in several monographs (I, NL) and illustrated by all of them.

A large variety of alternative organisational solutions are constructed upon the same technology and task design. Some are clearly biased by old organisational ideologies and are sources of ineffectiveness. Others are alternative solutions to the same problems in which different goals and variables are optimised. In particular, a more or less close consideration of QWL dimensions and shorter or longer range time perspective of economic considerations can lead to different designs that seem compatible both with task requirements and immediate socio-economic constraints.

As an illustration, we will review some alternative responses to the same task requirements.

The problem of idle time is variously approached in the different cases. The need for a quantitative appreciation of time use is indicated in USSR 1. It is reported that the elimination of a task formerly absorbing 70 per cent of the operators' time justifies the perception of lost status. In P it is shown how the time left free by automation has been transformed purely into idle time or work external to the factory. The same happened in H, which stresses the problem of in-factory unemployment.

A different strategy was adopted in UK 1. Several existing jobs were merged into a single job after automation, as in the case of moulders or oven jobs. Multi-skilling is the solution reported in the cement industry by F, rotation between maintenance and production being started up in two groups.

A radically new job design was adopted in N and I and is recommended in the Soviet Union national background: new jobs with a good mix of both control tasks and manual tasks. This was done not by designing an individual specialised job but by creating a work group with full internal task rotation. Separate job titles were eliminated, as each operator belonging to a work group had a similar job.

Problems of preparing workers to respond to unpredictable or stochastic events are often reported. Fadem in particular recalls Blauner's prediction of a shift from able craftsman to reliable worker.[9] High pay and human relations techniques are often used to encourage the latter, but perceptional charge and strain (J), contradiction between tedious work and readiness for rapid intervention (J), loss of status (USSR 1) and loss of participation (Y) play a negative role in encouraging and maintaining the characteristics of reliable workers. Solutions using work procedures that continuously require workers' intervention in the production process are described in I and N 2. The role of "member of work group able to control organisational and technical variances" is illustrated in I. Understanding the "analytical aspect of the work under the new conditions", i.e. creative use of opportunities to analyse, compare and record "various combinations of means and actions", is considered in USSR 1 as conducive to readiness as well as to motivation of operators.

5.3.4 Communication and co-operation

The complexity of information systems in automation leads to two opposing strategies. One is to automate the information system and to concentrate planning and decisions in a few specialised technical and supervisory jobs. This is described in FRG (programming displaced to supervisors and senior technicians), in the NC example in the United States national background (design of "push-button" operator jobs with resulting grievances and difficulties of minimising technical disturbances), in H (information system became totally external to current operations), in UK 1 (supervisory ratios were increased and supervisory tasks became more complex) and in J (increase of automatic feedback in order to simplify the workers' jobs).

The organisational prerequisite of this first strategy is reliance on high levels of formalisation of process algorithms and on the effective functioning of automatic control with passive monitoring from the operator. Some cases make clear that when the process exhibits some uncertainty or unpredictability, an impossible or harmful situation arises. The Soviet Union national background describes computer-based management systems (CBMS) designed to work with partially formalised parameters. N 1 describes two similar bleaching departments in which the level of automation and the uses of human information and decisions are different. In the more automated department, the possibility of full computer control depends upon the absence of major disturbances such as stoppages, changeovers or relaunching of production. Such disturbances exist in the less automated department.

An occupational implication of the strategy of intensive displacement of human work to planning and programming is so-called workforce polarisation. At one extreme, a limited number of very skilled technicians design and program the automated control system. At the other, operators are given transitional manual tasks or passive monitoring jobs with consequent progressive de-skilling.

The second strategy, illustrated in some cases, is that of designing the control system in such a way that operators can participate in running and changing it. The use of work groups to implement this strategy results from the recognition that human control tasks are highly interdependent and that the best organisational boundaries are those encompassing sufficient scope for workers to control disturbances and to assure results. In the NC example in the United States national background (1.8.1), a self-maintaining unit was created whose members manage an information system internal to the unit, together with the relevant interfaces with other parts of the enterprise. This was done by giving people responsibilities related to results as soon as each worker felt able to take on such responsibilities. In I, a planned process was used to map out the information network internal to the group and variances to be absorbed. The workers were trained to understand and manage the functional interactions within the group, as well as with higher-level information systems and with the "technical memory". Members of the group were also requested to contribute to improvements in process control and in the existing equipment. As a result, the gulf between technical and manual workers is reduced and a career path from operator to engineer may be created (F).

Organisations are more likely to adopt such a widely diffused information system if the process algorithm is not fully formalised or can be improved. The occupational implications are that the status and grades of group members tend to equalise and that distinctions between operators and technicians are reduced. Foremen are no longer needed or have new roles (N 1).

5.3.5 New occupational patterns

In most of the case studies, the traditional relationship between qualification, trade, job and work station disappears. It is almost impossible, for instance, to understand occupations in automation just through observation of a sequence of actions: analysis of functions is required. In this sense traditional job descriptions are becoming useless. The events people must control take place at unpredictable times so that work can no longer be described in the language of trades in which the transformation is obtained by a skilled series of actions using tools and machines.

Some cases report elimination of traditional trades or craft jobs (F, I, USSR 1) or of work stations with an identified relationship to a machine cycle (H, P). The effects on formal grading are less traumatic, but many cases suggest that workers are given higher grades than would be justified by traditional criteria. This is because of the above-mentioned need to reward personal reliability.

Some authors report that new patterns of qualifications, occupations and careers sometimes emerge in a more or less planned way. N 1 stresses that the notion of a specialised individual job disappears in work group arrangements. In its place is found the occupational notion of group member. N 1 describes how management and union got rid of traditional job descriptions and job titles. They agreed upon the use of roles as an analytical way to describe and regulate group membership and related personnel matters. The number of job titles decreased dramatically in all these cases while task rotation and internal mobility increased.

A trend away from the individual job as the basic occupational unit seems to be taking place in the cement industry of F. There, an approach is being developed consisting of a separation between jobs and qualifications. "Qualification is no longer linked to the post but to the individual." As the traditional craft disappears, high grades are maintained, but these are based upon qualifications and know-how rather than upon jobs.

In USSR 1 and 2, and P, it has proved difficult to preserve the existing grades of workers whose jobs were changed by automation. Cases of qualifications attached to the person and poorly associated with the work content appear in the United States national background. Here high grades, together with job redesign and training programmes for manual workers, were unsuccessful because the barriers with managerial staff in decision-making and in problem-solving were not crossed even though the new technology required it. As a result, a "reliable worker" is created at a high grade but with no clear responsibilities and with a weak job content.

5.3.6 New knowledge and skills

Knowledge and skills in automation could be associated either with new substantial roles or with transitional occupations waiting to be eliminated. In either case, the different nature of knowledge and skills in automation is recognised by all, as has already been frequently noted.

Some cases show that knowledge and skills are not the mere consequences of functional task requirements but are a matter of design, both in relation to the design of work itself (in the form of posts, jobs, roles, organisation units, etc.) and concerning the design of trades or crafts (including formal identification and certification of the knowledge and skills required, training, careers, rewards and mobility patterns).

Training in such new trades must be related to actual work performance. The failure of training programmes which were not connected with a real modification of work roles is reported in the NC example in the United States national background. Conversely, lack of training can prevent the transition from old to new occupations. In F, recruitment of younger personnel with no previous experience who were trained outside the enterprise meant that existing personnel were excluded from automated workshops.

A new occupational system is described in I, where new algorithms and old manipulatory skills are both included in new roles. Theoretical and on-the-job training were designed so that continuous career progression is planned and sanctioned by union/management grade assignments. On the other hand, the choice was made to select a sample of reliable workers capable of highly abstract learning, i.e. young workers not representative of the labour force.

One conclusion is that even though occupational systems designed according to both functional requirements of technical systems and workers' expectations exist, good examples of designs accessible to the majority of the existing workforce are lacking.

Table 9 gives a synopsis of the above description of conventional and new design patterns and of influencing or facilitating factors.

Many, if not all, the conventional work designs resulted in serious problems of effectiveness from the social, and often also from the economic, point of view. Although there is a scarcity of hard evidence that new design models always have positive results, they may at least be considered as examples of "the possible" in automation. Elements (or principles) of these new models include the following:

- jobs which combine computer system control, direct control and manual tasks in ways that avoid idle time;
- roles designed on the basis of functions performed, results achieved, variances controlled and interactions required;
- polyvalence and multi-skilling;
- work groups which are semi-autonomous units where people can carry out their roles co-operatively;
- new occupational patterns which differ from both fragmented jobs and traditional trades;
- training in both theory and practice;
- continuous careers which reduce the gulf between operational workers and technicians.

From table 9, the conclusion could be drawn that the design of jobs and occupations is largely a matter of choice based upon values and other characteristics of the "designer system". The range of choices available is limited to two sets of constraints: the design of the technical system and human tasks on the one hand, and the occupational system - including the labour market, industrial relations and education - on the other. Within these constraints, there are (a) good choices, in which joint optimisation is achieved; (b) poor choices, which lead to short-run ineffectiveness or inefficiency; and (c) ambiguous choices that optimise one variable at the expense of others and whose effects are longer run. In the latter case, disregard of QWL considerations may delay or displace positive effects.

Table 9: Conventional and new design models: characteristics and facilitating factors

Characteristics of control tasks in automation which influence work design	Job and occupational characteristics according to traditional organisational principles	Factors facilitating traditional design (in addition to cultural preferences)	Job and occupational characteristics according to new organisational principles	Factors facilitating design according to new principles (in addition to cultural preferences)
Limited time required	Monitoring jobs with much idle time; or merging of different jobs	Preservation of traditional demarcation among jobs. Closeness of the tasks in time and territory	Polyvalence between maintenance and production. Jobs include computer and direct control tasks together with manual tasks. Work groups	Management and union agree on internal mobility. Existence of enough manual tasks (not the case in chemical processes)
Readiness in responding to stochastic events required	High pay and grading. Human relations management techniques	Economic possibility of paying for reliability in place of skills. Social possibility of maintaining good human relations in spite of low work involvement	Procedures requiring continuous action. Role design in work groups. Training in analysis of process variables and models	Human control technically required. Operators responsive to goals and responsibilities
Functional interdependence implied	Computer control maximised. Heavy supervision. Specialised planning activities	Predictability of process algorithm. Social possibility of workforce polarisation	Work groups whose boundaries are sufficient to ensure results	Technical feasibility of designing appropriate group boundaries. Social acceptability of new roles for foremen, narrowed grade systems and few distinctions between operators and technicians
Defined by function performed rather than prescribed activities	Use of unskilled workers wherever possible. Fading away of traditional trades	Industrial relations protecting existing job titles and trades	Group work. Role design. Qualification attached to individual and not to post	Workers responsive to new roles and responsibilities. New occupations negotiable
Extended and abstract knowledge required	Recruitment of educated young workers with strong abstract reasoning skills	Short time to put systems into operation		Extended time available

5.4 Automation and organisational structures

Apart from the consideration of the roles of foremen and of change processes described above, few case studies pay special attention to changes in the general organisational structure of the firms where automation is introduced. One reason for this is that automation of limited scope - as reported in many of the cases - is unlikely to have an immediate impact on organisational structures, as P illustrates. Instead, it is encapsulated within the existing structure.

When automation has a wider scope, it changes the pattern of integration within the enterprise as in F, I, N 3, UK 1, UK 2, USSR 1 and Y. Then two different things seem to happen. First, a shift takes place away from an administrative or personal type of control: the I, USSR and Y cases give evidence that managerial control is partly incorporated into automatic data processing. Communications, instructions, decision-making, interfaces, etc., become more formalised and routine, as Simon has shown.[10] The co-ordination process as a consequence relies less upon hierarchical structures and more upon the integrating mechanism built into the automatic information system. A decline in the functional need for the hierarchical bureaucracy seems to appear. Secondly hierarchical structures, on the other hand, do not seem to be substantially weakened or altered. Exceptions are in F and USA, where the design of new factories incorporated a purposely designed pattern of non-hierarchical organisation. On the contrary, in I new forms of work organisation were limited to the boundaries of the immediate production process and affected only the personnel on the shop floor; new design ideas did not infiltrate the formal hierarchical structure. The upper- and middle-level authority structures, roles of managers, managerial models, etc., remained similar to the previously existing ones.

This phenomenon could be explained in various ways. One theory is that changes induced by automation must be managed by one or more power structures. The technocratic power of systems engineers, technicians, etc., is incorporated in existing power structures based on old models which are experienced in administering power. Hierarchical structures are seen as power structures for controlling technological change and its implications in their own interest. Automation would thus reinforce hierarchical structures.

A second, parallel explanation is that hierarchical structures have three functions: the co-ordination of the work process, the organisation of power and the stratification of people. Automation alters task relationships and profoundly affects co-ordination needs, as described by Thompson.[11] The decline of functional requirements of hierarchical structures is therefore just a limited consideration. The need to protect existing industrial power relations comes from the logic of social relations of production at societal level (Tannenbaum),[12] and the need to differentiate people in status and decision-making capacities comes from the logic of social stratification in society.[13]

Whatever the underlying reasons, it can be concluded that automation reduces functional requirements for hierarchical, formal organisational structures but that actual change depends heavily upon social variables both at enterprise and at societal levels.

This helps clarify the insistence of many socio-technical specialists on the association between work design and industrial democracy exercises at plant level.[14] It also supports those who maintain that the design of new technologies must come together with changes in labour market policy, educational systems, managerial training, industrial relations and similar societal variables.[15]

6. Automation and the quality of working life

Thus far, it has been shown that there are few direct effects of automation on QWL. Instead, effects come from alternative choices about technology, task structures, jobs and work groups, organisational structures, work time, administrative arrangements and social processes. Alternatives relating to each of these aspects can be wider or narrower according to choices made concerning the previous ones. For instance, technology and task structure in automation can be designed in many different ways as the cases show, but as soon as a specific choice is made concerning them, the remaining choices concerning jobs or occupational profiles are considerably reduced.

6.1 Whose quality of working life? Job security and segmentation of the labour market

The case studies clearly show that one of the main QWL problems in automation is that an increasing number of people will be excluded from automated industries. Although the cases were not expected to address the question of the effect of automation on overall employment, it is clear that occupational shifts will take place. It is demonstrated that older workers, less educated workers unused to abstract thought and less reliable personnel tend to be excluded sooner or later from automated work. Thus, a dramatic change not only in the total size but also in the composition of the industrial workforce could result, unless a special effort is made in the design of jobs and in training and selection procedures. However, the most advanced organisational designs on automated processes take it for granted that the workers in the automated systems will be different from traditional workers. This lack of "reproducibility" of the organisation in automation is reported in I: the new organisation, however well designed it is for its members, is suitable for only a limited segment of the working population. A related phenomenon (F) is the "export" of undesirable work from the big automated firms to secondary firms (subcontracting).

An effort to keep existing occupations and workers in the new automated system is made in several cases, as in H, P, USSR 1, etc. These traditional jobs are reserved during a period of transition for workers who cannot or do not adapt.

While overall employment effects are beyond the scope of this project, it should be noted that job security, which is an important dimension of QWL, is always challenged by automation in one way or another. This is a major reason for the tendency towards general social regulation of the effects of automation, as the national backgrounds and case studies note (see Chapter 2).

The recomposition of work into broader, more integrated occupations is encouraged by automation. There is evidence in most of the cases that pressures from the functional requirements of automation are not followed because traditional organisational philosophies based on job differentiation and segregation prevail. The creation of homogeneous social structures at work is a matter both of occupational design and of management of the labour market. Segregated jobs, lack of "reproducibility" of the organisation, and lack of labour market planning result in the frequently reported polarisation of the workforce into a few skilled technicians and many unskilled or transitional workers.[16]

6.2 Physical and psychological working conditions

An improvement in physical working conditions is reported in all the cases where strenuous tasks are shifted to machines or where the new work stations are remote from hazardous or polluted work environments. Reduction of accidents is reported in F, I, N and other cases.

Remote control can radically change conditions in industries where work has always been synonymous with danger and illness. Examples are found from the cement industry, steel works, aluminium plants and pulp and paper plants, in which automation has drastically reduced the workers' exposure to heat, fumes, glare, noise and physical safety hazards such as heavy objects in movement. The worker is thus released from difficult conditions and from worry about accidents and occupational diseases.

However, as is indicated above, it sometimes emerges that the workers' identity and status had been associated with the skills required to survive in a risky and unhealthy environment. F and I report that the change in working conditions meant that some segments of the working population became weak in the labour market. For instance, strong and healthy but unskilled peasant workers, who are accustomed to physical risks, were once the most sought-after labour for the steel industry in Italy. Now they are no longer recruited unless they have additional qualifications. USSR 1 reports that, before automation, kiln operators and their assistants had been entitled to special monetary compensation for the unfavourable conditions of their work, retirement at an earlier stage, free additional food, extra days of leave and other benefits. Their work status had been the highest in all categories of workers. After the change, when working conditions were substantially improved, the kiln operators' status shrank in importance. Kiln operators' work seemed now much easier to their fellow-workers. In the socio-cultural environment of the plant, the prestige of work was largely determined by its arduousness.

These negative social factors do not prevent the overall evaluation of remote control as positive. It should be noted that the improvements are not due strictly to automation but to appropriate design of control rooms, protective equipment, non-polluting production equipment, etc.

It is also possible to introduce remote control or other forms of protection in work environments where strong economic motives or easy technical feasibility do not exist. Policy statements that automation should be used for improving working conditions are found in the Soviet Union national background (1.4). In the national background on Italy (3), industrial robots are mentioned as a means for moving workers away from hazards, pollution and monotony. However, no case reports automation introduced with this as the main purpose. This is in line with other findings that very few cases are known of automated equipment being introduced for pure considerations of working conditions. Even entire industries where the physical working conditions are difficult (e.g. the wax industry, tanneries, explosives, etc.) have not placed emphasis on such equipment.

As far as psychological and psychosomatic working conditions are concerned, there are totally different effects according to the type of tasks. Tasks which are auxiliary to machines due to incomplete automation, as reported in FRG, H and J, imply a serious worsening of psycho-physiological working conditions.

In the automated assembly of FRG -

> considerable muscular tension (arms and neck) [was] observed in [the previous work system] due to unfavourable sitting and reaching positions; in [the new system], however, workers report intense mental stress due to monotony and lack of variety in their work. The new bonus wage system also leads to pressure of time, so that the worker tends to remain permanently in the working position ... [The new working system] is subjectively perceived as particularly stressful, probably because the operations and the relevant times are dictated by the machines. Those affected experience the restriction of individual freedom of organisation and action particularly intensely ... (Federal Republic of Germany, 5.5.)

Some women have lower performance than before: "the causes must be sought in the lack of acceptance of the system and the consequent low motivation" (ibid.).

In the Japan national background, where reference is made to industries manufacturing discrete items, it is said that -

> energy expenditure of new jobs has been markedly reduced and the working environment has generally improved. Further, these changes have accelerated standardisation of work and higher productivity under a more and more strictly organised division of labour. Thus extreme division of work, an increased rate of work and simplified work demands have created new problems of monotonous, repetitive tasks. Many workers perform such jobs at the periphery of or between automated machines. The result is a "gear-like" feeling as if the workers

> themselves have become part of the mechanical processes, which has spread among both automated machinery operators and repetitive task workers. The results are increased nervous strain in processing a large amount of information, locally concentrated load by repetitive manipulations, postural strain by restricted working movements, eye strain, and night and shift work in a greater number of workplaces. The psychological effects of monotony and boredom and of alienation from working life are becoming increasingly serious. All these conditions have a cumulative effect on a worker (Japan, 1.2).

J presents data on these consequences in different plants. In automatic production of fluorescent lamps where the human tasks are machine feeding, inspection and packaging, fatigue is concentrated in certain parts of the body and a high heart beat rate is observed, as well as compensatory movements irrelevant to the task itself, and feelings of monotony. As a result, product quality has decreased. In a bottling plant, inspection work is paced by conveyor speed and requires dealing with several bottles every second. Eye strain, fixed posture, high monotony, a decline in the Critical Fusion Frequency of Flicker (CFFF, an indicator of the activity level of the cerebral cortex) and absenteeism are reported as associated with conveyor speed, short work cycle, lack of variety of tasks and insufficient rest.

The above examples recall the evidence reported in over 40 years of studies about stress and strain in highly mechanised production systems. With the increased speed and the further reduction of work content in computer-aided mechanised systems, there is cause for alarm.

The psycho-physiological effects of automated control systems are very different. Pure vigilance tasks with no physical activity and no feedback system produce nervous strain, J reports. In an automated power plant, operators felt irritable and uneasy to a greater extent than in other work settings. This came from increased job strain, larger responsibility and uncertainties about what they would do in the immediate future. Stress due to the "awareness of high cost of the machines" (P) or "mental load" due to "uncertainties concerning official rules about equipment needing multiple unforeseen adjustments" (F) is also reported. However, when control consoles are properly designed and roles are built up in such a way to empower the worker to "understand the laws of the system" (I), no mental stress or mental charge is reported.

Negative feelings of isolation are reported whenever the workers' location in a remote control room is not accompanied by the creation of a work group and of a communication system with other people in the plant, as pointed out in P. Boredom caused by idle time is reported: the "long hours of monitoring are a tedious process ... and this becomes the main criterion for evaluating the strenuousness of work" (P).

6.3 Complexity of work, skills and new occupations

Jobs constructed from auxiliary tasks in highly mechanised computer-aided processes show a dramatic decline in complexity of work and related skill requirements. FRG reports that polarisation of the workforce is rooted in the polarisation of job content: a few skilled technicians on the one hand, many unskilled assemblers on the other. In UK 1, a similar phenomenon is described: "increased complexity of technology placed new demands" upon maintenance and supervisory functions and upon a few operators' jobs. The author measured along a nine-point scale the change in ten job attributes contributing to job satisfaction (variety, [non-] repetitiveness, mental absorption, discretion, control, goal attainment, perceived contribution, social opportunities, interdependence and skills). His conclusions are that "those jobs scoring at the mid-point or above on many of the job attributes were further enhanced as a result of the technical changes" (United Kingdom, 2.4). Other jobs became more repetitive, less mentally absorbing, had less direct feedback and were lower in skill demands. These examples further confirm the tendency towards progressive de-skilling of less qualified workers in the face of advanced mechanisation.

A totally different situation is reported when the main job content is human intervention in the automated control process. In these cases, a change in the type of complexity and in the pattern of skill requirements is always reported.

In F, a thoughtful analysis of the radical change in workers' know-how is reported. Workers' know-how was traditionally heuristic knowledge and skills developed through direct contact with the production process and through experience of the product/machine interface. This know-how was acquired during a long apprenticeship, and the workers eventually developed considerable autonomy and co-operation. Automation destroys this system. Workers and foremen must have algorithmic know-how which draws on scientific and engineering knowledge. The source of qualification is changed from internal (learning of a trade, accumulation of know-how through practice) to external (formal education).

Change in the nature of skills required is confirmed in other cases (FRG, I, N 2). Examples are given (I) and suggestions are provided (Soviet Union, 1.4) about possible combinations of manual and control tasks, of heuristic and algorithmic skills. This point is strongly made by S.

New jobs requiring skills at a higher level of sophistication are described frequently in the case studies: foremen (N), supervisors (UK), technical office engineers (I), maintenance people, new specialists (USSR). In the preceding section, we discussed organisational and information system designs which can either fragment work or create an integrated co-operation pattern among different roles. New work roles can be institutionalised into new occupations, including not only well-defined role content but systems of training, formal qualification procedures, and advancement and career patterns as well.

Examples of this are scattered through the case studies: I describes an innovative system for role description around which union and management can negotiate; in N 3 a new occupation, the process operator, emerges; quantitative increases and qualitative improvements in training are reported in F, N and USA; multi-skilling and job rotation are frequently reported; continuous careers based upon qualifications (F) or upon refinements in role performance (I) are contrasted with traditional industrial career paths; and the trend towards equal grades for all members of work groups is described in I and N. However, these examples only suggest a trend: in no case, with the possible exception of F, is the design of a new coherent system of occupations under automation part of a planned process.

It is difficult to be confident about this trend because there are numerous institutional variables (educational system, labour market structure, industrial relations, etc.) and many different types and impacts of automation. Even within the same firm the occupational effects of automation seem contradictory. As UK 1 notes, in partial automation there are varied changes in operating tasks and no coherent new pattern emerges.

Some data seem to confirm predictions of the rise of a new working class in which the work content and the occupational status of manual and non-manual workers becomes similar.[17] Other data, on the contrary, seem to confirm theories of polarisation between a few specialists and many unskilled workers.[18] Finally, there is some support for theorists of the uninterrupted de-skilling of human work and the dwindling of trades.[19] These contrasting findings suggest that a shift is needed from a search for undimensional explanations towards a contingent appreciation of the interplay of the three main variables affecting human work in automation: automation of technology, institutional settings and policies, and design. So far we can conclude that different types of automation, different societal settings and different design of technology and organisation produce different types of "workers in automation".

6.4 Autonomy, co-operation and social interaction

It is noted above that in cases of increased mechanisation associated with computer control, many jobs decrease in autonomy, discretion and social interaction. J finds that the negative effects of using robots appear at intersections in which the human role remains to perform unmechanised tasks, and that the pace of work at these intersections may be determined by that of robots. It concludes that the problems of mechanical automation that have been seen so far will continue to exist also in the case of robot technology. In FRG, no discretion, autonomy or social interaction is left to the assemblers. In Y, a physical scattering of workers, dispersion of their roles and increased binding of every worker to his workplace is reported. It is pointed out that this depends on choices: "the new organisation, together with technological demands, did not allow the functioning of communication well enough".

In contrast, a release from pairing of workers with machines and fixed work stations is reported in most cases where the work is based primarily on control tasks. In the majority of cases where the main work activity is monitoring in a central location, such as a control room, an increase in social contacts among workers is reported (N, P). An extreme example is described in P, where the workers spend most of their time reading or studying for activities outside the workplace. In I, a high level of operator intervention in automated control has permitted operators to identify numerous instances where they can make relevant decisions and to map out a complex communication network among different roles. As a result, work autonomy and co-operation in process control are fairly high.

In process control the range of human decisions and the information system are parts of a larger system. The design of roles, work groups and required skills and knowledge permits both higher and lower levels of decision-making and co-operation among operators. Integration of processes and increased technical complexity give an opportunity to designers to offer higher autonomy and co-operation to the workers, but this is far from inevitable. The NC example in the United States national background points out that the constraints on this approach are social rather than technical.

6.5 Control

The case studies present varying consequences in terms of workers' authority or control over production processes. Many describe automated factories as having impersonal systems of control, where workers become auxiliary gears in a larger mechanism. For example, workers feel they no longer make important decisions in USSR 1. The computer runs the process, and the computer operator asks for the intervention of manual workers only when disturbances occur.

Different consequences are reported when the human tasks include supervision of the computer. USSR 1 reports the use of an autopilot, i.e. the operators were given the final responsibility to decide whether to use computer control or to operate manually.

In some cases, the operators' responsibility increases, at least for the negative consequences of their errors. Moreover, a wider knowledge of plant functioning and properly designed control roles can empower workers to make a positive contribution not only to ongoing processes but also to improvement of the system. Classical syndromes both of powerlessness and of a sense of influence over the production process are cited. It appears that automation offers opportunities both for estrangement from production processes and for new forms of participation and positive influence on the techno-organisational life of the firm. The final result depends largely upon the control ideologies of management and trade union. What seems certain is that the traditional division of work (management controls the whole organisation, skilled workers control the work process, and unions control conditions of work and employment) tends to be lost.

6.6 Pay

Almost all cases report increased wages, sometimes unrelated to improvement of skills, responsibility and influence. It seems reasonable to explain this phenomenon as an effect of the management desire to reduce the negative reactions of workers to the technical changes and to encourage the idea of the reliable employee. On the other hand, the impact of wages on the final cost of the product is usually so small that this does not have important economic consequences for the firm.

The disconnection of pay from the content of work can create serious, long-range problems, even though compensation strategies are often adopted for reducing the maladjustments of workers to technical changes. However, there is considerable evidence that compensation strategies unrelated to strategies of work design are unlikely to have a durable impact on the attitudes and technical performance of the workers.

7. Summary and final remarks

The above analysis of case studies leads to the conclusion that the configuration of human work in automated systems is the result of several choices affecting both the technology and the organisational elements of the work system.

7.1 The impact of technology on work organisation: absence of determinism

A variety of alternative designs are technically possible and economically feasible concerning basic technology, secondary (or peripheral) technology, jobs and occupations, organisational structures and conditions of employment. Various procedures have been used in designing technical and organisational systems with QWL goals, in cases both of planned design of entire new units and of purposeful design of single elements of work systems.

It may be concluded, then, that design is both a technical endeavour and a social exercise. The need to develop purposeful design processes which are both technically and socially appropriate appears strongly from the case studies (see sections 4 and 5, above). The following factors influence different parameters, procedures and participants in design:

- the timing of design: the research and development phase of a new technology, the application of a new technology in the investment phase and the introduction of the new technology within a complex socio-technical unit of the design phase;

- the type of unit design: for example, an automated chemical plant has different requirements from a car plant;

- the scope of design: an entire computer-controlled plant has different requirements from a single NC machine;

- the societal goals of design, including economic and social aspects;

- the institutional setting: industrial relations, government, participation of engineers, scientists, etc.

The need for new processes of design comes from the nature of automation itself. Automation is a new phenomenon representing a rupture in the industrialisation process. The content and perhaps even the notions of work, jobs, skills, training, careers, organisation, etc., require review. The boundaries between work and life, organisations and institutions, and production and services are changing.

The design of automation is too complex and too important to be left only to technologists. Interdisciplinarity and participation are needed. A communication network is necessary among universities, research centres, institutes, enterprises, trade unions, etc.

7.2 The constraints of design

Basic conversion technology has an impact upon the quantity and gross quality of production ("aggregate work"). This is shown by the human tasks eliminated by the new technology. There are positive cases (elimination of heavy, hazardous tasks), negative cases (elimination of tasks requiring skills and process control) and very negative cases (elimination of control and skill components of tasks). The adoption (and not the content) of technology is in question here. At production unit level, a decrease in the quantity of work needed and the disappearance of traditional jobs and trades are frequently witnessed.

Similarly, there is an impact of automation upon the nature of tasks and their relationships. Human operational tasks, where not abolished, tend to become in most cases ancillary to the machine and poor in content.

Tasks associated with the control of production processes (control tasks) are of various types ranging from passive monitoring to human control via computerised control. All these types contain the following properties that suggest a change in the nature of human work:

- they occupy a limited amount of workers' time;

- they require response to stochastic events (i.e. events necessitating immediate action at unpredictable times);

- they imply a high level of functional interdependence and communication among the task holders;

- they demand patterns of activity which vary according to productive and organisational events; and

- they require knowledge beyond the most frequent activities performed, including some understanding of the principles behind the production process.

Within these general characteristics, tasks can encompass more or less skill, actual control, autonomy and interaction: i.e. they can be more or less rich "building blocks" for further job and organisation design.

An association can be seen between varying task structures and the following four types of automation,[20] indicating the differing functional requirements that different technical systems place on the nature of work:

- Type 1: electronic mechanisation, i.e. localised forms of automation in poorly integrated production systems with a predictable input/output state; includes computer-assisted isolated machines associated mainly with so-called "transitional human tasks of transformation";

- Type 2: automation with low uncertainty, including situations approximating to the ideal of the "automatic factory"; among operators, passive monitoring tasks prevail;

- Type 3: automtion with some degree of uncertainty but with little manual work; includes the most frequent cases of advanced automation, with frequent disturbances in input, conversion and output; the main tasks required are control tasks, including the capacity to operate manually or to modify the process;

- Type 4: computer-assisted integration with survival of relevant manual activities and with some degree of uncertainty; similar to Type 3, but a considerable amount of manual work remains.

Types 2-4 automation require "control tasks", which in all cases change the nature of human work. New tasks demand little or no physical intervention in the production process, and a more abstract work pattern emerges. The work shifts from a heuristic to an algorithmic-based mode of operating. Tasks allow choices within alternatives displayed by the electronic interface, unless it is possible to operate manually or to modify the algorithm. They imply not continuous activities but quick reactions to disturbances and problems relating to variances which occur from time to time. Within these broad characteristics, task content and relationships differ according to the type of design choice.

This classification of automation is further developed in Chapter 4, section 2.1.

7.3 The emergence of new design principles

Secondary (or peripheral) technology has a primary role in determining content and relationships of tasks in automation. The design of instrumentation, operational information systems, layout, etc., in one way or another can result in radical changes in the content and relationships of control tasks. When there are social system requirements on the nature of work, the object of design must be tasks and consequently secondary technology.

The nature of control tasks in automation makes the traditional philosophies and procedures of job and organisation design untenable. The nature of tasks is no longer compatible with prescribed job specifications, full use of work time in measurable activity, clear-cut boundaries among jobs, information and co-ordination concentrated in supervisory roles, segmentation of the workforce or narrow training.

From this some possible principles for a new paradigm of work in automation seem to emerge. These principles are set out in Chapter 4, section 3.

The actual design of occupations is influenced by three main factors: (a) the requirements of tasks; (b) the labour market; and (c) the organisational philosophies and strategies of the parties concerned. The best designs emerge from exercises where -

- a model exists for understanding the micro-macro relationships;
- a procedure exists for detecting socio-technical requirements;
- a means of integration is found between design and industrial relations and management of labour markets; and
- a method is put into practice for goal-setting and training by the parties concerned.

The overall organisation structure is not influenced by partial automation but is partly influenced by advanced automation. A shift away from administrative or personal control and a decline of hierarchical bureaucracy seems to be a functional requirement of automation. But hierarchical structures do not seem to be substantially weakened or altered. New forms of work organisation are limited to low organisational levels. Variables such as the organisation of power and the social stratification may finally have a stronger impact than the forms of co-ordination by mutual adjustment as functionally required by most types of automation.

The effects of automation on QWL depend upon the nature of work design. Thus no direct relationship can be found between advanced technology and QWL. However, some general concerns about QWL in automation arise, emanating both from the general nature of automation as a "rupture in the industrialisation process"[21] and from control applications which do not utilise the opportunities for taking QWL into account. These concerns include -

- possible job loss by older, less educated workers, unused to abstract control work, because the "automated organisation" is not designed for them;
- polarisation of the workforce as a result of segregation of large sections of workers in marginal or poor jobs or occupations;
- an increase in nervous strain and alienation in "transitional jobs";
- the dwindling of manual skills through loss of sensory-motor skills and "heuristic know-how";

- a sense of powerlessness when the nature of tasks and lack of training does not give workers control over the production process.

In conclusion, as far as the content of work is concerned, the possible negative impact on QWL derives not from automation in general but from Type 1 and Type 2 automation (electronic mechanisation and full automation). The issue of QWL leads us to the problem of design: good or bad QWL can be seen not as a consequence of automation but as a goal for designing automated units, a goal with a well-defined and measurable set of indicators. Thus the requirements of economic growth and the need to improve QWL can be optimised in concrete processes of design where various participants can realise their legitimate expectations.

Notes

[1] M. Duverger: Methodes des sciences sociales (Paris, Presses Universitaires de France, 1961). Most of our case studies are of the first type, as opposed to the tradition during the sixties when the second predominated in this field. See M.M. Towy-Evans: "The value and applicability of case studies", in Labour and Automation (Geneva, ILO), Bulletin No. 1, 1964, pp. 37-46. In Labour and Automation, Bulletin No. 2, 1965, 160 cases dealing with the social impact of technical change written between 1945 and 1963 are listed. Cases of changes in product, materials, machinery and control devices are taken into account: effects on employment, occupations, adjustment of workers and wages are tabulated.

[2] F. Emery and J. Marek: "Some socio-technical aspects of automation", in Human Relations, 1 Feb. 1962, pp. 17-26. Concerning generalisations from case studies, see also Towy-Evans, op. cit., and P. Naville: "Problems involved in measuring the effects of automation in case studies in France", in Labour and Automation, Bulletin No. 1, op. cit., pp. 47-69.

[3] See Ch. 2.

[4] Emery and Trist call this the "internalised environment". See F. Emery and E. Trist: "The causal texture of organisational environments", in Human Relations, 2 Feb. 1965, pp. 21-32.

[5] J. Child: "Organisational structure, environment and performance: The role of strategic choice", in Sociology, Jan. 1972, pp. 2-22.

[6] L.E. Davis and J.C. Taylor: "Technology, organisation and job structure", in R. Dubin (ed.): Handbook of work, organisation and society (Chicago, Rand McNally, 1976). They illustrate socio-psychological effects of technology and organisation design, and distinguish whether effects stem from the job configuration or from the organisational unit design.

[7] L.E. Davis: "Optimising plant and organisation design", in Organisational Dynamics, Autumn 1979.

[8] L.E. Davis: "Coming crisis for production management: Technology and organisation", in L.E. Davis and J.C. Taylor (eds.): The design of jobs (Santa Monica, Goodyear Pub. Co., 2nd ed., 1979).

[9] R. Blauner: Alienation and freedom (Chicago, University of Chicago Press, 1964).

[10] H. Simon: The shape of automation (New York, Harper and Row, 1965).

[11] J.D. Thompson: Organisations in action (New York, McGraw Hill, 1967).

[12] A. Tannenbaum: Control in organisations (New York, McGraw Hill, 1968).

[13] M. Maurice et al.: La production de la hiérarchie dans l'entreprise - Comparaison France-Allemagne (Aix-en-Provence, Laboratoire d'Economie et de Sociologie du Travail, 1978; mimeographed).

[14] F. Emery et al.: Form and quality of industrial democracy - Some experiences from Norway and other European countries (London, Tavistock, 1969).

[15] M. Maurice and M. Brossard: "Existe-t-il un modèle universel de structure d'organisation?", in Sociologie du Travail, Oct.-Dec. 1974, pp. 402-426.

[16] See a discussion on polarisation in FRG and N case studies.

[17] S. Mallet: La nouvelle classe ouvrière (Paris, Editions du Seuil, 1969).

[18] H. Kern and M. Schumann: Industriearbeit und Arbeitsbewusstsein. Eine empirische Untersuchung über den Einfluss der aktuellen technischen Entwicklung auf die industrielle Arbeit und das Arbeitsbewusstsein, Vols. I and II (Frankfurt am Main, Europäische Verlagsanstalt, 1970).

[19] H. Braverman: Labor and monopoly capital: The degradation of work in the twentieth century (New York, Monthly Review Press, 1974).

[20] This classification of automation, like the design principles mentioned below and explained in Ch. 4, is based on extensive research carried out by Federico Butera.

[21] See Ch. 5.

CHAPTER IV

THE IMPLICATIONS OF AUTOMATION FOR WORK DESIGN: SOME TENTATIVE CONCLUSIONS

1. Introduction

So far, the material contained in Part I of this volume has been carefully related to the results of an international project covering 12 countries and including a considerable amount of enterprise-level case-study and analytical material. Conclusions have been quite tentative where there were gaps in the information available or a lack of convergence in the analyses which emerged. Emphasis has been placed on connections or common points rather than comparisons.

This chapter takes a very different approach. While it is rooted in the same descriptive and theoretical information as earlier chapters, it is oriented towards more general findings which the authors believe have significant implications for policy and practice concerning automation at national, enterprise and shop-floor levels. Such an undertaking is necessarily somewhat speculative, not to say hazardous.

It is surprising how many modern institutions and policies - including industrial relations systems, systems for education and training, policies for employment and growth - are ultimately based on assumptions about the kind of work for which people are being prepared and the efficiency with which they will carry it out. These assumptions are being radically challenged by changes in the nature of work due to automation. The content and implications of these changes at shop-floor level and beyond are the subject of this chapter.

2. The content of work in automated production systems: types of automation and task characteristics

What happens to work when automation occurs? Some jobs disappear, others are partly changed, while still others are entirely new. Moreover, different combinations of these changes occur in different instances of automation. Even when the same automatic hardware is introduced in different locations, different results are found.

There are two main factors which complicate the relationship between automation and work. First, automation is in itself complex. It is necessary to consider different types of automation in order to discern the impact which is taking place. Second, the effects of automation are not directly on jobs but on tasks and task relationships. The tasks which are found in automated systems can be combined in various ways, so that different sets of jobs can be designed to operate the same production system.

This section presents four types or categories of automation, each with characteristic main tasks, as mentioned briefly at the end of Chapter 3. The kinds of task principally associated with each of the first three categories are analysed in terms of the desirability of jobs composed of such tasks. We then go on to discuss task relationships and problems of labour market segmentation. Section 3 then covers the design of work under automation, suggesting ways that both efficiency and human objectives can sometimes be met.

2.1 Four types of automation: characteristic tasks and characteristic problems

2.1.1 Type 1 automation

This type may be called electronic mechanisation. It includes isolated computer-assisted machines which require constant feeding, emptying or other operator interventions (automatic packaging, automatic assembly, NC machines in mass production, simple industrial robots, etc.). Most of the tasks left to people are residual and ancillary to machines, often located at the intersection of extremely fast and sophisticated processes. The other principal tasks include maintenance, repair and materials handling.

The fragmented tasks characteristic of Type 1 automation are rarely designed in advance; they are merely left over. Such tasks not only have fragmented content, but tend to be determined in their pace and timing by the "needs" of the machines which precede or follow. It thus becomes important to programme such tasks with strict specifications so that bottlenecks are not created. Alternatively, the need to fragment such tasks can derive from Taylorist efficiency considerations directed at minimising the time required for workers to carry them out. At the extreme, automation is used to minimise human thought through real-time instructions. Means have been devised to that instructions may be conveyed electronically and instantaneously to the worker, with the result that the traditional roles of people and machines are effectively reversed. The worker becomes a device to be "operated" by a machine over which he has no control.

Mechanised assembly-line tasks characteristic of Type 1 have long been studied, and the principal negative effects are well known. There are, however, differences in both degree and nature due to increased automation. As more and more work is automated, the remaining tasks are shortened in cycle (down to 1.3 seconds in one example). The discretionary parts of tasks, even if they are only the choice of the following correct part for assembly, are increasingly taken over by more reliable automatic control systems. Real-time control results, so that the worker cannot even get a few seconds ahead and momentarily escape from the requirements of rigidly and rapidly following the machine's instructions.

It has long been argued that the tasks characteristic of the first category of automation are "transitional", i.e. bound to disappear with advancing automation. It is true that from a technical point of view it should be relatively easy to incorporate any extremely repetitive, heavily machine-controlled task into automatic hardware. Experience has shown, however, that such tasks are being generated more rapidly than they are being "automated away".

2.1.2 Type 2 automation

This type of automation concerns operating conditions with low uncertainty. It includes situations approximating the ideal of the automatic factory: unmanned factories, automated power plants, automatically controlled railways, automated refineries, etc. It also includes isolated cases of automation of part of a production process when little active intervention from the worker is required (e.g. NC machines with long production cycles). The primary task found is passive monitoring, with operators often isolated from one another. The most important worker characteristic required is reliability rather than skillfulness although there may also be some more complex maintenance tasks. The tasks of initial design are highly complex, while those of management imply very great responsibilities but are often simple since control is built into the system. There are few active control tasks.

The passive monitoring tasks which are characteristic of Type 2 automation tend to be monotonous and, in the short run at least, not very purposeful. On the other hand, the worker is often held responsible if a problem occurs and he or she does not respond immediately and correctly. The combination of tasks which are extremely tedious and the responsibility for avoiding damage to expensive equipment or even for public safety disasters[1] places a heavy strain on the workers.

A second characteristic of Type 2 automation is that it represents the extreme case of elimination of jobs and obsolescence of skills. The development of the various skills which make up a craft or occupation is a major investment for a worker. At higher skill levels, this investment is in principle reflected in higher pay and greater job security. These economic benefits result from the increased bargaining power which derives from the scarcity of the skill. At the same time, skills are an investment in a more personal way. Highly skilled workers enjoy considerable status among their fellow-workers and can take pride in the results of their work. They can generally count on interesting jobs and on comfortable conditions at work.

Automation is capable of brutally eliminating the value of skills which have taken an entire career to develop. An NC machine - to say nothing of more integrated manufacturing processes - can be programmed to closer tolerances and can achieve much more rapid production than a skilled machinist. There would seem to be no well-defined (and therefore programmable) manual skills which are exempt from the implications of robotics or other automation developments, except for the skills required to maintain the automatic equipment itself.

2.1.3 Type 3 automation

This type of automation concerns operating conditions with moderately high uncertainty but little manual work. It includes the most frequent cases of advanced automation: both the machines and the unit are run automatically but disturbances are frequent in input, conversion or output. Examples are paper mills, printing shops, automated transfer lines with a varied product mix, etc. The greater uncertainty requires a higher level of intervention from operators. The main tasks required are control tasks, which usually include a capacity to operate manually or to modify the process, as well as a large variety of comparisons and adjustments. Tasks relating to planning and programming become as important as those concerning maintenance. Task relationships tend to be interdependent: co-operation is often necessary or useful among those carrying out planning, maintenance or control tasks.

The task most characteristic of Type 3 automation, the full control task, has significant advantages over ancillary or monitoring tasks. However, work consisting entirely of control tasks is radically different in terms of knowledge and skills required from earlier forms of industrial work. This results partly from the fact that automatic control systems eliminate the need for direct human observation and intervention at many stages of production processes. More important, the integration of control functions due to automation results in the need for decisions based on an entirely different kind of knowledge from non-automated systems. Once control systems begin to affect large, inter-related stages of the production process, human decisions need to be based on an understanding of the principles of the control system rather than on experience with a particular machine or job. These principles are reflected in algorithms, i.e. logical representations of processes such as that included in computer programs.

A negative possibility is when a job is designed in such a way that the worker may need neither manual skills nor experience concerning the specific industry or product. Thus the worker is not only physically remote from the product, but his mental work is based on abstract understanding of the control system rather than experience. Ultimately, it is more than skill which is lost: it is the entire concept of craft-based occupations and career patterns. Different consequences result from role designs encompassing a high degree of control over the computer and over the physical process: a good deal of communication and decision making, possibly within work groups, contribution to maintenance and innovation, and planned training and continuous career design.

2.1.4 Type 4 automation

Type 4 automation is similar to Type 3 in terms of operating uncertainty, but the need for manual work remains. The production of seamless steel pipe, chemical fibres or aluminium, for example, though they utilise considerable automatic control, require manual intervention (refitting, setting up, change-overs, etc.). Maintenance repair, transport and cleaning operations are also quite extensive in many processes. In Type 4 automation, there are numerous opportunites to combine manual and control tasks during job design. Interdependent task relationships provide a certain amount of pressure for multi-skilling and group work. There is, however, nothing inevitable about the appropriate combination of tasks into jobs. It requires both technological choices and careful design.

2.2 Task characteristics and roles

The types of automation listed above have been described in terms of the tasks which tend to accompany them. Each of these kinds of tasks is in fact an entire category containing a range of skill requirements. These ranges may be described as follows:

- manual tasks: from simple, unskilled labour (e.g. machine operators in Type 1 automation, cleaning or janitorial work in Types 2, 3 and 4) to skilled maintenance, materials handling and setting up tasks;

- monitoring tasks: from passive monitoring of a machine in order to stop it and notify a specialist in case of breakdown to passive monitoring of an entire production system with responsibility for intervention in a number of differing contingencies;

- control tasks: from direct manual control of a simple system to supervisory control of complex systems.

In order to approach the question of design it is necessary to consider the fact that workers may have comprehensive roles to carry out which include specific tasks. They may, for example, be responsible for providing or acquiring information by interacting with other workers, for the quality or quantity of outputs or for other results, or for ensuring that certain problems are taken care of when they arise, and that equipment or products are not damaged. This entails contingencies which may not be precisely foreseen: the exact time at which well-known problems arise may not be predictable, or action taken may depend on events at other stages in the production process, and so on. Thus roles place a whole set of expectations upon individuals or groups, and comprise a building block of the socio-organisational system in which task results and relationships are component elements. Such roles cannot be readily summed up. In section 3 below, on design, both tasks and roles are considered, because they determine what workers or groups actually do, the knowledge and skills required, the degree of co-operation among workers, and the career patterns and occupations which are possible.

2.3 Polarisation and segmentation

Polarisation of the workforce refers to the separation of workers into two categories, one consisting of highly skilled technicians and the other of unskilled manual workers. Of course, it is not necessary to restrict the idea of polarisation to two groups. Further divisions are possible, in which case it is more appropriate to refer to segmentation. Early fears about automation and polarisation were expressed in terms of the hypothesis that the latter was an inevitable result of automation. There are now numerous examples which contradict this hypothesis, at least in its strongest formulation. On the other hand, the association between types of automation and the types of tasks described above obviously opens opportunities for creating specialist groups of workers to carry out specific tasks. While this creates functional problems in advanced forms of automation - communication, co-operation, flexibility and response to emergencies become more difficult - there are countervailing pressures which ensure that work designs involving polarisation and segmentation are very common. Some of these pressures are socio-cultural, as in the case of occupational stereotypes for women or immigrants. Others are technical or economic, as is often the case in decisions to apply Type 1 automation. Still others are based on old-fashioned managerial philosophies derived from Taylorism.

There is an additional source of segmentation which relates to the possible separation of an enterprise or plant from the labour market as a whole. Even when the design principles listed in Chapter 3, section 7.3, are followed, which would imply that there would not be a problem of polarisation or segmentation within the enterprise, there is no guarantee that the jobs being created are appropriate for the local or national labour market. There is considerable evidence, in fact, to the contrary: especially in the case of new factories with a high degree of automation, work designs often provide no jobs outside a very narrow segment of the labour market, consisting essentially of young, well-educated and often male workers. Alternatively, job opportunities for others are available only through subcontracting arrangements such as janitorial services providing poorly paid, insecure and otherwise undesirable employment.

3. The design of work in various types of automation: constraints, principles and processes

The purposeful design of work in automated systems emerges as a critical necessity. In every instance of unplanned change reported in the project, there have been serious negative consequences requiring remedial action. In many of the cases where some planning was done, it was insufficiently early or thorough.

3.1 Constraints

Design, unfortunately, is not easily optimised. Serious limitations on the potential for appropriate design have been found in three broad categories: economic, technical and social.

The economic constraints on design are the most pervasive. Appropriate design is slow and needs to be individualised to the specific work situation. This can be done without causing delays if an entire new production facility is being designed - a process which in any case takes considerable time - and if the social side of design is considered as early as the technical side. In most cases, however, automation is introduced piecemeal. This approach is based on introducing limited automation where the greatest immediate economic advantage is to be found, to be followed if necessary by further steps of "automating away" any problems which emerge. The long-run economic and social costs of such an approach may be very large, but each individual decision seems economically "rational".

Technological constraints appear at various stages of the design process. Mass production technologies, especially light assembly, consistently result in Type 1 automation. This may not be inevitable, but the only alternative seems to be to carry automation so far that no assembly or machine-operation tasks remain. Both technological difficulties and very strong employment effects - not to mention the difficulty of circumventing the piecemeal approach to automation - mitigate against this alternative. Differing basic technologies have also tended to result in Types 2, 3 and 4 automation with their varying proportions of monitoring, control and manual tasks. In many cases, however, the design of peripheral (secondary, control) technologies can alter the proportions and the relationships among such tasks.

Social constraints tend to be expressed in a more subtle fashion, but they none the less place real limits on design. The respective roles occupied by managers and workers, for example, are necessarily connected to the relative power and status of each in a given culture or community. If the design of jobs, occupations and organisational structures do not take these social constraints into account, the result may be unacceptable to one or the other party.

3.2 Design principles

3.2.1 Type 1 automation

In this type of automation, economic and technological pressures drive towards fragmented, machine-paced jobs with few skills and with workers under constant threat of replacement by machines. Some limited practices have been developed (e.g. decoupling of machine pacing, job rotation, job enrichment) which attempt to provide a modicum of variety. This is difficult, however, and the results are likely to be limited. More advanced techniques (e.g. assembly performed by semi-autonomous groups) tend to require extensive changes in equipment and layout, and are usually associated with new factories. Thus the most important design principle concerning Type 1 automation is early, comprehensive consideration of the nature of the work which is being created for people. The problems of this type of automation come from basic technology, usually as chosen and designed in detail by engineers.

3.2.2 Types 2, 3 and 4 automation

The nature of the control tasks and of the resulting overall roles in these forms of automation is incompatible with the traditional philosophies and procedures of job and organisation design, which are based on -

- job specifications prescribing a limited number of well-defined tasks;
- use of all available work time in measurable activity;
- clear-cut boundaries among jobs;
- narrow, job-specific training;
- segmentation of the workforce;
- information and co-ordination roles restricted to supervisors.

These old design principles assume that the bulk of the tasks which need to be carried out by the workers are simple, manual and repetitive. No roles exist which exceed the sum of elementary tasks formally specified. Control tasks and roles do not fit any of these assumptions.

The following design principles have been demonstrated as possible, and have shown enough success to suggest that they could be adopted on a widespread basis:

- job specifications which combine manual, monitoring, computer control and direct control tasks;
- role specifications complementary to task specifications which give responsibility for functions, results, control of variances and co-ordination;
- use of multiple skills;
- work groups as internally managed units;
- career patterns which link manual and technical work and which overcome the boundaries between trades;
- training which combines both theoretical understanding and practical experience of large parts of the production process.

These design principles, while related to the merging nature of work under automation, do not occur inevitably. While there are costs of dysfunctional design, these costs are borne in part by workers and by society. Moreover, there is no guarantee that enterprises will recognise the opportunities presented by better design or will choose appropriate approaches and techniques from the many competing schools of thought on the subject.

3.2.3 Two possible types of work in automation

Two different types of work can result from automation: auxiliary work which takes care of imperfections in automated equipment, and process control work which involves actual control of technical processes and relatively high knowledge and status. The first type of work will be found more frequently in Types 1 and 2 automation, while the second is more often associated with Types 3 and 4. Purposeful design can extend the possibilities for process control work.

The human implications of polarisation and segmentation, as discussed above (section 2.3), are severe. The restriction of the desirable aspects of work to privileged groups and the relegation of others to undesirable jobs without even much hope for the future is one of the least attractive aspects of working life. It is difficult, however, to go much beyond the design principles already listed. What is really needed is occupational design comprising both societal and enterprise-level considerations. The principles of such design are not yet clear, though certain procedures can be discerned.

3.3 Principles concerning design processes at enterprise level

In order to encourage appropriate design, greater emphasis should be placed on the design process. The design process at present tends to be unplanned, it tends to avoid social considerations and it rarely involves in a serious way all the parties who are affected by design decisions. The alternatives - careful planning, joint consideration of both social and technological choices and participation of all those concerned - can be encouraged.

The key figure in the design of automated production systems is of course the engineer, who is always a participant in design, and often the only one. In many cases, however, engineers are anonymous and removed from the production process, designers of hardware and systems which are applied in places they have never seen, perhaps after adaptation by yet another engineer of merely installed without modifications.

There is an increasing evidence that design by engineers is seen as insufficient. From the technological - or technocratic - point of view, this is reflected in increasing recourse to systems analysts, who are seen as having a wider view than engineers. However, there exists a need not merely for a wide view but for a different view. People cannot be engineered in the way that machines can. Solutions imposed from outside remain external impositions even if their purpose is humanistic. If the changes due to automation have to take into account all desirable opportunities, they must be designed and introduced in a participative way.

In spite of the technical rationality that might be expected to accompany non-participative change, much automation is introduced in a haphazard way and results in unpleasant surprises. Moreover, piecemeal introduction of automation, even if individual decisions are carefully taken, is unlikely to build up into a coherent system. However obvious the point may sound, it is necessary to insist that design be planned. This applies to remedial measures as well as design of entire new systems. Such planning necessarily must begin at the earliest possible stage in the design process: each decision forecloses later choices.

There is a temptation in practical design work to separate the questions of participation and planning. The extent to which this can be done without foreclosing essential options is very limited. In Type 1 automation, the choice of the basic technology makes the preponderance of certain tasks inevitable. In Types 2, 3 and 4, the choice of the basic technology does not predetermine the nature of tasks, but the design of control systems and other peripheral technologies can foreclose most of the real opportunities for improving QWL.

4. Automation and work design in a wider context: the roles of public institutions and other interested parties

Engineering developments concerning automation spread very quickly through multiple applications of the same hardware and system designs. Work design techniques which allow consideration of QWL travel a much more circuitous route. This, combined with the disparity between the resources available for engineering applications of automation versus those available for QWL improvements leads to a rather wide gap between the intent of national legislation, policies and programmes, and practices at enterprise level. While planned, participative introduction of automation may be the ideal, remedial action or no action at all is the rule.

Implicit in nearly all the results of this project is the complexity and difficulty of appropriate design. There is a need for detailed analysis of the design options presented by various combinations of economic, technical and social factors. Planned, participative change processes require additional knowledge and experience. It is therefore not surprising that there is a lack of appropriate expert resources in most of the enterprises where automation is being applied.

4.1 Focal problems for societal concern

If the spread of the various types of automation is outpacing the ability to cope with change at enterprise level, it is necessary to identify the resulting problems before proposing solutions. The problems emphasised above may be summarised as follows:

(a) problems concerning QWL -

(i) monotonous, stressful, dead-end jobs (especially in Type 1 automation);

(ii) devaluation of existing skills and experience during the introduction of automation, with implications for both personal losses (de-skilling) and collective losses (reduction of bargaining power based on skill shortages);

(iii) minimum barriers to "automated jobs" which cannot be met by many workers, both older and young;

(iv) adjustment and "insertion", polarisation and segmentation problems;

(b) problems of direct concern to managers -

(i) automated systems which do not operate as engineered (e.g. worker rejection of Type 1 automation, emergencies in Type 2 automation);

(ii) management roles in advanced automated systems;

(c) problems of broader industrial relations or social concern -

(i) the challenge to educational, training and adjustment systems;

(ii) possible repercussions underlying assumptions of industrial relations structure and practices;

(iii) the need to link principles with actual practice.

The above listing, which is neither exhaustive nor particularly detailed, suggests the scope and seriousness of the problems which are to be faced.

4.2 The challenge for trade unions

The first category of problems listed above suggest that the trade union agenda is particularly heavy. From a trade union perspective, the protection of workers and the improvement of their QWL is especially difficult in the context of the problems raised by automation. The following three problems for trade unions deserve special emphasis:

(1) The problems of work content posed by the task characteristic of automation - especially Types 1 and 2 - derive from choices which are outside the usual scope of trade union influence or action. Trade unions have traditionally had little access to the managerial decision-making process or to the even more remote decision processes of engineers. While they can react to obvious problems once automated technologies have been installed, the scope for remedial action is usually quite limited.

(2) One of the most important bases of collective power is undercut by automation at precisely the same time that energetic action is needed. Acquired skills and experience are usually an asset in bargaining

because they represent human capital from which the enterprise benefits and which is difficult and expensive to replace. Yet such acquired skills and experience may find their value to the enterprise suddenly diminished and can even be seen as a liability. When enterprises could otherwise engage workers at lower pay to carry out new tasks, senior workers needing retraining are in difficulty.

(3) Polarisation or segmentation may reduce the solidarity of trade union effort. The workers who benefit in terms of status and pay may not be those with the highest seniority and may even come from outside the enterprise. Moreover, subcontracting or other practices may mean that the most disadvantaged workers do not belong to the union at all.

Most trade unions have developed very efficient techniques for dealing with matters such as contract negotiation or grievance settlement. Neither of these areas presupposes the kind of decentralised shop-floor expertise which would be required to participate in work design. Centralised expertise which is useful in contract negotiation or in assisting workers with grievances could be used to develop principles and procedures, but it may not in itself constitute a means for decentralised understanding or participation. Even in areas like manning and job classification, where some trade unions have developed extensive expertise in a complex field, the approach taken is essential <u>ex post facto</u> - that is, based on previous management decisions. Participation in design work which takes place prior to the installation of automated technologies is more difficult. In a few countries, trade union oriented research efforts have demonstrated an important potential role for trade unions in the field of work design. However, the information and training programmes which would be required to allow widespread, concrete action are still well below the level required by the complexity of the task of design.

The greatest danger for trade unions would seem to be treatment of the issues of automation and work design in the same way as other, more familiar problems. For example, technology agreements have sometimes included prior notification provisions if new technologies are to be introduced. This kind of provision can be very valuable in terms of job security: bargaining can then be opened on any possible redundancy problems. However, the kind of work design which could optimise efficiency and human considerations is unlikely to emerge from a bargaining process, especially one carried out at a high organisational level. The issues are much more complex, and appropriate design procedures require much more knowledge and participation at shop-floor level.

4.3 The challenge for management

Work design is the traditional role of management and it would not be inapporopriate to attribute to managers the major share of responsibility for both successes and failures. Nearly all the cases of introduction of automation or new work designs reviewed by this project have been the result of management initiatives. Where undesirable effects have occurred, this has usually been due to lack of planning or a management failure to explicitly consider the social implications of decisions. More positive results have been associated with extensive, early and planned consideration of social implications. While the number of positive results varies from country to country, the challenge to the professional capabilities and responsibilities of management seems evident.

Among the many difficulties faced by managers in trying to make production systems more effective in both production and human terms is a less than total control over technology. The design of automatic equipment rarely takes place under the control of the management where it is applied. The design and even the local layout of hardware, and the conception that programming of software may be done by engineers, systems analysts and programmers whose concerns are far from those of operational management.

Before considering what management might do in order to make better use of the opportunities afforded by automation, it is perhaps worthwhile to review the changes which have been observed in management practices as a result of automation. One notable change is that the traditional role of the foreman has tended to disappear. Programmers and other technicians have taken over some functions, while others are incorporated in automatic control systems themselves. While new functions in the area of technical work or training may be found for foremen, this is often an uncomfortable change for them. Although higher management choices affect the nature of work on the shop floor, and may sometimes entirely determine the content of workers' jobs (as in the case of purchase of machines or equipment with only one possible way of operation), the middle and upper levels of management may not themselves be changed by the functional requirements of production and social changes resulting from automation. Technical integration does not necessarily entail a drive towards social integration, or at most its influence is comparatively weak.

A factor of obvious interest to managers in considering the applicability of alternative work design practices is the economic costs and benefits involved. If measurements are made in the short run from a narrow point of view, little evidence may be found that work design based upon social as well as engineering/economic criteria is superior. However, the failure to use social criteria leads to at least two forms of negative results. First, in the case of advanced mechanisation and "transitional tasks", the undesirable job characteristics and their effects such as stress and dissatisfaction result in absenteeism, high labour, turnover, poor quality of work and other cost or productivity-related problems. Second, in more advanced forms of automation, poor design is related to the inability of the human system to react properly to unusual but possibly disastrous events with enormous economic and safety consequences.[2]

In view of the world-wide interest in work organisation improvements as exemplified in some Scandinavian and Japanese practices, and the obvious relevance of such improvements to the introduction of automation, it is tempting to ask why the extent of improvements in practice has been so low. Part of the reason may lie in the limitations of the practices themselves,[3] in resistance from workers and trade unions, or in the inherent constraints to organisational and technical change. The strong changes in first-level supervision coupled with resistance to change at higher levels suggests, however, that additional factors are at play. The objections voiced by first-level supervisors suggest that they were particularly unhappy with what they perceived as a loss of status and power. They were no longer unfettered decision-makers but instead found themselves required to collaborate with their subordinates due to the interdependence of work roles. As the design principles listed earlier in this chapter (section 3.2) necessarily require co-operation and power-sharing, the most serious challenge to managers may be the need to overcome those decision-making traditions and habits which are no longer appropriate.

4.4 Public policy and action

There is considerable evidence that automation has outpaced the ability of managements and trade unions to control, much less to optimise, its implications for QWL, at least for very large numbers of enterprises. Moreover, when the implications of automation - especially when coupled with poor work design - are aggregated, they go far beyond the concerns of a single enterprise. This final section lists the concerns which seem likely to require the attention of public policy in the near future. In many countries, public institutions, academic bodies and others are already in the process of developing appropriate responses.

4.4.1 Development of automated technologies

Most countries have recently made a considerable investment in the development of new technologies and automation. The policies, institutions, research programmes, etc., which have been established have both functional and social reasons for taking QWL into consideration, a fact that tends to be agreed with in principle but ignored in practice. If this shortcoming can be overcome, the dominant theme of technology may serve as both a focal point and a catalyst for the articulation of policies in many fields, which could be particularly advantageous if the policies are influential in scientific and engineering research. In practical terms, the most difficult problems to overcome is the application in many places of technologies which have been developed by a research group or vendor enterprise that is unfamiliar with the specific enterprise-level or shop floor problems involved. This is especially the case when the vendor is large and effective at marketing while the customer is small and weak at adaptation.

4.4.2 Adjustment, retraining and education

The changing content of skills, and especially the problems of de-skilling will require adjustment at national as well as at enterprise level. It is extremely difficult to predict the rate at which automation will proceed and dislocation result, but all observers agree that the rate is likely to accelerate in the near future. Two factors will need to be taken jointly into account by national policies and programmes: (a) the need to protect the existing investment of workers in education, skills and experience; and (b) the need for educational and training policies which make workers familiar with algorithmically oriented work (e.g. programming), both for young entry-level workers (or order to avoid employment barriers) and as part of continuing education and readjustment training. In addition, the need for management training and workers' education are considerable and growing.

4.4.3 Industrial relations

Negotiation on the implications of automation has usually been limited to those effects which are easily quantified, and to broad principles. The implementation of improvements at shop-floor level requires a different sort of approach: joint influence by parties with differing goals in a common exercise in order to design something new. Such an approach seems extremely difficult to attain the face of four limiting factors -

- the conflictual situation arising from opposing interests of management and workers in areas such as job security and wage levels;
- the destruction of craft-based, seniority-linked career systems which provide both a stable occupational structure and a future for workers, and a source of first-level supervision with links to the workers;
- the lack of necessary design expertise on the part of both shop-floor level management and workers' representatives; and
- the lack of accepted procedures of workers' participation in design that threaten neither management prerogatives nor trade union autonomy.

4.4.4 The emergence of a design community

The functional requirements of integration when designing automated systems, the difficulty and complexity of appropriate work design, and the widening participation in the design process provide us with hints that a community of interested institutions and individuals has begun to develop around automation and work design. The development of such a "design commuity", should it occur, could provide a means of overcoming the gap which has been identified between public policy and shop-floor practice.

The characteristics of a design community are not completely clear and will undoubtedly vary from country to country, from industry to industry and from enterprise to enterprise. None the less, at least at national level, it seems possible to identify the following important points:

- the design community will often be able to rely on a national policy enunciating design principles and procedures. These principles and procedures will usually be the subject of legal obligations;

- it will place emphasis on its scientific basis: objectivity and careful research will be more important than ideology;

- it will contain heterogeneous groups with differing and sometimes conflicting goals, but will, however, develop consensual rather than conflictual methods of decision making; and

- it will encompass public and private institutions outside the enterprise which may bring research results, engineering developments, information about new products and techniques, consultancy services and sometimes financial support to activities at enterprise and shop-floor level.

Notes

[1] See the discussion of the Three Mile Island accident in Ch. 16 on the United States (section 1.9.1).

[2] While there was considerable evidence of economic motivation for the introduction of automation in this project, the actual economic results were mixed. See the relevant sections in Chs. 2 and 3. See also A. Hopwood: "Economic costs and benefits of new forms of work organisation" in ILO: New forms of work organisation, Vol. II (Geneva, 1979), pp. 115-145. As regards disastrous events, see note 1.

[3] For limitations and problems concerning quality circles and zero-defeat groups, see Ch. 10 on Japan (section 1.6).

CHAPTER V

AUTOMATION AND WORK: CONCEPTS AND MODELS FOR ANALYSIS AND DESIGN

1. Introduction

The conceptual and empirical bases of the approach to automation and work design adopted in this book relate to extensive published research in several fields over a period of many years. While it is not possible to do full justice to such a considerable body of knowledge in a single chapter, it is none the less worthwhile to provide a summary description of the most relevant literature. In most cases, this is done in a very brief way, though the interested reader can find full bibliographic references in the classified bibliography at the end of the chapter.[1] For topics which relate very closely to the substance of the book, such as the definition of automation, a more complete review is provided.

In addition, the final section of this chapter provides a summary of a technical model which may help to organise or illuminate the design of work under conditions of automation.

2. Social causes and consequences of technical developments: related fields of study in the sixties and seventies

First of all, it is necessary to set up clear boundaries to our subject within the wider and perhaps imprecise field of the social effects of technology. The phrase "social effects" has been variously interpreted to include employment, privacy and other considerations which are outside the scope of this book. Rather than social effects, this chapter concentrates on the QWL and more particularly on the nature of work. At the same time, it is important to draw a strong distinction between automation and "new technology". New technologies very often concern changes in conversion processes themselves - for example, a new way of making steel or refining oil. Moreover, the concept of technology goes beyond conversion techniques and equipment to include the context and purpose inherent in the technology itself. For example, Woodward (3.3, 1970) defines technology as "the collection of plant, machines, tools and recipes available at a given time for the execution of production tasks and the rationale underlying their utilisation"(page 4). The concepts of QWL and of automation will become clearer in the sections below which deal with their definitions and relationships.

Important contributions to our particular topic come from neighbouring fields of study. Many of the reported consequences of automation have been studied under different technical conditions (Council for Science and Society, 3.1, 1981). For example, the process of fragmentation of jobs, the withering away of occupational skills, the reduction of workers' control over production process, etc., are associated with a work design philosophy born in a previous stage of mechanisation. Both this previous work design philosophy and former technology are still present in most cases of automation. In addition, reference to the history of the division of labour and of the traditional work organisation philosophies may be useful.[2] Studies on the effects of automation on employment, labour market and social structures have to be kept in mind as well.[3]

The rise of an impressive technical debate and a set of co-ordinated research programmes on automation between 1958 and 1966 has provided most of the available substantial literature. These developments were stimulated mainly by governmental and societal concern with automation's impact on the level of employment, on educational and on industrial relations. It is difficult to isolate the findings of that period from its own context. Variables which can affect a typical social invention such as the organisation. Such considerations as technology, size, location, market and environmental instability were examined in relation to business functions, hierarchical structure, supervisor/technician/worker ratios, communication patterns, decision-making processes, job specialisation, etc. Some schools of thought were more interested in problems of manoeuvring the top organisation structure (Woodward, Pugh et al., etc.), others in problems of work design (Touraine, Davis, Blauner, Meissner, etc.), but both moved decisively away from the normative "one best way" approach to organisation design. This area of organisational studies is very pertinent to our topic. Many propositions about technological influences on organisation must be seen in the context of a discipline trying to overcome a dogmatic approach to the organisation rather than an expression of positivistic bias or technological determinism.[4]

The literature of the seventies is dominated by the idea of QWL. It encompasses and integrates various dimensions -

- a revised consideration of workers' needs, attitudes and expectations going far beyond the paradigm of the "economic man" upon which the organisational paradigms were built (Walton, 5.1, 1976);

- the identification and measurement of parameters of quality of work itself extended beyond the traditional ones of working conditions (Seashore, 5.1, 1976a);

- the stress on the change in work itself: tasks, jobs, work groups, communication and decision systems, internal mobility, training, career as influencing QWL, in addition to physical work environment, safety, job security, pay, good human relations, etc. New principles of work design, new paradigms, new formulas and experiments are the outcome of this changed attitude to work design (Klein, 3.4, 1975; ILO, 3.4, 1979); European Foundation for the Improvement of Living and Working Conditions, 3.4, 1978);

- the increased attention paid to the social procedures through which work is designed: workers' participation, interdisciplinarity, trade union involvement, new forms of social accountability of firms, changes in industrial relations procedures, government involvement through social planning and special agencies, etc. (Emery and Thorsrud, 3.4, 1969);

- the rallying of social research institutions and knowlege regardless of disciplinary boundaries in order to contribute to designing a better world of work.

Sometimes QWL ideas and experiments were linked with advanced technologies; more often they were not. However, basic ideas were developed regarding socio-technical systems analysis and design (Emery, 3.3, 1959), the making of organisational choices permitted by technology (Trist et al., 3.3, 1963) and of the anticipatory design of technology itself with human values (Davis, 5.2, 1979).

Because of their non-deterministic view of the effects of technology, QWL experiences are not described in terms of effects, impacts, trends and consequences; on the contrary, they develop the idea of technology subjected to social options by stressing design, joint design, joint optimisation, anticipation, etc. The QWL experience (or movement, as it is sometimes called) offers two sets of different contributions: (a) a series of findings leading to scientific knowledge of relationships between work and technology; and (b) a different view about how to develop options for the future.

3. Automation

3.1 Automation in the popular sense of the term

The word "automation" was first used by John Diebold (2, 1952), who wrote the first authoritative book on the subject (Automation: The advent of the automatic factory) and by D.S. Harder, Vice-President of manufacturing of the Ford Motor Co., who used this term to describe the new concepts encompassed in automatic handling within transfer lines (so-called "Detroit automation").

The popular use of the word automation has an evaluative nature, stressing the particularly high level of either: (a) the technical sophistication of the automatic control of a machine or set of machines; or (b) the integrative mechanisms of a productive system built into its technical components; or both. Popular language gives to automation both the meaning of a technology and of a system of production.

The word automation is usually associated mainly with engineering developments -

- continuous flow production, with advanced mechanisation both of individual machines and materials handling so that "the product moves untouched by human hands" (Detroit automation). It is largely the outcome of mechanical engineering thinking (de Bivort, 2, 1955);

- automatically controlled machines: the development of automatic control functions, recently mainly through electronic devices. Control comparisons in modern automation concern not only fixed expected performance (as in traditional servo-mechanisms) but a programmed, varied set of expected performances. NC machines fall into this category. Electrical engineering knowledge and techniques played a major role in this field, together with mechanical engineering and computer science;

- computerised control of flow production processes (both continuous chemical processes, as in an oil refinery, and continuous or semi-continuous mechanical processes, as in a steel rolling mill). Computer science and process engineering have fostered most of these developments.

Advanced mechanisation of materials handling, sophistication of automatic machine control and computerised process control are the basic "automatic techniques" most people have in mind when labelling a technical development as "automation". In this respect, automation certainly includes recent developments such as automated packaging, machining centres, computer-integrated manufacturing, robots, flexible manufacturing systems, unmanned factories, centralised and decentralised remote control, etc. Computerisation of research, engineering, marketing, production planning and control, quality control and administration imply both the use of similar techniques and equipment and the integration of the overall information processes.

3.2 Why so many definitions of automation?

Automation has been defined in a number of different ways, but there are two main tendencies relating to its underlying meaning.

First, a group of authors used the term simply to describe an advanced stage of technology development, encompassing in a single concept technical evaluation, organisational complexity, elimination of workers from production processes, etc. This idea of automation also indicates a stage or a quality of other phenomena which have long been the subject of scientific investigation: technology; technical change; systems of production; process control; organisation of work; worker/machine allocation; labour/capital intensity, etc. Jaffe and Froomkin (2, 1968) provided some definitions -

- technological changes; any and all changes which affect the final product or the process of producing the final goods or services;

- invention: a paritcular combination of technological changes regarded by the public as a new device or contrivance;

- mechanisation: any technological change which increases output per worker (or work-hour);

- automation: this term should be reserved for that type of production process utilising the automatic or feedback principle, in which a control mechanism triggers an operation after taking into account what has happened before. The ultimate in automation is the closed-loop process, a method of operation which requires no human interference from the time the raw material is inserted into the machine to the time the finished product is stored or stacked at the end of the production line.

For these authors automation would be nothing more than a particular type of technological change. Sultan and Prasow (2, 1964) draw similar conclusions. For them "the word 'automation' is employed loosely, and there seems to be little advantage in distinguishing it from 'advanced' forms of technical change" (page 32). However, although this general definition is frequently used in an advanced technological age, we shall not consider it in further detail but will move on to more specific uses of the term "automation".

A second larger group of authors has accepted that automation concerned a specific phenomenon, and has reached basic agreement that the core of automation is the unprecedentedly high degree of displacement of human work by machines. These authors have gone on to discuss the phenomenon from various theoretical perspectives, notably in terms of levels of conceptualisation and of key elements used to view the production system.

3.2.1 Levels of conceptualisation

Levels of conceptualisation of automation may be seen in terms of the following four dimensions:

- comprehensive concepts of automation, encompassing interpretations of relationships between technology and society. They usually tend to provide a theoretical framework, with propositions about industrial societies, their dynamics and their overall trends. They may also be interpretative models. For instance, if we need to evaluate whether automation (or new technology) is a continuation or a rupture in the industrialisation process, we must operate at this level;

- relative definitions, intended to explore through empirical research or action programmes whether some newly detected problems may be created by the new phenomenon that we can call automation. For instance, if we need to analyse the profound change in the skill requirements occurring in modern industry, we must operate at this level. Sadler (4, 1968) writes:

  If there is a case for social science research directed specifically at problems associated with automation, as distinct from problems associated with technological change in general, it must rest on the assumption that there is something special about automation which makes it give rise to problems that are different from those generated by other forms of technical progress. To state this assumption more clearly, it is that in recent years there have been certain developments in technology which are qualitatively different from previous ones and which, in consequence, render past experience of handling technological change a less and less reliable guide to future policies (page 2).

Most definitions given in sociological, psychological and epidemiological research on automation are of this kind;

- socio-technical models of analysis and design involving organisational units affected by advanced technology. If a joint design of technology and the social system in a productive unit is planned, it is necessary to map out the relevant connections among technology, organisation and social system within a meaningful model of the specific unit under study (Emery and Marek (4.2, 1962; Davis, 1979). The "design model" presented later in this chapter uses this kind of conceptualisation;

- analytical dimensions of automation, intended to develop typologies and to test organisational and QWL correlates of automation. In order to test whether work groups are facilitated or inhibited by automation, we need to use this level of conceptualisation (Crossman, 2, 1965; Bright, Herbst, Meissner, Blauner).

There is, of course, an interplay among these levels. Societal studies could be improved if analytical dimensions were sorted out and tested. Empirical research on specific problems could benefit from a higher level of conceptualisation. Authors may have an all-embracing model - or, more frequently, a "style" - allowing them to move across the different levels, but they usually concentrate on one or two of them. Research at each level is highly beneficial, but comparisons crossing the different levels create confusion. It is inappropriate, for example, to oppose Richta's pessimistic view of automated society bringing about a new social order with Diebold's representation of new manufacturing philosophies for the education of managers or with Crossman's taxonomy contributions to an analytical understanding of the impact of new technologies on skills.

3.2.2 Key elements

The various key elements relating to automation seen as displacement of human work by machines are as follows:

- control systems: automation would be characterised by closed-loop feedback control mechanisms, whether mechanical, pneumatic, electronic, etc.;

- technology: automation would be associated with electronic computer technology as the mechanisation of sensory, control and thought processes. The terms "new technology" or "EDP technology" are used more frequently in this case;

- organisation or production: automation would consist of continuity of production flow and of integration among different machines in a unitary control;

- organisation of the enterprise: automation would be the integration of different phases of the industrial process and of different business functions, through an appropriate compter-assisted information and decision-making system;

- economics: automation would be a particular type of technological change;

- historical: automation would be a step in the history of industrialisation concerning not only technology and production but the structure of society at large.

In the following section we review the literature on automation, as used in the second sense, in terms of these key elements.

3.3 Perspectives on automation as the replacement of human work by machines

3.3.1 The functional/technological approach

This approach, to which Bright and Crossman made a major contribution, identifies automation on the basis of the human functions replaced by machines. Other authors also subscribe to this view: "Automation is the use of machines to run machines" (Drucker);[5] "the central idea [of automation] is that mechanical or chemical processes are directed, controlled and corrected within limits automatically, that is, without further human intervention once the system is established" (Dunlop, 2, 1962); and again, "automation is the mechanisation of sensory, control and thought processes" (Killingsworth, 2, 1963).

Some authors indicate automatic control as the main characteristics of automation, thus maintaning the continuity between mechanisation and automation. Bright (2, 1958) offers a classification of levels of mechanisation within which the shift of functions between worker and machinery appears to take place in a continuous trend (table 10).

Yet others, who also stress automatic control, maintain the discontinuity of automation because of the appearance of a new and powerful computer technology. Crossman (4.2, 1960) writes: "Automation differs from mechanisation in that it involves <u>machines</u> which have been built in the capacity to supply process information ... through general purpose and flexible device like <u>computers</u>" (our underlining). He goes on -

> Automation is the second major stage in the historical process of replacing human labour by machinery ... the replacement of human information processes by mechanical ones ... Under this definition

Table 10: Levels of mecahnisation and their relationship to power and control sources

Initiating control source	Type of machine response		Power source	Level number	Level of mechanisation
From a variable in the environment	Responds with action	Modifies own action over a wide range of variation	Mechanical (non-manual)	17	Anticipates action required and adjusts to provide it
				16	Corrects performance while operating
				15	Corrects performance after operating
		Selects from a limited range of possible prefixed actions		14	Identifies and selects appropriate set of actions
				13	Segregates or rejects according to measurement
				12	Changes speed, position, direction according to measurement signal
	Responds with signal			11	Records performance
				10	Signals preselected values of measurement (includes error detection
				9	Measures characteristic of work
From a control mechanism that directs a predetermined pattern of action	Fixed within the machine			8	Actuated by introduction of workpiece or material
				7	Power-tool system, remote controlled
				6	Power tool, program control (sequence of fixed actions)
				5	Power tool, fixed cycle (single function)
From man	Variable			4	Power tool, hand control
				3	Powered hand tool
			Manual	2	Hand tool
				1	Hand

Source: J.R. Bright: Automation and management (Boston, Harvard University, Graduate School of Business Administration, Division of Research, 1958).

> automation is a change of operating method which can be introduced in any field where brain-power is an input to an economic production process, whether in manufacturing, distribution, finance, insurance, banking, agriculture, medicine, education, law, government, public utilities, the arts or elsewhere ... All the human mental processes involved in guidance and control, operating machinery, remembering, calculating, solving problems, planning, decision making, discussion and so forth, can be seen as various modes of manipulating and using information; in a word, they are all information processes, which can be carried out equally well by mechanisms other than the human brain. It is precisely the substitution of the former by the latter means that is automation ... (pages 161, 162, 164).

The following features of technical systems connote automated technology:

- inclusion of indirect labour in machines: "Automation includes the mechanisation of more indirect labour activities ... material movements ... control activity ... testing and inspection activities ... and the mechanisation of data-processing through the computer" (Bright, 2, 1958);

- incorporation of sensory capacities into the machines: "[Automation] contemplates the wholesale reproduction of the sensory and mental functions of human operators in production systems which go far beyond the fixed sequence-fixed operation variety of automatic production" (Rogers, 2, 1958);

- closed-loop feedback (or self-correcting) control mechanism. Diebold group (2, 1964) distinguishes cases of open-loop control pertaining to old technologies (single program control with variable program built in; variable but separate program) and closed-loop control pertaining to new technologies (see Chapter 3, section 5.14) for an explanation of closed-loop control).

It is chiefly electronic devices like computers which have the capability to perform these work functions.

To summarise, the above definitions of automation stress the technological nature of automation, being a further displacement of human work by machines. Due to the nature of the displaced work and the machines, definitions and classifications of automation do not concentrate on the physical properties of the machines (hardware and software) but rather on their functional performances described by means of cybernetic concepts. The latter allow the allocation of human functions to the electronic machines. In addition, analysis and design of human work take place in new terms: control performed rather than physical action carried out; skills in place of procedures; goals instead of time use, etc.

This cybernetic way to describe work is helpful in every situation: it makes clear that physical work has almost always been inextricably connected with control (Butera, 4.2, 1978) and that the removal of the human control function from production occurred prior to automation. On the other hand, other more complex control activities concerning a wider segment of transformation processes arise, which need to be allocated to the same or different people, or to machines.

The report by the Council for Science and Society (3.1, 1981) comments on the difficulties of a clear-cut separation between mechanisation and automation. It suggests that definitions of automation and mechanisation abound and so do disputes about their application. Roughly, they correspond to the view that in the early industrial revolution, muscle power was replaced by mechanical power - water, stream and electricity. Later, human guidance and control was replaced by automatic control (i.e. automation). The report contends that the distinction between the two is often based on the absence or presence of feedback: upon whether the result of an action is sensed and used to correct the performance of the action. The control of temperatures, pressures and flows in an oil refinery is automation, because the temperatures, etc. are measured and compared with their desired values, and action is taken to correct any error. Sometimes complicated examples of mechanisation are called "Detroit automation" after the very complex machines used in the motor industry during the forties and fifties.

The report goes on to say that difficulties abound in these definitions: a robot may have feedback in its control drives, but may have no way of sensing the parts it is handling. Then in its external function we should have to call it mechanisation, though it is usually thought of as an example of advanced automation. In addition, feedback is largely in the eye of the beholder: almost any stable process can be regarded as having internal feedback. It is a question largely of the way in which we choose to write the mathematical equations that describe it.

On the other hand, continues the report, there is this justification for the distinction between mechanisation and automation. Until the forties, the imitation of human feedback, that is, of corrective action determined by observing what is happening, was difficult and poorly understood. Understanding, and skill in application, developed increasingly over the period 1950-70. The development of electronic devices rapidly increased our ability to sense and to measure and to construct systems in accordance with the theory.

The report concludes that mechanisation may be defined as conversion tasks displaced by people to machines. Automation is seen as both: (a) automatic feedback systems permitting a further displacement of conversion tasks from people to machines; and (b) automatic feedback systems replacing similar human systems in cases where transformation is already performed by machines.

Wild (5.1, 1975), who emphasises the continuity between mechanisation and automation, calls our attention to the co-presence of both in many concrete instances and to the unity of the historical process producing both -

> It seems reasonable to view both mechanisation and automation as trends rather than states, in both cases the trends being associated

> with the replacement of human activities by the activities of inanimate objects ... Mechanisation, for our purpose, is therefore seen as an aspect of, or component part of, automation, mechanisation being concerned with activities, whilst automation of such activities implies the use of control procedures ... Automation subsumes mechanisation (page 31).

The functional technological approach, as we have seen, provides illuminating criteria for positioning automation within the field of technical developments: it is fundamental and offers the basis for analytical taxonomies. However, it fails to provide a model of the configuration of the unit affected by automation, unless it is fully automatic: the adoption of terms like "partial automation" makes the situation even less clear. Nor does it incorporate any explanations as to why automation is introduced or what are the restraining and facilitating factors. The impression remains that only technical and conceptual advancements and limitations are taken into consideration. Finally, this approach does not fully account for the place of automation within the history of industrialisation, in spite of the declaration that it is "the second industrial revolution". Elements like occupational patterns, workforce composition, labour market, capital accumulation, administrative institutions, education systems, industrial relations, etc. are treated by this approach as mere consequences instead of as contituent elements of the historical process of industrialisation.

3.3.2 Automation as a form of organisation of productions and of the enterprise

The idea that automatic machines and computers were only a component step towards automated systems was first advanced by Diebold (2, 1952) and later supported mainly by people concerned with mechanical processes: "Automation denotes both automatic operation and the process of making operation automatic, with a resultant emphasis of self-regulation of the entire production process."

Continuity of production flow as a component part of automation is also frequently described -

> ... automation means continuous automatic production, linking together more than one already mechanised operation with the product automatically transferred between two or among several operations. Automation is thus a way of work based upon the concept of production as a continuous flow, rather than processing by intermittent batches of work (Northrup, 4.6, 1958, pages 35-36).

Integration among different machines in a unitary control system is also stressed by Diebold.[6] He maintains that it is no longer necessary to think in terms of individual machines or even in terms of groups of machines; instead, for the first time, it is practical to look at an entire production or information handling process as an integrated system and not as a series of individual steps. Diebold contends that automation is more than a series of new machines and more basic than any particular hardware: it is a way of thinking, as much as it is a way of doing.

Pollock (2, 1956) has a similar view, defining automation as "the integration of discontinuous or partial production processes into a co-ordinated process which associates advanced machine tools under the direction of electronic equipments".

Buckingham (4.1, 1961) provides a distinction along these lines between mechanisation and automation, stressing the change in the basic conception of the factory. In his description, mechanisation came first, creating the factory system and separating labour and management in production. Then mass production brought the assembly line. Finally automation, since the Second World War, has added the elements of automatic control and decision making, turning the factory from a haphazard collection of machines into a single, integrated unit. According to Buckingham, mechanisation is a technology based on forms and application of power, while automation is a technology based on principles of production organisation. Automation is more than a technology: it is a concept of manufacturing.

Integration of the different phases of industrial process through different business functions is also part of the concept of automation (Diebold, 2, 1952). He sees an enormous difference between a process which merely makes use of automatic controls and a process which is truly automatic. Diebold makes it clear that redesign of product, or of process or of machinery - and sometimes of all three - is often necessary in order to take full advantage of the new electronic technology.

An integrated view of the whole production process is implicit in automation and the scale of its application, whatever intermediate step we might observe. A full description of this view is summarised in Sultan and Prasow (2, 1964) -

> The elements common to these views of automation are:
>
> (a) the integration of production planning to fuse purchasing, production and distribution activities, and in the technical sphere the linkage of one machine activity to another;
>
> (b) the application of instrumentation techniques that simulate human skills through both open- and closed-loop control systems. Both input and output behaviour are communicated to control systems which in turn induce necessary changes in the production process;
>
> (b) the integration of informational technology involving market variables with process variables to influence production (page 15).

A constituent part of this approach is the notion of "degree", "completeness" or "level" of automation. In this view, cases of real automation are those of an unmanned mechanical factory or of a fully computer-controlled chemical plant and the like. Others need to be qualified by some adjective which suggests the relative distance from the ideal stage of "manufacturing fully performed and managed by machines", "a system autonomous from men" (Naville, 2, 1963): partial automation, incomplete automation, initial automation, etc. These adjectives refer both to the degree of autonomy of a technical contrivance from man and the degree of automation or production and business functions like conversion, transportation, maintenance, scheduling, accounting, etc.

This approach relies not on the idea of advanced technology but of advanced organisation. In technology social aspects are consequences, whereas in organisation social paradigms are built in as goals. Technological definitions are the main responsibility of engineers, while production organisation definitions involve production systems designers.

Naville (2, 1963) clearly points out that automation is the autonomy of functioning of a material, not human system, a property of a system not of a machine. He does not believe that automation exists in the absolute (the "automatic factory" being a particular case) but that there are degrees of automation. He concludes that automation is not technical in character but that, like every technology, it is primarily a system of concepts, and its technical aspects are results rather than causes.

These definitions and approaches have the advantage incorporating the following dimensions in the concept of automation:

- hardware deployed;
- forms of control;
- scope of applications (as in the technical/functional approach);
- continuity of work flow;
- integration among machines;
- integration among the production stages;
- integration of manufacturing and information processes;
- design of processes;
- integration between business functions, and especially among manufacturing and marketing;
- organisational philosophy.

However this could also result in a technocratic approach based on production engineering instead of the process or system engineering approach. In fact, alternative paradigms of organisation, of internal social systems and of society are not usually indicated in these definitions.

In conclusion, it is necessary to ask whether organisation paradigms will be traditional mechanistic/Taylorist, organic (Burns and Stalker) or something different. Which human ideas will automation encompass (Davis)? Who will design these complex systems and how?

The notion of automation just presented is neither evaluative nor analytical. How should different instances of automation be classified within this scheme? What are the organising principles, the main criteria for setting up an analytical grid out of so complex a picture?

3.3.3 Automation as an aspect of the economics of technological change

Another possible key set of variables for defining and explaining automation are economic considerations. Technological progress, with all its disparate elements, could be seen as the deployment of new knowledge to produce in different ways or to produce different products with the aim of improving productivity and capital formation.[7] This kind of economic approach, however, fails to incorporate an explanation of the complex structural and organisational change determined in the enterprise and in the economy (Reyneri and Ricciardi, 3.5, 1972), with economic consequences greater than "output per worker". No explanation is given about what socio-economic factors make automation grow. An explanation of automation in terms of the process of intensification of fixed capital (with no immediate relationship with productivity or innovation) is proposed by Braverman (3.1, 1974), but no evidence is given as to how this single factor could produce such a multifaceted phenomenon.

The economic dimension cannot be considered as a mere "cause" but as a component element of the automation itself. Economic parameters and features of the production system are interlocked in any technical investment. Economic measurements of elasticity, flexibility, readiness, time, innovativeness, preservation of resources and other facets of organisational performance are becoming increasingly more important than actual output per work-hour, particularly in new design (Davis, 4.1, 1971).

3.3.4 Automation in the history of industrial society

Old and new historians of industrialisation have pointed out - according to a large variety of interpretative models of human civilisation - the complexity of societal changes due to industrialisation: machines, systems of production, economy, institutions, patterns of life and cities merge to form a total picture.[8]

Limited space does not permit an exploration of the wide range of literature that deals with the history and the perspectives of mankind and with the totality of complete systems of thought. Besides, authors who have considered automation as a historical event usually fail to give operational definitions that could allow empirical analysis. We would merely point out that most of the discussion about the social effects of automation is understandable outside a broader historical perspective.

Many historical approaches tend towards statements that are descriptive or normative, rather than interpretative. A historical perspective is - at the least - a powerful complementary tool, because it takes into account important elements not observable in the unit under study, such as labour market, history of ideas, societal goals, etc.

3.3.5 Conclusions and final remarks about a comprehensive notion of automation

A first conclusion is that it is worthwhile to distinguish automation from neighbouring phenomena and concepts such as technical change, advanced technology, rationalisation of production and the like. We have seen that according to the key elements of automation there is sometimes an emphasis on continuity, sometimes on rupture and sometimes on both approaches. This is illustrated in table 11.

Table 11: Key elements of automation: relationship with other forms of technical development

Key element	Continuity	Rupture
No distinction with other forms of technical development	X	-
Functions performed (closed-loop automatic feedback control system)	X	X
Technology (computer technology)	-	X
Organisation of production ("automatic factory")	X	X
Organisation of enterprise	X	X
Economic	X	-
Historical	X	X

First, automation is the continuation of a historical process of the incorporation of human work into machines. Human work is seen as mental and physical effort to achieve a transformation of physical objects and information from one state to another, and also as control over the technical and social conditions for achieving the transformation. However, this has never been a mere process of shifting tasks from people to machines. New transformation tasks that people never had to or could accomplish were created for machines; new tasks for invention, regulation and co-ordination were created for people.

Nevertheless, automation is also a rupture in that historical process because the scope and speed of this incorporation of work into machines produces a qualitative change in the components of the industrial units, in industrial work and in society itself. The ultimate is neither new control systems, nor computer technologies, nor integration of production flow, nor integration of business functions, but rather a combination of all of them.

The rate of introduction of automatic processes and systems has been impressive, affecting both production processes and products in everyday use. Data on the penetration of electronic devices associated with automation in the various fields of application show that their quantitative growth and their rate of innovation in the last 20 years is higher than any other technical device in the last 200 years.

We also see a rupture in many relevant elements of production system -

- in performance: a production is faster, cheaper and of better quality, and above all it allows production of new products or levels of quality which were previously unattainable;

- in production methods: computers are able to incorporate almost all human control systems;

- in the type of work incorporated in machines: beyond the absorption of trade skills into machines, there is the incorporation in computers of human communication and decision-making abilities which formerly required people in a social context (the organisation). The systems of production, of information and of co-operation are the subject of change, rather than machines and jobs;

- in the knowledge used: while the sciences leading the first Industrial Revolution (mechanics, chemistry, economics, organisation) were partly drawn from the existing experience of workers, and their applications could be intuitively understood through practice, the sciences of automation (applied mathematics, cybernetics, computer sciences, operational research, modelling) are based on abstract elaborations and are understandable only through learning algorithms;

- in the content of work left to humans, both residual (as previously) and totally new in nature;

- in the quantity and structure of employment inherent in automated processes;

- in the process of integration and exchange with society as a whole: knowledge, hardware, programmes and people have an unprecedented level of interaction outside the automated unit itself.

In the following pages we will see that this general level of discussion about automation does not allow forecasting of any specific work situation or the QWL in a concrete unit (from a work group to an industry). However, our discussion so far is enough to allow us to say that models of work and related social procedures are likely to change.

The revolution is not only conceptual but real: practically any sort of work content can now be performed by automatic devices, any production process can be automated, and enterprises of any size in any industry can be involved in automation because of the availability of microprocessors and the sophistication of EDP applications. Automation is never "production systems without men" but production systems where people choose, implement, modify and run a highly formalised joint technical and social system.

Integration among machines, phases, business functions and productive units, and among productive units and society's institutions are more than functional consequences of automation; they are inherent functional components. Therefore, automation is an appropriate word to indicate a specific phenomenon. It can be defined as the displacement - within the physical system itself - of goals, models, procedures and languages for controlling and integrating meaningful conversion processes: thus not only work but also typical social functioning and structures enter the machines. The conversion processes may be more or less automatic; they may operate on materials or information, or both; or they may require more or less human work. In all cases we have automation when the three following conditions are respected:[9]

- the adoption of highly formalised languages through devices (mostly electronic) providing automatic feedback;

- the representation, in the form of a viable algorithm, of the "law of the system" and of a rationale for its design and control; and

- the boundaries of the unit under consideration are such that they provide a meaningful result from a technical, economic and social viewpoint.

At this level of generality, we can say that automation promotes the process of transformation from a constellation of distinct jobs and professions to an aggregate of human work in which high social integration is required and totally different criteria of social differentiation are applicable. As we shall see, automation does not have deterministic social effects; it merely sets up some functional requirements and constraints that can be interpreted accordingly in organisation design. These general concepts serve mainly to represent scenarios about the nature of work and the composition of the workforce and to draw implications concerning social policy.

Automation, as we have defined it, imposes functional requirements on the restructuring of work: when jobs, organisation units, communication and decision-making processes are fragmented, they may become inconsistent with production objectives. Recomposition is also required between work and knowledge, work and organisation, work and life.

This can not lead automatically to a restructuring of the occupational structure, i.e. the elimination of unnecessary differentiation among workers regarding job titles, grades, wages, etc. Cases have been reported of a decreased number of job titles, compression of grade scales, full task rotation, new broad production roles (for example, the process operator) and a blurring of sharp distinctions between blue- and white-collar workers. However, other cases bear witness to the opposite trend, the polarisation of the workforce in automation: a few workers are highly skilled while a greater number are given simple, repetitive, unskilled jobs. In the next section we shall indicate the reasons for this alternative occupational design in different types of automation, different institutional settings and different philosophies of design. In particular, the push toward segregation and segmentation of the workforce is still operating within the labour market and is reinforced by increasing unemployment. Old labour market patterns and technological unemployment lead toward a polarisation of the occupational structure which contrasts with the functional requirements of work restructuring originating from most cases of automation.

In conclusion, there are two basic implications of the definition of automation as a "societal phenomenon" -

(1) The content and perhaps even the notions of work, job, skills, training, career, organisation, etc., require review. The boundaries between work and life, organisation and institutions, production and services will have to be reset, and their relevant transactions will take new forms. Research is needed on the basic aspects of work, organisation and society, and interdisciplinarity and participation are an essential component. A communication network should be set up among universities, research centres, institutions, firms and trade unions which conduct research on the changing world of work; at the same time, a new international scientific community is emerging.

(2) The design of automation is too complex a business to be left to technologists alone. In many countries potential social implications are eliciting suggestions that industrial management promote techical change in closer connection with public institutions. Trade union/management relations in most countries are focused on problems of technological design with new sophisticated approaches. At the same time, new agencies for studying and assisting these developments are being created, and "composite technology designer" seems to be emerging in many industrialised nations.

This is one particular view of automation. Others might be preferable within a different theoretical framework. Above all, those who are engaged in research and work in the field of social aspects of automation must give up the idea that the use of a single concept (or a single conceptual framework) might be helpful.

A new type of study concerning a more limited set of relationships is now flourishing (e.g. microprocessors and employment, VDUs and stress, robots and operators jobs, etc.) (Commission of the European Communities, 3.2, 1980). The next section is intended to provide a point of reference for these kinds of studies.

4. Analytical dimensions of automation and typologies for design and comparative studies

What are the differences among the various technical developments? What is the impact of these developments on work organisation and QWL? How can we forecast their specific impact? How can we design?

Analytical dimensions and typologies are provided in most of the literature already mentioned. The concrete developments included in automation differ greatly according to three basic dimensions: technological, production system and state of the system.

The technological and production system dimensions are fixed and inherent for any type of development of the same kind: conversion and transportation equipment used; product produced; hardware of feedback systems; software employed; information system structure; logical functions performed; business functions; industry areas of application, etc.

The state of the system dimension is a contingent dimension which varies according to some changing state of the unit under examination: scope of application within the unit; state of input and output; variances; functional requirements of information systems; economic and technical performance required, etc.

These dimensions are described in more detail below.

4.1 Inherent dimensions

4.1.1 Technological dimensions

1. Type of material handled: the nature of material (Perrow, Foster) is mentioned as a major source of the different consequences of automation. Mechanical and chemical process would permit mechanisation and automation (Reyneri and Ricciardi, 3.5, 1972; Council for Science and Society, 3.1, 1981), as well as information processing, to proceed in a different way.

2. Type of transportation and layout: the physical system of transportation is a determining factor in the type of organisation adopted (Emery and Thorsrud, 4.6, 1976) and in the penetration of automatic electronic control (Bright, 2, 1958). Integral mechanisation of

transportation shifts, for instance, a mechanical manufacturing system into a chemical type (Woodward, 3.3, 1969).

3. Technology of physical transformation: machines (lathes, drills, etc.), their various technological "generations" and their typologies (Meissner, Hage Hailken, Harvey, Perrow, Stinchombe) have different impacts on organisation and work.

4. Instrumentation (gauges, screens, display units, etc.) (Sholten and Keja, 4.4, 1977).

5. Technology of process control (mechanical, eletro-mechanical, hydraulic, electronic) (de Bivort, 2, 1955).

6. EDP technology: central process unit, mini-micro computers, peripherals, nets, software, etc.

7. Type of process control (Bright).

8. Type of information process (Crossman, Foster).

4.1.2 Production system dimensions

1. Industry of application: the same NC machine employed in a large aerospace industry and in a small factory producing parts makes a different type of automation.

2. Systems of production (prototype, batch, mass, continuous processes (Woodward, 3.3, 1970) have different impacts on organisational configuration.

3. Management decision systems (Woodward, 3.3, 1965).

4. Business functions of application and their integration (extension of automated processes on manufacturing, R & D, administration, production planning, marketing, etc.) have differentiated influences (Diebold, 2, 1952).

4.2 Contingent (state of the system) dimensions

1. State of input and output: the level of permanent or contigent uncertainty in input or output presentation (material or data) and decodification are determinant in the type of control system needed and actually implemented (Herbst, Susman, Galbraith).

2. Disturbances or variances occurring during the production process: the number and gravity of disturbances occurring during the production process affect the shape and the effectiveness of human and automatic control systems (Davis and Engelstadt, 5.1, 1966; Butera, 4.2, 1979).

3. Functional requirements of the information system: the number of elements to be linked together, frequency of information exchange and uncertainty in the content of information place functional requirements on information systems; all this can vary frequently, with major consequences for the effectiveness of the automated information process (Galbraith).

4. Scope of application of automated devices within the unit: a unit with a robot at the end of an assembly line and a unit consisting solely of robots are totally different cases of automation.

5. Technical and economic performance required (Udy, 3.3, 1970).

4.3 Typologies for comparative studies of automation

The above-mentioned authors have provided parameters for measuring these dimensions and have built up typologies. However, few consider all three basic components of automation (technological, production system, state of the system). We would therefore suggest that researchers and designers simultaneously use different typologies when they have problems of comparison, even though they may partly overlap.

A classification of any single case along all or some of the major typologies can help systematic comparison. This is illustrated in table 12, which suggests a classification scheme for cases of automation according to dimensions of the three main components.

Each column refers to a suggested typology, which is briefly reviewed below.

4.3.1 Technological and functional dimensions

1. Type of material handled;

   (a) energy;

   (b) chemical;

   (c) mechanical;

   (d) information.

2. Type of workplace technology (Meissner):

   (a) handling;

   (b) hand work;

   (c) machine work;

Table 12: Typologies for cases of automation

Cases	Technology and control system				Production system				State of the system
	1. Type of material handled (4 types)	2. Type of workplace technology (Meissner, 8 types)	3. Type of control (Bright, 17 types)	4. Type of informa- tion systems (Crossman, 432 types)	1. Type of production process (Woodward, 11 types)	2. Degree of indepen- dence of the unit and the enterprise (from Anfossi)	3. Manage- ment systems (2 types)	4. Labour character- istics of the unit (6 types)	Checklist (4 items)

(d) machine work sequence;

(e) assembly line;

(f) hand and machine line;

(g) remote control;

(h) automation.

Meissner proposes a classification of workplace technology on the basis of two dimensions -

(a) type of conversion (no conversion; hand tools; hand and machine tools; manual tools; steered automatics; steered and self-regulating automatics; and self-regulating automatics); and

(b) type of transfer (hand transfer; auto-transfer; dead line; steered line; live line)

adopted in the particular stage of the production process where a worker or group of workers perform their tasks. He finds that actual technologies can be grouped into only eight categories, as listed above, because some combinations are not in use or are sufficiently similar to be combined for analytical purposes.

3. Type of control (Bright, 2, 1958). As we have reported, Bright classifies the different stages of mechanisation through to automation chiefly according to an increasing level of control performed by machines. Four basic types of control are identified -

(a) hand control: the control is all in the operator's hand;

(b) mechanical control: the machines embody in a fixed manner some amount of control of the operation to be carried out;

(c) machines have variable control according to variation of the environment, to which they respond with signals;

(d) machines have variable control and respond with action.

Within these four broad categories, 17 types of control are identified (see table 10).

4. Type of information systems (Crossman, 4.2, 1960). In spite of his limitations to technological functional aspects, Crossman exhibits the most complete typology, including both what we have called the inherent and the variable components of automation. He develops a taxonomy of information processing technologies on the basis of their abstract logical properties, rather than their concrete features, and lists 432 distinct types of information processes, distinguishing a unit by -

(a) the format, or language, in which data is input or output;

(b) the nature of the transformation applied to the input and output;

(c) the amount and type of memory;

(d) the relative fixity of the transformation applied;

(e) whether or not there is any random component in the output;

(f) whether or not there is feedback from output to input.

Table 13 explains the categories included in each analytical dimension.

4.3.2 Production system dimensions

As we have said before, we need other dimensions for understanding automation. These are related to the production system rather than to the machine.

1. Type of production process (Woodward, 3.3, 1970). Woodward classified 11 types of production processes, according to three main groups (figure 3).

2. Degree of independence of the unit and the enterprise (from Anfossi):

   (a) technically and administratively independent;

   (b) technically independent but interconnected with larger administrative procedures (planning, scheduling, accounting, economic control, etc.);

   (c) technically and administratively interconnected within a productive cycle.

3. Management decision systems:

   (a) centralised;

   (b) decentralised;

   (c) centralised and decentralised.

4. Characteristics of the unit in respect to labour:

   (a) capital intensive;

   (b) labour intensive;

   (c) mainly skilled labour;

Figure 3: Woodward's typology of production processes

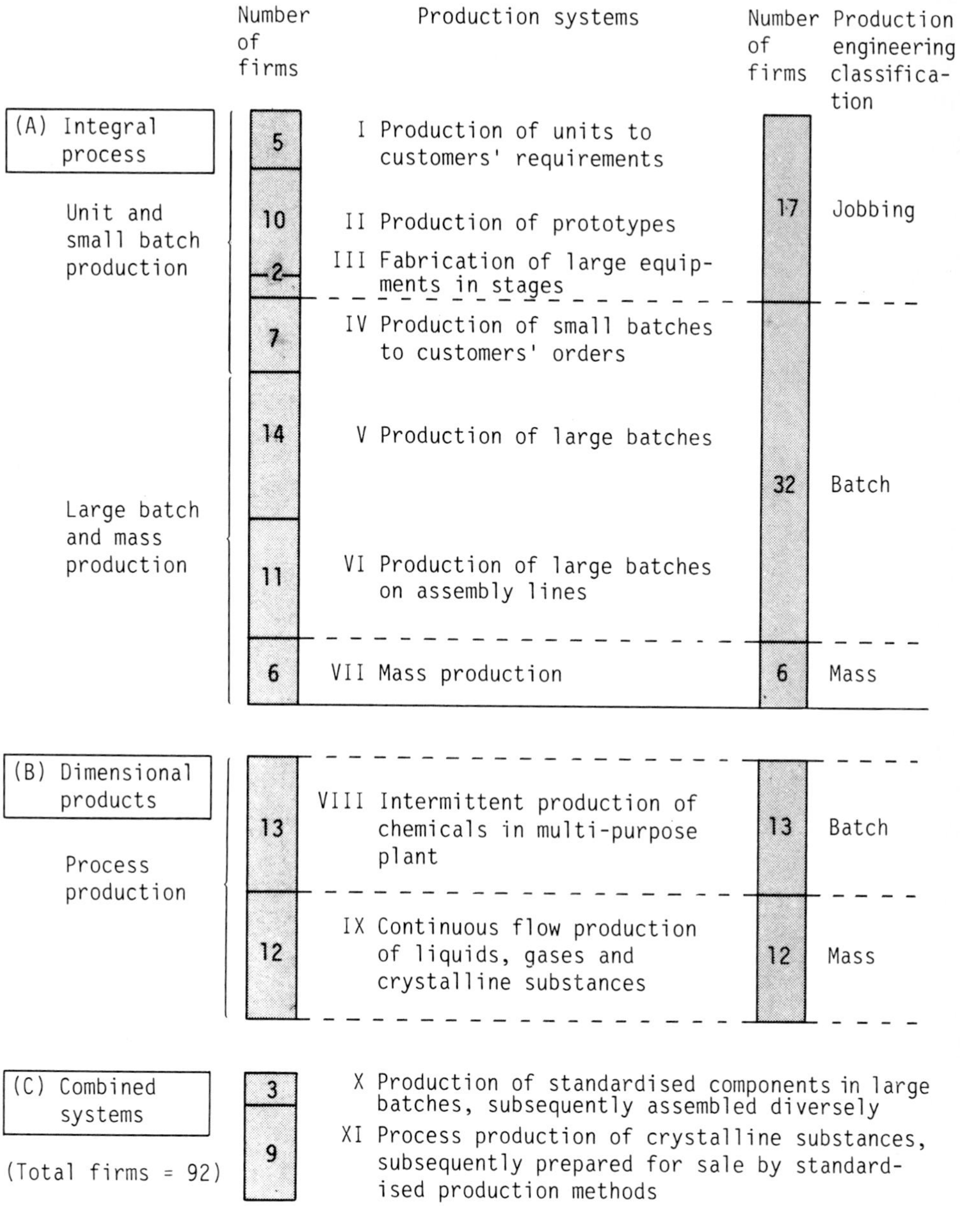

	Number of firms	Production systems	Number of firms	Production engineering classification
(A) Integral process				
Unit and small batch production	5	I Production of units to customers' requirements	17	Jobbing
	10	II Production of prototypes		
	2	III Fabrication of large equipments in stages		
	7	IV Production of small batches to customers' orders	32	Batch
Large batch and mass production	14	V Production of large batches		
	11	VI Production of large batches on assembly lines		
	6	VII Mass production	6	Mass
(B) Dimensional products				
Process production	13	VIII Intermittent production of chemicals in multi-purpose plant	13	Batch
	12	IX Continuous flow production of liquids, gases and crystalline substances	12	Mass
(C) Combined systems	3	X Production of standardised components in large batches, subsequently assembled diversely		
(Total firms = 92)	9	XI Process production of crystalline substances, subsequently prepared for sale by standardised production methods		

Source: J. Woodward: Industrial organisation: Behaviour and control (London, Oxford University Press, 1970).

(d) mainly unskilled labour;

(e) mainly senior personnel;

(f) mainly new personnel.

4.3.3 State of the system dimensions

We propose a comparative analysis to check whether or not -

(a) there is a high degree of uncertainty in the input;

(b) there are serious disturbances whose control could not be cheaply performed by machines;

(c) all the meaningful information needed goes beyong the possibilities of EDP; and

(d) the automatic control is localised in particular phases of the cycle.

4.3.4 Conclusions

Starting from these different typologies, one can outline the "relative" typologies of working definitions of automation which are useful in the analysis of specific social correlates. In Chapter 4, for instance, we searched for a possible association between automation on the one hand, and positive/negative consequences on the content of work and related QWL on the other. To this end we developed the typology of automation (Types 1-4), as presented in Chapter 4, section 2.1.

5. Some effects of automation on work and working conditions: a map of the literature

Many empirical studies have been conducted on the effects of automation on work organisation, working conditions and workers' behaviour. Few unequivocal conclusions can be drawn from the literature for reasons we will discuss at the end of our short review. In fact, various studies -

- explore different instances of automation;
- report effects upon different aspects of organisation, working conditions and behaviour;
- have different representations of the cause/effect chains among the various elements; and
- have different interpretative models and views of the alternative choices.[10]

Table 13: Crossman's typology of information systems

Dimension of classification	Category
(a) Type of data input or output	Discrete ("digital") data: messages coded as sequences of choices from a limited set of alternatives (e.g. the alphabet, the dictionary, the digits 0 and 1, the digits 0-9, etc.)
	Continuous ("analogue") data: messages coded as mathematically continuous variables (e.g. voltage, pressure, duration, angle, etc.)
	Patterned data: multidimensional arrays of data, whether continuous or discrete, posssessing structural features (e.g. engineering drawings, weather maps)
(b) Function performed	Control: determining the state or sequence of states of a system or process, and/or the way in which power is employed to effect changes of state
	Communication (data transfer): transferring data on the state of sequence of states of a system to a distant point
	Data process: recoding, manipulating or otherwise transforming input messages
(c) Amount and type of memory	No memory: past inputs are entirely disregarded, each message being processed and output as received
	Fixed memory: a memory element stores fixed information (e.g. player piano roll, IBM card, cam, gramophone record)
	Variable memory: input information can be stored and output at a later time, the storage element being reusable (e.g. tape-recorder, desk calculator)
(d) Programming	Fixed program: the same operations are applied to all incoming information (e.g. desk calculator)

Table 13 (cont.)

Dimension of classification	Category
	Preset program: a new set of processes can be set up readily, remaining the same for all input information (e.g. Hollerith card-sorter)
	Self programmed: the mode of information processing is determined by the incoming information itself (e.g. digital computer)
	Adaptive and heuristic program (learning): the information processes adapt themselves to optimise some external criterion of success (e.g. chequer-playing programme)
(e) Determinacy	Determinate: the output is wholly determined by present and past inputs (e.g. desk calculator)
	Probabilistic: the output is determined in part by chance (e.g. random search processes, communication in noise)
(f) Feedback	Open loop: the input to the process does not depend on its own recent past output (e.g. teleprinter)
	Closed loop: the input or functioning of a process depends at least partly on its own recent past output

Source: E.R.F.W. Crossman: Automation and skill (London, HMSO, 1960).

A wise organising principle to consider for the review of these studies could be the distinction between: (a) objective effects of technology on work organisation (jobs, occupations, skills, etc.) and working conditions (fatigue, stress, etc.); and (b) subjective effects (perceptions, attitudes, etc.). However, most studies which focus on one aspect offer contributions to the others. Recent QWL approaches stress the need for an integrated diagnosis and design of both objective and subjective correlates of technical developments.

5.1 Skills

The effects of automation on skills first attracted attention when the so-called "Detroit automation" was introduced in the motor vehicle industry. Moreover, the implications of this development for formal grading and training programmes gave an added impetus to this attention.

In 1955, Friedmann (3.1, 1955) had described some new skill requirements for operators based upon line supervision. In the same year, a young French researcher published his doctoral dissertation on the evolution of manual work at Renault factories (Touraine, 4.2, 1955). Alain Touraine had found a correlation between three stages of technical development and the nature of work organisation. In phase C, which includes extensive use of automatic machines such as transfer lines, the operators' work content would be based more on responsibility than on direct manipulations. Qualification would be based upon role in the production process, consisting of surveillance, data collection and control. Operators would no longer have relationships mainly with materials, but with the entire production system. Workers who did not lose their jobs would be upgraded.

A more concentrated analysis of automation was made by Bright (2, 1958), with rather pessimistic conclusions about skills and consequent grading. He noticed that after a certain level of mechanisation, skill requirements for operating labour (and occasionally maintenance labour) were often reduced. The need for physical dexterity and mental ability tended to increase while moving from the lowest levels of mechanisation, but to decrease before the highest levels of mechanisation were reached. In the field of supervisory labour, higher skills and education were required, due to the complexity of control mechanisms. However, Bright concluded that this demand would also probably fall off after a certain level.

In maintenance work, Bright pointed out that modern control systems and electronic techniques required entirely new skills, but, here too, the numbers required were not expected to be large. It was inaccurate to expect that maintenance workers would increase proportionately with the complexity of automatic machinery.

Referring to a similar type of automation, Faunce (4.5, 1958) focused on workers' attitudes. He found that a large proportion of workers felt that the greatest advantage of new working conditions was the decrease in materials handling, and hence automation was considered as an upgrading of work. The general feeling of the workers was that greater skill was called for in the new process. However, the analysis of what the new processes involved revealed no significant skill differences in the new work compared with the old, though the demand for maintenance workers increased. The psychological impact of working with highly complex machinery raised the average worker's self-image. Faunce also reported that more attention and responsibility, but less control over the work pace, were required.

Crossman (4.2, 1960) suggested that different types of automation had different consequences on work. Continuous flow production (transfer machines and automatically controlled process) plants -

> ... have a common pattern of work and skill. The work comprises control and communication ... The first of these elements requires "control skill" ... Its chief component is decision, selecting the best control action for each combination of circumstances that arises ... The better operators tend to work ... by an intuitive appreciation of the state of affairs, possibly using a mental "working model" of the process. New maintenance skills will probably emerge (page 57).

Blauner (4.5, 1964), in a secondary analysis of a Davis study of a chemical plant, writes that in automation -

> ... workers characterise their jobs as being more "mental" or "visual" than physical ... Traditional craft skills have been completely eliminated from the productive process ... The major requirement is responsibility ... (page 134).
>
> ...The most critical feature in automation is that it transfers the focus of emphasis from an individual job to the process of production ... It is a shift of emphasis from job to process, the workers' role changes from providing skills to accepting responsibility. His scope of operations increases ... (page 143).
>
> ...Modification of duties ... demand workers of fairly high ability ... companies have generally limited this hiring to high school graduates, in an attempt to get workers who are intelligent and adaptable ... (page 158).
>
> ...Skill in automated plant is new and vaguely defined (page 156).

Davis (4.1, 1971) virtually concludes the long controversy about the upgrading or downgrading effects of automation, observing that traditional concepts of skill are not applicable in automated systems where -

> ... humans ... are interdependent components required to respond to stochastic, not deterministic, conditions: i.e. they operate in an environment whose "important events" are randomly occurring and upredictable. Sophisticated skills must be maintained, although they may be called into use only very occasionally. This technological shift disturbs long-established boundaries between jobs and skills (Davis and Taylor, 4.2, 1976, page 69).

5.2 Automation and job/organisation design

Davis' conclusions on the skill discussion are the result of a stream of research focused upon another pressing policy question: What are the principles, models and procedures for job and organisation design?

Walker (4.2, 1959), in his study of an advanced steel mill, found that work roles are highly interdependent, the levels of supervison are reduced and the role of supervisor shifts its attention from men to machines.

Mann and Hoffman (4.5, 1960) found that centralisation in the control room of an automated power plant fostered contacts among workers, thus increasing their identification with the group and with final results.

Trist et al. (3.3, 1963) proposed influential conclusions about job and organisation design in advanced technology: the extreme division of labour of the previous stages of industrial development would be replaced by situations with fewer and less specialised work roles. Multi-skilling and shared responsibility would be required, possibly giving rise to "composite or autonomous work groups" with less or no internal supervision.

Blauner (4.5, 1964) stated that process technology reversed the historical trend towards increasing division of labour and specialisation.

Woodward (3.3, 1965) found that in continuous processes advanced technology control becomes more impersonal. The ratio of technical and supervisory personnel to operatives increases, but the latter have more responsibility and the work group has greater autonomy. Strict job specifications and close supervision as organisational models for getting things done are substituted by technical procedures and by the definition of responsibilities for roles and work groups.

While new methods of work and organisation design were developed incorporating a greater consideration of skills, communication, participation and control of results (Davis, 5.1, 1966), there was growing interest in a better definition of the analytical components of work and organisation in new technology.

Herbst (3.3, 1974) made his contribution to the impact of technology on tasks and task relationships, providing analytical tools for the description and design of work not necessarily encompassed in an individual job.

Meissner (4.2, 1969) analytically described relevant aspects of the inner nature of work and organisation, in particular in automation: communication, co-operation and influence. In spite of ambiguously calling them "behavioural adaptations", Meissner does explain how different technological situations require, permit or impede a variety of ways of working that cannot be described by the elliptical language of job and organisation formal structures.

Susman discussed the connotation of group autonomy in automation (4.2, 1970) and the uncertainties of the state of the system that might require such an autonomy in order to achieve productive results (4.2, 1975).

In conclusion, the literature suggests that advanced technologies would set up the conditions for overcoming traditional job design criteria based upon maximum fragmentation, time saturation and personal control (Davis et al., 5, 1972). New models and procedures for design would arise. A strong move away from the "one best way" to design organisation is advocated both by authors concerned with work organisation (so-called "micro-organisation") and by those concerned with organisation of production (so-called "macro-organisation").

5.3 Automation and workers' attitudes and behaviour

How will workers react to automation? How will they adjust to the new situation? These kinds of questions gave the impetus to most American studies in the field and further contributed to the more basic issues on the impact of automation.

Mann and Hoffman's Automation and the worker (4.5, 1960) is now a classic in this perspective. As we previously noted, the authors first identified specific changes in organisation and work context due to automation, and then tested the subsequent behavioural reactions. They write -

> Within the new power plant the reduction in the manner of levels of supervision was seen to have effects on the relative influence of different groups and on the degree of communication achieved in that plant. There was an increase in the satisfaction with the amount of communication from the top of the organisation to non-supervisory employees and a greater sharing of the decision-making power ... (page 64).
>
> ... In building the new power plant, technological changes and management decisions resulted in the redesigning of operating jobs ... The combination of job enlargement and of job rotation increased for many of these men - at least temporarily - the intrinsic satisfaction which could be derived from the work itself. Some negative effects in the form of somewhat increased tension were also found among these plant operators (page 103).

The issue of the relationship between job content and worker motivation is the subject of numerous works of the sixties and seventies. Vanplew (4.5, 1973) summarises an entire series of studies with this sentence: "Taken as a whole, our data confirm those other findings which characterise automated process work as interesting and rewarding intrinsically", but "explanation of workers' attitudes and behaviour cannot restrict its focus to either the one or the other side of the factory gates" (page 428). Finally, the author questions - together with Herzberg and Daniel - the very idea of "overall job satisfaction".

The expectation of finding sound correlations between work situations and workers' attitudes was frequently confronted by authors insiting on the significance of workers' values and perceptions. Grossin comments after a recent empirical analysis (4.5, 1979): "The attitude toward the technological change seems to rely mainly upon personal orientations which are expressed also in the work situation, orientations influenced by dominant ideas, major cultural preferences, main ideological streams in a country" (page 100).

5.4 Working conditions and quality of working life

The literature produced on the subject in recent years is vast. We will briefly recall the pioneering works produced since the fifties.

Since 1957, when the ILO had warned against the possible negative repercussions of automation on safety and health, the possibilities of using automation for moving workers away from hazardous and heavy conditions have been more frequently advocated than studied. Studies have nevertheless been made of accident risks due to the relaxation of workers' vigilance and reliance devices (Seeger, 4.4, 1960; Hatch, 4.4, 1962). The possible increase in nervous strain in automation is indicated by the World Health Organization (4.4, 1959) and discussed in depth by Murrel. Problems of mental health are described by Kornhauser (4.4, 1965). In particular, problems of isolation (Weybrew, 4.4, 1961), stress due to excess responsibility for catastrophic events and the worsening of working conditions due to increase of shift work in automated plants have been the object of study.

The concept of alienation is often used when negative effects can be experienced, even though no clinical disease is certified. Blauner has observed the alienating effects of powerlessness in automated systems, when lack of control over production processes is experienced. Fullan (4.5, 1970) reviews the literature and maintains that social integration can mitigate the effects of objectively alienating work. Smith (4.5, 1968) asserts that work roles are important but that the subjective significance of the work done is no less so. He talks of the "self-identity/work-identity complex".

The QWL developments have given rise to a more complex set of parameters. Seashore (5.1, 1976a) identifies four basic dimensions of QWL based upon the notion of integrity (self, body, social growth, life roles). That approach integrates the physio-psychological aspects with attitudes and "alienation" concepts. Gardell and Gustavsen (4.4, 1980) and Oddone (4.4, 1981) recall the practical developments of QWL in Sweden and Italy, which consist mainly of social exercises where workers have had an active role in evaluating their own experience and have avoided total dependency on experts.

5.5 Conclusions

The literature on the effects of automation on work does not lead to unequivocal conclusions as to the cause/effect terms of the positivistic sciences. The reasons for this, which are fourfold, are considered below.

5.5.1 Different instances of automation

The findings of case studies are often generalised beyond the classes of technological change which the case study might represent. Such was the case with Mann and Hoffman's "Automation and the worker" (4.5, 1960). The authors concluded that "the prospect of automation in the future seems to offer a tremendous potential for the worker and companies involved. There are, however, sizeable obstacles to be overcome and numerous precautions which must be taken before these benefits can be achieved (page 213). The apparent empirical evidence of these optimistic conclusions was over-generalised for many years.

5.5.2 Consideration of different aspects of automation or work

Studies that explore the relationships between technological typologies and selected elements of socio-organisation systems have sometimes reported quantitative evidence, whereas at other times they have not. As pointed out by Davis and Taylor (4.2, 1976), these typologies mostly "rely on more gross categories of technology, making difficult any further replication" (page 391).

Sometimes there are problems emanating from the fact that different meanings are used for the same word. As an illustration, after a decade of polemics between the supporters of the prevailing influence of the technology on the organisation (Woodward, 3.3, 1965) and those maintaining size as a more important variable (the Aston group, Hickson, et al., 3.3, 1969), others (Child and Mansfield) finally made it clear that Woodward's concept of technology encompassed some dimensions of organisational structure while the Aston group's definition of organisation was based upon formal structure and not - as in Woodward - on the configuration of the organisation (supervisor/employee ratio, span of control, etc.).

In the layman's world these problems may occasionally become serious: for instance, popular discussions about the effects of automation on skills indiscriminately borrow the contrasting findings of studies on substantial skill requirements and on formal grades. However, we have seen that the worsening of skill requirements often goes along with upgrading.

Finally, it is frequently difficult to identify the legitimate concrete unit of investigation (job, work system, organisation, socio-technical system, etc.) and to select the relevant level of observation (group, department, enterprise, system of enterprise, industry, educational system, labour market, etc.). Conclusions about the negative effects of robot technology on individual jobs - for instance - cannot be extended to its impact on the work system.

5.5.3 Simple cause/effect relationships

Unlike physics or chemistry, it is almost impossible to test a cause/effect relationship of technology upon work when a mediating variable - like social action and behaviour - plays a major role in the final results. The existence of an organisational choice which is always available under certain technological constraints was pointed out by Trist (3.3, 1963). Child (3.3, 1972) identified the relevance of managerial choices and ideologies on the structure and configuration of the organisation; others (Della Rocca) described the effects of trade union action on the work organisation. On a more general level, Gouldner (3.3, 1954) and Dubin (3.1, 1958) researched the influence of cultural patterns on the organisation; social action (Touraine, 3.3, 1965) was proved to be very important in the making of an organisation. Others (Maurice and Brossard, 3.3, 1974) attribute the major impact on organisation to societal dimensions. Crozier and Friedberg (3.3, 1977) have defined the organisation as "a concrete system of action" in which human structures set rules of game, and regulatory mechanisms are built in. Reynaud (3.3, 1979) stresses the joint regulation within which social actors choose the "recipes" available as represented by forms of organisation.

5.5.4 Interpretative models of social structures and social behaviour: alternative goals for analysis and design

In the social sciences two approaches can be distinguished. One relies on observation of the "objective" conditions of people while the other is concerned with more subjective questions about what people know, feel, desire, how they react, etc. The two approaches are of course linked together, but many of the interpretative models are more sensitive to one or the other. In our subject some authors deal with structural issues like objective working conditions, employment conditions, educational structures, social stratification, etc. Others deal on the contrary with behavioural issues such as satisfaction, motivation, perception, etc. This is why it is difficult to compare the results of Braverman (3.3, 1974), Popitz (3.3, 1976), Habermas (3.1, 1968), Touraine (3.3, 1965) and Maurice and Brossard (3.3, 1974), which deal for the most part with the first question, with those of Faunce (4.5, 1958), Mann and Hoffman (4.5, 1960), Walker and Guest (3.3, 1962), etc., which deal with the second.

Those authors who are concerned mainly with the first question deal with independent variables influencing human conditions by means of objective structures produced by people, such as economic structures, power structures, labour market, technology, organisational structures, etc. Those concerned with the second approach tend to attribute the major influence to variables that are a component part of the current generation and a manifestation of the behaviour itself: at the individual level, values, emotions, tensions, knowledge, perception, interests, etc.; at the societal level, interaction, communication, defensive and co-operative associations, influence and power, etc.

Another tendency is to look for "ecological influences"[11] on human activity and behaviour, that is, variables outside individual and group functioning. In the study of industrial workers, variables like technology, organisational task, organisation and physical work environment have been studied as potential explanations of human behaviour. Many of these variables are the same as those appearing in researches about social structures. This line of research attempts searches about social structures to look for inclusive models (or wide-range models as Merton calls them) explaining both social structure and functioning. In this field, models are medium and short range. For this reason, the influences of technology upon the occupational structure of a country or upon co-operation and communication at shop-floor level pertain not only to two different levels of reality but also to two different areas of scientific concern.

In our subject area, the opportunity to take into account both structural and behavioural views is offered, among other factors, by the rise of QWL as a concept encompassing evaluation and purposeful action both on objective working conditions and on subjective concern on the part of workers. However, the formulation of comprehensive interpretative models is still in the future.

6. Some concepts and models for designing work in automation

The present review of the literature still leaves two questions open. What components of work are more important in the work design process in automation? How can we take into account the complex nets of influences coming from inside and outside the unit to be designed? In the following pages two answers will be provided: (a) the critical component of work in automation is the task structure; and (b) the notion of work organisation must be expanded into a more complex model. To this end we provide some definitions, some directions for the sequence of analysis and design, and some suggestions for links to be considered.

6.1 Technology and quality of work: task structure as the main object of design

The choice of a technology is never a purely technical one. The same concept of technology encompasses an idea of purpose, as researchers on production systems have frequently pointed out.

Parsons (3.3, 1970) writes that technology should be conceived as modes in which knowledge is put to instrumental uses in the interest of goals and purposes, the significance of which is not given in the body of knowledge itself. Woodward (3.3, 1969) defines technology as "the collection of plant, machines, tools and recipes available at a given time for the execution of the production task and the rationale underlying their utilisation" (page 73). Meisner (4.2, 1969), on the same lines, writes: "We would say that the technology of the workplace consists of pieces of steel, stone, wood in certain shapes. But that is not what is meant ... These material things ... are the product of designs, the manifestations of ideas of those who planned a process and the means of facilitating it" (page 14).

Technology design has always encompassed definite models of man, society and economy in addition to a representation of the physical world. Udy (3.3, 1970) presents a model of the mutual influences between technology and socio-economic variables (figure 4).

Figure 4: Interaction between technology and socio-economic variables

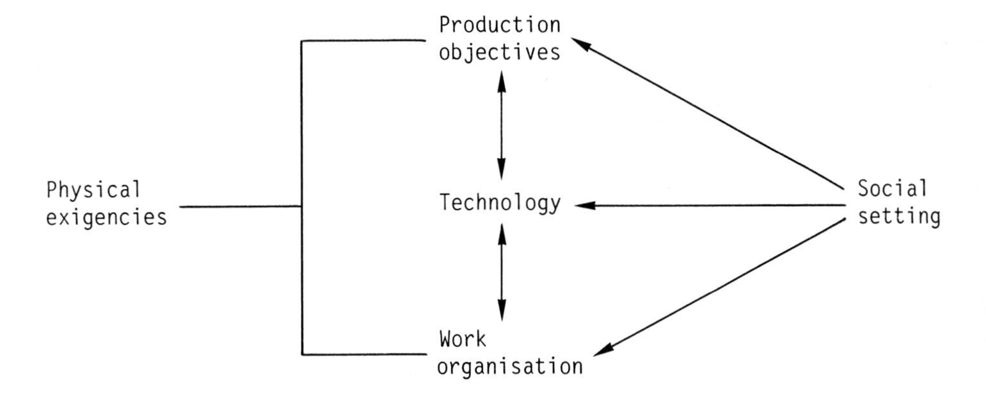

In the socio-technical approach of the Tavistock Institute (Emery and Trist, 3.3, 1960), the interdependencies between the technical and the social systems are examined mainly in a concrete unit. For the Tavistock researchers, technical systems have been designed encompassing ideas and purposes concerning social systems throughout the entire history of industrialism. Now the traditional social parameters of such joint designs are often no longer acceptable to workers and society. In addition, over the last 40 years the tendency to optimise the technical system alone has been very great, and the already poor equilibrium between the technical and social systems has deteriorated. To improve one of the two systems would thus imply a joint design of the other. In conclusion, in each industry there are always different degrees of freedom for technological choices that jointly optimise economic and social parameters.

The selection of a technology is a social choice because it encompasses inherent goals, assumptions and accepted influences concerning the quantity and quality of work needed and the relative modifications in the social system. In contrast, there is the so-called "technological determinism" which assumes that technology has its own logic and mode of growth. The majority of the literature agrees that once a technology is selected, the quantity and often the type of related human work change.

The introduction of NC machines, for instance, independently of any further analytical design of the organisation, reduces the number of work-hours per unit of production. This means that, using the new technology, the final product requires less human work at shop-floor level. The development of such a technology, on the other hand, required a great deal of new work in R & D. In NC machines, the set-up of tools and the correction of the ongoing process are made automatically, regardless of the job design solution adopted. Thus, the core content of the skills of an entire trade which had been developed since the appearance of machine tools, were then shifted into the machinery. We call this a change in aggregate work.

6.1.1 Aggregate work

Although the manning and configuration of the organisation have usually not been developed when the decision to buy an NC machine is made, the first component or stage of the "overall work" has already been decided. The prevailing "language" at this stage is economic, concerning investment decisions: the return on investment, the cost amortisation of new equipment compared to the cost of wages, and the increase of fixed capital usually guide the evaluation of that investment. Explicit social considerations are also taken into account, such as the opportunity to abolish some jobs in which workers are unreliable or recruited with difficulty. Sometimes these considerations bear more weight than the economic ones.

Bright's classification (table 10) gives a clear picture of the transfer of aggregate work from people to machines in various phases of industrial development.

The great changes in quantity and quality of work during industrialisation have been the main concern of authors dealing with the transformation of economy and society: from Marx to Weber, from Dobbs to Galbraith, from Dubin to Braverman. Many studies on the social effects of automation deal with this level. For instance, Richta (3.1, 1969) suggests that automation frees people from transformation work and will restore the unity of work activity: he takes the optimistic view that there exists potential for the resolution of major societal problems. Wiener (3.1, 1964), on the other hand, takes a radically pessimistic view, forecasting a dramatic reduction in the human work needed because of automation, with the resulting threat of unemployment and power concentration.

All these views identify gross changes in aggregate work. However, opposite predictions may result from different interpretative models. Up to this point, the content of jobs, organisation, occupations, etc., can usually be designed in a wide range of alternative patterns. Nevertheless, the design of tasks sharply reduces this range.

The classical organisational theories like scientific management were in principle dogmatic and normative. They conceived organisations as self-contained entities having a limited interaction with the external world and organised through the generation and application of rational absolute rules. In contrast, the "task approach" (Trist et al., 3.3, 1963; Woodward, 3.3, 1965); Katz and Kahn, 3.3, 1966); Perrow, 3.3, 1967; Miller and Rice; and others) stresses that what is to be accomplished by organisational units, workers or machines has a determining impact on the structure and functioning of the organisation. The adaptation of the organisation to its tasks largely explains the differences in organisation design.

The choice to have a task accomplished is social in nature, but once defined it takes the form of a structured relationship between the organisation as a social invention and some objective "ecological" elements, e.g. the physical world, the equipment, the other organisations and institutions, the market, etc. The organisational invention is no longer free, but its survival depends upon positive response to the chosen task(s). This involves the application to organisations of the "principle of reality" developed in psychology.

6.1.2 Decomposition of aggregate work into tasks

A task is a structured relationship with some portion of reality. The reality can be the physical properties of a material, the functioning of an information system, the figure of goods distribution, etc. The structured relationships comprise not only instructions but also "hard" system, etc. In abstract terms, a task is "a process whereby a distribution of input is convered (transformed) into a constrained distribution of output by the application of technology(ies)" (Abel and Mathew, 5.2, 1973, page 164).

Tasks are not synonymous with activities, with situations or contingencies, with instructions, with jobs, with goals, with technology or with control, even if they overlap with some of them (for an in-depth discussion, see Hunt, 5.1, 1976). Instead, they are something objective, related to the external world: the incorporation of a social choice into a concrete, operating structure.

Different levels of tasks correspond to the different levels of organisational reality. Primary tasks (Trist et al., 3.3, 1963) are what an entire organisation has to accomplish in terms of products, sales, technological innovation, etc., they should not be confused with organisational goals and missions. The global task (for example, an enterprise) is an intuitive concept. Its decomposition (Abel and Mathew) into an analytical set of tasks is a matter of different possible designs. In the production organisation, it is a question of technological rather than organisation design.

At a more restricted level, each unit (department, office, individual role, etc.) has its own tasks, all of which involve a defined transformation process of some objective physical or abstract properties.

6.1.3 Tasks and work for people

In the design of work for people, we deal mainly with the analytical tasks that become the content of work ("the work itself") performed by workers. Tasks can be material or informational. Human tasks are what one or more persons have to do, deriving from instructions, from a required performance or from physical properties of the process; how to do them can be externally specified or left to the judgement of the task holder - it can be imposed, agreed or chosen. In any case, task denotes the functional requirement of a work activity (what to do for what purpose).

Herbst (3.3, 1974) illustrates how human tasks and machine cycles in industry are two faces of the same coin -

> An event can be analysed with respect to its role within the technological frame of reference and with respect to its role within the social behaviour frame of reference; and, by implication, every technical event has a corresponding behavioural representation, and every behavioural event has a corresponding representation within the technological frame of reference ... If operations, independently of who carries them out, are looked at in terms of their sequential and mutual dependence relationships, we arrive at a representation of the production task structure. If activities are analysed with respect to the sequence of an inter-relation of material changes, we arrive at a representation of the material production processes (pages 114-115).

The task structure is the notion which includes both the content of human tasks in production processes (in terms of physical and psychological load, skill content, importance, etc.) and the type of relationships among tasks. It does not include the notions of who performs the tasks, how they are assembled into a job, how workers are paid, etc.

The content of tasks is what is called the work itself, the task attributes, the "immediate job": what has to be done and its necessary demands on people.

Regarding task relationships, the following types are described by Herbst (3.3, 1974) -

- structure of time dependence (sequential, interdependent, independent);
- operational relationships (isolated tasks, collective tasks);
- goal dependence (goals of task: independent, dependent, interdependent, conflicting or not);
- informational relationships;
- absorption-transmission variances.

The task structure is largely designed by activities usually classified as technological design (worker-machine allocation, layout, transportation, instrumentation, physical communication control, feedback systems, etc.). The work relationship structure logically takes place before the design of jobs, or job allocation systems, of organisation unit boundary definitions, of co-ordination systems and of interaction patterns, etc.

The task concept has been utilised in two ways, task as ecology of work organisation and task as ecology of human behaviour.

6.1.4 Task as ecology of work organisation

This first approach was initiated by the studies of the Tavistock Institute, and definitions provided emanate mostly from that school of thought. Tasks are generated together with technology. They contain the amount of psychological and physiological charge, complexity and skills, control, autonomy, etc., that can be placed in jobs (usually without fully determining the job content). Fragmented simple tasks in the assembly line provide limited content for designing rich jobs, while complex control tasks in an automated rolling mill offer the opportunity to design skilled jobs and occupations.

Task relationships influence communication, co-operation and the formation of work groups. Tasks affect working conditions and human behaviour directly by their very existence, or indirectly through organisational design. The task structure has been defined (Butera, 4.2, 1979) as the "skeleton of human organisation" and "the intermediate variable between technology on the one hand and organisation and working conditions on the other" (page 91).

6.1.5 Task as ecology of human behaviour

Authors who consider tasks as the ecology of human behaviour have studied their direct effect on a variety of behavioural aspects.

March and Simon (3.3, 1958), discussing the behavioural effects of close supervision, found that they vary according to the complexity of the task involved. This finding was confirmed by Likert. Hackman found that certain task factors control up to 50 per cent of the variance in behaviour and output in small laboratory groups.

The effect of task properties on work performance have been studied by Brown et al.; O'Brien and Owens; and Maylor and Dickinson. Even though they do not mention tasks, a similar concern was expressed by authors like Bell, Ohman and Swadows, and summarised by Strauss and Sayles in their review Technology and job satisfaction.

In 1965, Turner and Lawrence (4.5, 1965) wrote the most influential book about "response to task attributes". They criticise the assumption that "modern technology condemns the majority of the workforce to a routine and unchallenging life at work" and that "attention has to be turned to ways in which this might be counteracted through more meaningful leisure activities" (page 1). During the peak popularity of the human relations approach, following Elton Mayo's studies at Western Electric, the authors write -

> The Western Electric studies produce a powerful and justified trend in favour of social and interpersonal consideration. But we believe that some of the interpretation of the Western Electric research as downgrading the relative impact of job attributes on behaviour has been mistaken (page 3).

The authors concentrate their research on "the attributes or technologically determined characteristics of tasks" in order to understand how "industrial workers respond to various characteristics of the intrinsic job" (page 25) (our underlining).

Turner and Lawrence identified four main "intrinsic task attributes" as variety, autonomy, responsibility and interaction. These variables were foudn to be positively associated with two behavioural responses: work attendance and job satisfaction. Other variables affecting behaviour had been covered in the data analysis as "situational factors" (such as individual characteristics, perception of task attributes and subcultural predispositions).

Herzberg (4.5, 1966) is the most famous supporter of the idea that improvement in the work itself will result in higher satisfaction and motivation, via the positive response to human needs of self-actualisation. He provides empirical evidence that motivation factors (i.e. associated with positive feelings of satisfaction) are those intrinsic to "job content" or the work itself. He distinguishes them from factors whose absence is a source of dissatisfaction; these are extrinsic to the job and are called "job context" factors.

6.2 Models for analysis and design of work: the different coexisting organisations

Some final questions need to be answered. What should be designed and when should work be designed in automation? Which notion of work, work organisation or organisation should be adopted? Which component elements should be designed, in which sequence? Which inter-relationships should be taken into account?

In this subsection, models for analysis and design indicated by the present author are briefly reported (Butera, 4.2, 1979).

Organisation can be conceived first as a social invention bringing dispersed actions into a whole. Critical points for analysis and design of organisations pose the following questions. Who is the decision-making authority for design? What are the boundaries and homogeneity of the object designed? Which disciplines are used to provide conceptual tools? In this perspective we must conclude that in any concrete unit different "layers" of organisation coexist, each with its own boundaries and internal coherence. Different authorities establish and run each organisation layer and different negotiations take place. Various disciplines are involved.

The classification outlined below was drawn from an empirical study of a steel works (Butera, 4.2, 1979).

6.2.1 Formal organisation

This is the system of written rules oriented to regulate the actions of the unit's members. Job descriptions, organisational charts, official norms, procedures, etc., are included. The boundary to which they apply corresponds to the firm's premises. The source is management authority, with the constraints coming from the law and contracts. The content is a potentially coherent "system of government". The disciplines involved are mainly the administrative sciences.

6.2.2 "Technical" organisation

In productive organisations, most human actions and relationships are regulated by procedures which are meant to give directions for running the equipment or technical procedures. A work cycle in an assembly line, a machine cycle in a lathe shop, a production schedule, a computer program, etc., make up this layer of organisation. Custom and practice, even if oral, should be included in this category. The boundary is that of a production system, and can be broader or narrower than an enterprise according to circumstances. The source is technical authority, with the constraints of cost and workforce availability. When effects on the labour force are more evident (e.g. changes in the work cycle or new equipment), negotiation becomes the major source or constraint of this layer of organisation. Its content is a set of potentially coherent rules for production management. Disciplines involved are production management, industrial engineering, materials management, cost control, systems analysis, etc.

6.2.3 Professional institutions

In some units the main regulatory activity is provided by large institutions having branches within the unit (Gouldner, 5.1, 1959). For instance, in craftwork, research-oriented companies or hospitals the main rules are set by the professional institutions having formal status and sanction in the society concerned. Boundaries are usually societal boundaries. The source is institutional authority, as formulated in professional communities, etc. The content is a set of professional rules defining the role definitions/relationships and how the work has to be done. Disciplines involved are not organisational disciplines but those of the institutions (medicine, engineering, physics, etc.).

6.2.4 Perceived organisation

Organisation is perceived differently by members of the organisation, according to their formal position within it, their cultural preferences, their main interests, etc. People with similar perceptions have different interpretations of the same organisation which often become rules. The source of perceived organisation is various subcultures. Its object is sets of coherent systems, or representations based upon beliefs, values and culture. Disciplines usually include social psychology, cultural anthropology, etc.

6.2.5 Informal organisation

The human relations school has developed this cateogry. As Gouldner says, the notion of informal organisation is derived from a number of vague concepts. The only feature they have in common is that of being a marginal addition to formal organisation, oriented to achieving goals that are different from the institutional ones. Popular usage represents informal organisation mainly as cultural structures, such as models of beliefs, feelings, etc., that may supplement or contrast with the goals of the formal organisation. We suggest that informal organisation be used only to indicate social structures that are not planned for in the formal organisation. (e.g. primary groups, cliques, counteracting groups). Informal organisation is, by definition, particularised in the sense that its content and scope are limited to the defense of special interests. The source of informal organisation is the general social structure and functioning.

6.2.6 Overall organisation

Each of these layers of organisation or sub-organisation described above has its own internal coherence. Sometimes they regulate different behaviours; sometimes their patterns can conflict with each other but still be complementary. For instance, a hierarchical formal organisation and highly prescriptive work cycle specifications can coexist with cordial social relations between foreman and workers, or with informal output restriction among groups of workers. Another phenomenon is that of an individual who exceeds production norms and thus becomes a social outcast. Informal organisation is supplementary to, not in contrast with formal organisation.

The various layers provide rules for keeping together the human and technical resources within the particular unit. We call this the overall organisation, which always encompasses a more or less recognised model that provides a framework for any modification in any organisation layer.

6.2.7 Real organisation

In some cases people do their work, follow rules and manage their relationships in striking contrast with formal and technical organisation, that is with the formalised, explicit will of management. In some cases, this could stem from counter-organisations. As a result, the unit either collapses or changes its mission, in spite of an increasing formalisation. In other cases, the source of these contrasts is in fact adaptation by people in order better to achieve the goals of the unit, as well as individual goals: for instance, in a steel mill, the goals of producing better quality at lower cost with less fatigue and risk (Butera, 4.2, 1979). In these cases the de facto organisation is an unwritten set of rules intended to adapt to well-identified goals and constraints. This adaptation is mainly a co-operative effort developed by the people themselves: workers, foreman, managers. Its object can be task definition, task allocation (Susman, 4.2, 1975), role definition, authority structure, decision-making processes, etc. The process of adaptation affects not only what are called organisation processes, but also "hard" elements like organisational mechanism and organisation structures.

When the contrast with the formal and technical organisation is wide and when a paradigm of organisation emerges which is based upon different principles, we have called this real organisation. It often indicates: (a) a signal that a profound organisational change is under way which is not visible within formal sources: the adaptation taking place indicates a reactive response of work to goals, and a new trend towards different formalisation, even though unwritten; and (b) a set of new principles for further design in the same or a similar unit: these are sometimes similar to principles developed in planned designs (Davis, Cherns).

6.3 The functioning of the unit

An organisation unit can be conceived as an open system, that is a system of transactions with the outside (Katz and Kahn, 3.3, 1966). The main features of these exchanges concern input (materials or information) converted into output through the adoption of work and related technology. In every case, the necessary work has four functions to perform -

- conversion from input to output;
- co-ordination and integration of various elements;
- maintenance, allowing the system to preserve its pattern; and
- adaptation.

Uncertainties of input, insufficient knowledge of conversion procedures and input different from that expected or accepted - all push towards a continuous adaptation of the functioning of work and technologies. This creates the need for innovation. As soon as adaptation takes place, the different kinds of work must be redesigned in terms of content and distribution.

Input and output of material or information whose conversion is the main mission of the system is not the only transaction: information on co-ordination processes is input and output in the form of programs, directions, etc., while information on maintenance work is input or output in the form of instruction, knowledge, etc. Innovation can also be imported or exported in the form of goals and know-how.

A picture of the unit in motion is shown in figure 5. There follows from this two implications: (a) the degree of uncertainty suggests the appropriate configuration of work and organisation; and (b) the analysis of cases where values of input, conversion and output exceeds the expected range (variances) may explain the reason for the adaptation of de facto work organisation.

Figure 5: The unit in motion

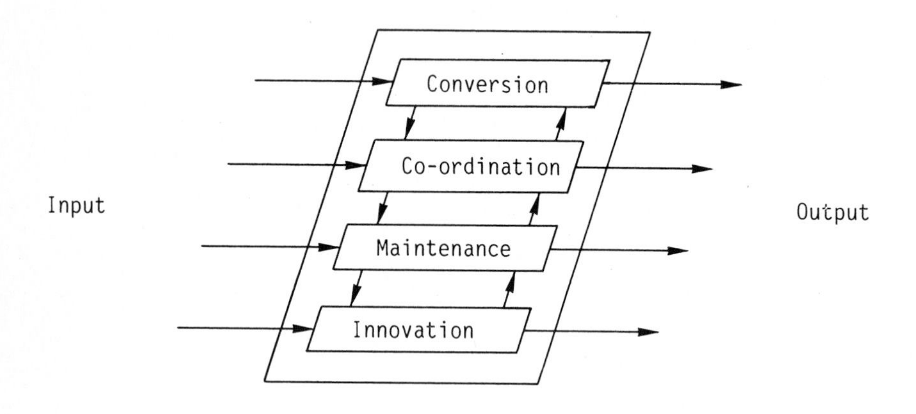

6.4 The anatomy of organisation: configuration, structure and performance

What is the object of work design? What concrete element of work should be designed first? What limits will it place on the design of other elements? Which structural features of the unit (such as size, technology, population) or expected performance will influence the choices? How will external variables directly or indirectly affect work design? In this section, a model is presented as a map for providing the answers.

Human work may be defined as structure relationships between people on the one hand and physical and social production processes on the other, resulting in a physical or symbolic transformation having economic implications. The work to be designed can be described as made up of various component elements, or overall work configuration, as summarised in table 14. Consequently its configuration is more extensive than the body of formal rules usually associated with the idea of work organisation. The body of rules is sustained by a "skeleton" made up of the necessary work, the control system and the task sructure, and gives life to the "nervous system" made up of the living social system including norms, group and interpersonal processes, roles, expectations, perceptions, values, beliefs, goals, etc.

These different components of the same object have historically been analysed as different phenomena subjected to different disciplines. However, interdependence, the effects on people and the needs of planned design in automation now suggest an integral overview.

Components (or stages) have two properties: (a) each can be described as a "stage" in the formation of the overall work, in the sense that all are in a logical sequence whereby the design of each stage more or less limits the possible choices for the other stages; and (b) each component is influenced by different structural and performance dimensions of the unit, that is by different "internalised environments" (Emery and Trist, 3.3, 1960).

The following example may serve to illustrate these two properties. Firstly, the task structure, as we have seen, influences the range within which jobs and organisational units can be efficiently designed, while job design in turn defines the object to which personnel rules can be applied. All this sets up constraints for the social system. Secondly, the aggregated work is influenced mainly by economic and social dimensions, the task structure by the technology adopted, the personnel rules by societal factors, etc.

This rapidly sketched view challenges both the theories of technological and societal determinism. Technological determinism regards technology as the major source of influence on the nature of work, occupations and social relationships of production. Authors like Woodward, Mallet, Richta and Blauner are frequently included in this category. Societal determinism attributes greater influence on work structures to power structures, social conflicts and institutions. Authors such as Touraine, Reynaud and Maurice are usually considered as exponents of this school of thought. Nevertheless, the distinction seems to be a rather artificial one, and the authors themselves correctly refuse these labels. In fact, in most cases they make studies of specific variables within enterprises and make no claim to explain how society is made up. Sometimes those polarised positions seem to hark back to theories of past years (ecological versus sociological thought, Kårl Marx versus Max Weber, etc.), whose opposing explanations of the nature of social reality are still operating.

Our view challenges any real or hypothetical technological determinism because it stresses that technological choices are ultimately socio-economic ones, and that there is no deterministic cause/effect relationship among the set of variables described. In any stage we can encounter the decisive impact of social action that, for good or for ill, could trigger off a flow of consequences. Underlying the model, action by people is indicated as by far the most important intervening variable in the process of design because it could affect both the external or contextual variables, as well as the aspects of the configuration.

On the other hand, this view also challenges any real or hypothetical societal determinism because it makes clear that there are stages of

Table 14: Overall work configuration

Stages	Corresponding to	Areas of conventional concern
1. Aggregated work (amount and gross quality of work required for a given output with a given technology)		Production engineering
2. Control system (pattern and architecture of the production system)		
3. Task structure (content and relationships of component tasks of stages 1 and 2)	- Cycles - Transformation - Information system - Technical procedures	Industrial engineering
4. Formal organisation (organisation structure, procedures, jobs, formal communications, etc.)		Administrative sciences, sociology of work, personnel management, industrial relations
5. Task assignment (to individuals and segments of the work-force)	Work organisation	
6. Conditions of employment (wages, qualifications, work time, training, careers, etc.)		
7. Social system (professional and social roles co-operation and coalition systems, values, beliefs, social goals, etc,)	Perceived organisation, informal organisation, social interaction behaviour, motivation, etc.	Psychology of work, industrial sociology, etc.

formation of work (e.g. personnel administration, social systems) whose content is more essentially social in nature and whose object is a matter of social relationships (like industrial relations, legislation, institutional actions, etc.). Here, however, the social choice is not free of constraints; the alternative pattern of personnel administration is more or less limited by the work content influenced by previous technological choices.

There is evidence that jobs, roles, work time arrangements, qualification patterns, payment systems, career patterns, social rules at the workplace, values about work and attitudes, as demonstrated in similar technical systems but in different countries or regions, differ enormously according to societal dimensions Maurice and Brossard, 3.3, 1974; Reynaud, 3.3, 1979; Gallie, 4.3, 1978); Pizzorno and Crouch, 4.6, 1978). Our point is that these are minor differences as compared with those observed between work situations with different content; however different the pay, training, recruitment and treatment might be between a French and Japanese assembly line operator with a 45-second cycle, their different conditions seem marginal if compared with those of the skilled craftsman developing prototypes in any part of the world.

The major question of the seventies has been whether or not the decline in the content of human work can be halted and possibly reversed. From our perspective, the social exercise consisting of industrial relations, human relations, subjective social actions, etc., as shown in stages 6 and 7 of our scheme (table 14) can be effective in changing the content of work only if the previous stages of work formation have also been affected. The type of control system and task structure needed for enlarging options on organisation, occupational structure, social systems, etc., must be designed for the most part together with technology at an early stage.

If the opposition between technological and social determinism seems rather artificial, two contrasting theses remain: those authors who claim that work content has a limited role in shaping occupations, social stratification and social division of labour, as well as in influencing behaviour, and those who assume that the work content is the strategic, even though by no means exclusive, variable affecting the structure of the world of work. The first group, of course, does not place great importance on design of the technology, while the second stresses that a technology more conducive to the improvement of the work content must be designed in most industries with a social orientation in mind. Our model falls within the second point of view.

The content of work is assumed to have an influence on both social structure and social behaviour. As with any middle-range model, it focuses only on a limited set of phenomena; both the dynamics of the phenomena at the societal level and the "non-ecological" determinants of social behaviour are excluded. Our model also takes into account the fact that the content of work is only one component of the overall work configuration. The expansion of the definition of work organisation is not sufficient for understanding and forecasting fully the conditions of work or the QWL of people included in a specific unit. We therefore need to insert two more dimensions into our model of work.

The first is the structure underlying the overall work, which falls into three categories -

- actual technology and physical hardware (know-how, production system model, equipment, physical environment, buildings, layout, etc.);

- actual nature of the unit of production (size, location, administrative relations, financial characteristics, etc.);

- actual labour force employed (young/old, male/female, skilled/unskilled, etc., workforce composition).

This dimension is to the work configuration as the table legs are to the table top: it includes the concrete visible structure which supports that configuration, and comprises the "internalised" segments of the environment affecting the overall work configuration. Alternative features of the structure also have a direct impact on QWL. Light, noise, size of company, rural/urban location, age or skill composition of the workforce - all influence working conditions, both indirectly (via work configuration) and directly.

The final dimension of the model of work is performance: economic, technical and social. This dimension expresses the reasons for which the work system is running. The relative importance given to each class of performance and the relative goals and measurements could influence the functioning, configuration and structure of the system.

Herbst wrote: "The product of work is people". Joint optimisation of social and techno-economic criteria is an idea proposed in the seventies. The definition and measurement of QWL criteria (see Chapter 1) is not only an evaluation tool but also a cause of "teleological" change, i.e. goals become variables concretely affecting the overall work configuration and structure, people's perception and, finally, the actual QWL.

A general scheme emerges from the preceding discussion (figure 6). The three faces of the cube mutually influence each other, but the main cycle of influces is: performance (expected ⟶ structure ⟶ configuration ⟶ performance (achieved) ⟶ (required change in) ⟶ structure / configuration and so on.

The shape and dynamics for a single unit take place within a similar shape and dynamics for the enterprise, as represented in figure 7.

The relationships with the environment come through -

- structure, as a process of internalisation of the environment;

- performance, as market responses in a market economy or attainment of plans in planned economies; and

- the work configuration as pressures from the labour market, educational system, social system, political system and technology available.

Figure 6: Overall work model (single unit)

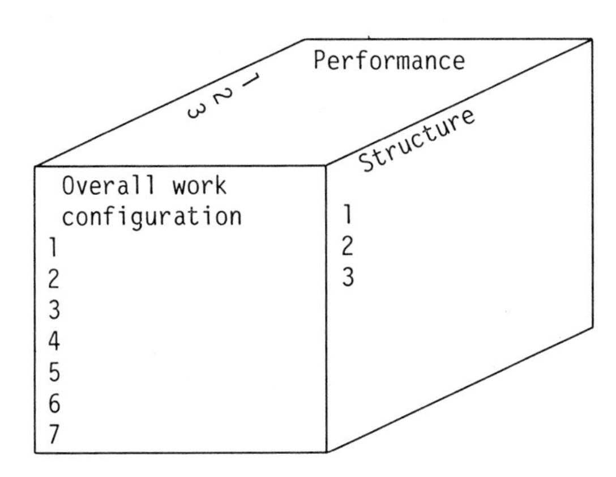

Figure 7: Overall work model (the enterprise)

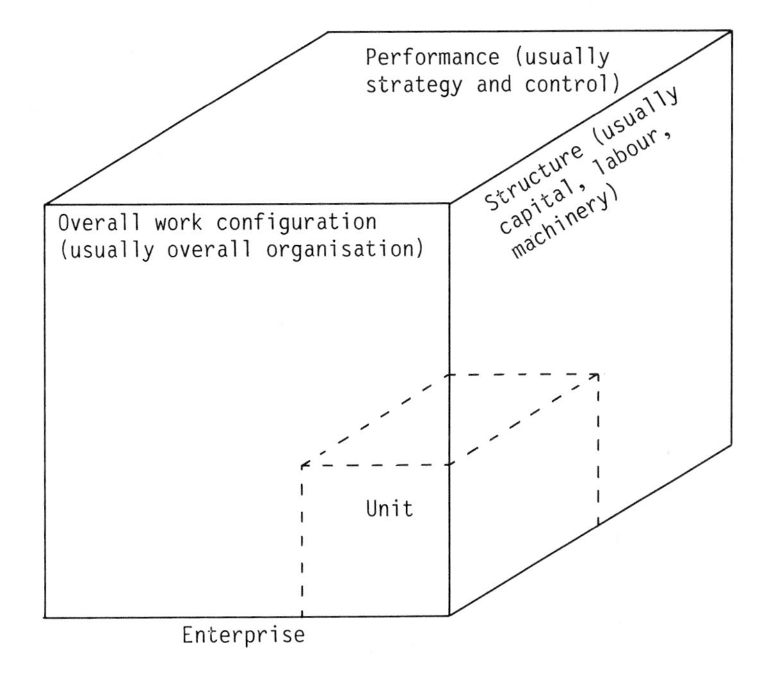

Our model suggests, in substance, that a phenomenological notion of work - however expanded - is not enough. Two more dimensions are added to this, which indicates how relationships among people in the face of production processes can be described and designed): a structural dimension (concrete equipment, productive units and people) and a functional dimension (goals and results).

The model accounts for the uniqueness of the organisations claimed by socio-technical theorists, in which social and technical systems are combined in various ways. It also incorporates the societal effects on organisation claimed by the proponents of sociological explanations. In addition, it seems to offer a conceptual framework for both designers at unit level and policy-makers at societal level. To company managers and trade union leaders, it offers a map of interactions to be considered when particular aspects of the system are designed, and suggests moving away from any dogmatic "best way" to organise. To high-level policy-makers, it provides an idea of how societal programmes can have different effects on various types of units, and advocates participation, experimentation and diffusion of learning processes.

In conclusion, figure 8 illustrates the sequence of the main variables to be considered in work design with QWL goals.

Figure 8: Variables in work design with quality of working life goals

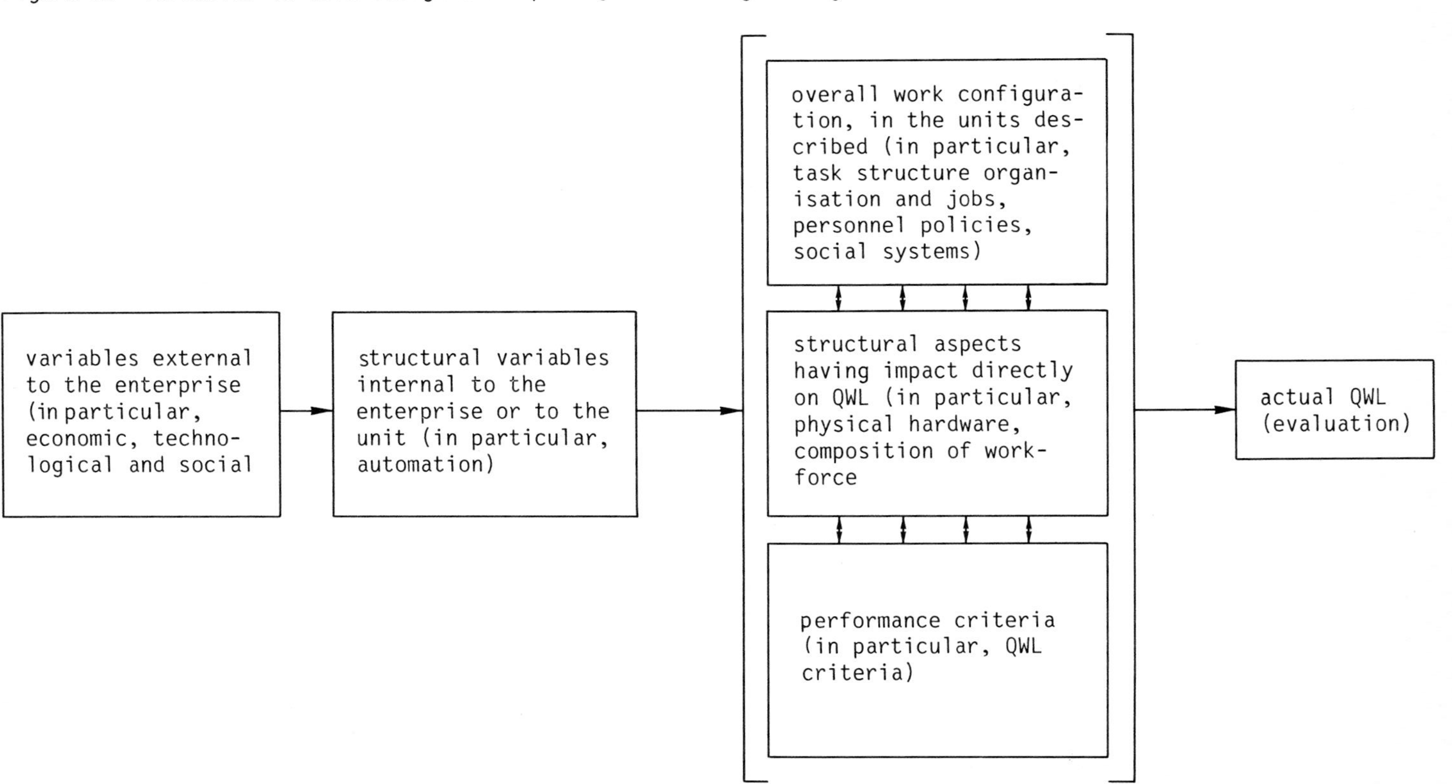

Notes

[1] References to the bibliography within the text take the form: author's name; section of bibliography; date of publication. They should be distinguished from general notes, which are indicated by numbers in the text.

[2] See bibliography, section 3.1.

[3] See bibliography, section 3.2.

[4] See bibliography, section 3.3.

[5] Quoted in Sultan and Prasow (2, 1964).

[6] ibid.

[7] For a detailed analysis, see M. Blaug: "A survey of the theory of process innovation", in Economics, Feb. 1963; W. Nordhaus: Invention, growth and welfare (Cambridge, Massachusetts, MIT Press, 1969); and Reyneri and Ricciardi (3.5, 1972).

[8] See bibliography, sections 3.1 and 3.2.

[9] The presence of the computer by itself is not necessarily automation, as for example in isolated NC machines or in EDP use for calculation or printing.

[10] See also, in the most exhaustive review of studies on the impact of technology on job structure, L.E. Davis and J. Taylor: "Technology, organisation and job structure", in Dubin (2, 1976).

[11] The term was used in this sense by Stanley Udy: "The comparative analysis of organisation", in J.G. March (ed.): Handbook of organisations (Chicago, Rand McNally, 1965).

CLASSIFIED BIBLIOGRAPHY

Categories of publications covered

BIBLIOGRAPHY

1. New technologies and automation: types of technology and trends

Advisory Council for Applied Research and Development. 1978. Industrial innovation. London, HMSO.

Barker, M.J. (ed.). 1979. Industrial innovation: Technology, policy, diffusion. London, MacMillan.

Commission of the European Communities. 1979. Surveying secondary and tertiary changes in technology and society due to micro-electronics-based innovation. FAST Expert Meeting, Brussels, 1979.

International Metalworkers' Federation. 1979. Micro-processors and robots: Effects of modern technology on workers. Discussion papers of IMF Central Committee, Vienna, 1979. Geneva.

McLean, J.; Rush, H. 1978. The impact of micro-electronics on the UK: A suggested classification and illustrative case studies. Brighton, University of Sussex, Science Policy Research Unit.

Nora, S.; Minc, A. 1978. L'informatisation de la société. Paris, Documentation Française.

OECD. 1976. Proceedings of the OECD Conference on Computer/Telecommunication Policy. Paris.

Proceedings of the Third International Symposium on Industrial Robots. 1973. Zurich, 1973. Munich, Verlag Moderne Industrie.

Rada, J. 1980. The impact of micro-electronics: A tentative appraisal of information technology. Geneva, ILO.

Rempp, H. 1982. Introduction of CNC machine tools and flexible manufacturing systems: Economic and social impact. Maastricht, European Centre for Work and Society.

United Kingdom, Department of Industry (National Engineering Laboratory). 1980. Industrial Robots, Vols. I and II. London.

United States Department of Labor (Bureau of Labor Statistics). Technological change and manpower trends (in 16 industries). Bulletin Nos. 1981/1977, 2005/1979, 2033/1979, 1658/1970. Washington, DC.

2. Automation: general studies, classifications, analytical dimensions

Bouchot, Y. et al. 1980. Automatisation. Lyon, Presses universitaires de Lyon.

Bright, J.R. 1958. Automation and management. Boston, Harvard University, Graduate School of Business Administration, Division of Research.

Crossman, E.R.F.W. 1965. "Taxonomy of automation", in OECD. The requirements of automated jobs, North American Joint Conference, Paris, Dec. 1964. Mimeographed, 1965.

De Bivort, H. 1955. "Automation - some social aspects", in International Labour Review (ILR), July-Dec. 1955, pp. 468-485.

Diebold, J. 1952. Automation: The advent of the automated factory. New York, Van Nostrand.

Diebold Group. 1964. Criteria and bases for a study on the extension of automation in American industry. A study prepared for the United States Department of Labor, Office of Manpower, Automation and Training. Washington.

Dubin, R. (ed.). 1976. Handbook of work organisation and society. New York, Rand McNally.

Dunlop, J.T. (ed.). 1962. Automation and technological change. Englewood Cliffs (New Jersey), Prentice Hall.

Ellul, J. 1978. Le système technicien. Paris, Calmann-Lévy.

Foster, D. 1963. Modern automation. London, Pitman.

Jaffe, A.J.; Froomkin, J. 1968. Technology and jobs - Automation in perspective. New York, Frederick A. Praeger.

Killingsworth, C. 1963. "Automation, jobs and manpower", in United States Department of Labor. Nation's manpower revolution. Washington, United States Government Printing Office.

Lilley, S. 1957. Automation and social progress. London, Lawrence and Wishart.

Margulies, F. et al. 1977. Automation und industrielle arbeitmehmer. Vienna, Sweitlige Studies.

Michael, M.D. 1962. Cybernation: The new conquest. Santa Barbara, Center for Study of Democratic Institutions.

Nathusius, K. 1968. Die Automation als Gesellschaftliche und Wirtschaftliche Herausforderung. Boppard am Rhein, Harald Boldt Verlag.

Naville, P. 1963. Vers l'automatisme social?. Paris, Gallimard.

Nora, S.; Minc, A. 1978. L'informatisation de la société. Paris, Documentation Française.

Pollock, F. 1956. Automation: Materialien zur Beurteilung der Okondhishen und Sozialen Folgen. Frankfurt am Main, Europaische Verslangsaltalt.

Rijnsdorp, J.E. 1979. "Levels of automation: A multidimensional approach", in Rijnsdrop, J.E. (ed.). Case studies in automation related to humanisation of work. Oxford/New York, Pergamon Press.

Rogers, J. 1958. Automation: Technology's new face. Los Angeles, University of California, Institute of Industrial Relations.

Simon, H. 1969. The science of artificial. Cambridge (Mass.), MIT Press.

Sultan, P.E.; Prasow, P. 1964. "Automation: Some classification and measurement problems", in Labour and Automation (Geneva, ILO), Bulletin No. 1, 1964, pp. 9-33.

Weizenbaum, J. 1975. Computer power and human reason, from judgement to calculation. San Francisco, W.H. Freeman.

3. Background areas of study

3.1 History of the division of labour and technology

Braverman, H. 1974. Labor and monopoly capital: The degradation of work in the twentieth century. New York, Monthly Review Press.

Council for Science and Society. 1981. New technology: Society, employment and skills; Report of a working party. London.

Dubin, R. 1958. The world of work: Industrial society and human relations. Englewood Cliffs, Prentice Hall.

Ellul, J. 1964. Technological society. New York, Vintage Books.

Friedmann, G. 1955. Industrial society [Problêmes humains du machinisme industriel]. New York, Free Press.

---. 1964. The anatomy of work [Le travail en miettes]. Glencoe, Free Press.

Giedion, S. 1948. Mechanisation takes command. London, Oxford Univeristy Press.

Ginzberg, E. 1964. Technology and social change. New York, Columbia University Press.

Habbakkuk, H.J. 1962. American and British technology in the XIXth century, London, Cambridge University Press.

Habermas, J. 1968. Technik und Wissenshaft als Ideologie, Frankfurtlak, Suhrkamp.

Landes, D.S. 1969. The unbound Prometheus. London, Cambridge University Press.

Moore, W.E. 1965. The impact of industry. Englewood Cliffs, Prentice Hall.

Mumford, L. 1934. Technics and civilisation. New York, Harcourt Brace Co.

Ochainine, D.A. 1971. L'homme dans les systèmes automatisés. Paris, Dunod.

Richta, R. 1969. Civilisation at the crossroads - Social and human implications of the scientific and technological revolution. New York, International Arts and Sciences Press, 3rd ed.

Rosenbrock, H. 1979. The redirection of technology. Bari, International Federation for Automatic Control symposium.

Rustant, M. 1959. L'automation, ses consequences humaines et sociales. Paris, Editions ouvrières.

Strassmann, W.P. 1972. "The response to automation and advanced technology: A comparison of developed and developing countries" in Automation in developing countries, Round-table discussion on the manpower problems associated with the introduction of automation and advanced technology in developing countries, Geneva, July 1970. ILO, Geneva, 1972, pp. 17-36.

Walker, C.R. (ed.). 1962. Modern technology and civilisation. New York, McGraw Hill.

Wiener, R. 1964. God and Golem, inc.. Cambridge (Mass.), MIT Press.

3.2 Technology, labour market and social structure

Bessant, J.R.; Dickson, K.E. 1980. Computers and employment: An annotated bibliography, British Computer Society. Birmingham, University of Aston.

Commission of the European Communities. 1980. Microelectronique et emploi en Europe. Series Le Dossier de l'Europe 16/80. Brussels.

---. 1981/2. Technology and employment. Bulletin Nos. 1, 2 and 3. Brussels.

Foster, D. 1980. Innovation and employment. Oxford, Pergamon Press.

Freeman, C.; Curnow, R. 1978. Technical change and employment: A review of post-war research, Manpower Services Commission. Brighton, University of Sussex, Science Policy Review Unit.

Jaffe, A.J.; Froomkin, J. 1968. Technology and jobs - Automation in perspective. New York, Frederick A. Praeger.

Killingsworth, C. 1963. "Automation, jobs and manpower", in United States Department of Labor. Nation's manpower revolution. Washington, United States Government Printing Office.

OECD. 1966. Manpower aspects of automation and technical change. Paris.

Piore, M.J. 1968. "The impact of the labor market upon the design and selection of productive techniques within the manufacturing plant", in Quarterly Journal of Economics, Nov. 1968, pp. 602-620.

Ranis, G. 1970. "Technology, employment and growth: Lessons from the experience of Japan", in Automation in developing countries, Round-table discussion on the manpower problems associated with the introduction of automation and advanced technology in developing countries, Geneva, July 1970. ILO, Geneva, 1972, pp. 37-52.

Sorge, A.; Harmann, G. 1980. Technology and labour markets. Berlin, Wissensdraftszentrum Berlin, mimeographed.

3.3 Technology and organisation

Aldrich, H.E. 1972. "Technology and organisational structure", in Administrative Science Quarterly, Vol. 17, 1972, pp. 26-43.

Blau, P.M.; Shoenner, R.A. 1971. The structure of organisation. New York, Basic Books.

Butera, F. 1976. "Environmental factors in job design", in Davis, L.E.; Cherns, A.B. The quality of working life. New York, Free Press.

Child, J. 1972. "Organisation structure, environment and performance: The role of strategic choice", in Sociology, No. 6, 1972, pp. 1-22.

Crozier, M.; Friedberg, E. 1977. L'acteur et le systême. Paris, Editions de Seuil.

Davis, L.E.; Taylor, J.C. (eds.). 1972. Design of jobs. Harmondsworth, Penguin Books.

Dobrov, G.M. 1979. "La technologie en tant qu'organisation", in Revue internationale des sciences sociales, Vol. XXXI, No. 4, 1979, pp. 628-648.

Emery, F.E. 1959. Characteristics of socio-technical systems. London, Tavistock Institute, mimeographed. Reprinted in Davis L.E.; Taylor, J.C. (eds.). Design of jobs. Harmondsworth, Penguin Books, 1972, pp. 177-198.

---; Trist, E. 1960. "Socio-technical systems", reprinted in F. Emery (ed.). Systems thinking. Harmondsworth, Penguin Books, 1960, pp. 281-296.

Friedmann, G. 1950. Où va le travail humain?. Paris, Gallimard.

---; Naville, P. 1961. Traité de sociologie du travail. Paris, Librairie Armand Colin.

Gasparini, G. 1975. Technologia, ambiente, struttura, Milan, Franco Angeli.

Gouldner, A.W. 1954. Patterns of industrial bureaucracy. New York, Free Press.

Harvey, E. 1968. "Technology and the structure of organisations", in American Sociological Review, No. 33, 1968, pp. 247-259.

Herbst, P. 1974. Socio-technical system design. London, Tavistock.

Hickson, D.J. et al. 1969. "Operations technology and organisation structure: A reappraisal", in Administrative Science Quarterly, Vol. 14, No. 4, 1969, pp. 378-397.

Katz, D.; Kahn, R.L. 1966. The social psychology of organisations. New York, John Wiley and Sons.

Lawrence, P.R.; Lorsh, J.W. 1976. Organisation and environment. Boston, Harcourt Brace.

March, J.G.; Simon H.H. 1958. Organisations. New York, Wiley and Sons.

Maurice, M.; Brossard, M. 1974. "Existe-t-il un modèle universel de structure d'organisation?, in Sociologie du travail, Oct.-Dec. 1974, pp. 402-426.

Maurice, M. et al. 1978. La production de la hiérarchie dans l'entreprise - Comparaison France-Allemagne. Aix-en-Provence, Laboratoire d'Economie et de Sociologie du Travail; mimeographed.

Parsons, T. 1970. "The impact of technology on culture and emerging new modes of behaviour", in International Social Science Journal, Vol. XXII, No. 4, 1970, pp. 607-627.

Perrow, C. 1967. "A framework for the comparative analysis on organisation", in American Sociological Review, No. 32, 1967, pp. 65-105.

Popitz, H. et al. 1976. Technick und Industrierbeit: Soziologische Unterschungen in der Hüttenindustrie. Tübingen, Mohr.

Pugh, D.S. et al. 1968. "Dimensions of organisation structure", in Administrative Science Quarterly, Vol. 13, 1968, pp. 65-105.

---. 1969a. "The contexts of organisation structures", in Administrative Science Quarterly, Vol. 14, 1969, pp. 91-114.

---. 1969b. "An empirical taxonomy of structures of work organisation", in Administrative Science Quarterly, Vol. 14, 1969, pp. 115-126.

Reynaud, J.D. 1979. "Conflit et regulation conjointe: Esquisse d'une theorie de la regulation conjointe", in Revue française de sociologie, Vol. XX, 1979, pp. 367-376.

Taylor, F.W. 1951. Scientific management. New York, Harper.

Taylor, J.C. 1971. "Some effects of technology in organisational change", in Human Relations, No. 24, 1971, pp. 105-123.

Touraine, A. 1965. Sociologie de l'action. Paris, Editions du Seuil.

Trist, E. et al. 1963. Organistional choice. London, Tavistock.

Udy, S.H. 1970. Work in traditional and modern society. Englewood Cliffs, Prentice Hall.

Walker, C.R.; Guest, R. 1962. The man on the assembly line. New York, McGraw Hill.

Woodward, J. 1965. Industrial organisation: Theory and practice. London, Oxford University Press.

---. 1970. Industrial organisation: Behaviour and control. London, Oxford University Press.

3.4 The quality of working life

Butera, F. 1972. I frantumi ricomposti. Venice, Marsilio.

--- (ed.). 1981. Le ricerche per la trasformazione del lavoro industriale in Italia: 1969-1979. Milan, Franco Angeli.

Clifford, B. 1980. Quality of work life developments in Sweden - A review of the literature. London, Work Research Unit, mimeographed.

Cooper, C.L.; Mumford, E. 1979. Quality of working life in Western and Eastern Europe. London, Associated Business Press.

Davis, L.E.; Cherns, A.B. (eds.). 1975. The quality of working life. Vols. I and II. New York, Free Press.

Emery, F.E.; Thorsrud, E. 1969. Form and content of industrial democracy. London, Tavistock.

European Foundation for the Improvement of Living and Working Conditions. 1977. L'analyse de changements dans l'organisation du travail (4 volumes). Dublin.

---. New forms of work organisation in the European Community. 1978. (9 volumes). Dublin.

International Council for QWL. 1979. Working on quality of working life: Developments in Europe. Boston/The Hague, Martinus Nijhoff Pub..

International Institute for Labour Studies (IILS). 1978. Social aspects of work organisation, Research series No. 33. Geneva.

ILO. 1975. Making work more human: Working conditions and environment, Report of the Director-General, International Labour Conference, 60th Session, Geneva, 1975.

---. 1979. New forms of work organisation (2 volumes). Geneva.

Klein, L. 1975. New forms of work organisation. Cambridge, Cambridge University Press.

OECD. 1973. Advances in work organisation. Final Report, International Management Seminar, Paris, Apr. 1973.

Swedish Employers' Confederation. 1975. Job reform in Sweden. Stockholm.

Takezawa, S.I. 1976. "Quality of working life: Trends in Japan", in Labour and Society (Geneva, IILS), Jan. 1976, pp. 29-48.

Trist, E.L. 1981. The evolution of socio-technical systems, Occasional papers No. 2. Ontario, Quality of Working Life Center.

UNESCO. 1975. Revue internationale des sciences sociales, Vol. XXVII, No. 1, 1975. Issue devoted to "Les indicateurs socio-economiques: Theories et applications".

United States, Executive Office of the President. 1973. Social indicators, Washington, DC, United States Government Printing Office.

Wilson, N.A.B. 1973. On the daily quality of working life, United Kingdom, Department of Employment. London, HMSO.

3.5 The economics of automation

Abernathy, W.J.; Utterback, J.M. 1975. Innovation and the evolving structure of the firm. Cambridge (Mass.), Harvard University Press.

Blaug, M. 1963. "A survey of the theory of process innovation", in Economics, Feb. 1963.

Heller, W.H. 1951. "The anatomy of investment decisions", in Harvard Business Review, Mar. 1951, pp. 95-103.

Hetman, F. 1973. Society and the assessment of technology: Premises, concepts, methodology, experiments, areas of application. Paris, OECD.

Jaffe, A.J.; Froomkin, J. 1968. Technology and jobs - Automation in perspective. New York, Frederick A. Praeger.

Meyer, D. et al. 1979. Automation, travail et emploi: étude empirique des principaux automatismes avancés et eléments d'approche macro-économique. Paris, Université IX, Institute de recherche et d'information socio-économique.

Nordhaus, W. 1969. Invention, growth and welfare. Cambridge (Mass.), MIT Press.

OECD. 1973. Society and the assessment of technology. Paris.

Reyneri, E.; Ricciardi, C. 1972. La variabilita degli effetti dell' automazione sulla struttura del lavoro: Problemi di metodo per una ricerca sulla realtà industriale italiana. Milan, Istituto Lombardo per gli studi economici et sociali.

United Nations Economic Commission for Europe. 1971. Aspects économiques de l'automatisation, New York.

4. Research on automation, work and working conditions

Abbott, L.F. 1976. Social aspects of innovation and industrial technology: A survey of research, United Kingdom, Department of Industry. London, HMSO.

Brady, R.H. 1961. Organisation, automation and society. Berkeley, University of California Press.

Cherns, A.B. 1980. "Speculations on the social effects of new micro-electronics technology", in ILR, Nov.-Dec. 1980, pp. 705-721.

Emery, F.E. 1981. Social aspects of automation internal to the firm. Harrogate, British Conference on Social and Economic Effects of Automation, 1981.

Greenberg, L.; Weinberg, E. 1964. Automation: Nationwide studies in USA. Geneva, ILO.

Haug, M.R.; Dofny, J. 1977. Work and technology, International Sociological Association. London, Sage Publications.

ILO. 1972a. Automation in developing countries, Round-table discussion on the manpower problems associated with the introduction of automation and advanced technology in developing countries, Geneva, July 1970. Geneva, 1972.

---. 1972b. Labour and social implications of automation and other technological developments, Report VI, International Labour Conference, 57th Session, Geneva, 1972.

Lilley, S. 1957. Automation and social progress. London, Lawrence and Wishart.

Naville, P. 1964. "Problems involved in measuring the effects of automation in case studies in France", in Labour and Automation, Bulletin No. 1, 1964, pp. 47-69.

Sadler, P. 1968. Social research on automation, Social Science Research Council. London, Heinemann.

Sheppard, C.S.; Garrol Socnal, D. (eds.). 1980. Working in the twenty-first century. New York, Wiley and Sons.

Walker, C.R. 1959. Toward the automatic factory. New Haven, Yale University Press.

4.1 Automation and organisational structures

Bright, J.R. 1958. Automation and management. Boston, Harvard University, Graduate School of Business Administration, Division of Research.

Buckingham, W. 1961. Automation: Its impact on business and people. New York, Harper and Row.

Burack, E.H.; McNichols, T.J. 1968. Management and automation research project - Final report. Chicago, Illinois Institute of Technology.

Davis, L.E. 1971. "The coming crisis for production management: Technology and organisation", in International Journal of Production Research, 1971, Vol. 9, No. 1, pp. 65-82.

Della Rocca, G. 1983. Il ruolo delle parti nella progettazione e applicazione di nuove forme di organizzazione del lavoro. Dublin, European Foundation for the Improvement of Living and Working Conditions.

Drucker, P. 1970. Technology, management and society. London, Heinemann.

Galbraith, J.R. 1977. Organisation design. Reading (Ma.), Addison Wesley Pub. Co.

Haine, P. 1979. Computers in business. London, Macmillan.

Lupton, T. 1981. "The effect of automation on organisation and human relationships within the firm", in British Conference on the Social and Economic Effects of Automation, Harrogate, 1981.

Rosenbrock, H. 1977. "The future of control", in Automation, Vol. 13, 1977.

Simon, H. 1965. The shape of automation for men and management. New York, Harper and Row.

Tannenbaum, A. 1968. Control in organisations. New York, McGraw Hill.

Thomas, H.A. 1969. Automation for management. London, Gower Press.

Thompson, J.D. 1967. Organisations in action, social science bases of administrative theory. New York, McGraw Hill.

Woodward, J. 1970. Industrial organisation: Behaviour and control. London, Oxford University Press.

4.2 Automation and the content of work and occupations (tasks, skills and jobs)

Bonazzi, G. 1975. In una fabbrica di motori. Milan, Feltrinelli.

Bright, J.R. 1958. "Does automation raise skill requirements?", in Harvard Business Review, July-Aug. 1958, pp. 85-98.

Brown, A.A. 1966. "Artefacts, automation and human abilities", in Lawrence, J.R. (ed.). Operational research and the social sciences. London, Tavistock.

Butera, F. 1979. Lavoro umano e prodotto tecnico. Turin, Einandi.

Chapanis, A. 1965. "On the allocation of functions between men and machines", in Occupational Psychology, Jan. 1965, pp. 1-11.

Clegg, H. 1981. "The effects of automation on employment, wages and working hours", in British Conference on the Social and Economic Effects of Automation, Harrogate, 1981.

Cotgrove, S.F. 1958. Technical education and social change. London, Allen and Unwin.

Council for Sciene and Society. 1981. New technology: Society, employment and skills. London.

Crossman, E.R.F.W. 1960. Automation and skill. London, HMSO.

---. 1965. "European experience with the changing nature of jobs due to automation", in OECD. The requirements of automated jobs, North American Joint Conference, Paris, Dec. 1964. Mimeographed, 1965.

Davis, L.E.; Taylor, J.C. 1976. "Technology, organisation and job structure", in Dubin, R. (ed.). Handbook of work, organisation and society. Chicago, Rand McNally, pp. 379-419.

Durand, C. 1959. "L'evolution du travail dans les laminoirs", in Revue française du travail, Jan.-Mar. 1959, pp. 1-18.

Emery, F.E.; Marek, J. 1962. "Some socio-technical aspects of automation", in Human Relations, No. 15, 1962, pp. 17-25.

Jordan, M. 1963. "Allocation of functions between man and machines in automated systems", in Journal of Applied Psychology, No. 47, 1963, pp. 161-165.

Lucas, Y. 1973. Codes et machines. Paris, Presses universitaires de France.

Meissner, M. 1969. Technology and the worker - Technical demands and social processes in industry. San Francisco, Ghandler Publishing Co.

Popitz, H. et al. 1976. Technick und Industrierbeit: Soziologische Unterschungen in der Hüttenindustrie. Tübingen, Mohr.

Susman, G.I. 1970. "The impact of automation on work group autonomy and task specialisation", in Human Relations, Dec. 1970, pp. 567-577.

---. 1975. "Task and technological prerequisites for delegation of decision-making to work groups", in Davis, L.E.; Cherns, A. The quality of working life: Problems, prospects and the state of the art, Vol. I. New York, Free Press.

Touraine, A. 1955. L'evolution du travail ouvrier aux usines Renault. Paris, Editions du CNRS.

Walker, C.R. 1959. Towards the automated factory. New Haven, Yale University Press.

Weir, M.; Mills, S. 1974. "The supervisor as a change catalyst", in Industrial Relations Journal, Winter 1973-74, pp. 61-69.

4.3 Automation, stratification and careers

Acquaviva, S.S. 1958. Automazione e nuova classe. Bologna, Il Mulimo.

Boguslaw, R. 1965. The new utopians. Englewood Cliffs, Prentice Hall.

Christensen, E. 1968. Automation and the worker. London, Labour Research Department Publications.

Dubois, P. 1978. "Techniques et division des travailleurs", in Sociologie du travail, Vol. XX, No. 2, 1978, pp. 174-191.

Gallie, D. 1978. In search of the new working class; Automation and social integration within the capitalist enterprise. Cambridge, Cambridge University Press.

Kern, H.; Schumann, M. 1970. Industriearbeit und Arbeiterbewusstein (2 volumes). Frankfurt am Main, Suhrkamp/Kno.

Mallet, S. 1969. La nouvelle classe ouvrière. Paris, Editions du Seuil.

Sandberg, A. 1979. Computers dividing man and work: Recent Scandinavian research on planning and computers from a trade union perspective. Stockholm, Arbetslivscentrum.

Stinchcombe, A.L. 1965. "Social structure and the organisation", in March, J.G. Handbook of organisations. New York, Rand McNally.

Wells, J.A. 1970. Automation and women workers. Washington, DC, US Women's Bureau.

Whyte, W.F. 1956. "Engineers and workers: A case study", in Human Organisation, Winter 1956, pp. 3-12.

Zweig, F. 1952. The British worker. Harmondsworth, Penguin Books.

4.4 Automation, working conditions and quality of working life

Gallino, L. 1983. Informatica e qualità del lavoro. Turin, Einaudi.

Gardell, B. 1980a. "Sweden - Production techniques and working conditions", in Current Sweden, No. 256, Aug. 1980, pp. 1-16.

---; Gustavsen, B. 1980b. "Work environment research and social change: Current developments in Scandinavia", in Journal of Occupational Behaviour (Chichester, UK), Jan. 1980, pp. 3-17.

Hatch, T. 1962. "Human-factors engineering and safety research", in Journal of Occupational Medicine, Jan. 1962, pp. 1-5.

ILO. 1957. Automation and other technological developments: Repercussions on safety and health, Report of the Director-General, International Labour Conference, 40th Session, Geneva, 1957.

---. 1979. Policies and practices for the improvement of working conditions and working environments in Europe, Report III, Third European Regional Conference, Geneva, 1979.

Kornhauser, A. 1965. Mental health of the industrial worker. New York, Wiley.

Marri, G. 1981. "L'ambiente di lavoro in Italia", in F. Butera (ed.). Le ricerche per la transformazione del lavoro industriale in Italia; 1969-79. Milan, Franco Angeli.

McLean, A.A. 1979. Work stress. Reading (Ma.), Addison-Wesley.

Murrell, H. 1978. Work stress and mental strain: A review of some of the literature, Work Research Unit Occasional Paper No. 6. London, United Kingdom Department of Employment.

Oddone, M. 1981. "Il problema dell'ambiente di laboro in Italia", in F. Butera (ed.). Le ricerche per la trasformazione del lavoro industriale in Italia: 1969-1979, Milan, Franco Angeli.

Rijnsdorp, J.E. 1979. Case studies in automation related to humanisation of work, Proceedings of the IFAC workshop, Enschede, Netherlands, 1977. Published for the International Federation of Automatic Control. Oxford, Pergamon Press.

Rose, J. 1970. Anatomie et physiologie de l'automatisation. Paris, Dunod.

Seeger, O. 1960. "Occupational safety in automated production", in Werkstatt und Betrieb (Munich), Mar. 1960, pp. 125-128.

Sell, R.G. 1980. Microelectronics and the quality of working life. London, United Kingdom Department of Employment, Work Research Unit.

Sholtens, S.; Keja, A.J. 1979. "Jobs and VDU's model approach", in Rijnsdorp, J.E. Case studies in automation related to humanisation of work, Proceedings of the IFAC workshop, Enschede, Netherlands, 1977. Oxford, Pergamon Press, 1979.

Teknologisk Institut Gregersensvej. 1981. VDU and work environment - A literature review. Dublin, European Foundation.

Welford, A.T. 1960. The ergonomics of automation, London, HMSO.

Weybrew, B.B. 1961. "Human factors and the work environment: The impact of isolation upon personnel", in Journal of Occupational Medicine, June 1961, pp. 290-294.

WHO. 1959. Mental health problems of automation, Technical Report Series No. 183. Geneva.

4.6 Automation, attitudes and behaviour

Ancona, L.; Iacono, G. 1960. "Workers' attitudes to technological change: Some observations", in Chapanis, A. L'automation: Aspects psychologiques et sociaux. Louvain, Publications universitaires.

Blauner, R. 1964. Alienation and freedom. Chicago, University of Chicago Press.

Chadwick-Jones, J.K. 1969. Automation and behaviour - A social psychological study, London, Wiley.

Faunce, W.A. 1958. "Automation in the automobile industry", in American Sociological Review, 1958, No. 23, pp. 401-407.

Fullan, M. 1970. "Industrial technology and workers' integration in the organisation", in American Sociological Review, Dec. 1970, pp. 1028-1039.

Grossin, W. 1979. "Attitudes des travailleurs à l'égard du changement technologique", in Revue française des affaires sociales, July-Sep. 1979, pp. 65-127.

Herzberg, F. 1966. Work and the nature of man. Cleveland, World Publishing Co.

Jessup, G. 1978. Behavioural considerations of production technology: Technology, employment and job satisfaction. London, United Kingdom Department of Employment, Work Research Unit.

Mann, F.C.; Hoffman, L.R. 1960. Automation and the workers: A study of social change in power plants. New York, Henry Holt.

Marcson, S. 1970. Automation, alienation, and anomie. New York, Harper and Row.

OECD. 1968. Adjustment of workers to technical change at plant level - Final report, International Conference on Adjustment of Workers to Technical Change at Plant Level, Amsterdam, 1966. Paris, 1968.

Perroux, F. 1970. Alienation et société industrielle. Paris, Gallimard.

Sell, R.G.; Shipley, P. 1979. Satisfaction in work design: Ergonomics and other approaches. London, Taylor and Francis.

Smith, M.A. 1968. "Process technology and powerlessness", in British Journal of Sociology, Mar. 1968, pp. 76-88.

Thurman, J.E. 1977. "Job satisfaction: An international overview", in ILR, Nov.-Dec. 1977, pp. 249-269.

Turner, A.M.; Lawrence, P.R. 1965. Industrial jobs and the workers. Cambridge (Mass.), Harvard University Press.

Vamplew, W.C. 1973. "Automated process operators: Work attitudes and behaviour", in British Journal of Industrial Relations, No. 11, 1973, pp. 415-430.

Wedderburn, D.; Crompton, R. 1972. Workers' attitudes and technology. Cambridge, Cambridge University Press.

4.6 Automation and industrial relations

Ahlin, J.E.; Svensson, L.J.P. 1980. "New technology in mechanical engineering industry: How can workers gain control?", in Economic and Industrial Democracy (London), Nov. 1980, pp. 487-521.

Bradley, V.A. 1972. "Informing workers on management use of computers", in Management and Productivity (Geneva, ILO), Jan.-Mar. 1972, pp. 22-38.

Cooley, M. 1978. Architect or bee: The human/technology relationship. Slough, Langley Technical Services.

Elliot D.; Elliot, R. 1976. The control of technology, London, Wykeham Pub.

Emery, F.E.; Thorsrud, E. 1976. Democracy at work: The report of the Norwegian industrial democracy programme, Leiden, Martinus Nijhoff.

Kassalow, E. 1972. "Automation, technological change and unionism in the less-developed countries", in Automation in developing countries, Round-table discussion on the manpower problems associated with the introduction of automation and advanced technology in developing countries. Geneva, July 1970. ILO, Geneva, 1972.

Northrup, H.R. 1958. "Automation: Effects on labour force, skills and employment", in Annual Proceedings of Industrial Relations Research Association, 1958. Washington.

Pizzorno, A.; Crouch, C. 1978. Conflitti in Europa. Milan, Eras.

Rezler, J. 1969. Automation and industrial labour, New York, Ramdom House.

Shils, E.B. 1963. Automation and industrial relations. New York, Holt, Rinehart and Winston.

Taylor, G.W. 1961. "Collective bargaining", in American Assembly (ed.). Automation and technological change. Englewood Cliffs, Prentice Hall.

5. Categories and models for analysis and design

5.1 Tasks, jobs, organisation dimensions and quality of working life indicators

Abel, P.; Mathew, D. 1973. "The task analysis framework in organisational analysis", in Warner, M. (ed.). The sociology of the workplace. London, Allen and Unwin.

Birchall, D.W. 1975. Job design: A planning and implementation guide for managers. London, Gower Press.

Boguslaw, R. 1972. "Operating units", in Davis, L.E.; Taylor, J.C. Design of jobs. Harmondsworth, Penguin Books.

Burbidge, J.L. 1975. Introduction of group technology. London, Heinemann.

Centre de Recherches Economiques Pures et Appliqués. 1975. Les performances sociales des organisations. Paris, Université de Paris, IX Dauphine.

Davis, L.E. 1966a. "The design of jobs", in Industrial Relations, Oct. 1966, pp. 21-45.

---; Engelstadt, P.N. 1966b. Unit operations in socio-technical systems. London, Tavistock.

Davis, L.E. et al. 1972. "Current job criteria", in Davis, L.E.; Taylor, J.C. Design of jobs, Harmondsworth, Penguin Books.

Emery, F.E. 1959. Characteristics of socio-technical systems. London, Tavistock Institute, mimeographed. Reprinted in Davis, L.E.; Taylor, J.C. (eds.). Design of jobs. Harmondsworth, Penguin Books, 1972, pp. 177-198.

Gallino, L. et al. 1976. "La qualità del lavoro", in Quaderni di sociologia, 2/3, 1976, pp. 297-322.

Gouldner, A.W. 1959. "Organisational analysis", in Merton R. et al. Sociology today. New York, Basic Books.

Herbst, P. 1974. Socio-technical systems design. London, Tavistock.

---. 1976. Alternatives to hierarchies. Leiden, Martinus Nijhoff.

Herrick, N.Q.; Quinn, R.P. 1971. "Working conditions survey as a source of social indicators", in Monthly Labor Review, Vol. 94, No. 4, 1971, pp. 15-24.

Hunt, R.G. 1976. "On the work itself: Observation concerning relations between tasks and organisational processes", in Miller, E. (ed.). Task and organisation. London, Wiley and Sons.

Laboratoire d'Economie et Sociologie de Travail. 1971. Recherches d'indicateurs sociaux concernant les conditions de travail. Aix-en-Provence, mimeographed.

Meissner, M. 1969. Technology and the worker - Technical demands and social processes in industry, San Francisco, Chandler Publishing Co.

OECD. 1975. Indicators of the quality of working life. Paris, mimeographed.

Régie Nationale des Usines Renault. 1976. Profils de postes: Méthodes d'analyse des conditions de travail. Paris, Masson et Sirtes.

Research Committee of the National Living Council 1974. Social indicators of Japan. Tokyo, Japanese Economic Planning Agency.

Seashore, S.E. 1976a. "Defining and measuring the quality of working life", in Davis, L.E.; Cherns, A.B. (eds.). The quality of working life. New York, Free Press.

---. 1976b. "Task force on evaluation", in Davis, L.E.; Cherns, A.B. (eds.). The quality of working life. New York, Free Press.

Susman, G.I. 1976. Autonomy at work; A sociotechnical analysis of participative management. New York, Praeger.

Walton, R.E. 1976. "Criteria for the quality of working life", in Davis, L.E.; Cherns, A.B. (eds.). The quality of working life. New York, Free Press.

Wild, R. 1975. Work organisation: A study of manual work and mass production. New York, John Wiley.

5.2 Experiments, approaches and models concerning joint design of automation and work

Aguren, S.; Edgren, J. 1980. New factoris: Job design through factory planning in Sweden. Stockholm, Svenska Arbetsgivareforeningen.

Archer, J. 1976. "Toward a more satisfactory work group: Aluminium Co. of Canada", in Davis, L.E.; Cherns, A.B. (eds.). The quality of working life. New York, Free Press.

Davis, L.E. 1979. "Optimising plant and organisation design", in Organisational Dynamics, Autumn 1979, pp. 3-15.

Farrow, H.F. 1979. Computerisation guidelines: Guidelines for managers, other employees and trade unions involved in the introduction and use of computer-based systems, Manchester, National Computing Centre.

Hedberg, B. 1980. Design for a camping steelwork. Stockholm, Arbeitslivscentrum, mimeographed.

Mumford, E.; Henshall, D. 1979a. Participative approach to computer design: A case study of the introduction of a new computer system. London, Associated Business Press.

---. Weir, M. 1979b. Computer systems in work design: The ethics method; effective technical and human implementation of computer systems. London, Associated Business Press.

Taylor, J.C. 1982. "Integrating computer-based information systems and organisation design", in National Productivity Review, 1982, No. 2.

6. Other studies

Cohen-Hadria, Y. 1978. Analyse bibliographique du terme de l'automatisation. Paris, Centre de Recherche en Gestion, Ecole Polytechnique.

Harrison, A. 1968. Bibliography on automation and technical change and studies of the future. Santa Monica, Rand Corporation.

ILO. 1965. "A tabulation of case studies on technological change", in Labour and Automation Bulletin, No. 2, 1965, pp. 1-87.

Rijnsdorp, J.E. (ed.) 1979. Case studies in automation related to humanisation of work, Proceedings of the IFAC workshop, Enschede, Netherlands, 1977. Oxford/New York, Pergamon Press, 1979.

United States, Department of Labor, Bureau of Labor Statistics. 1962. Implications of automation and other technological developments: A selected annotated bibliography. Washington, United States Government Printing Office.

Author index

	Date	Section of the bibliography
Commission of the European Communities	1979	1
---	1980	3.2
---	1981/2	3.2
Cooley, M.	1978	4.6
Cooper, C.L.	1979	3.4
Cotgrove, S.F.	1958	4.2
Council for Science and Society	1981	3.1, 4.2
Crompton, R. see Wedderburn	1972	4.5
Crossman, E.R.F.W.	1960	4.2
---	1965	2
---	1965	4.2
Crozier, M.	1977	3.3
Curnow, R. see Freeman	1978	3.2
Davis, L.E.	1966a	5.1
---	1966b	5.1
---	1971	4.1
---	1972	3.3
---	1972	5.1
---	1975	3.4
---	1976	4.2
---	1979	5.2
De Bivort, H.	1955	2
Della Rocca, G.	1983	4.1
Dickson, K.E. see Bessant	1980	3.2
Diebold, J.	1952	2
Diebold Group	1964	2
Dobrov, G.M.	1979	3.3
Dofny, J. see Haug	1977	4
Drucker, P.	1970	4.1
Dubin, R.	1958	3.1
---	1976	2
Dubois, P.	1978	4.3
Dunlop, J.T.	1962	2
Durand, C.	1959	4.2
Edgren, J. see Aguren	1980	5.2
Elliott, D.	1976	4.6
Elliott, R. see Elliott, D.	1976	4.6
Ellul, J.	1964	3.1
---	1978	2
Emery, F.E.	1959	3.3, 5.1
---	1960	3.3
---	1962	4.2
---	1969	3.4
---	1976	4.6
---	1981	4
Engelstadt, P.N. see Davis	1966b	5.1
European Foundation for the Improvement of Working and Living Conditions	1977	3.4

	Date	Section of the bibliography
Iacono see Ancona	1960	4.5
ILO	1957	4.4
---	1965	6
---	1972a	4
---	1972b	4
---	1975	3.4
---	1979	3.4
---	1979	4.4
International Council for QWL	1979	3.4
International Institute for Labour Studies	1978	3.4
International Metalworkers' Federation	1979	1
Jaffe, A.J.	1968	2, 3.2, 3.5
Jessup, G.	1978	4.5
Jordan, M.	1963	4.2
Kahn, R.L. see Katz	1966	3.3
Kassalow, E.	1972	4.6
Katz, D.	1966	3.3
Keja, A.J.	1979	4.4
Kern, H.	1970	4.3
Killingsworth	1963	2, 3.2
Klein, L.	1975	3.4
Kornhauser, A.	1965	4.4
Laboratoire d'Economie et Sociologie de Travail	1971	5.1
Landes, D.S.	1969	3.1
Lawrence, P.R. see Turner	1965	4.5
---	1976	3.3
Lilley, S.	1957	2, 4
Lorsh, J.W. see Lawrence	1976	3.3
Lucas, Y.	1973	4.2
Lupton, T.	1981	4.1
Mallet, S.	1969	4.3
Mann, F.C.	1960	4.5
March, J.G.	1958	3.3
Marcson, S.	1970	4.5
Marek, J. see Emery	1962	4.2
Margulies, F.	1977	2
Marri, G.	1981	4.4
Mathew, D. see Abel	1973	5.1
Maurice, M.	1974	3.3
---	1978	3.3
McLean, A.A.	1979	4.4
McLean, J.	1978	1

	Date	Section of the bibliography
Reynaud, J.D.	1979	3.3
Reyneri, E.	1972	3.5
Rezler, J.	1969	4.6
Ricciardi, C. <u>see</u> Reyneri	1972	3.5
Richta, R.	1969	3.1
Rijnsdorp, J.E.	1979	2
---	1979	4.4
---	1979	6
Rogers, J.	1958	2
Rose, J.	1970	4.4
Rosenbrock, H.	1977	4.1
---	1979	3.1
Rush, H. <u>see</u> McLean	1978	1
Rustant, M.	1959	3.1
Sadler, P.	1968	4
Sandberg, A.	1979	4.3
Schumann, M. <u>see</u> Kern	1970	4.3
Seashore, S.E.	1976a	5.1
---	1976b	5.1
Seeger, O.	1960	4.4
Sell, R.G.	1979	4.5
---	1980	4.4
Sheppard, C.S.	1980	4
Shils, E.B.	1963	4.6
Shipley, P. <u>see</u> Sell	1979	4.5
Shoenner, R.A. <u>see</u> Blau	1979	3.3
Sholtens, S.	1979	4.4
Simon, H. <u>see</u> March	1958	3.3
---	1965	4.1
---	1969	2
Smith, M.A.	1968	4.5
Sorje, A.	1980	3.2
Stinchcombe, A.L.	1965	4.3
Strassman, W.P.	1972	3.1
Sultan, P.E.	1964	2
Susman, G.I.	1970	4.2
---	1975	4.2
---	1976	5.1
Svensson, L.J.P. <u>see</u> Ahlin	1980	4.6
Swedish Employers' Confederation	1975	3.4
Takezawa, S.I.	1976	3.4
Tannenbaum, A.	1968	4.1
Taylor, F.W.	1951	3.3
Taylor, G.W.	1961	4.6
Taylor, J.C.	1971	3.3
--- <u>see</u> Davis	1972	3.3
--- <u>see</u> Davis	1976	4.2
---	1982	5.2

PART 2

NATIONAL SITUATIONS AND CASE STUDIES

CHAPTER VI

JOB DESIGN AND AUTOMATION IN THE FEDERAL REPUBLIC OF GERMANY

(by Wolfgang H. Staehle)

1. Automation: its development and study in the Federal Republic of Germany

It is a common procedure in the Federal Republic of Germany to distinguish certain stages in the replacement of human labour by machines, following Bright;[1] these stages, which are as follows, can then be discussed in the relevant historical and social context:

1. In the premechanisation stage, the worker uses more or less complicated tools for the job, which usually demands good training plus experience and skill. Besides the skilled worker, less highly trained assistants (labourers) are also required.

2. In the mechanisation stage, the worker is freed from the need to use tools. Actual work on the object in question is done by a machine; the machine merely has to be set up, fed and operated. The worker has less freedom of action; other qualifications are needed.

3. In the automation stage, the human operations still required in stage 2 are also taken over by machines, leaving only tasks of control, monitoring and maintenance. In the final stage of full automation, even these tasks can be handed over to machines, with the help of EDP-based automatic feedback control.

Although all the above stages of mechanisation are still to be found coexisting in the Federal Republic of Germany today, a definite trend towards stage 3 has been observed since 1966.[2] These tendencies first appeared most clearly in branches of industry with an acknowledged advanced state of technological development, such as petroleum and natural gas extraction and processing, and tobacco processing. However, other industries (mining, chemicals, metals, electrical engineering) also showed high degrees of mechanisation and automation, judging by the growth of their productive equipment and labour productivity. Agriculture and forestry should not be forgotten, either: in the sixties these sectors leaped from stage 1 to stage 3, thereby ranking among the leading industries as far as the development of productivity is concerned.

The main drive towards automation in the Federal Republic of Germany has been in the development of electronics, particularly micro-electronics. Neither specialisation of activity nor smallness of business organisation are a bar to the spread of EDP, which supports and promotes: (a) automation in production, including process computers for controlling manufacture; NC machine tools; industrial robots; automated handling equipment; CAD (computer-aided design) in the design field; and fully automated warehouses; and (b) automation in administration, commerce, banking and insurance, including VDUs; optical document readers; automatic word-processing; telecommunications; and computer-aided information systems.

In the Federal Republic of Germany, a whole range of scientific disciplines are concerned with the lines of development of automation outlined above. In the context of this paper, industrial sociology is especially important. Besides early work by the Dortmund Social Research Agency,[3] the work of the Sociological Research Institute of Göttingen (SOFI) has played a trend-setting role in the automation debate.[4] Kern and Schuhmann's central thesis is that technological development (especially automation) leads to a differentiation of work situations, which in turn produces a polarisation of the level of qualification of workers. According to this polarisation theory, as automation advances it produces a small number of highly trained automation workers juxtaposed with a large number of unskilled (or de-skilled) people responsible for the remaining work, which is simple in nature. Whereas Kern/Schuhmann see a certain natural inevitability in this development (technological determinism), Mickler et al. think that there is an intervening variable, in the form of an enterprise's interest in making use of its capital, between technological development and activities/qualification/requirements. New opportunities for freedom of action, which flow to a certain extent from automation, are used - in SOFI's view - not to humanise work but to make better use of capital (intensification of work). The work of Fricke[5] should be mentioned here: he argues for a strategy of humanisation which would use opportunities for freedom of action provided by new forms of work organisation to secure greater self-determination (autonomy) for people at work. In his view work should be organised as a learning process, in the course of which new capabilities and skills are developed.

Besides the efforts of industrial sociologists to explain the phenomenon of automation, we should mention the recent research contributions in the field of work psychology. The most important studies here are those carried out as part of the interdisciplinary project called "Automation and qualifications" at the Psychological Institute of the Free University of Berlin. So far the Automation and Qualifications Project Group has prepared a very informative review of the development and spread of automation (Automation in the Federal Republic of Germany)[6] and has worked out a classification scheme (qualification, co-operation, autonomy) for a work activity analysis of automation.

In addition, work science (in the sense of ergonomics) is concerned with worker-machine systems, especially from the point of view of an optimum adaptation of machines to the man or woman at work. The principal themes dealt with in this connection are the load and demands placed on people at work, and strategies to reduce such burdens.[7] Apart from the traditional bias of ergonomics in the direction of engineering and economics, a new behavioural approach is emerging. Under the influence of the trend towards humanisation of work, the ergonomists (and the REFA Institute for work study organisation, which is active in a similar field) now regard themselves as having two equally important central objectives, namely economic efficiency (improving efficiency in order to increase profitability) and human considerations (satisfying the needs of employees at their place of work). The REFA Institute is of particular importance for the development of new ideas in the field of work organisation since it offers additional training in work study, time measurement and job design to a large number of engineers.

The science of business administration, which has explicitly chosen the above-mentioned economic interest as a selection principle for its areas of research, has also been undergoing changes for several years.

Formerly, organisational problems in the functional areas of organisation, production and personnel used to be analysed almost exclusively from a cost angle,[8] primarily using the tools of work science, work psychology and industrial and organisational psychology (provided these were relevant to costs). In recent years, however, as a result of the shift towards the behavioural sciences and greater theoretical consideration, importance has also been attached to a non-economic approach.[9] The worker is no longer regarded solely as "homo economicus", as a production or a cost factor, but as a social being with his or her own desires and needs which must be taken into account when decisions about organisational and personnel matters are made.[10] As a result of the trend towards an interdisciplinary, behavioural approach and the debate about the so-called work-orientated theory of the firm,[11] signs of a comprehensive analysis of the idea of automation and work organisation are beginning to appear in business management studies.

The above-mentioned discplines have a contribution to make about the following aspects of automation and their consequences:

- economic advantageousness: which elements of the work can be performed more cheaply by people and which by machines? Wage costs versus capital costs;

- qualifications: consequences of automation for the structure of qualifications (e.g. the polarisation thesis); process-dependent versus process-independent skills;

- freedom of action and autonomy: new opportunities for freedom of action opened up by automation; use of this freedom for the benefit of the employer (intensification of work) and/or the employee (QWL);

- co-operation and communication: influence of automation on social interaction at the place of work; tendency for machines to cause isolation;

- workloads and effects on health: transition from physical/muscular to mental/nervous demands; responsibility for machines; increase in psychosomatic illnesses;

- working time and leisure time: consequences of automation for arrangement of working time; increase in shift work; social isolation in leisure time.

What is urgently necessary, in our opinion, is to pay greater attention to an interdisciplinary analysis of the factors influencing industrial work situations, as this should be a prerequisite for a relevant assessment (i.e. one that has due regard for the particular situation) of strategies of change. In the American literature the problems of (human) work structuring are dealt with primarily from the standpoint of job design and job structuring (or redesign and restructuring); in other words, a narrow analysis relating to the individual isolated workplace is favoured, although recently this approach has been broadened through the introduction of situational influence factors. A comprehensive approach, based on an analysis of the background conditions of the work organisation, is to be found not so much in the United States as in the United Kingdom and Scandinavia, where research by the Tavistock Institute (the socio-technical approach), by Aston University in Birmingham (the Aston approach) and by the advocates of industrial democracy have encouraged an approach to technical/organisational, human/social and community relationships that does justice to the complexity of industrial work structures, that takes into account situational factors and that has an integrated effect.

In the Federal Republic of Germany, situational approaches based on these ideas have attracted particular attention.[12] At the Institute of Business Administration at the Technical University of Darmstadt and at the Free University of Berlin, the situational approach has been further developed and applied to questions of work organisation; to this end a system of indirect (macrostructural) and direct (microstructural) factors influencing the work situation has been developed. This is intended to enable work situations to be described as operationally as possible.

The implementation of the demand that new technologies should be handled in a human way and, more particularly, the chances of successful introduction of corresponding forms of work organisation are determined at national level through the interplay of the interests and demands of the social partners and of the opportunities and constraints of the overall social conditions that have come into being over the course of time.

As shown in figure 9, a certain historical situation in society may offer possibilities (e.g. technical developments, market demands) for or constraints (e.g. lack of skills, social acceptability) to changes in the work situation. On the other hand, societal institutions, like political parties, trade unions, churches and the mass media, put pressure on management to organise work in an acceptable way. In the Federal Republic of Germany, management's freedom of action in this field is narrowed down by a multitude of laws, rules and regulations. Trade unions, employers' associations and state agencies interact in a variety of ways (e.g. collective bargaining, joint consultation) which produce the framework for future management decisions (see the industrial relations triangle in figure 9).

2. The macrostructure of the work situation

In recent years the socio-political climate in the Federal Republic of Germany has been characterised on the one hand by economic recession and the resulting labour market problems (unemployment), and on the other by legislative initiatives by the Federal Government in the field of social and labour policy, and action by unions and employers in collectively negotiating working conditions.

Following the three major groupings in the German industrial relations system (i.e. Government, employers and their organisations, and workers and their organisations), attitudes and activities in relation to automation and its consequences are differentiated accordingly.

Figure 9: Macrostructure of the work situation

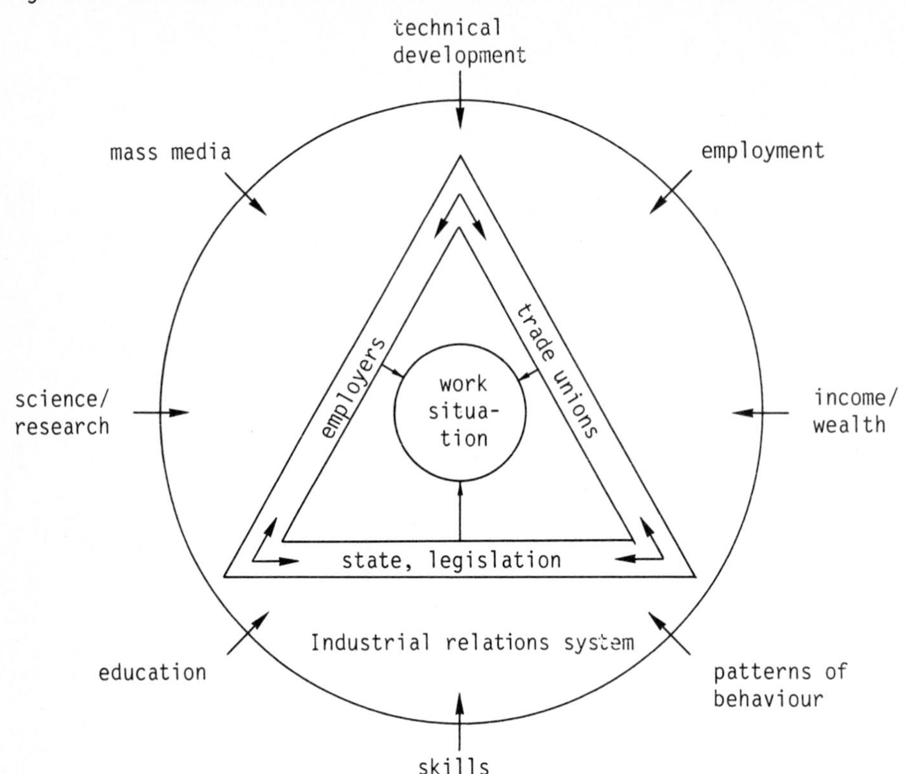

2.1 State, government and legislation

The establishment of humane and safe working conditions has been the concern of legislators for more than 100 years. Regarding industrial relations in the Federal Republic of Germany in recent years, greater attention has been paid to questions of work organisation.

The Industrial, Civil and Commercial Codes have been supplemented in recent years by the revised Works Constitution Act (1972), the Work Safety Act (1973), the Workplace Regulations (1976) and the Co-determination Law of 1976.[13] The outstanding event in this connection was the introduction of the Works Constitution Act (1972). Although earlier legislation had given employees' representatives rights of co-determination in the most varied areas, the new Act now extended works councils' rights, which had previously been of an informative and advisory nature, to give them a say in the shaping of the workplace, working procedures and the working environment. Paragraphs 90 and 91 of the Act embody an explicit demand by the legislators and are significant to the extent that in implementing these measures, accepted work/science principles about humanised job design must be observed.

The Work Safety Act obliges firms above a certain size to appoint medical officers and safety experts; the latter are also required to assist the employer in matters to do with humanised job design. Paragraph 3 of the Workplace Regulations requires employers to organise and operate the workplace in accordance with generally accepted safety, medical and hygiene rules and in accordance with other accepted work/science principles.

At the trade unions' instigation, a commission on economic and social change, made up of trade union and employers' representatives and scientists, was set up by the Federal Government in 1971. Its tasks were to examine the repercussions of technological, economic and social change with a view to further development of social policy and to state the options open in the field of economic, social and educational policy planning. However, the non-committal nature, from the political and practical point of view, of the opinions expressed by this commission, and consequent irritation, produced an increasing desire for concrete reforms, reaching a peak towards the end of 1973.

In May 1974 the Federal Ministries of Labour and Social Order and of Research and Technology presented an action programme entitled "Research into the humanisation of work". The programme's general objective is to investigate ways in which working conditions can be better adapted to the needs of working people than has hitherto been the case. The results of the research and development work are intended to take the form of practical proposals for solutions.

The action programme has the following detailed goals:

- compilation of safety specifications, standards and minimum requirements for machinery, installations and workplaces;

- development of humanised working technologies;

- preparation of exemplary proposals and models for work organisation and job design;

- dissemination and application of scientific findings and plant experience.

The funds invested in this action programme by the Federal Government are very modest, i.e. 2 per cent of the whole research budget, compared with the support it gives to technology (especially research in EDP and nuclear reactors). In the first year, 1974, 13 million Deutsche Marks (DM) were made available; thereafter the amount was raised to DM30 million in 1975, and DM40 million in 1976. The sum of DM134 million will be provided in 1981. The current research programmes cover -

- measures for work organisation and factory organisation;

- ergonomic measures;

- technological measures; and
- other measures.

Organisational and technological measures in the production field will be clearly in the foreground. In the future, however, increased emphasis is to be placed on projects to improve working conditions in the clerical, administrative services sector.

Trade union representatives consider this Federal Government programme rather as a state-subsidised rationalisation programme for management purposes than as a humanisation programme for the worker. Therefore, they rely much more on laws and collective agreements. Indeed, the industrial relations system in the Federal Republic of Germany relies heavily on institutionalised and legal provisions, and this seems to have much effect on QWL than research programmes without any obligation on the side of management.

2.2 Trade unions

The German trade unions, particularly the individual union with the largest membership, IG Metall, were already at an early stage (the beginning of the sixties) pointing out possible dangers from automation, such as mass unemployment, the devaluation of traditional occupational qualifications and industrial concentration. It will be seen that against the background of these fears the unions' classic goals of raising earnings, securing fair pay, improving safety at work and obtaining shorter working hours were increasingly joined by qualitative objectives concerned with the actual content of work. The shift of emphasis in trade union concern is demonstrated by the paper published by the DGB (German Trade Union Federation) under the title Humane job design (first ed., 1972). The preparation of a corresponding programme was given a major boost at congresses such as the IG Metall Congress on "The quality of life" at Oberhausen in 1972 and in speeches on Labour Day (1 May) 1973; it was demanded that the quality of life of all workers should be further improved, that economic power should be controlled by co-determination and that the working world should be humanised.

The importance to the trade unions of the new goal of humanisation of work became clear when, in the course of the negotiations on Outline Collective Wage Agreement II for the metal industry in North Württemberg/North Baden, claims connected with the actual content of work became the central issue in an industrial dispute for the first time. The reduction of extreme forms of division of labour, increased intellectual involvement through greater freedom of action, improvement of opportunities for social contact and reduction of monotonous tasks are today regarded as valid trade union aspirations in addition to the traditional demands for improvements in working conditions through reduction of primarily physical demands, greater safety and higher wages.

In March 1978, further-reaching demands concerned with job and pay security were successfully fought for by industrial action in the printing and metal industries. The collective agreement for the printing industry gives protection to type-setters against the effects of new printing technologies. In addition to qualitative job security guarantees for skilled workers, pay guarantees and retraining assistance, concrete provisions on the structuring of work and workplaces were obtained. The metalworkers' union, IG Metall, also won guarantees against the consequences of technological change in the Collective Agreement to safeguard classification and to safeguard pay in the event of declassification. Protection against declassification for the individual employee was introduced: in future, blue-collar workers could not be declassified by more than two grades and white-collar workers by more than one, and declassification could be effected only where it was impossible to transfer the employee to or retain him for another appropriate job of equal grade within the firm. In addition, a pay guarantee in the event of declassification was obtained.

A negotiating breakthrough was also made in the cigarette industry, involving for the first time a reduction in working hours to below 40 hours per week for employees who were aged over 60 years and had worked for the firm for at least 10 years. Either working hours could be reduced to 20 hours per week on full pay, or the employee could retire completely and receive 75 per cent of gross pay.

2.3 Employers' representatives

Unlike the trade unions, the employers take a predominantly optimistic view of automation. The relief of the worker from physically demanding and even degrading activities is stressed, and faster economic growth (new products and types of production) and increasing capital investment are usually adduced as the consequences of "technological change".

The employers, too, emphasise humanisation of work as a socio-political goal. There is an endeavour to provide working conditions which meet the employee's desire for an opportunity to develop skills, and for recognition and social contact. In the debate on humanisation of the working world and on the introduction of new forms of work organisation, the employers tend to emphasise criteria connected with work safety or with good ergonomic job design. Alongside these aims they think that employees' motivation and satisfaction at work should be enhanced by suitable measures. In this way they hope to reduce wastage, labour turnover and absenteeism. Employees' needs ought to be met, in the employers' view, by partial restructuring of the work situation, in order to combat the "motivation crisis" among employees and thereby maintain or improve the economic efficiency of the production process.

Apart from ensuring profitability, changes in work organisation are intended primarily to secure increased flexibility in the production process. Only if profitability and flexibility are improved or at least maintained do employers turn their thoughts to redesigning jobs. For this reason they have been accused of pursuing cost-cutting rationalisation under the guise of humanisation of work.

One argument put forward by the employers crops up repeatedly in the humanisation debate: that work organisation has always been a traditional field of action for employers or management and that the search for optimum forms of work organisation has always been one of their main objectives. Humanity and economic efficiency are complementary, not conflicting goals, according to the employers. What is today called humanisation of work has been striven for and attained over decades by continuing reorganisation and by improvement of working conditions, the employers claim. After having discussed the macrostructure of the work situation with special reference to the industrial relations system of the Federal Republic of Germany, we now take a closer look at the microstructure in order to examine what happens inside the circle called "work situation" (figure 10).

3. The microstructure of the work situation

Besides an analysis of the macrostructure, a profound knowledge of the actual work situation is a necessary condition for describing and assessing new forms of work organisation.[14] This involves making a greater effort to analyse the factors influencing industrial work situations, which is a prerequisite for any reform of existing work organisation systems if such a reform is to take sufficient account of the prevailing situation and of the desired objectives. To do this we employ an action-oriented approach, the aim of which is to explain the action and behaviour of people in work situations (figure 10). Following this approach, by work organisation we understand relatively stable relations between people and machines, i.e. worker-machine systems. The primary concern of work organisation (as a function) is hence the situationally appropriate allocation of tasks to workers and machines.[15]

The actions and behaviour of people carrying out tasks (as shown by the findings of empirical surveys that have so far become available) are influenced mainly by two classes of variables: technology and organisation. Although the majority of independent variables analysed in empirical surveys can be included in one of the two categories, this classification is certainly not intended to be an exhaustive list of all the influencing factors involved, but only of the dominant ones. As intervening variables the size and location of the organisation, the product range, the legal form and the goal system can be cited. Different combinations of these factors result in different forms of work organisation, which in turn, together with the working environment (the physical situation at the workplace) and the objective task, influence the perceived work situation. By work situation we mean the entirety of working conditions observed and experienced by the individual. The perceived work situation, therefore, is the central empirical unit under investigation in explaining the actions and behaviour of workers. Actions and behaviour represent goal-oriented reactions towards the perception of stimulus material, instructions about operations and goals (objective task) in specific task situations.

According to the model presented, the objective work situation is defined primarily by the contextual variables. The impact of these

Figure 10: Microstructure of the work situation

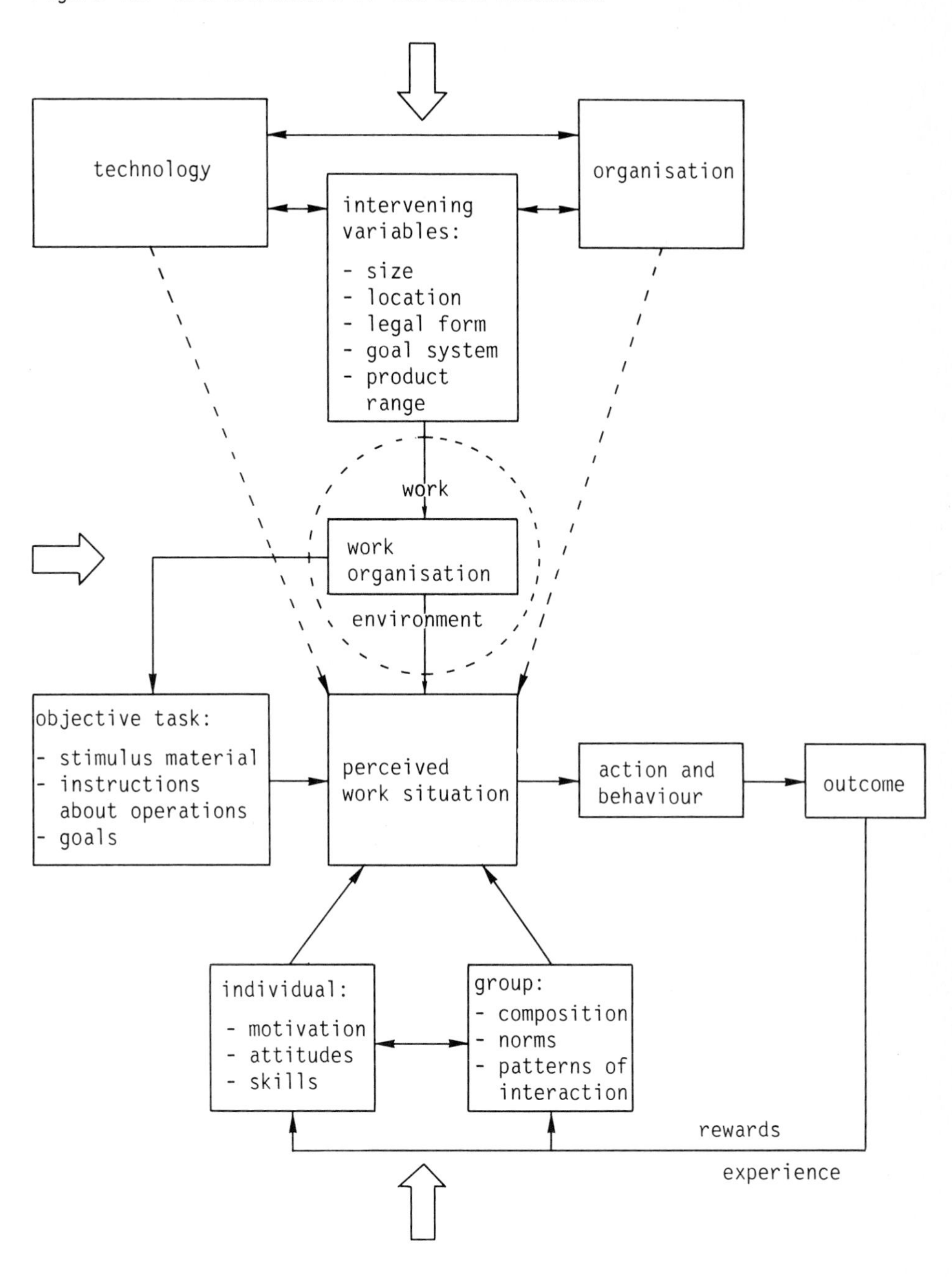

contextual variables on the subjective work situation is on the one hand influenced directly by the definition of the objective task and the nature of the work environment. On the other hand, contextual variables will also be perceived indirectly because of subjective feelings and formation of attitudes towards social reference groups, means of production, work objects or organisational goals (see dotted lines in figure 10). The process of perception varies from individual to individual and from group to group. Motivational conditions, attitudes and skills on the one hand, and those conditions specific to groups such as norms, patterns of interaction and group composition on the other, are analysed as influencing factors in the perceived situation.

The influence of macrostructural and microstructural factors on the perception of the person in question at his or her place of work (subjective work situation) is thus exercised in the first place directly through the definition of the concrete work task and the nature of the actual working environment, and in the second place indirectly through the attitudes towards such reference points as technological change, co-determination and method of wage payment.

Steps to change the way work is organised must be based on an analysis of the objective and subjective situation. According to the three large arrows in figure 10, there are at least three different directions of planned organisational change -

1. In the classical top-down approach (arrow pointing downwards), management starts the design process by changing the technical system first (e.g. automation) and subsequently adapting the work organisation, tasks and jobs to it.

2. In the human relations oriented, bottom-up approach (arrow pointing upwards), management first analyses the existing social system, looking for the needs, qualifications and attitudes as well as the perceived possible fears of the workers concerned. Change processes will then be designed and implemented jointly by management and workers (and their representatives).

3. In the integrative socio-technical approach (horizontal arrow), the aim of design is a joint optimisation of the requirements of the social system <u>and</u> the technical system. To reach this goal, one has first to design a meaningful work organisation with relevant tasks and jobs which meet human as well as efficiency needs.

Such redesign strategies cannot be applied on an ad hoc basis - as the case study shows - but must take account of all relevant background factors and individual and group-related process circumstances in a comprehensive organisation and personnel development plan. To achieve this it is advisable to build a project team in which the interest groups underlying the direction of the arrows are represented (i.e. management; workers; works council members; safety engineer, etc.).

4. Theoretical approach to the case study

The case study presented in the next section of this chapter is indicative of some of the problems as discussed above, relating to the causes and consequences of automation, but is in by means representative. The research process follows the model described in figure 10 by attempting to conceptualise and investigate the different variables.

Depending on the form assumed by the variables, there will be different forms of work organisation, constituting the objective work situation. The subjective work situation comprises the elements of the objective situation as perceived by the individual. Therefore, in a dual work situation analysis it is necessary to investigate not only the objective but also the subjective work situation (figure 11).

A third factor decisively influencing the perception, behaviour and action of the worker in a new work situation is the process of organisational change (redesign of work organisation associated with the introduction of new technology). Through introducing the process of change, the predominantly static character of the model is overcome. Besides the change process itself, data on all the variables must be collected twice, before and after the change.

Figure 11: Federal Republic of Germany: theoretical framework of case study

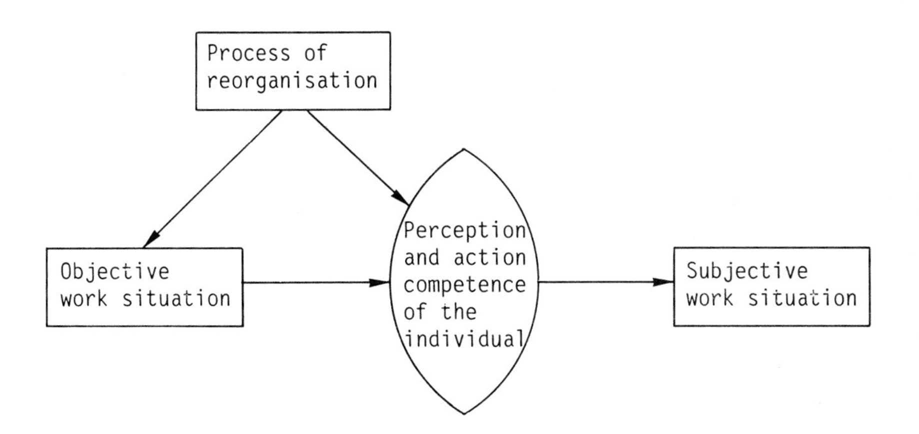

Intensive interviews are the chief method used to survey the three aspects as shown in figure 11 -

- objective work situation (interviews with department heads, job descriptions, structured observations, work flow analyses);

- subjective work situation (interviews with workers);

- process of reorganisation (interviews with managers or organisation and planning department, content analysis of reports, memoranda, records and minutes documenting the change process).

Following this research design, the case study is divided into the following main sections:

- technology;
- company (organisation);
- department affected by the reorganisation (work organisation);
- reasons for and objectives of the change;
- description of the change (process of reorganisation);
- assessment of the change.

5. Case study - Job design and automation in production: printed circuit board assembly

5.1 The technology

The printed circuit board (PCB) with its complement of various active and passive components now constitutes the central functional part of every piece of electronic equipment, whether in the consumer field (e.g. entertainment electronics) or in commercial applications (e.g. instruments).

A PCB consists of a piece of substrate material (e.g. laminated paper or rigid epoxy resin/glass fabric) cut to size and having a conductive track on at least one side. The holes to take the mechanical and electrical components are drilled or punched. When the components have been fitted and all operations such as soldering and protective lacquering have been completed, the result is a complete circuit board or flat module. Assembly refers to the insertion of components in the circuit board. These components may be -

- plugs, soldering terminals, bridging wires (insulated or bare), with varying bend-off dimensions;
- resistors of various ratings and tolerances with different resistance values;
- capacitors of a wide variety of types, with differing capacitance values and with axial and radial connections;
- coils with transformers of different types;
- relays with make and break contacts to a variety of specifications;

- diodes of different types and sizes;
- transistors of different types and with different housings;
- integrated circuits in the form of dual in-line packages 14- or 16-pin, of a variety of types;
- insulators;
- heat sinks and mounting hardware;
- handling rails and reinforcements;
- labels.

The main problem of PCB assembly is to ensure that the correct components are selected from this plethora of different types and properly positioned in the correct polarity.

This type of assembly work lends itself particularly to examination in terms of automation and organisation in that all three stages of mechanisation and/or automation may still be encountered in practice. In addition to conventional manual assembly from an assembly diagram (capital cost about DM1,000), there is semi-automatic assembly, where the correct component position is indicated by a light ray coupled to the bin feed (capital cost, including programming, about DM47,000), and the new fully automatic radial PCB assembly machine (capital cost, including programming facility, about DM270,000). The economic advantage of the use of alternative technologies is almost entirely dependent on the quantity and diversity of the components to be fitted and on the number of identical circuit boards to be assembled (batch size). A clear trend towards complete automation is unlikely in the very near future for the following reasons:

1. Integrated circuits are increasingly superseding the conventional PCB for certain types of electronic instruments (in particular, small, high-grade instruments for commercial applications). These are tiny chips containing up to 2,500 circuits; they are combined with or bonded to a "spider", the chip connections being linked to the pins and solder tags by wires having a thickness of 1/20,000-1/30,000 mm. None of these operations can be carried out at the types of assembly positions hitherto used. To protect them from mechanical agencies, the circuits are incorporated in steel, ceramic and plastic packages. This technology has the following advantages:

(a) reduction of soldered joints and hence prolongation of the unit's life;

(b) compact dimensions;

(c) high reliability;

(d) light weight;

(e) high packing density; and

(f) even temperature distribution in operation.

2. The traditional PCB will continue to be used in consumer electronics, for example, in the entertainment field (television sets, stereo equipment, etc.). Because these units are relatively bulky and the electronic components have been miniaturised, it is not at present necessary for microminiature modules to be used. In addition, not all components can yet be made in the form of chips. When constructed in integrated form, these components are also often more expensive than traditional discrete units. Even so, the conventional PCB will change in appearance. It will be made smaller and its packing density will be higher. Components such as transistors, diodes, resistors and capacitors are also becoming smaller. Integrated circuits with up to 24 pins are already being manufactured. For this purpose, the relevant conductive tracks must be made thinner and thinner and the spacing between the conductors must become closer. The conductive tracks are then located on both sides of the circuit board, transversely on the component side and longitudinally on the solder side. This is an obstacle to fully automatic assembly and imposes particular demands on the assembler (in particular, precision motoricity, skill and concentration).

In the case study described below, the assembly department of the company concerned was converted from manual to semi-automatic assembly.

5.2 The company

The company is a medium-sized industrial enterprise (currently employing some 700 people). The firm was established in 1950 as an undertaking specialising in electronic instruments; after expanding at a fast pace, it was transformed in 1955 into a limited liability company with members able to call upon more substantial capital resources. The company's products are marketed throughout the industrialised nations of the world. In view of the constant expansion of the specialised field of electrical measurement of mechanical parameters and enlargement of the product range, rationalisation and technological restructuring measures are constantly in progress.

The company is in the throes of a process of profound restructuring, affecting both the organisational and technical fields. The instruments were originally produced individually; following a period of small-series production, the important stage of conversion to large-series production is now about to be embarked upon. Since the company makes 150-250 different instruments, series size cannot be standardised. Batch sizes range from 30-100 PCBs.

Production is planned in advance on a quarterly basis in accordance with the expected sales of the instruments. Each instrument includes an average of three PCBs, and there are a total of about 500 types of PCB with a wide variety of component densities.

In view of the need for constant adaptation to the evolution of technology, expansion of the market and enlargement of the range of products, the organisation is relatively flexible and open (see figure 12).

Figure 12: Printed circuit board assembly: functional chart

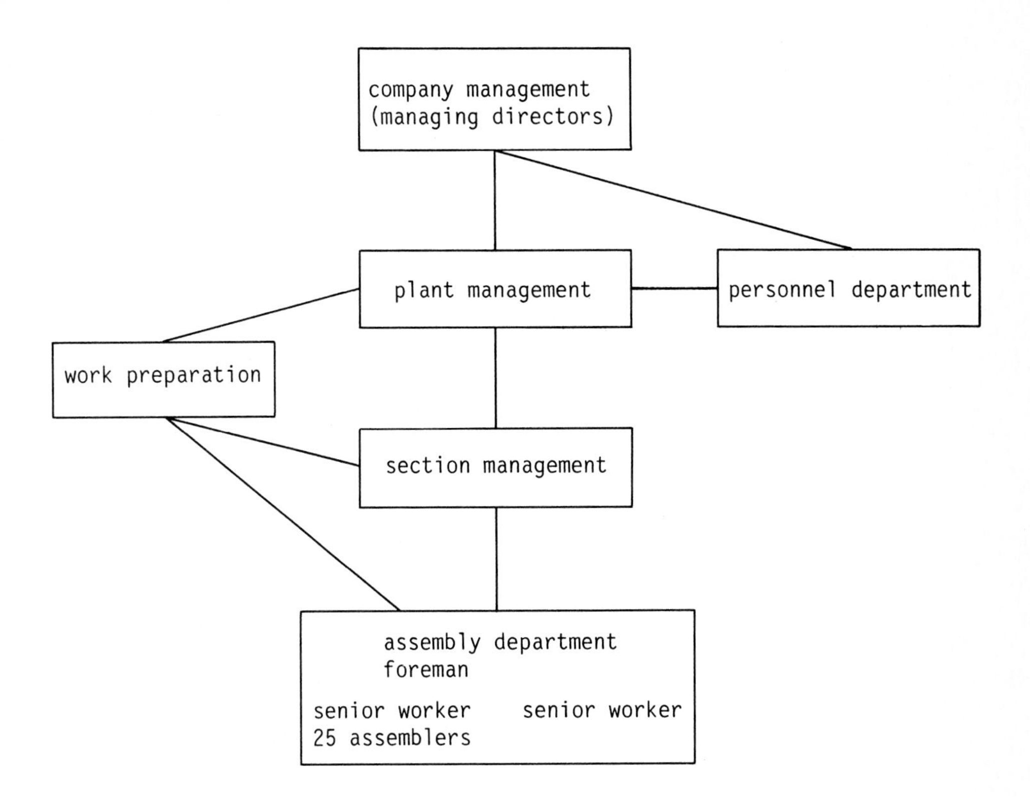

5.3 The assembly department

5.3.1 The workforce

A total of 28 persons is employed in the department as a whole, which is headed by a foreman and two senior workers (all male). Apart from a man responsible for preparation of the components to be assembled, the entire labour force is female (12 Germans, eight Yugoslavs, two Italians, one Greek, one Turk and one French woman).

The foreman and one senior worker have worked in these positions in this department for about ten years, and the other two men six or seven years. The average length of service of the assemblers is about ten years, although some have worked for the firm for 12 or 15 years. Length of service with the firm and time worked in the department are substantially identical. Four women are over 50 years old, all the others being between 28 and 40. Most of the employees work a standard 8-hour day with some flexibility of hours, although two women work mornings only for six hours per day. Breaks are standardised: a breakfast break from 9.00 to 9.15 hrs and a lunch break from 12.00 to 12.30 hrs.

There is very little trade unionisation, only 10 per cent of the labour force being members of the appropriate trade union, IG Metall. Nevertheless, the workers feel themselves well represented. One of the (male) workers is a member of the works council and is also the workers' representative for the department. There did not seem to be any conflicts due to trade union activity or industrial relations problems.

5.3.2 Reasons for and objectives of the change

The change was necessitated by problems in the assembly department, which was regarded as a production bottle-neck. The reasons stated were deficient quality and quantity of output and high staff absenteeism. Management's objectives were -

- improvement of quality (reduction of sources of error);
- improvement of flexibility in terms of product and labour deployment;
- reduction of absenteeism; and
- reduction of training and work familiarisation times.

The workers in the assembly department were asking for -

- better ergonomic design of their jobs;
- better organisation of work flow; and
- higher pay.

The change-over was planned at the beginning of 1979 and implemented within a year. The Plant Management and Work Preparation Department were involved, but the works council was not consulted. The first test run with a semi-automatic assembly table took place at the end of 1979, further tables then being successively introduced.

The process of introducing the change was recorded by our research team through interviewing all parties concerned. According to this information, the change process followed the top-down management approach described earlier (section 3). Possible alternatives to the existing work organisation were only discussed in terms of other technologies for doing the job. Alternative job content or changes in the social system were not even considered. The criteria for the success of the change were defined by management alone, using the stated objectives of change as guide-lines.

5.4 Description of the change

5.4.1 Work system 1 (before conversion)

The equipment of the individual work position comprises the following items:

(a) work table (earthed);

(b) two rotating bin holders;

(c) 72 storage bins for components (colour-coded);

(d) tweezers;

(e) soldering iron and solder;

(f) simple wooden frame to take the PCBs;

(g) additional lamp;

(h) foot-rest;

(i) working documents (assembly diagram of the relevant PCB, parts list, order stating batch sizes).

The components to be assembled are prepared and issued in a separate room, the subsidiary store, where four people cut the ends to length or bend the connections and put together the quantities required for each series. (The subsidiary store is supplied from the central or main store and its capacity is to be substantially increased to save movement time.)

The women in the assembly shop receive the components for an entire commission ready cut and bent. Their job begins with the preparation of the work position, i.e. placing the tools in the correct positions and mounting the PCBs in the assembly frame; the remaining operations comprise fitting the components to the PCBs, transporting them to the solder bath, resoldering and inspection. This sequence of operations is illustrated in figure 13.

Figure 13: Printed circuit board assembly: block diagram of material flow

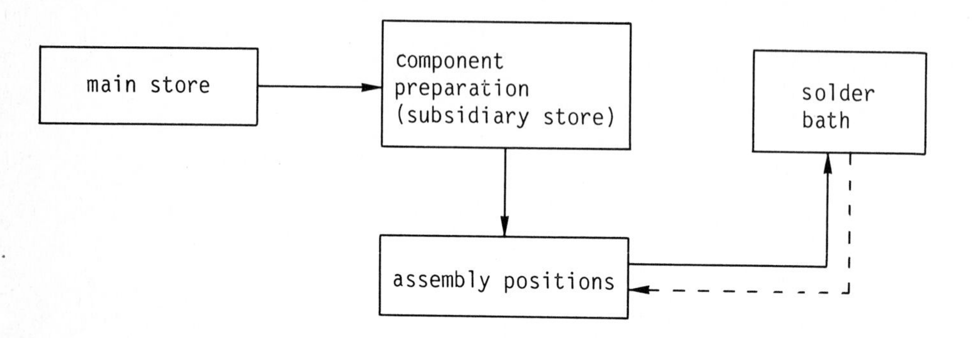

The components are accommodated in 72 bins positioned in front of the assembler and the assembly frame, and on two swivelling frames to the right and left of her seat on four levels, each row being further back than the one below it. The assembler has an assembly diagram drawn in black and white on a scale of 2:1 and clearly showing the correct polarity of the components; this diagram is also printed in yellow on the actual board to which the components are to be fitted. The number of components to be fitted totals 200-260, and they may be of 100-120 types or configurations, average batch size being 50-60 boards. The configuration may, for example, be characterised by different connection lengths and forms. Components not accommodated in the 72 bins are placed on the table in additional boxes.

The components are inserted either with the tweezers or with the fingers. The assembler takes individual components from a dispenser with five magazines, which also stands on the table, using a punch-like tool. For high packing density, the PCBs are "tiered", i.e. the components are positioned vertically according to the rule "from small to large" and "from bottom to top", in several planes. The problems of exact assembly are exacerbated with increasing density and decreasing size.

If a box contains fewer components than needed, the worker must go to the subsidiary store and have the correct parts located, cut to length and bent. When all components have been fitted, the assembler takes the frames with the PCBs to the solder bath. The solder bath is operated by one person, only 30-50 per cent of whose time, however, is employed in this work. On the way back from the solder batch, the assembler collects new frames to take further boards for assembly. After passing through the solder batch, the PCBs are taken back to the assembly position, where they undergo finishing operations, as follows:

- exposure of protected connections;
- removal of solder lugs;
- nipping off wire ends (in about 20 per cent of all wires);
- visual inspection;
- cleaning; and
- fitting of any additional components.

In the penultimate operation, the assembler places the PCBs on a trolley. A senior worker takes the trolley to the test area, where (as far as possible) the components undergo functional testing (electrical functional testing prior to assembly is carried out only on a statistical sample basis); otherwise a visual inspection only is performed.

If mistakes are not observed until the stage of final inspection, the whole instrument may sometimes have to be completely dismantled. The error rate in PCB assembly is between 0.2 per cent and 0.5 per cent. To complete the order, the assembler must fill in her wages sheet and an account sheet as a timing cross-check.

Work system 1 is perceived by the workers as demanding a considerable amount of technical skill and knowledge about the whole assembly process. This makes the job difficult but interesting. Complaints were voiced not about the work content (the objective task) but the work environment, especially design of tools and chairs, lighting, heating system and pay.

5.4.2 Work system 2 (after conversion) (computer-controlled magazine with light ray as assembly guide)

The equipment of the individual work position now comprises the following items:

(a) programmable assembly table with -

  (i) PCB holder;

  (ii) presentation recess (movable component bin);

  (iii) repeat button;

  (iv) "next component" button;

  (v) digital display (for serial number of component bin); and

  (vi) foot-switch;

(b) tweezers;

(c) soldering iron and solder; and

(d) additional lamp.

This assembly comprises a programmable semi-automatic machine with electronic memory and a dynamic display.

The unit is programmed in the department by a specially trained senior worker. For this purpose the parts list for the PCB must be rewritten in order of assembly. The size and module dimensions are the determining factors. A sample of the PCB or a 1:1 scale model is also required for programming. The senior worker guides the program to all points to be displayed; the program is then electronically stored and can be called up at any time. If it is required at a number of assembly positions, the program can be duplicated. Programming of the unit for a basic PCB, including revision of the parts list, takes about six hours.

The content of an assembler's work in work system 2 is modified as follows: she no longer has to find the position of each component herself, using the assembly diagram as in work system 1, but has at her disposal a light ray which acts as a positioning guide, by projecting a spot of light on to the board where the relevant component is to be inserted. The light ray and the appropriate component bin are controlled in synchronism by a computer, the signal being triggered by a foot-switch or manually operated switch.

At the maximum cycle speed, a component could in theory be fitted every 1.3 seconds; however, since the assembly of predominantly small boards (up to 16 of them at a time on one frame) requires intense concentration, this cycle speed is not achieved. The setting up of the machine is particularly time consuming. Since the position and polarity are in each case indicated by spots of light, precise registration of the boards in the frame is essential where component packing density on the small circuit boards is high. This registration of the boards prolongs the set-up time, as two registration points must be correctly positioned for each board. The two assembly points are then displayed by the light spot. The display is triggered by the manual switch or foot-switch, the component polarity being indicated by flashing.

The components are accommodated in 96 compartments of the 12 magazines; there are also facilities for placing trays on the work surface. The magazines are filled with components in the work preparation department, which greatly reduces set-up time in this respect.

The operations to be performed by the assembler are then as follows. Without looking, she takes a number of components out of the component bin with her left hand and transfers them to her right hand. The light spot moves to the positions where the components are to be fitted; she returns excess parts to the bin and then presses the button "next component" with her left hand. Where components are to be taken not from the table but from additional boxes, the presentation recess remains closed. A digital display indicates the serial number of the component bin.

This system differs from work system 1 in that the components are always taken from the same position, a recess in front of the assembly frame, and hence immediately in front of the assembler. She thus need no longer stretch out awkwardly to reach inaccessible bins, and the number of backward and forward movements of the upper part of the body is minimised. Owing to the exact indication of components and assembly positions, mistakes are extremely unlikely and tend towards zero. Training time in these work positions is also practically nil. Only the registration of the small boards in the frame prolongs the set-up time and demands some degree of dexterity.

5.5 Assessment of the change

A comparison of the costs of the two systems shows that while the fixed costs have increased substantially, the variable costs have fallen drastically owing to the considerable fall in quality costs combined with higher hourly output. Table 15 gives a cost comparison of the two systems.

According to the company, about 1 million components per year can be fitted at a type 1 work position and about 1.5 million at a type 2 position. When used to full capacity, system 1 thus entails total costs per working position of DM69,290 and system 2 DM46,058. This convincingly demonstrates the economic advantage of work system 2. However, the result of the comparison is significantly less favourable in terms of skill requirements and onerousness of the work.

Work system 1 calls for familiarity with details of the construction of the product of the work (the PCB) and higher-order aspects, to allow the main activity (fitting and combination) and the subsidiary activities (preparation and fetching, preparation of working facilities and the products, correction and retouching) to be performed. The relevant component must be identified from the details printed on the board, the module dimension and the layout of the holes drilled in the board, and then located in the bins. The colour code of the resistors must also be known by heart. If the information on the board is insufficient, the workers must be able to obtain this information from the assembly diagram associated with each order. These skills can be learned, and sooner or later become technical routine. However, to achieve an optimum sequence of assembly of components on the PCBs, system 1 requires not only technical routine but also empirical-adaptive and system-optimising thought processes in the form of technical intelligence. This thinking requirement can be replaced by technical routine only after long experience, if at all.

The long experience (or systematic analysis of the working process) required for rapid assembly is separated from the individual workers in work system 2 and is incorporated in the programming. Since the light ray on the semi-automatic assembly table dictates a substantially optimum order of assembly of the components, no thinking is necessary on this point. As a result, work at the assembly tables becomes appreciably more monotonous and hence mentally more burdensome. One worker who had previously worked in system 1 characterises the new system as follows:

Table 15: Printed circuit board assembly: cost comparison of the two systems

Cost item	Work system 1	Work system 2
Capital cost (DM), including programming	900	47 000
Amortisation (years)	1	8
Fixed costs:		
Capital costs (DM/year; depreciation factor 0.1874)	990	8 808
Program modification costs (DM)	-	4 050
Repairs and maintenance (DM/year)	-	800
Total (DM)	990	13 658
Variable costs:		
Wages (DM/hour)	8.46	8.86
Ancillary wage costs (DM; 1980 = 70 per cent)	5.92	6.20
Quality costs (DM/hour)	22.86	3.49
Total (DM/hour)	37.24	18.55
Assembly time (quantity/hour)	545	857
Variable costs (DM/unit)	.0683	.0216

- = nil or negligible.

> The light spot shows the part that has to be fitted. You can't say that the light spot is not as good as fitting the components freely by hand. But I have usually fitted the component first and switched on the light spot afterwards, because when you have been there so long, you know the boards.

It must, however, be borne in mind that existing qualifications, skills and abilities which are unused for a long time become rusty and are no longer available. In general, it can be affirmed that this division of labour encourages the withering away of skills. There has also been an unmistakable shift in the physical and mental stresses occasioned by the work. Considerable muscular tension (arms and neck) were observed in system 1, due to unfavourable sitting and reaching positions; in system 2, however, the workers report intense mental stress due to monotony and lack of variety in their work.

The new bonus wage system also leads to pressure of time, so that the worker tends to remain permanently in the working position. Thus the change in the wage system was perceived negatively, although the level of pay increased. Before the change the workers were paid in wage group 2 (i.e. DM8.46 per hour); after the change everybody was moved to wage group 3 (i.e. DM8.86 per hour). In addition, through the introduction of a new bonus system (rewarding quantity of output) in work system 2, the average pay level increased by another DM.60 per hour.

Work system 2 is subjectively perceived as particularly stressful, probably because the operations and the relevant times are dictated by the machine. Those affected experience the restriction of individual freedom of organisation and action particularly intensely.

The statements of some women whose performance in the new system is inferior to what it was in the days of manual assembly are consistent with this picture. Since this cannot be explained by the working facilities, the causes must be sought in the lack of acceptance of the system and consequent low motivation. Management's considerations of economic profitability are based on a "theoretical output" which can only be translated into "actual output" if the workers co-operate on a basis of joint responsibility. For this purpose, the workers' learning requirements must be taken into account. In other words, the possibilities and chances of learning skills, interacting and communicating must be discussed before new working systems are introduced or the organisation of work is modified.

6. Conclusions

Overall, this case study reveals the following consequences of automation:

- increasing fixed costs and falling variable costs;

- more involved preparation (programming, set-up time), unskilled performance of work, short cycle times;

- reduced physical stress but greater mental stress;

- lower skill requirements, shorter training times, higher concentration demanded;

- monotonous, mechanical work without prospects of acquiring further skills and promotion.

Management-initiated changes in work organisation usually tend to be identical to work division measures. In the past, the assemblers themselves put together and checked the material from the store and also bent and cut the components, which was regarded as "more interesting than the present system", but this work is now carried out by "preparation" staff. Checking, automatic-mechanical bending and cutting of the components, and determination of the quantities of each - all take place at a working position which calls for different skills owing to the diversity of tasks and experience required. The organisational separation into assembly and work preparation thus gave rise to working positions calling for little skill at the preparation stage. This division of labour was also associated with the loss of individual skills at the assembly positions.

A similar situation was brought about by the introduction of the semi-automatic assembly table. The technical intelligence hitherto required at each assembly position for development of an assembly order consistent with the relevant PCB (in turn a prerequisite for rapid assembly) is now only required of the senior worker who programs the assembly tables. This could be regarded as an indication of a trend towards polarisation of skill levels, i.e. a few people acquire higher skills while the majority are de-skilled.

As far as the initial management objectives are concerned, besides considerable cost advantages, improvements in quantity and quality of output could be achieved. On the other hand, no reduction in absenteeism could be reported and, in addition, overall satisfaction with tasks and jobs decreased. One reason for these negative results may be found in the process of reorganisation. As we learned from our investigations, the workers concerned were not involved in the deliberations and decisions relating to the change. The works council was not consulted, which constitutes in this case a violation of the Works Constitution Act.

This holding back on information and participation on the part of management reflects the predominant technological and engineering orientation of most managers in the Federal Republic of Germany, and their fear of resistance to change when plans for automation should become known at an early stage. On the contrary, we suggest a participative team approach to organisation redesign, as outlined in section 3.

New forms of work organisation which try to consider QWL aspects must inevitably remain fragmentary, as long as machines and technical requirements alone take priority, rather acceptance of the basically variable character of techniques and technology. We must break away from the idea of optimising the technical system and then adapting the social system to it by applying appropriate techniques, with the object of considering work organisation as a socially determined system (socio-technical system approach). People, not machines, are of primary concern. Techniques and technology are there to serve human beings and should be conceptualised as variables, not as unchangeable facts in the situation model. New concepts in work organisation should induce new technical developments, and not vice versa.

Therefore, it is not of interest to our approach to examine how people can be adjusted to technically and/or organisationally determined tasks. We should rather attempt to determine the actions and behavioural consequences to which certain tasks may lead and, as a result, the kinds of stimuli with which tasks must be equipped in order to produce a favourable work situation.

What in our opinion is urgently necessary is increased research on and analysis of influencing factors in industrial work situations as a prerequisite for socially acceptable changes in existing work organisation.

Notes

[1] J.R. Bright: "Does automation raise skill requirements?", in Harvard Business Review, Apr. 1958, pp. 85-98; idem: Automation and management (Boston, Harvard University Press, 1958).

[2] Projektgruppe Automation und Qualifikation (Automation and Qualifications Project Group) (F. Haug et al.): Automation in der BRD (Berlin, Argument-Verlag, 3rd ed., 1979).

[3] O. Neuloh and H. Widemann: Arbeiter und technischer Fortschritt (Köln/Opladen, Westdeutscher Verlag), 1960; U. Jaeggi and H. Widemann: Der Angestellte im automatisierten Büro. Betriebssoziologische Untersuchung über die Auswirkungen elecktronischer Datenverarbeitung auf die Angestellten und ihre Funktionen (Stuttgart, Kohlhammer, 2nd. ed., 1963).

[4] H. Kern and M. Schumann: Industriearbeit und Arbeitsbewusstsein. Eine empirische Untersuchung über den Einfluss der aktuellen technischen Entwicklung auf die industrielle Arbeit und das Arbeitsbewusstsein, Vols. I and II (Frankfurt am Main, Europäische Verlagsansatalt, 1970); O. Mickler et al.: Technik, Arbeitsorganisation und Arbeit. Eine empirische Untersuchung in der automatisierten Produktion (Frankfurt am Main, Aspekte-Verlag, 1976).

[5] W. Fricke: Arbeitsorganisation und Qualifikation. Ein industriesoziologischer Beitrag zur Humanisierung der Arbeit (Bonn and Bad Godesber, Verlag Neue Gesellschaft, 1975).

[6] Projektgruppe Automation und Qualifikation (F. Haug et al.): Automation in der BRD, op. cit., 1st ed., 1975; idem: Theorien über Automationsarbeit (Berlin, Argument-Verlag, 1978).

[7] W. Rohmert and J. Rutenfranz: Arbeitswissenschaftliche Beurteilung der Beanspruchung und Belastung an unterschiedlichen industriellen Arbeitsplätzen (Bonn, Bundesminister für Arbeit und Sozialordung, 1975).

[8] cf. E. Grochla: Automation und Organisation. Die technische Entwicklung und ihre betriebswirtschaftlich-organisatorischen Konsequenzen (Wiesbaden, Gabler, 1966); H.J. Drumm: Automation und Leitungsstruktur (Berlin, Duncker and Humblot, 1970).

[9] See, for example, M. Gaitanides: Industrielle Arbeitsorganisation und technische Entwicklung, produktionstechnische Möglichkeiten qualitativer Verbesserungen der Arbeitsbedingungen (Berlin/New York, de Gruyter, 1975); and B. Schiemenz: Automatisierung und Produktion (Göttingen, Vandenhoeck and Ruprecht, 1980).

[10] See, for example, W.H. Staehle: "Die Stellung des Menschen in neueren betriebswirtschaftlichen Theoriesystem", in Zeitschrift für Betriebswirtschaft, No. 1975, pp. 713-724.

[11] Projektgruppe im WSI (WSI Project Group): Grundelemente einer arbeits - orientierten Einzelwirtschaftslehre (Köln, Bund-Verlag, 1974).

[12] W.H. Staehle: "Der situative Ansatz in der Betriebswirtschaftslehre", in H. Ulrich: Zum Praxisbezug der Betriebswirtschaftslehre in wissenschafts-theoretischer Sicht (Bern/Stuttgart, Haupt, 1976), pp. 33-50; ibid: "Die Arbeitssituation als Ausgangspunkt von Arbeitsgestaltungsempfehlungen", in G. Reber (ed.): Personal - und Sozialorientierung der Betriebswirtschafts-lehre, Vol. 1 (Stuttgart, Poeschel, 1977), pp. 223-249.

[13] For details, see A.L. Thimm: The false promise of codetermination (Lexington/Toronto, Lexington Brooks, 1980).

[14] W.H. Staehle: "Die Arbeitssituation als Ausgangspunkt von Arbeitsgestaltungsempfehlungen", op. cit.

[15] M. Gaetanides, op. cit.

CHAPTER VII

AUTOMATION AND WORKING CONDITIONS IN FRANCE

1. National background

1.1 Introduction

In France, the problems of industrial automation falls within the framework of a wider debate on new technologies that has taken on increasingly large proportions over recent years.

Automation and its effects on employment have long been the subject of study and debate among a number of French investigators (Friedman, Naville, Touraine, etc.) but until now concern had been limited to research circles. Recently, however, the stakes at play in these new technologies in general and in automation in particular have abruptly been brought to the attention of the public at large. This is in sharp contrast to the major public debate on the limits of growth, in which little attention was given to these new technologies.

Parallel to this, intensified competition, stimulated in particular by the expansion of international trade and progress in automation technology, has made a powerful contribution to the introduction of these technologies into manufacturing industries. As their performance has risen and their components have been miniaturised and fallen in price, they have been used not only in heavy plant to improve productivity but also in low-price equipment for cost-effectiveness. However, it is the actual or potential effects of automation on employment in general and on numbers of jobs in particular that have launched a widespread debate on this topic and, further, on the overall structure and functioning of our cultural system which could be endanged by computerisation and production automation.

The first part of this chapter, which deals with data on the automation of French industry, aims to give a general overview at a national level, and to investigate the degree to which automation has currently penetrated industry and the relationship between automation and working conditions. Finally, a schematic picture is given of action taken by the authorities and of attitudes adopted by employers and employees in this context.

1.2 The demographic and cultural scene

The profound effects and wide-ranging uses of computerisation and automation must be viewed in relation to the size and structure of the workforce, its system of values and its attitudes towards technical progress and the changes that will ensue.

1.2.1 Size and structure of the workforce

The size and characteristics of the workforce have undergone considerable changes since the sixties, and these changes are expected to continue.

(1) Changes in the number and composition of workers

The increase in the number of workers is due first to demographic growth. Total population size, which had been virtually stable since the start of the century, increased by around 25 per cent between 1954 and 1980, rising from 43.1 to 53.7 million, an equivalent of a rise in the workforce of about 20 per cent. This increase is attributable mainly to natural demographic growth and, during the period 1954-68, to foreign immigration (repatriations and the influx of foreign workers).

This growth disparity between total population and working population is attributable to trends in employment rates. These were driven down by the upward movement in age at first employment following the raising of the school-leaving age and the lowering of the retirement age, and driven up by the growing employment of women. Finally, immigration, which increased strongly as from the sixties, has reached a stage of zero growth over recent years.

A second factor is the increase in age at first employment. The rise in the standard of living, the importance attached in France to the possession of diplomas and the difficulties that a large number of young people have had in finding jobs in recent years have all contributed to an increase in full-time school attendance, resulting in first employment occurring at a somewhat higher age (table 16). Consequently, employment rates for young people aged 15-25 have fallen significantly (table 17). For women, this trend, together with that in school attendance rates, has led to a marked fall in the size of the non-working population not in full-time education. This rise in the school-leaving age has, of course, helped to improve the educational level of the workforce (see below).

Thirdly, there has been a sharp reduction in employment rates for the elderly, even though this trend has been reversed somewhat in recent years (table 18). This movement is attributable, on the one hand, to a fall in the proportion of the total workforce accounted for by non-salaried workers (who usually retire later than wage earners), and to the introduction of retirement incentives, especially for elderly agricultural workers. Among wage earners, on the other hand, the trend is due to the effects of pre-retirement or early retirement systems introduced to counteract high unemployment levels.

The generations that are retiring earlier were, of course, in full-time education for shorter periods than subsequent generations and, in particular, than the generation now starting work. Consequently, the early retirement movement is accentuating the trend towards a workforce with a higher average level of education.

Fourthly, in common with many industrialised countries, France has, over recent years, had recourse to a large number of immigrant workers. The percentage of the total population accounted for by foreign workers rose from 4.7 per cent in 1962 to 6.9 per cent in 1980; as a percentage of the workforce, the figure rose from 5.6 per cent to 6.3 per cent (6.8 per cent in 1978).

The influx of foreign workers, following a sharp increase starting in the sixties, has fallen progressively due to the downward trend in the business cycle and immigration control measures. It is estimated that net immigration was zero over the period 1975-80.

Table 16: France: percentages of young people aged 17-24 in full-time education, 1968-75

Age (years)	Males		Females	
	1968	1975	1968	1975
17	54.3	67.4	62.6	73.7
18	40.3	50.4	47.1	58.4
19	30.2	33.9	32.4	39.8
20	22.6	24.9	22.3	26.7
21	17.3	19.0	15.4	18.8
22	12.4	14.4	9.9	13.4
23	9.7	11.4	6.8	9.4
24	7.1	8.7	4.5	6.2

Source: Institut National de Statistique et des Etudes Economiques (INSEE), 1968 and 1975 population censuses.

Table 17: France: employment rates of young people aged 15-25, 1968-75 (percentages)

Year	Males		Females	
	Age group			
	15-19	20-24	15-19	20-24
1968	43.0	81.7	32.5	63.6
1974	29.3	82.3	21.8	65.8
1975	29.7	81.3	23.0	65.7
1976	27.5	81.9	21.3	67.3
1977	27.1	80.9	20.9	67.5
1978	25.2	80.1	18.9	66.9
1979	25.9	80.2	19.4	67.7
1980	25.7	80.2	18.1	66.9

Source: INSEE, employment surveys.

Table 18: France: employment rates of persons aged 60-64, 1968-80 (percentages)

Year	Males	Females
1968	65.9	35.3
1974	60.4	31.2
1975	54.4	29.3
1976	50.3	27.1
1977	47.6	27.2
1978	43.6	23.9
1979	43.7	23.9
1980	46.9	27.1

Source: INSEE, employment surveys.

Finally, a significant factor has been the growth in the employment of women. Statistics on reserves of available labour show that, over the period 1962-80, women accounted for some 70 per cent in the growth of the workforce (i.e. a rise twice as rapid as that for male workers). The percentage of women in the workforce rose from 35 per cent in 1962 to 41 per cent in 1980, and this figure should continue to rise over coming years. This trend is attributable to the rise in employment rates for women and, in particular, for married women.

The rise in the employment rates of married women is linked, on the one hand, to fertility patterns (cause or effect) and, on the other, to the rise in the educational level of women.

Employment rates tend to increase with the level of scholastic achievement, and it is women with technical qualifications who have shown the highest increase in employment rates between 1968 and 1975. In addition, the growth of the tertiary sector has, without any doubt, favoured the expansion of female employment.

Thus, the rise in the number of working women - involving as it does those women with the highest educational level - tends to raise the qualification level of the workforce as a whole.

(2) Rise in the educational level of the workforce

The amount of education acquired by workers has increased considerably since the beginning of the twentieth century. Between the early 1900s and the 1970s, the time spent in full-time education by the workforce as a whole increased, on average, by a half. Between the generations born in the periods 1906-10 and 1941-45, the percentage of male workers who left school at 15 years of age or over rose from 22 per cent to 59 per cent and the percentage of female workers from 23 per cent to 68 per cent. The relatively higher rate for women, which was already apparent for the earlier generation, became increasingly marked for those born between 1941 and 1945.

The increase in the number of years spent in full-time education has resulted in a significant rise in the level of education as measured by the highest qualifications obtained. Table 19 shows the rise between 1973 and 1979 in the educational level of young people entering the labour market.

Table 19: France: percentages of new labour market entrants by level of qualifications, 1973 and 1979

Sex/year	Below vocational proficiency certificate (CAP)	Vocational proficiency certificate to baccalaureate	Above baccalaureate	Total
Males				
1973	52	34	14	100
1979	42	42	16	100
Females				
1973	46	34	20	100
1979	33	45	22	100

Source: INSEE.

Table 19 shows that the educational level of young people in general, independent of sex, has risen, and that the educational level of women starting work for the first time is higher than that of men. A comparison of the average educational level of persons entering with those leaving employment shows that the proportion of those with the equivalent of the French vocational proficiency certificate is three times higher among those entering than among those leaving the labour force.

Attention should also be drawn to a number of trends in the types of courses studied. Over the last ten years, the number of students attending courses between the second cycle of secondary education and the highest level of university education has risen by more than 30 per cent; however, the increase for female students has been twice as rapid (+42 per cent) as that for male students (+22 per cent). Over the whole course range, the number of students taking vocational, technical or scientific courses has increased more rapidly (+35 per cent) than those taking other courses (+27 per cent). In the case of male students, the disparity has increased markedly: an increase of 32 per cent in students taking vocational, technical or scientific courses as against around 10 per cent for other types of courses. The number of female students taking such courses has risen by approximately 40 per cent for women as against 35 per cent for men, but not as rapidly as the number of female students taking other types of courses (+44 per cent). Among students in vocational, technical or scientific education over a given period, the number of those taking courses for the French vocational proficiency certificate has fallen by nearly 20 per cent whereas the number of those in higher studies (engineering schools, university technical institutes, scientific university courses) has risen by over 30 per cent; however, the most remarkable development has been in intermediate-level education, where the numbers enrolled have virtually trebled in the space of ten years: a 3.6 per cent increase in those studying for a vocational diploma and a 2.5 per cent increase in those studying for a technician's diploma or a technical baccalaureate. Over a ten-year period there has been a 35 per cent increase in vocational, technical or scientific courses as a whole; however, within this total the portion accounted for by intermediate-level courses has risen from 18.5 per cent to 39.5 per cent.

1.2.2 Values and attitudes

Attitudes and behaviour with regard to work and technological progress are to a large extent conditioned by a complex system of values which only partially straddle the natural generation gaps.

If attitudes to technological progress are to be understood, they must be interpreted within the context of a given culture and thus made more specific. Advanced industrialisation in northern Europe, the United States or Japan has had only a moderate standardising effect on existing cultures, and general attitudes to technological progress probably continue to differ markedly from one country to another.

A feature peculiar to France or, at least, very pronounced there is the very wide disparity between the ways in which different groups regard technological progress. One approach which still accounts for a significant current of opinion although presented here in an over-simplified manner is to view technological progress very positively since it is seen, in itself, as a neutral phenomenon. In such an approach, it is the uses made of technology that are important and give rise to debate, conflict and the need for choices to be made. Another approach views technology as a product of science; its proponents assert that, since it results from the interaction of social forces, it cannot be neutral, and claim that the effects of technology are determined largely by the balance of power.

An interpretation halfway between these two extremes would probably be the closest to reality. Nevertheless, these two concepts, in some form or other, are reflected in statements made in particular by the membership of the major French trade unions. Research in the human sciences has emphasised the lack of certainty rather than rigid determinism both in the choice of technology and its consequences, but there is little awareness that choices exist and, in fact, behaviour and practice tend not to take advantage of them.

Wage earners in secondary industry seem to have attitudes to technological progress which are very different from those working in the tertiary sector. However, both groups do have a common point of view: they both see technological progress as something which upsets an equilibrium acquired at great cost, and also as a cause of employment.

Going beyond this common ground, wage earners in the tertiary sector express much greater apprehension than their colleagues in secondary industries about the changes that computerisation has brought to their work. Machines have been introduced into a world which up until that point had, to a large extent, been spared its influence. Although data processing machines are not "dirty", and call for the same type of theoretical reasoning that wage earners in the tertiary sector commonly use in their work, it is still true that many of those involved feel that, in fact, the job content has been diminished; they rightly see themselves as servants to machines, and just as much, in their eyes, as workers in secondary industry. The difference in status between blue- and white-collar workers, which is so strong in France, is being undermined, and the majority of tertiary sector workers view the spread of new technologies with profound apprehension.

Workers in secondary industry see the arrival of new technologies in a somewhat different way. They are more accustomed to technological change than are their fellow-workers in the tertiary sector. A recent public opinion survey carried out for the French Ministry of Labour[1] shows, however, that their appreciation of technological progress is far from being totally positive: although 62 per cent of these workers think that technological progress has improved their working conditions, 59 per cent believe that it has reduced workers' status and, in particular, has done away with team spirit, mutual assistance and time-proven worker solidarity. The less skilled the workers, the greater this fear of automation, no doubt since it is their jobs which are, in fact, the most endangered.

Less is known of the opinions of management about automation. It seems, however, that managerial staff are also highly apprehensive, and many company directors emphasise the very marked resistance encountered at this level, particularly to the computerisation of managerial tasks.

Public opinion surveys indicate that the age group between 30 and 40 years is currently the one least apprehensive about new technologies. Young people have a less positive attitude. However, in any interpretation of this attitude, it would be necessary to make due allowance for the particularly unfavourable position of young people in the labour market. Their spontaneous reaction is to link computerisation and automation more with the idea of unemployment than with an improvement in working conditions.

On the whole, attitudes and behaviour in the face of new technologies seem relatively divided, although negative opinions seem currently to predominate over positive ones.

1.3 Economic data on the automation of French industry, and its effects on working conditions

The aim of this section is first to outline some general data on the economic background against which automation is taking place, and then to bring together available data on the current status of automation in France and its repercussions on working conditions.

1.3.1 Selected overall economic data

Over the last ten years, the French economy has continued to undergo profound changes which started after the Second World War. From 1968 to 1979, gross national product (GNP) rose by 60 per cent. The opening up of the economy internationally which started in the sixties has continued, and exports and imports have, over the same period, grown around 2.8 times. Exports have risen from 17 per cent to 30 per cent of GNP.

This has involved a far-reaching restructuring of the country's production capacity stimulated, in particular, by domestic and foreign demand and by international competition in the form of financial, economic and technical rationalisation[2] and a redistribution of labour. This redistribution has had a profound effect on the number of people in each of the various social and occupational categories, and the relative size of these categories within the working population as a whole (table 20).

It may be seen from table 20 that the number of wage earners has risen twice as fast as the total working population, while the number of self-employed persons has fallen by more than 20 per cent. When these figures are broken down by social and occupational category, it is found that the number of persons in the liberal professions and in top management increased by nearly two-thirds, the number in middle management by nearly 50 per cent and the number of salaried staff by nearly one-quarter. On the other hand, the number of manual workers increased only slightly (+4 per cent).

Table 21 gives a comparative breakdown, by major industrial sector, of the structure of the workforce in 1968 and 1979 and the growth of the workforce by sector (excluding non-commercial services).

The level and growth of fixed productive assets per head in 1968 and 1979 are shown in table 22. The areas of greatest growth, apart from agriculture, have been capital goods, building and civil engineering and consumer goods.

1.3.2 Level and means of automation in French industry[3]

The aim of this section is to review the main available data on the current status of automation in French industry. However, any attempt to describe, even in an approximate fashion, the extent and forms of automation

Table 20: France: structure and growth of the workforce by social and occupational categories, 1968-79

Social/occupational category	Breakdown (%)		Index of no. of workers in 1979 (1968 = 100)
	1968	1979	
Farmers	11.5	7.5	70
Agricultural workers	2.7	1.4	54
Industrialand commercial employers	10.7	8.1	81
Liberal professionals and top management	5.1	8.0	169
Middle management	10.3	14.0	146
Salaried employees	15.0	17.1	123
Manual workers	36.6	35.3	104
Service workers	6.2	6.7	117
Others	1.9	1.9	114
Total	100	100	100
Numbers (millions) of which:	19.9	21.5	108
- wage earners	15.2	17.8	117
- self-employed	4.7	3.7	79

Source: From *Enquêtes emploi* (INSEE Collection D.68 and D.73).

in the French industrial manufacturing system encounters numerous obstacles due, in part, to common factors which are worth mentioning.

In France, as in other countries, the first obstacle is the absence of adequately clear and precise definitions for the multitude of forms in which automation is implemented. The second is due to the fact that what in numerous cases is commonly termed "automation" is part of the continuum of long-term development in various technologies and, in particular, mechanical engineering, electronics and data processing. The problem stems from a sort of halo phenomenon resulting from both fashion and myth, which confuses the factual data. For these reasons, the validity and interpretation of these data, where they exist, should be subject to numerous precautions.

Table 21: France: structure and growth of the workforce by major economic sectors, 1968-79

Economic sector	Breakdown (%)		Index of no. of workers in 1979 (1968 = 100)
	1968	1979	
1. Agriculture, forestry, fishing	18.5	11.0	61
2. Agricultural and food industries	3.6	3.4	98
3. Energy	2.1	1.7	83
4. Intermediate goods	9.2	9.2	102
5. Capital goods	9.1	10.5	119
6. Consumer goods	9.5	8.3	89
7. Building and civil engineering	11.5	10.6	94
8. Commerce	14.0	15.2	112
9. Transport and telecommunications	7.1	7.8	113
10. Commercial services	15.4	22.3	148
Total	100	100	103
of which:			
industry (4+5+6)	27.8	28.0	103
sectors not subject to foreign competition (7+8+9+10)	48.0	55.9	103

Source: From data published in Economie et Statistiques, No. 127, Nov. 1980, p. 9.

Table 22: France: fixed productive assets per head by major economic sectors, 1968 and 1979

Economic sector	Level of fixed assets (to nearest FF'000, 1970)		Index of fixed assets in 1979 (1968 = 100)
	1968	1979	
1. Agriculture, forestry fishing	29	71	245
2. Agricultural and food industries	82	135	165
3. Energy	387	664	172
4. Intermediate goods	91	152	167
5. Capital goods	48	94	195
6. Consumer goods	43	76	177
7. Building and civil engineering	20	38	194
8. Commerce	36	62	174
9. Transport and tele-communications	131	226	173
10. Commercial services	30	50	168
Total	54	96	177
of which:			
industry (4+5+6)	61	108	178
sectors not subject to foreign competition (7+8+9+10)	44	75	173

Source: From data published in Economie et Statistiques, op. cit.

The degree to which a sector is open to penetration by various automated technologies can be assessed by an analysis of the disparity and variety of the manufacturing processes employed in different industries. Table 24 attempts to present this situation schematically. In the following pages we review the progress of automation by process, follow up with some quantititative data and, finally, assess the degree of dependence on foreign equipment.

Figure 23: France: penetration of production processes by automated technology

Process \ Technology	Machining centres	NC machine tools	Robots	Integrated manufacturing
Continuous			xx	xx
Semi-continous			xx	x
Batch		x		
- large-batch assembly	x		xx	xx
- small-batch assembly			≃	
- long-run machining	x	xx	x	x
- short-run machining	x	x		
- packaging			xx	x
- handling			x	

≃ = very weak penetration; x = susceptibility to penetration; xx = high penetration.

Source: Agence Nationale pour l'Amélioration des Conditions de Travail (ANACT).

(1) Automation in continous and semi-continuous processes

The continuous and semi-continuous processing industries currently account for about half of all the automated equipment in French industry. These processes have been a prime target for automation, due in particular to the type of products they handle, the processing they carry out and the mass production nature of their markets. By increasing productivity and allowing real-time information flow, widespread use of computerised techniques has raised plant performance through greater reliability and flexibility.

Automation in the processing industries is far in advance of that in other areas and involves mainly, and to varying degrees, the following sectors: petroleum; fluid and energy flows (water, gas, electricity, etc.); chemicals; rubber; pharmaceutical production; iron and steel; minerals; non-ferrous metals; glass; paper; transport; building materials; agricultural and food industries; and electrical engineering.

The rise in energy and raw materials costs, the rationalisation and expansion of production capacity, labour problems (especially labour management in the continuous operation of large plant, and workers' rejection of arduous, dangerous or dirty work) and the rise in wage costs have necessarily helped maintain the drive towards automation, and this has been facilitated by improved technology, especially in the field of sensors. However, there are many technical problems to which no solution has yet been found, and to these difficulties must be added the problems of the business cycle, the high cost of installing such systems and a whole range of social problems among the workforce.

Table 24 shows the trends in level of automation in the main processing industries between 1973 and 1978.

(2) Automation in batch processes

Machining operations, in the eyes of workers involved in them, are the very basis of manufacturing. To the external observer, machining (whether for short or long production runs) is the domain of the machine tool and skilled operator. There are, of course, major differences depending on whether short or long production runs are involved. Still typical of this area of production is the immense workshop with manually controlled mass production blanking and forming presses. This is the realm of both the machine and technology, but also of the specialised worker glued to his machine, whose freedom of action, even in the best of cases, is only slightly more than that of his workmates on the assembly line who are controlled by a specific production tempo.

It was only to be expected that the machine shop would be a prime target for automation. The first step was to equip conventional machines with relatively simple automation which helped to lighten the operator's workload while better regulating the machine's output. This was followed by NC lathes and capstan heads which, in addition to increasing precision, made it possible to integrate a series of operations into a single machine, thus eliminating intermediate storage and transfer from one work station to another. Nowadays, some machining centres with tool wear and breakage detectors are able to work continuously without any supervision. Efforts are also being made in the field of machine feed and ejection. Two or three French firms have installed a central robot which directly feeds several machines located around it, but this is still the exception.

NC machine tools operating under direct or computer control offer significantly improved productivity due not so much to reductions in machining time as to savings in down-time, non-machining time and adjustment time which may be considerable on conventional machines. With its short production times and, in particular, the rapid program changes it allows, this type of machine is ideal for short production runs; it also offers the further advantage of product diversification, even in factories which had previously specialised in mass production work.

Whereas metal machining evokes a picture of the machine tool, installation, assembly and finishing operations are closely linked to the production line and to the repetitive movements of the factory's least-skilled workers. Over the last ten years, very experimentally at first but in a significant manner nowadays, the robot has made its appearance to replace the human worker in repetitive movements in the workshop. Programmable manipulators are now being manufactured in small quantities in France, and this makes it possible to give an approximate assessment of their industrial importance. In this field too, micro-electronics have led to considerable improvements in the performance of these machines which are currently used mainly in two types of work: welding and painting. Screwing and bolting operations

Table 24: France: breakdown of investment of main processing industries by level of automation, 1973 and 1978 (percentages)

Industrial sector	Level of automation							
	1973				1978			
	0	1	2	3	0	1	2	3
Chemicals	25	55	20	0	24	52	22	2
Electricity	15	55	25	5	12	53	27	8
Petroleum	18	60	19	3	15	58	22	5
Paper	47	43	9	1	37	45	14	4
Agricultural and food industries	58	28	12	2	44	34	19	3
Building materials	20	65	15	9	12	66	20	2
Water treatment and distribution	45	38	17	0	34	47	19	0
Ferrous and non-ferrous metals	20	65	15	0	11	63	24	2

Key: Level 0: local operation (single machines or work stations); level 1: linked systems (covering more than one unit or work station); level 2: technical optimisation (complete technical systems); level 3: economic optimisation (covering overall economic structures).

Source: Bureau d'Informations et de Prévisions Economiques (BIPE).

have still not been affected by automation for purely technical reasons, and special assembling machines are still very rate in assembly shops.

Painting robots with up to seven movement axes are now capable of highly complex configurations. They are employed for the most delicate operations and their use has proved particularly suitable for short production runs.

In addition to the use of robots for painting, other major operations in surface treatment have also been transformed by the new technologies. Efforts here have been concentrated mainly on mechanical handling, the movement of equipment and the duration of vat and bath operations. In this field, greater accuracy and regularity have led to very significant quality improvements. On the other hand, vat and bath supervision, the mechanical equipment for vats and component transporters are still largely untouched by new technology. The loading and unloading of components in surface treatment are still carried out manually, due mainly to the high level of investment entailed by automation.

The automation of machining and assembly has affected product finish and quality: retouching has been significantly reduced. Moreover, there is no doubt that developments in automation will in the long term lead to radical changes in product design, with unit assembly giving way to modular assembly, for example. However, such foreseeable developments are still far from widespread implementation.

Both in semi-continuous and long- or short-run production work, packaging and handling were and still are the areas in which the least effort has been made towards automation. Traditionally, packaging has been carried out by an abundant unskilled female labour force, and handling by immigrant workers.

Technological change in packaging started first in the agricultural and food, and pharmaceutical industries where hygiene requirements and the need to combine large volume and improved productivity led, initially, to the mechanisation of the main handling procedures. Automation has made it possible to integrate different machines along the packaging line. Nowadays, what is seen is no longer machine integration but a transformation of the packaging methods themselves by an integration of different operations within a given machine; this has been made possible primarily by the development of micro-electronics. Quality control is also carried out by high-reliability automated equipment.

A study carried out for the General Delegation for Scientific and Technical Research (DGRST)[4] concludes that mechanical handling is under-automated in France, and that the obstacles to automation in this area are, in part, of the same type as those in packaging. Activities are subordinate to the production processes themselves are usually more easily overlooked. The realisation that significant productivity gains could be obtained by optimising manual handling operations is now leading to efforts to automate the handling of individual loads by conveyors and wire-guided trolleys. Handling equipment for bulk goods is still not easy to automate. Closely linked to handling, storage is an area where, due to cost reasons, efforts at automation will be more significant. Bufffer stores to even out production flows, intermediate storage and final warehousing are being automated by the use of manually, then computer-controlled materials handling systems. However, this still affects only an extremely small number of firms, whereas inventory management as such is more and more often computer-controlled.

(3) Quantitative data

Starting in the sixties, there has been progressive growth in the use of NC machine tools. Reports[5] estimate that there were 5,800 NC machine tools in French industry in 1980, in comparison with 2,200 in 1974 and 1,000 in 1970; a figure of 10,000 for 1985 is forecast. Table 25 compares the percentage breakdown by industry of stocks of conventional and NC machine tools in 1974.

The same report estimates that, at the time of the survey (1974), 25 per cent of NC machine tool stocks was in the hands of firms employing less than 200 workers (10 per cent in firms employing less than 100 workers) and that, by 1980, the corresponding figure should have risen to between 30 and 35 per cent.

Table 25: France: number and breakdown of stocks of conventional and NC machine tools, 1974

Total breakdown by industry	Conventional machine tools	NC machine tools
Total number	985 000	2 200
in the following industries:	(%)	(%)
mechanical and metalworking industries	38	35
motor vehicle	20	22
aeronautic	6	20
armaments	2	8

Source: O. Pastre and J. Toledano: "Automation et emploi: L'hétérogénéité des conséquences sociales du progrès technique", in O. Pastre et al.: Emploi et système productif, Collection Economie et Planification (Paris, Commissariat Général du Plan, 1979).

As far as CAD machines are concerned, Pastre and Toledano estimate that, in 1978, there were 700 in service, to which should be added 1,000 mini-computers for long-run production work (1977 figure).[6]

The term "robot" covers machines of widely differing degrees of automation, ranging from programmable manipulators to advanced robots (fitted with sensors, co-ordination systems and operating arms) and even as far as the "intelligent" robot capable of taking situation-dependent decisions. The Pastre and Toledano report estimates that there are less than 200 robots currently in service. However, an official report, although confirming that France is behind in robot use, estimates that there are 300 high-level robots in France in comparison with 1,000 in Sweden, 3,000 in the United States and 5,000 in Japan.[7]

Finally, two flexible workshops are currently operating and five are under construction.[8]

The number of CAD systems in operation in France is estimated by Pastre and Toledano at around 100, and these are used in design offices in the following sectors: mechanical engineering; electrical and electronic engineering; motor vehicle construction; shipyards; aeronautics; armaments; architecture; and plant and design engineering.

Tables 26 and 27 show, on the one hand, the relative importance of each type of automated process for the main industries in France in 1976, and, on the other, the main industrial sectors involved or likely to be involved in different types of automation.

Table 26: France: relative importance of each type of automated process in main industries, 1976
(as a percentage of total automated equipment)

Industrial sector	Continuous or batch process control	Machine control	Data transaction control	Environment and testing
Chemicals, petrochemicals, pharmaceuticals, plastics, rubber	65	30	-	5
Refinining, petroleum transport and distribution	84	-	6	8
Electricity production, transport and distribution	92	-	-	8
Iron, steel and non-ferrous metals	46	49	-	5
Pulp, paper and board	31	65	-	4
Agricultural and food	62	35	-	3
Glass	44	45	-	11
Building materials	44	48	-	8
Mechanical engineering, motor vehicles, aeronautics and shipyards	10	82	-	8
Textiles	35	56	-	9
Transport	1	14	70	15
Electrical and electronic engineering	-	69	-	31
Other sectors	37	36	10	17
Average, all sectors	46	38	4	12

- = nil or negligible.

Source: Société d'Etudes pour le Dévelopment Economique et Sociale (SEDES); BIPE.

Table 27: France: breakdown of advanced automation by industrial sector

Industrial sector	Type of automated equipment			
	Process automation	NC machine tools	Mass production automation	CAD systems
Chemicals, rubber, glass, plastics, pharmaceuticals	++		++[1]	
Oil	++			
Iron and steel	++		++	
Agricultural and food industries	++		+	
Mechanical engineering industries		++	++	++
Textiles, clothing	+		+	
Electrical engineering	++	++	++	++
Electricity, gas, water	++			
Paper, board	++		+	
Building materials	++		+	+
Minerals and non-ferrous metals	++		+	
Semi-finished metal products	++			
Motor vehicles and bicycles		++	++	++
Shipbuilding, aeronautics, armaments		++	++	++
Leather			+	
Wood			+	

[1] The main industries involved in mass production automation are plastics and glass.

+ potentially suitable for the use of automated equipment; ++ automated equipment currently being used.

Source: O. Pastre et al.: Automation, travail et emploi, Report drawn up for the Mission à l'Informatique du Ministère d'Industrie (Paris, Ministère d'Industrie, 1979).

As has been pointed out above, these quantitative data should be viewed with circumspection owing both to the lack of precision in the terminology and the limits of the statistical information available. Finally, from the point of view of small and medium-sized industries, a recent survey of 600 firms employing less than 500 workers showed that, in 1980, 14 per cent of these firms had automated machines (13 per cent for manipulators and robots, 41 per cent for automatic machines), and that the penetration rate increased as expected with the size of the firm.

Another survey[9] carried out amongst a representative sample of 800 workers confirmed that automation currently affected only a limited number of workers and sectors. At the time of the survey, only 13 per cent of workers carried out tasks which were totally or highly automated; 30 per cent had a job that was somewhat automated, and 56 per cent a job which was not automated at all. Throughout their working life, 23 per cent of workers had been employed on a job which was totally or partly automated.

(4) Degree of dependence

Finally, it is necessary to look at the degree to which the French economy depends on the supply of foreign automated equipment, and especially how this affects freedom to specify that working conditions be taken into account when equipment is still at the design stage.

The data given here are taken from a report dealing specifically with this question, which contains interesting concepts that cannot be developed within the framework of this study.[10] Table 28 gives the parameters of the situation.

The authors of the report reach the following conclusions:

(a) basic components: French industry seems very dependent on foreign semi-conductors; in view of increasing semi-conductor integration, this may, in the long run, constitute a threat to products which are traditionally downstream;

(b) automation components: France seems relatively weak in the field of analysers, control valves, industrial calculators and peripherals. This weakness is a disadvantage in the face of foreign manufacturers who offer more comprehensive product ranges;

(c) automation equipment: the situation is satisfactory in the field of process controls, but tends to be less satisfactory in machine controls (in particular numerical controls and robots), production management and CAD. However, the Renault state-owned company is well advanced in this field since it manufactures its own robots and is planning to market them. Moreover, there are a number of firms which buy in chiefly foreign components or units and adapt them or build them into sub-assemblies that are currently not picked up by the statistics.

Table 28: France: position of the French automated equipment industry in the French market (percentages)

Components	Total Demand	French manufacture		Imports	Imports + production under licence	Exports	Manufacture + exports
		Total	of which under licence				
Sensors, transmitters	100	65	20	35	55	25	80
Analysers	100	40	15	60	75	10	50
Indicators, recorders	100	70	10	30	40	15	25
Regulation	100	75	10	25	35	20	95
Speed changers	100	60	13	40	55	15	75
Regulating valves	100	55	20	45	65	15	70
Teletransmitters	100	90	5	10	15	22	112
Relays	100	80	5	20	25	20	100
Programmers	100	70	10	30	40	20	90
Programmed automated equipment	100	70	10	30	40	17	87
Industrial calculators	100	55	10	45	55	25	70
Analogue calculators, simulators	100	80	5	20	25	20	100
Interfaces	100	60	5	40	45	15	75
Peripherals	100	30	5	50	75	10	40

Source: SEDES/BIPE: Recherche technologique et indépendance industrielle, Study carried out for DGRST et CORDES (Paris, 1979).

1.3.3 Automation and working conditions

This subject raises two basic questions to which two supplementary comments will be added.

The first question returns to earlier remarks contrasting the simplicity of "automation" as a concept with the extreme diversity of its forms in practice.

As to the second question, experience shows that different enterprises seldom use the same technology in the same way or in the same form of work organisation; technology is not necessarily the determining factor in production organisation and working conditions. Moreover, there is no doubt that the aim, or at least the importance, of the ILO project has been to adopt a case-study approach. The monograph published by the French National Agency for the Improvement of Working Conditions (ANACT) has the same scope, and this is in line with comments that ANACT makes in fields other than automation.

Finally, it must be added that this study does not cover the relationship between automation and employment levels since this cannot, of course, be dealt with without considering the effects of numerous other factors. Nevertheless, the very fact that workers perceive or even fear that there is a link between automation and reduced levels of employment - especially at the micro-economic level of their own plant - is a factor to be borne in mind since employment and employment conditions are for the worker a major aspect of working conditions.

It is also necessary to make two comments here. The first is to emphasise that, whatever its cause or nature, change within a firm produces effects due both to the difference between the initial and final situation, and also to the way in which the change was designed, prepared, introduced and carried out.

The second comment relates to the extent of the lessons that can be drawn from a study of the past, or even of the present. The effects of automation on working conditions that can be detected by an examination of this kind also result from past choices in automation design and implementation. Naturally enough, this type of study is incapable of reporting on the consequences of the various possible choices which were not adopted.

Over recent years, the literature on the impact of automation on working conditions published by investigators has been extensive; however, no clear picture has been formed as to the consequences that automation may already have had or those that it is likely to have in the future. Leaving aside the fully automated process industries, it seems, as already indicated, that we are currently embarking upon a phase of profound technological change in which it will prove particularly difficult to distinguish between effects attributable to the change itself and those which are more durable because they are specific. Furthermore, although it is sometimes possible to make a scientific and objective analysis of "material determinants", the way workers view their employment situation is certainly influenced by their appreciation of the changes taking place.

Without wishing or claiming to be exhaustive, an attempt may nevertheless be made to list the main points related to working conditions emphasised both by employers, employees and investigators when various types of automation are being or have been introduced.

(1) Health and safety

For quite simple reasons, new technologies seem to have a positive effect on health and safety in the strictest sense of the term. In many cases, they tend to separate workers from the potentially harmful industrial product and from the machines for which they are responsible by increasing the number of safety devices and reducing the interventions required on the machine whilst it is operating. Machines which are more integrated and which require less direct human intervention - the NC machine tool comes particularly to mind here - can often be enclosed more effectively and are therefore less noisy than first-generation machine tools. Paint-spraying robots make it possible to eliminate the exposure to harmful products common to this type of work; but since robots operate in an explosive environment, particularly severe safety requirements have to be met. The use of wire-guided conveyor trucks fitted with anti-collision devices and sectorised circuits so that a truck cannot enter a zone already occupied by another should result in a significant fall in occupational accidents. Doubtless, further progress will be achieved in this field and, in view of the number and severity of handling accidents, serious hopes for an improvement in this field seem well founded.

Although remote-controlled manipulators have not changed basic occupations in the iron and steel industry, they have made a considerable contribution to improving physical working conditions: separation of the worker from the heat source and reduction in physical lifting and holding tasks. In all activities of this type, by moving workers away from the direct source of harmful agents, remote-controlled manipulators must necessarily have significant positive effects.

Without listing in detail all the relevant new technologies, one has the general impression that these technologies help to reduce what ergonomists call the workload and physical burden in mass production processes.

Leaving aside the above-mentioned effects, which will be dealt with in more detail in the case-study, control room work is one of the automated or semi-automated processes which are frequently cited as a source of significant stress; this applies in particular to windowless, first-generation control rooms. Furthermore, work with cathode-ray tube (CRT) screens presents the same problems as when this type of equipment is used in the tertiary sector.

In automated industries, it seems that the new technologies have environmental effects that are likely to appear in due course in mass production industries. Workers tend to complain more about noise, whereas the actual noise level has not increased: in older workshops, machine noise was closely mixed with human noise and activitiy which made the machine noise more tolerable although it did not reduce its harmful effects.

Machine noise seems to be perceived more acutely the more the individual is isolated in large areas. In pass production industries, the coexistence of machines of different technological generations and the relative abundance of labour at the workplace mean that this is a problem which is not frequently raised; however, it may become a major problem in highly automated industries where isolation tends to sharpen perception, and where harmful factors which were little noticed under other circumstances become intolerable. This question of physical working environment should be a major concern when new technologies are introduced.

This brings us to the subject of mental stress, which has received considerable emphasis by employers, employees and investigators. There are few precise, overall indicators which can be used to measure with any accuracy the content of this concept. So far it has been too little used in the analysis of conventional jobs to allow serious comparisons. Yet, for workers doing automated jobs, it seems that the strain of changing over from heuristic to algorithmic modes of work is likely to increase this stress considerably; passive supervisory work increases and active intervention decreases. In a foundry, for example, a feed hopper was left unautomated so as to allow the operator some physical activity since the rest of his work was basically to supervise the machine. Numerous other examples could be quoted: in short-run production work, instead of working on "their own" machines, skilled workers each supervise the operation of four machines which requires virtually no intervention on their part.

Finally, continuous or semi-continuous production processes often entail shift work, and mention should be made of the effects of shift work (especially night shifts) on workers' health and life-style. Since automation often increases machine utilisation time, it sometimes also helps to increase the number of workers doing shift work in relation to the total number of workers. Between 1957 and 1977, the proportion of units in processing industries (including building) in which workers did shift work increased from 13 per cent to 16.4 per cent, while the proportion of workers involved in shift work rose from 10.3 per cent to 29.9 per cent, with even steeper increases being registered in many sectors.

Consequently, the overall balance is perhaps less positive than it would seem at first sight; however, awareness of these problems is also more acute, as has been indicated by the example of the foundry mentioned above. It is also difficult to determine the relative significance of the learning and acclimatisation phases for these new technologies. The way in which new technology is introduced, and the effective and active involvement of the workers during the introductory phase are decisive factors in reducing this nervous fatigue, as has been shown in many cases.

(2) Work organisation, job content and skills

The new technologies used by automation are viewed positively when they actually make it possible to reduce the physical arduousness of work. They cause far more apprehension in the case of the mental stress that they are likely to provoke. However, leaving aside employment problems, which are not dealt with here, it is certainly the subjects of work organisation, job content and qualifications that have given rise to the most contrasting stances and liveliest discussions.

The first and most common question about job content and skills is that of de-skilling. This term, as used by investigators and trade unionists, is sometimes ambiguous since it has various implications.

First, it indicates the disappearance of certain trades linked mainly with the direct processing of materials and which consequently involve personal, specific experience, a special touch and know-how acquired through long apprenticeship; the new technologies take over this knowledge from the worker, who becomes a simple machine supervisor. For example, paint-spraying robots are used precisely for the most delicate items, and this is the case too with surface-treatment machines and certain firing jobs, etc.

Consequently, de-skilling also points to the concept that not only do these machines consume the least skilled jobs but also that, in return, they offer an exceptionally small number of low-content, supervisory jobs. Although this hypothesis is sometimes confirmed by the facts, it is not as widely proven as it may seem.

Without any doubt, certain crafts are doomed to disappear, but the reduction of the operator's jobs to a simple supervisory activity is perhaps less predetermined than it seems. Take for example the statement of a company director: "It is the difficulty nowadays of recruiting experienced craftsmen that makes direct numerical control so attractive". It is for this very reason that pre-programmed rather than continuously programmable NC machine tools are used. With the latter, the worker can modify the program from one work-piece to another without stopping the machine, in order to optimise the work cycle, increase tool approach speed or depth of cut, etc. Thanks to the worker's know-how, the initial program can be constantly refined. The ongoing improvement of new technologies may offer other possibilities for skills and, in this field, it is certainly unwise to lay down absolutes, when in fact there may be no more than a semblance of predetermination. As with earlier generations of technology, these new technologies offer and will continue to offer alternatives in which the choices may be more social than technological. On the whole, it seems that automation requires an increased number of skilled workers even though it does, in fact, tend to decrese or even eliminate a large number of unskilled jobs; in all cases, it transforms the trades that workers have learned.

The above considerations related to workers' jobs should, after all, be related to observation of the structure of employment, taking into account all categories of workers.

Table 29 shows the trend in job structure by level of skills, between 1952 and 1973, for two industrial sectors which widely use continuous or batch processes. Thus, when considering changes in skills, each category should be reviewed in relation to the change in its weighting in the employment structure. These data on two branches of industry are reflected in the lessons that may be drawn from a study of table 20 (section 1.3.1) on 1968 and 1979 for the population as a whole.

There is no doubt that the analysis that can be made of work organisation is more complex. Since they are more flexible than earlier technologies, the new technologies can promote the development of team or group work and allow major decentralisation, including that of management, but at the same time help to transform the role of the hierarchy, and even to

Table 29: France: changes in the structure of skills for two industrial sectors, 1952-73 (percentages)

Occupation category	Oil, liquid fuels		Chemicals, rubber	
	1952	1973	1952	1973
Engineers, managers	7.3	17.7	5.2	8.1
Tradesmen, technicians	9.7	27.8	7.7	19.8
Employees	27.1	24.1	16.4	16.5
Skilled workers	27.5	24.5	20.3	27.6
Unskilled workers	27.8	5.9	50.3	27.7
Total	100.0	100.0	100.0	100.0

Source: Pastre et al.: Automation, travail et emploi, op. cit.

suppress certain intermediate echelons whose main task was more one of worker supervision and the allocation of human resources than of true technical expertise. An analysis of automation in many firms also brings to light phenomena for which it is difficult to determine the extent of the consequences. Expertise from outside the firm is frequently required during both the design and installation stages of new technologies and, very often, even in their maintenance. At the other extreme, firms tend to contract out the most unattractive jobs or regulate them by significant use of temporary labour. The firms own core of workers tend to grow relatively smaller, while the workplace is occupied by workers from different firms with different bilateral agreements. The tasks of workers' representatives and the exercise of trade union rights are made more complicated, and it is sometimes difficult to take into account the extent of problems. Collaboration between employees from different firms but working on the same premises may not be without difficulties and promotion ladders, which are more direct, may contain stages which are more difficult for individuals to pass through.

In many cases, the appearance of new technologies has made existing classifications obsolete without, however, changing them: they may tend to cause splits in the firm's inherent unity, as well as widening the gap between the less skilled and the more skilled, making promotion from one level to another impossible. Unless they are truly negotiated and backed up with suitable training courses, these technological choices may, at this level, have extremely severe consequences on individuals, and these consequences are an integral part of working conditions. There is also the tendency, in certain cases, to substitute male workers for female workers, thus increasing the unemployment distortion for this category of the labour force. Finally, the new technologies may marginalise, or even exclude, older workers at all levels who have not been able to adapt to the rapid changes that have been brought about.

There is no determinism in any of these areas, and a number of experiences provide proof of this. However, extreme vigilance is necessary if we are to avoid the recurrence of what decades of Taylorism have taught us.

1.4 Action by public authorities, and attitudes of employers and employees

1.4.1 Action by public authorities

In the field of computerisation and automation, the action of public authorities may, in an oversimplified way, be seen as the pursuit of two major types of objective. The first was to make public opinion and social forces as a whole aware of the economic, social, societal and political stakes at play in the new technologies, and in computerisation in particular.

First, a number of public reports drafted at government request were widely disseminated. These includes -

- a report published in 1975 on the basis of the work of a "Data processing and liberties" committee;[11]
- a report published in 1979 with the title L'informatisation de la société [The computerisation of society].[12] This was the outcome of a study entrusted to Simon Nora aimed at "promoting thought on the ways of computerising society" and deals in particular with the problems of employment and working conditions;
- the publication in 1980 of the proceedings of a major international symposium, Informatique et société [Computerisation and society], which studied, in particular, work and employment problems;[13] and
- the publication in 1981 of the proceedings of an international seminar organised in June 1980 by the Minister of Economics which dealt with economic choices in the eighties.[14]

Moreover, mention should be made of work carried out at the instigation of the General Planning Commissariat in the preparation of the VIIIth Plan, and, in particular, La sociéte française et la technologie [French society and technology],[15] and L'impact de la microélectronique [The impact of micro-electronics].[16]

The second aim of public authorities in the area of computerisation and automation was to lay down priority objectives and to establish a means of achieving them. Briefly, their policy has been characterised by three priority approaches, all of which are interlinked and aimed at both supply and demand for computerisation: the first aims at expanding the use of computerised and automated systems as a means of improving the performance of France's production capacity; the second concerns with the development of a national computer and automation industry; the third has as its objective the integral consideration of working conditions as early as possible.

In view of the wide range of factors involved and of target situations, the framework of institutions that has so far been set up, but which may be subject to modification, is somewhat complex, since a number of ministerial departments are carrying out a group of activities intended to come together to achieve the authorities' general objectives.

Set up in 1979, the Steering Committee for the Development of Strategic Industries (CODIS), chaired by the Prime Minister and aided by a management committee chaired by the Director General of Industry, on the one hand lays down a number of major priority industrial objectives and, on the other, specifies ways in which help can be given for the achievement of these objectives. CODIS has adopted six major themes, three of which deal with computerisation and automation: office automation; consumer electronics; flexible workshops, robotics and components for automation.

The various ministerial departments each have their own means of action, including in particular the Ministry of Industry, the Ministry of Research and Technology, the Ministry of Posts, Telecommunications and Cable Broadcasting, and the Ministry of Labour.

Finally, since we are dealing with areas which often require experimental action and greater flexibility of intervention, a network of peripheral organisations has been set up.

Attached to the Ministry of Industry are the Directorates of Metallurgical, Mechanical and Electrical Industries and of Electronic and Data Processing Industries, the latter being linked to an Information Technology Commission[17] serviced by a technical computer centre and a Study and Research Centre for Information Systems. The Electronic and Data Processing Industries Directorate is responsible for the Automated Production Development Agency (ADEPA). ADEPA, which is a non-profit-making association initially dealing exclusively with machine tools, now covers the whole of automation. It administers the granting of subsidies to small and medium-sized producer or user industries and carries out assistance and training activities for these firms.

Under the responsibility of the Ministry of Industry, the Information Technology Agency (AD), another public body, has the task of sectoral dissemination of computer applications, regional action, research aid, education and training (equipment purchasing assistance for university technology institutes and the major educational establishments, and pilot studies in educational units.

The Ministry of Research and Technology has at its disposal a range of research resources and means of promoting innovation: the General Delegation for Scientific and Technical Research (DGRST); the National Centre for Scientific and Technical Research (CNRS),[18] a governmental agency which has a large network of specialised laboratories; the National Agency for Research Application (ANVAR); and the National Institute for Information Technology and Automation Research (INRIA).

The ANVAR has the task, on the one hand, of financing research by making innovation grants to firms and, on the other, of offering reimbursable innovation subsidies for pilot projects. In the same way as CNRS or DGRST, it has a wide range of responsibilities, and computerisation and automation form only one of its areas of research and action. Finally, INRIA is a government agency that carries out research, experimentation and information dissemination.

The various forms of assistance to industry for work on robotics granted, in particular, by CODIS or coming from ADEPA, ANVAR or ADI may be supplemented by soft "robotics" loans from specialised banks.

The Ministry of Education, independent of its direct action through its training projects, has just received the conclusions of a commission set up to study a "computer plan" for educational establishments. These conclusions foresee, on the one hand, a year's computer training for 200 teachers in all disciplines and, on the other, the prolongation and expansion of an experiment in which 58 secondary schools were equipped with computers. Finally, computer studies are to be offered as an optional subject experimentally in ten secondary education colleges. Moreover, the Study and Research Centre on Employment and Trade Skills (CEREQ), which comes under the responsibility of the Ministry of Education, is carrying out studies dealing in particular with the effect of computerisation and automation on jobs and trade skills.

The Ministry of Labour, and the Employmetn Study Centre which is attached to it, are carrying out research concerned in particular with the effects of computerisation and automation on work and employment. Through the Working Conditions Improvement Fund (FACT) and the Training Assistance Fund (FAAFE), the Ministry of Labour grants firms subsidies for research, investment and training courses which, from the point of view of working conditions, are innovative, exemplary or could be given wider application; certain of these are related to automation and computerisation.

ANACT, a public institution with a tripartite board of directors which includes representatives of employers' organisations, trade unions and various ministerial departments and experts, is under the supervision of the Ministry of Labour. In addition to the advice provided by its officials concerning all applications submitted to FACT and FAAFE, ANACT has as its mission: to collect and analyse information concerning all activities for the improvement of working conditions; to contribute to the development and promotion of research, experimental work or concrete projects concerning working conditions; to encourage manufacturers to design ergonomic machines and industrial buildings; to support training activities; and to develop methodologies for action. Within this framework, a number of projects deal with computerisation and automation, and these are among the priority themes of the current programme.

Together with the Ministry of Industry's Information Technology Commission and the Ministry of Labour, ANACT has been involved in the preparation of an "Automation and Working Conditions Plan" (PICT), as one of the principal instigators of this plan.

1.4.2 Attitudes of employers' organisations and company directors

Employers' organisations and company directors, with their common preoccupations and their differences, do not seem to have made any statements of principle about technical progress and automation; however, this is in line with their normal practice in matters of this sort.

Their attitudes are expressed more in relation to the conditions of operation, survival and development of industry and national production capacity, especially in the face of competition, and usually in actions rather than words. The aim is to exploit technical progress to improve company performance. Some feel that the penetration of new technologies meets with obstacles that they consider more psychological or social than technical.[19] However, certain company directors think it is necessary to find a compromise when introducing computerisation so as to exploit its effects with the aim of lightening and enriching jobs, facilitating the elimination of arduous jobs and promoting choices about work organisation which may or may not affect workers' skills.[20] Certain employers' organisations consider that works councils are the appropriate places for discussions on major projects.[21]

As far as small and medium-sized firms are concerned, a recent survey shows that the heads of such enterprises are far less reluctant than before about the possibility of installing automatic equipment and robots in their companies, and expect that it would lead to an improvement in worker skills and not a fall in employment.[22]

1.4.3 Attitudes of trade unions

The attitudes of the various trade union organisations to automation and to new technologies and technical progress in general show both similarities and differences.

In line with its long-term approach to technical progress, the General Confederation of Labour (CGT) asserts its "certitude that there is no fatality ... and it is possible to proceed in such a way that all applications of data processing will produce results which are beneficial to the workers and society".[23] Realising the problems involved in introducing computers into all areas of employment and the consequences of rapid computerisation, the CGT has formulated the following objectives for automation:

- the improvement of working conditions and QWL, in particular by the elimination of repetitive tasks, diminution of the workload and a reduction in working hours;

- the creation of jobs and the improvement of work skills, the utilisation of knowledge already acquired, expansion of opportunities for displaying initiative, the upgrading of training facilities and the reorganisation of work on the basis of co-operation between various categories of workers;

- the granting of new rights to information and self-expression to workers and their representatives, and the creation of workshop or office councils;

- the strengthening of rights and means of action available to works councils (right of appeal, access to all information, possibility of appeal to experts, right of inspection) and to the Safety and Health Committee (CHS), Improvement of Working Conditions Committee (CACT), occupational health services, etc.;

- the creation of a national computer and automation industry.

The attitude of the French Democratic Confederation of Labour (CFDT) to technical progress is a more questioning one, particularly as expressed in the organisation's publication entitled Les dégats du progrès [The price of progress].[24] It now views social change and technical innovation as inextricably intertwined and tends to challenge the idea that technical progress may be distinguished from its uses and be divorced from social change: according to the CFDT's Secretary-General, "technology in itself is not neutral".

While emphasising the risks and disadvantages of computerisation for workers (physical working conditions, effects on job content, skills and the very meaning of jobs skills as such, and the impact on work organisation, the reduction of scope for initiative and on mental workload), the CFDT does however consider "that it can nevertheless be used as a tool for progress".[25]

In Les dégats du progrês, the CFDT looks at the specific characteristics of the new generation of technology; it seems, in some manner, to have moved away from determinism and concede a certain degree of choice regarding the design and implementation of systems which may mould not only working conditions but, more widely, "categories, images, logics" and even management of the organisation of social relationships, i.e. power.

In September 1979, the Secretary-General of the CFDT, faced with the growth in computerisation and automation, put forward a number of proposals as to the role of works councils -

- consultation on the consequences of introducing new technologies and consultation with works councils and workers at the design stage of systems (these consultations should deal not only with the material aspects but also with systems of organisation and authority);

- for every investment in computerisation, management should submit to the works council a forecast on the economic and financial repercussions, the level and quality of employment and any modifications to the system of decision making;

- introduction of small-scale pilot projects for each new computerised system;

- arrangements for the works council to appeal to outside experts;

- allocation of time to workers and their delegates to analyse the files; and

- retrospective studies on computerised systems already installed.

Force ouvrière (FO) emphasised in a White Paper[26] that the computerisation of our society would continue and that the employment situation would deteriorate; it gave a vigorous reminder that employment is the priority factor. As far as equipment design was concerned, emphasis was placed on the need to incorporate ergonomic features into systems at the planning and design stage to ensure maximum protection of workers and users. A call was also made for the establishment of a body in which all questions of national significance concerning data processing and industrial automation would be the subject of consultation between the economic and social partners in question.

In line with its attachment to the system of collective bargaining, FO states that collective agreements should incorporate new forms of protection aimed at modifying social security provisions, especially in relation to job transfers and changes in job skills; this would entail the establishment of a policy on initial and continuing training that was really matched to the needs of those concerned.

A report presented by the French Confederation of Christian Workers (CFTC) at its 40th Congress in 1979 emphasised the importance of data processing and automation which are "currently and beyond any discussion one of the driving forces in technical progress". The CFTC believes that technical progress is a determining factor in economic and social change and points to the high priority it attaches to respect of the individual confronted with the changing situation. According to the CFTC, it has too often been forgotten that data processing is not an end in itself but merely a technique for the organisation and management of firms for administrations.

At times, there has been the impression that it was being used for the satisfaction of technical achievement of the pleasure of employing the latest fashionable gadget, and consequently its arrival on the scene has evoked hostility and even rejection; this has not had a favourable effect on the smooth running of companies or administrations, nor has it improved the employees' working conditions.

The CFTC reiterates its position in favour of true consultation which takes into account all questions raised by the impact of data processing on working conditions and methods in offices and factories.

In a recent White Paper, La Novotique,[27] the French Confederation of Executive Staffs (CFE, formerly CGC) examined the impact of new technologies on industry, and emphasised in particular the need for France to build up an independent data processing industry and to master the manufacture of electronic components.

Pointing out the contradictory effects to be expected in the field of employment, the final outcome of which the CFE does not think can be foreseen, the White Paper studies the main impacts of data processing on job development from the point of view of job skills, working conditions and work organisation. For these new technologies to be successful, the CFE considers it necessary to -

- demystify the subject among executives so that they are not overwhelmed by it;
- develop worker involvement;
- ensure that the introduction of new technologies does not result in polarisation of jobs but makes it possible to promote redistribution, decentralisation and new allocation of authority and opportunities for personal development;
- develop the training and upgrading of management;
- improve working conditions by taking these into account systematically in the design of new equipment (elimination of danger zones, improvement of safety and health by the use of automation, true reduction in the physical arduousness of jobs, etc.); and
- avoid and reduce the harmful consequences of new technologies on working conditions (noise from certain automatic machines, difficulty in deciphering display characters, nervous tension, monotony and the tedious nature of certain jobs, depersonalisation of work, reduction initiative and responsibility, downgrading of jobs, coercive supervision, etc.).

1.4.4 Review of the system of employer-employee relations

The French system of employer-employee relations has a formal and institutional framework and also a practical component which interprets and even transforms this framework, due in particular to the organisation and power of the employers' and employees' organisations and the privileged position given to agreement and state intervention.

Employers' organisations are grouped together in the National Council of French Employers (CNPF), the structures of which were reformed in 1969. However, this situation should not mask the extreme heterogeneity of the employers. Small companies are, for example, grouped together in the General Confederation of Small and Medium Enterprises (CGPME), which often keeps its distance from the social policies implemented by the CNPF in 1969, the large companies formed the Association of Large Enterprises making a call on small savers (AGREF). Finally, various associations such as the Centre for Young Company Directors, a power group within the CNPF, or the GEFACT, are attempting to promote more open social policies in the companies which form its membership.

The transformation of the CNPF in 1969 was aimed at providing the organisation with the means to carry out a more active policy, in particular by negotiating at three levels: between the various trades and professions, by industrial sector and by company.

Employees' organisations (trade unions) are notable for their small memberships and for the divisions wrought by their ideological differences. There is every reason to believe that the level of trade unionisation is between 20 per cent and 25 per cent, and this figure must fall far below 20 per cent if one excludes the public sector, where the level of unionisation is much higher. Only since 1968 have unions been permitted as such inside individual companies. This low level of membership, which also has repercussions on the state of the unions' finances, is further accentuated by the divisions in French trade unionism.

The Boards of Conciliation elections, which concern only private sector workers were held in December 1979. When abstentions (36.7 per cent) are taken into account, the voting gives an approximate indication of the size of the various trade unions as follows:

CGT	:	42.4
CFDT	:	23.1
FO	:	17.4
CFTC	:	6.9
CGC (now CFE)	:	5.1
Independents	:	4.6

The various trade unions (and in particular the five main ones recognised as being representative) have seats on numerous public bodies both at the national, regional and local level.

Within individual companies, employees' representatives are either elected (personnel delegates, since 1936, in firms with more than ten employees; members of the works councils, since 1945, in firms with more than 50 employees) or stem from the trade union branches (trade union delegates, since 1968, in firms with more than 50 employees). Very schematically, the division of competence is as follows: personnel delegates deal with individual complaints, trade union branches handle negotiations and works councils are responsible for consultation on a wide range of subjects.[28]

As far as the public sector is concerned (public administration and public companies), there is a special system of representative institutions, and the level of trade unionisation is relatively high.

Basically, collective agreements relate to industrial sectors, or even cover a number of occupational categories. However, it was not until the Act of June 1971 (amending the 1950 Act on collective agreements) that company agreements came under the statute of collective agreements.

The collective agreement in France is, by its very nature, halfway between a regulation and a contract. The content is stipulated by the parties, but the convention is valid as soon as it is signed by a single representative organisation and the text becomes compulsory even for those who have not signed. Moreover, and without exception, the duration is indeterminate and denunciation by a single party does not modify its validity.

Moreover, in general, collective agreements deal in most cases with minimum levels of protection designed mainly for the employees in the most marginal companies, and consequently scarcely mobilise employees as a whole. On the other hand, due to their continuity and the coverage that they give to employees, collective agreements have numerous advantages from the point of view of rights and protection.[29]

Within this highly regulated field of collective agreements is embodied another characteristic of the French system of employer/employee relations, i.e. the role of the State.

First, the State may extend the provisions of collective agreements to workers or firms belonging to the sector covered by the agreement but who, since they were not involved in the negotiations, were excluded. Furthermore, the provision of a collective agreement may be extended by the State to another geographical or occupational sector with similar economic conditions.

Wider still, there exists in France a link between the law and the collective agreement. In certain cases, legislation with social repercussions is in practice negotiated before being submitted to Parliament. In others, the agreement is one of the conditions in the implementation of the legislation and may even in some cases lead to its amendment. In the same way, the agreement may be the starting point for legislation whereas, in other cases, it is under the threat of legislative intervention that certain agreements are drawn up.

Finally, it should not be forgotten that in the operation of the system of employer/employee relations, the State can, as an employer, play a leading role.

2. Case-study - Automation and working conditions in the production units of a cement manufacturing group

2.1 Introduction

The analysis of a changing situation always raises theoretical and methodological questions which are very difficult to answer. "Pin-pointing" the significant event, on the basis of which one can attempt to structure an analysis in "before and after" terms, is particularly delicate when dealing with technological change. Discoveries and progress come together and combine to form equilibria which are in constant evolution, and the way in which they interact with the status of the workforce is often understood only later, whereas other changes occur which help in understanding the past but leave the future uncertain. In the case of automation and work organisation, these factors have led us to adopt here the observation and analysis of the development of changes that occurred in the establishments of a given group, rather than the specific observation of a given factory.

The Group studied forms part of the cement industry. The choice of this Group is partly pragmatic and results from the quality of the information and observations available on it. However, the choice of this Group and this industry brings us back to other factors related to the subject itself: the radical technological change that has occurred in the cement industry over a period of 20 years, and in particular the last ten years; the co-existence, within a single group, of factories at different stages of automation, of technologies at different stages of sophistication; and a spontaneity in social policy which has helped to earn this company the title of the "French employers' social laboratory".

2.2 The main characteristics of the cement industry

A very brief review will be given of the main characteristics of the industry; these must be constantly be borne in mind since they have had some influence on the changes that the industry has undergone.[30]

The cement industry is primarily a totally integrated industry and carries out a complete production cycle from raw materials to the finished product. The industry is required to process vast quantities of raw materials with a low initial cost but, in order to produce a high-grade product, requires high-cost industrial equipment. Moreover, not only is the raw material of low initial cost but the finished product is also of low commercial value (cement accounts for only about 2 per cent of the total construction cost of a dwelling); and high capital investment (the cost of an entirely new factory is around three times the amount of its annual turnover) explains in part the major mergers and rationalisations that have taken place in this industry.

The cement industry went through a period of major growth after the Second World War, when the reconstruction effort resulted in a considerable demand for cement. At this time, cement works were very numerous and, as a rule, were at a low level of mechanisation because no major investment had been made since the years 1925-30. The factories had small outputs and were very labour intensive since virtually all the quarrying, crushing and handling operations were manual. Working conditions were particularly hard. Workers were constantly in contact with the product in the quarry and the stone was broken with a 7 kg sledge-hammer. During manufacturing the kiln had to be charged and discharged, and it was often necessary to resort to sledge-hammers and shovels. During dispatch, the hopper and the jute sacks gave off clouds of dust. The sacks of cement were loaded on to lorries, railway wagons and barges, either by trolley or by hand over relatively long distances. With exposure to bad weather, dust, gases, heat and noise, the working conditions in a cement works were amongst the worst that existed at that time.

Increasing demand in the cement industry, as in many others, resulted in a process of rationalisation of production which gave rise to a major effort of mechanisation, then automation, a significant reduction in the number of employees and an economic and financial rationalisation without which it would have been impossible to meet the major investments considered necessary. Currently, there are 15 companies which are responsible for the totality of French production: four of them alone account for 84 per cent of production, with the other 11 producing the remaining 16 per cent. Cement production is widely distributed throughout France, and this decentralisation is explained by the need to locate the factory as near to the quarrries as possible and also by a desire to be close to consumption centres to avoid the high cost of transporting a heavy product.[31]

Figure 14 consists of an operational diagram of a cement works, which is a relatively simple one and should permit a better understanding of the various stages in the manufacture of cement. It should also help us to distinguish the various difficulties that must be overcome in automation.

Cement is produced by grinding the basic product, clinker, with a certain number of additives. A clinker production line comprises the following four major units:[32]

- the shop or shops in which the raw materials are prepared. This starts with natural quarry extraction, followed by the grinding of the raw materials, and finishes with the reconstitution of an artificial quarry in which the products are blended and proportioned. The basic operation in this shop is sampling. The final mix is around 80 per cent limestone and 20 per cent argillaceous materials;

- the raw-materials finishing shop. In the dry process, which is currently used in virtually all plants, this shop dries, grinds and stores the powder in large bunkers where it undergoes constant blending. A slight correction in the limestone content may be made during the grinding process;

- the kilning shop. The thermal part of the process entails the ground clinker passing through a series of counterflow kilns in which it is heated to 850°C, decarbonised and then calcined at 1,450^{o}C, which gives rise to the reactions that produce the clinker. The kiln is fired by

Figure 14: Operational diagram of a cement works

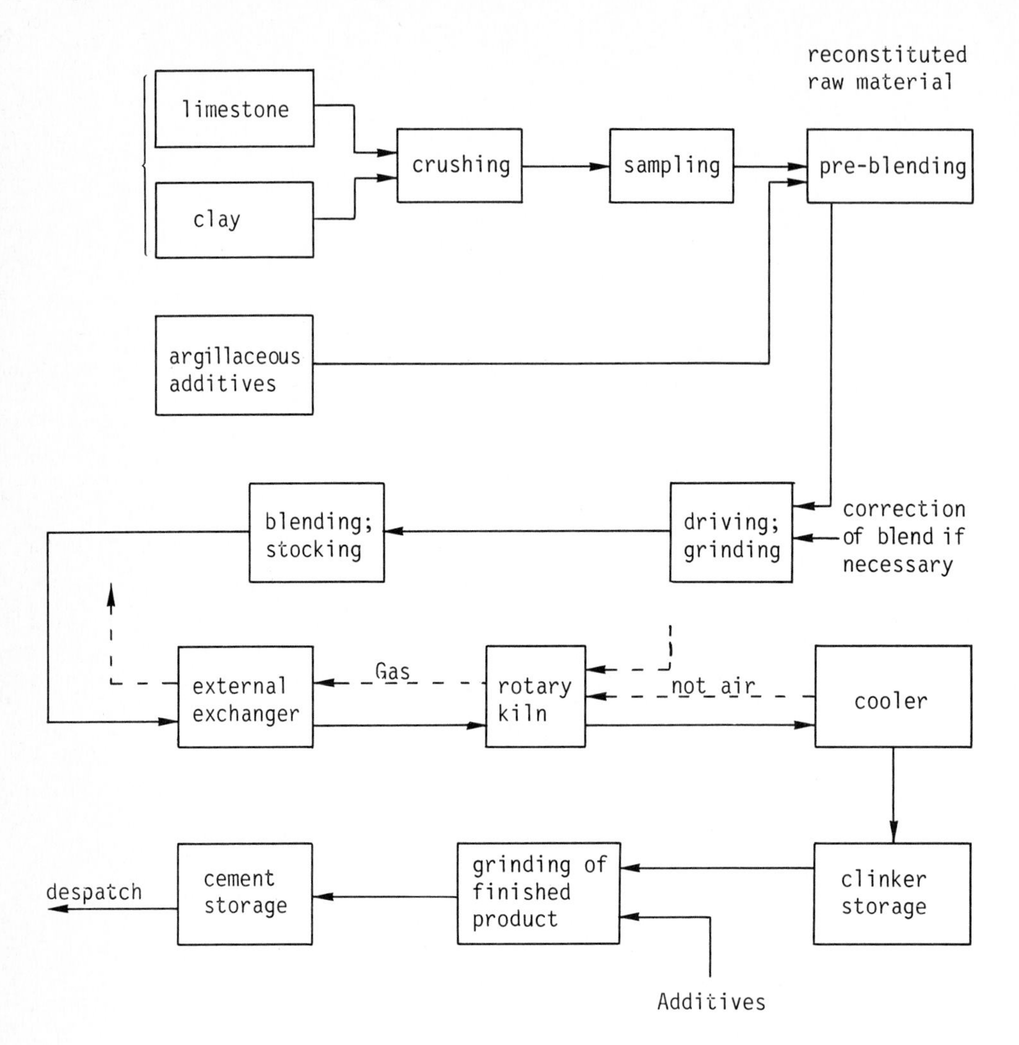

Source: Lafarge Conseils et Etudes: Une philosophie nouvelle en contrôle commande de processus (unpublished internal document).

heating oil, natural gas or crushed coal for which the combustion air has been preheated by the hot clinker. This process can be carried out by a variety of systems engineered in different ways;

- finished product manufacturing shop. Here the clinker is mixed with the required additives and ground to the correct particle size. It is not only the chemical structure of the clinker, but also the fine particle size which gives cement the required strength characteristics after a more or less rapid setting time.

Cement manufacture is as much a handling as a thermal industry. Its basic characteristics, when looked at by the automation specialist, are the long transit times and the coexistence of a fluid and a solid with fundamentally different thermal properties. In addition, it is an industry with a continuous, integral firing system.

Cement manufacturing techniques have developed over the years. In extraction and crushing, there has been an increase in the use of explosives and more and more powerful quarrying and crushing machines. Pre-blending has also been mechanised significantly. Research specific to the cement industry has dealt mainly with kilning techniques. Briefly, the kilning process has moved from wet process, which was previously the main one, to dry process. Research has aimed primarily at reducing energy consumption, since energy accounts for one-third of cement production costs. A sector contract has therefore been drawn up between the Ministry of Industry and Research and the National Federation of Cement and Lime Manufacturers with the aim of reducing consumption from 1,118 therms per ton of clinker in 1973 to 980 therms per ton.

Finally, bagging and dispatching have also undergone major changes. Automatic bulk delivery, bagging and palletisation have also helped to bring about significant changes in the situation in the factory yard and, in particular, gradually to eliminate piece-work specific to this sector.

2.3 The stages of automation[33]

Currently, ten cement works out of 22 in the Group studied in this report have a centralised production system.[34] Not all of the ten automated plants have achieved the same level of automation; however, the coexistence of newer and older forms of automation permits a particularly interesting analysis of the various stages of the automation process and the constraints they place on the workers or the opportunities they open to them. Before describing the various stages in automation, consideration should be taken of the situation described in section 2.2 which, in part, explains the various approaches taken. First, the need significantly to increase production and the predominantly low level of productivity led to to an initial phase of mechanisation aimed primarily at labour-saving and productivity improvements. When imbalances appeared in the cost of raw materials and, particularly, the cost of energy, the task was no longer merely one of reducing labour costs, since it was felt that the number of workers had already been reduced to a minimum level by optimising their operations and by producing a more consistent product. Table 30, which

Table 30: Cement works: trends in number of workers employed at different stages of technological development

Stage of technological development	Quarry	Factory	Production
	No. of workers	No. of workers	Tons/day
Before mechanisation	200-3 000	300	200-500
After mechanisation (1930-50)	20	200[1]	400-1 500
With central control room (1980)	3	80-120[2]	1 500-4 500

[1] 5-10 workers per work station.

[2] 4 workers per work station.

Source: Luttrin: "Historique de l'automatisation en cimenterie", in CRG and AFCET: Aspects humains de l'automatisation: Les cas de l'industrie cimentière (Paris, 1978), p. 13.

shows the trend in the numbers of workers employed in a production unit at different phases in technological development, will allow a better picture of the situation.

If the processes are taken in chronological order, it will be seen clearly that the quarry is where the reduction in the workforce has been the most impressive and that this reduction is, beyond any question, due initially to a very high level of mechanisation which has progressively been supplemented by the automation of certain work stations. The most recent crushers do not require continuous supervision.[35] The conveyor belts for moving the raw materials now have linked controls and therefore require far less stringet supervision. The jobs which the quarry workers are still required to carry out are mainly the driving of quarry machines, which is still often very arduous, and blasting. Since the team is small in size and the work arduous, each worker takes on a variety of jobs, a point which will be dealt with further below.

However, it is mainly in the factory itself that automation has had its greatest impact. In a conventional cement works, although the term "continuous process" refers to a continuous flow of materials, the manufacturing work is in effect a juxtapostion of worker-machine systems, each independent of the other and each geographically forming a production shop and carrying out one phase in the production process. Cohen-Hadria explains this description diagrammatically (figure 15).

Figure 15. Continuous process in a conventional cement works

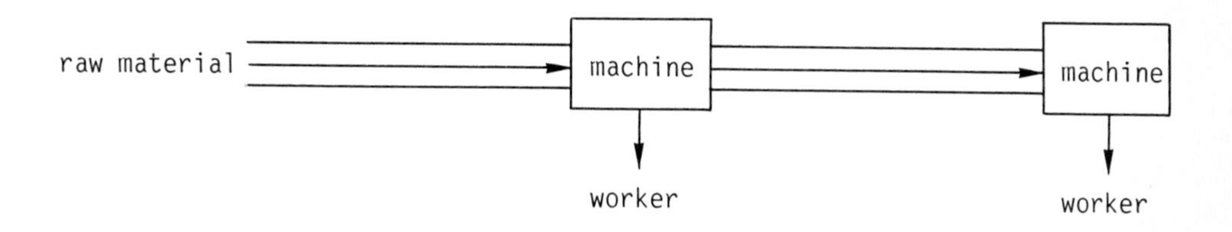

Source: Yves Cohen-Hadria: Automatisation et organisation du travail: L'exemple des cimenteries (Paris, Ecole Polytechnique, Centre de Recherche en Gestion (CRG, 1979).

It was customary to call the workers in a conventional factory "tradesmen" where the trades were named after the machines on which the men worked or after certain processes: e.g. crushers, bridgers, kilners (the latter being at the top of this professional hierarchy, which is widely encountered). The trade was learnt on the job, since it was made up more of empirical than of theoretical knowledge. Apprenticeship was a long process and was based on word-of-mouth transmission of knowledge. The division of the jobs in the production team reflected the technical division of the process into basic phases of raw material processing, and the hierarchy which was built up from the different jobs did not necessarily have any link with the complexity of the job but rather with the different production phases. The shift foreman, a senior worker who moved from job to job, was virtually the only one to have an overall picture of the manufacturing process.

Starting in the fifties, new equipment began to appear which deprived workers of a part of their direct intervention on the materials being processed (sensors which were able to make simple measurements, automatic scales which controlled materials flow) and moved them away from the machines by adding electrical control equipment.

This first phase of automation was accompanied by the suppression of certain jobs. Returning once again to Cohen-Hadria's diagram, for the remaining jobs, the worker-machine combination was moving towards a worker-control panel-machine grouping in line with the appearance of new sensors which were able to make more intricate measurements[36] (figure 16). It became evident that this phase of automation would be accompanied by a major change in maintenance activities. Whereas in the preceding sytem each worker was responsible for the maintenance of his machine, the situation would no longer be the same with the new installations which were to lead to automation of maintenance operations.

Figure 16. Cement works: first phase of automation - worker/control system/machine interface

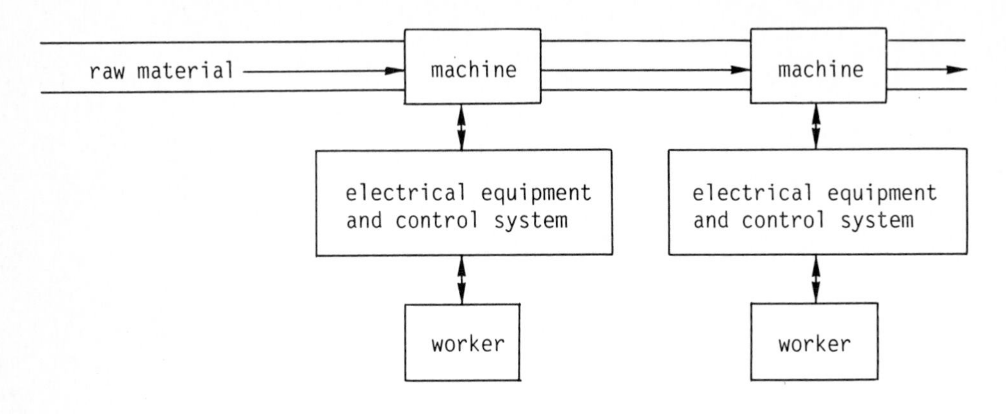

Source: Cohen-Hadria, op. cit.

The second phase of automation was to appear at the end of the sixties, with the progress made in remote control, measurement and data processing techniques. This did not mean total automation since human intervention was still a decisive factor in the control of hazardous situations, especially during kilning. However, this second phase placed a question mark over the technical separation based on different phases of materials processing. Cohen-Hadria emphasises that there remains only a single worker multipurpose machine interface instead of several; each operator is, in fact, in charge of the whole process allocated to him (figure 17).[37] This centralisation tended to eliminate the work of the shift foreman, individualised jobs and traditional know-how in favour of more abstract knowledge. This is an overall change in work organisation to which we will return later.

The phases defined by Cohen-Hadria are termed "the first type of automation", by Lebas,[38] and it is this which is currently dominant in automated plants. According to Lebas, this type of automation has basically a "labour-saving" function.

Currently, cement works are undergoing a new form of automation, although it is still too early to analyse the overall consequences; this new type of automation differs from the earlier one in that it is aimed at process "optimisation". A very clear description can be found in a paper by Yves Cottet on the modern concept of numerical industrial data processing.[39] Lebas notes in his research that the old type of automation is still subject to the "rationale of mechanisation"; it replaces human control by automatic control,[40] and its main functions is labour saving.

Figure 17: Cement works: second phase of automation - team/system interface

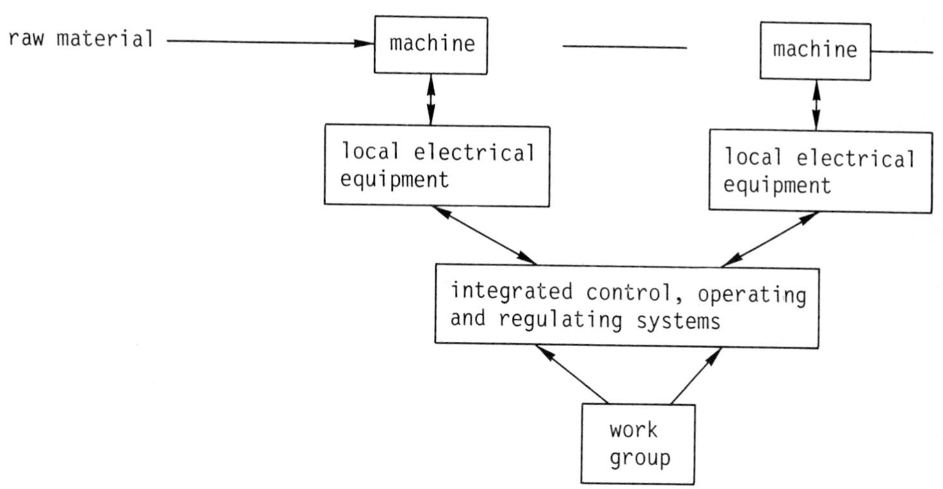

Source: Cohen-Hadria, op. cit.

The new type, on the other hand, is subject to a new model of the "worker process" relationship. Since the worker is no longer involved in either the monitoring or the control of the process, the computer carries out these operations and is programmed for "optimisation". This new form of automation will not be without repercussions on the operator's work. Cottet emphasises that we are faced with something new which will disrupt existing habits and change behaviour. Moreover, he argues, the very structure of factory management will obviously have to be revised. Operating control and process will doubtlessly form two independent groups which are able to deal with their own breakdowns, thus doing away with the real barriers between production and maintenance, and mechanisation and electrical installation. However, he goes on, it will also be necessary to be wary of the system's excessive power and not leave it to just anyone to do just any of the tasks. Access to the system will have to be strictly supervised, but on the other hand, everyone must be capable of reading the data on the screens.[41] It is possible to see here the implications on the work organisation of this last phase of "new automation".

Mechanisation and automation have transformed the appearance of the quarry and the factory, as well as life in the factory yard: an increase in bulk cement deliveries, automatic palletisation which eliminates loading jobs, and automatic bagging - all lead to a growth in supervisory activities with a possibility of job rotation; the cleaning work is usually subcontracted.

2.4 Changes in personnel management policy[42]

Technological change has also contributed to a permanent change in work organisation and the working conditions of cement manufacturing workers. However, these changes have taken place in a relatively specific context since, together with dockers, printing workers and certain public enterprise workers, cement workers are amongst those with the highest rate of trade unionisation. Various trade union sources indicate that nearly 9,000 of the 13,000 workers in the industry are union members. The CGT, with around 65 per cent of the union membership, is by far the leader, followed by the CFDT (around 20 per cent), the FO, the CFTC and the CGC (now CFE) which account for the remaining union members. The reasons for this high level of trade union membership are, without any doubt, to be found in the particularly arduous working conditions which were a feature of the industry in the fifties. This high level of trade unionisation and the desire expressed by the senior management of the Group under study have, beyond all question, allowed a level of dialogue and negotiation which is relatively unusual within the French system of employer-employee relations.

The first phase of automation was characterised, as has already been pointed out, by a major change in mechanical maintenance activities, with maintenance activities being directly eliminated by technological change. Milling machine and lathe workers, and boilermakers, who are all very specialised, appeared either less useful or too rigid in their specialisation. More and more frequently, the maintenance of the most sophisticated machines was being subcontracted.

Consequently, as from 1960, the emphasis was placed on flexibility and a wide range of skills among maintenance workers. This change of direction led, in 1963, to the signature of a "Charter for the promotion of maintenance workers" which was extended to cover the whole occupational category in June 1969 by the adoption of a collective agreement. Thus, as from 1963, in the Group under study, a method was introduced by employers and employees in each factory for the grading of jobs in relation to each other on the basis of their assessment on a number of criteria: physical workload; intellectual and vocational training requirements; responsibilities; and working conditions. The concept of multi-skilling was also recognised for workers who were able to carry out other jobs of the same level or a higher level. Skill was no longer linked to a job but to the individual, and all maintenance workers, provided that they were willing to take training courses and examinations, could rise through all the grades leading to the job of a highly skilled worker. Provision was made for a less strict promotion scheme for certain categories of age and seniority. As from 1964, the company set up a large training centre in which all the maintenance workers underwent training courses.

As has been pointed out by Mireille Dadoy, this ability to do different jobs had a double value: it motivated workers to acquire new knowledge and created the conditions for real promotion. Moreover, it allowed management flexibility which was absolutely essential with the reduction in the number of workers. However, Dadoy adds, this type of work flexibility has had ambiguous results: on the one hand, it has resulted, to a certain degree, in lowering the qualifications of maintenance personnel since the new knowledge acquired did not have a sufficiently sound basis to be used with skill, and the older knowledge was gradually being lost due to inadequate application; on the other hand, this flexibility created a dynamic approach which is in line with new types of worker interchangeability.[43]

These new forms of interchangeability appeared with the second phase of the first type of automation and the installation of the centralised control room. This affected not only maintenance workers but also aimed at creating a new category of employee: the operator-maintenance worker. In 1972, the company negotiated two specific agreements with the unions: one concerned the centralised production workers, the other the centralised production foremen.[44] These two agreements were intended to define the new functions and channels of promotion, based on principles similar to those laid down in the first promotion charter signed earlier in the company.

The introduction of the principle of multi-skilling was based on a specific form of work organisation with the rotation of seven work teams. The centralised production employees worked alternately on manufacturing and maintenance, based on the example of the rotation of seven work teams over a seven-week period:

21 days on 3 x 8-hour shifts;

14 days on maintenance (day work);

14 days rest.

Each work team is made up of a centralised production foreman and two centralised production workers. In practice, this model presented certain difficulties in distributing the jobs between the foremen and the workers, as has been emphasised by Cohen-Hadria in confirmation of the observations of other investigators to whom reference has been made: the role of the centralised production foreman became a relatively fragile one in relation to that of the centralised production workers in his team.[45]

The technical and organisation changes deprived from the foreman of an old privilege: that of being the only person with an overall picture of the sector for which he is responsible. Data centralisation also gave the process operator in the control room the ability to obtain an overall picture. Consequently, in practice and to maintain the hierarchy between foreman and worker, the foreman was given the job of control panel operator whereas the workers did the rounds. Nevertheless, the fact that such a problem could be raised showed clearly that the gap between the two jobs was being closed. In one plant, the foreman's job was even eliminated.

The non-automated factories, however, maintained a much more traditional organisation with a rotation system between five or six teams.

Over and above the questions raised by this "new work organisation", the move to seven teams made it possible for each production worker to work no more than 52 night shifts instead of 73 in a five-team rotation. A plant agreement was signed in November 1980 in which the main provision was the introduction of an Individual Point-Acquisition Plan. Each shift-worker is therefore able to build up, over his shift career, a "points capital" which, at the age of 55, allows him either to take an available day work job, or to take additional annual holidays.

Finally, the Group is pursuing a very active safety policy.

2.5 Factors in interpretation

The interpretation of the technical and organisational changes which occurred in the cement manufacturing company under study is complex and many questions about future developments remain unanswered. Interpretations in the literature are of varying degrees of pessimism depending on the investigators in question. However, all emphasise the same questions although the answers they give are different.

The first question, at the heart of all major technological changes, deals with skills - undoubtedly an ambiguous concept. Flexible manning - especially among maintenance workers - is at the basis of an extremely difficult debate which can only be outlined here.

All the authors emphasise that the flexible manning system introduced in automated factories is aimed at providing the firm with the flexibility required, as Davis has noted, by the existence of unforeseen events that technology does not yet control. Coriat writes that the object of setting up production teams by shop is to achieve characteristics of plurality, adaptability and interchangeability. However, this author emphasises that flexible manning finished by breaking down within teh factory the virtual monopoly of technical know-how exercised by workers in the specialised maintenance teams. What is being attacked here is the whole resistance and braking ability of each of these categories.[46] Dadoy has also pointed out, as mentioned above, that this flexible manning by maintenance workers is, to a certain extent, carried out "downwards". In all the original skills, each worker knows less. Nevertheless, it may be emphasised that, although it is true that maintenance workers lose some of their traditional ability to resist change, based on acquired monopolies, there is no doubt that the same workers find other means of action based on the new jobs that they hold; and as has been seen, these are as strategic as the earlier ones.

Over and above this debate on flexible manning, there is another acute question related to worker know-how and to what Lebas calls "the means of transmitting technical memory".[47] Production worker know-how in traditional cement works was essentially a heuristic knowledge acquired by direct sensory contact with the raw material and even, in the case of kilners, by an in-depth knowledge of the machine-material interface, based on experimentation. This know-how was transmitted from one worker to another by a long period of apprenticeship which is also related to a whole major system of social mythology in which the old know more than the young and the worker more than the engineer, and which makes it possible to build up a specific social equilibrium based on autonomy and interdependence.

Automation breaks down this system and - with the second phase of automation - breaks it down definitively. Workers and foremen are not only further and further removed from the raw material, but their know-how consists more and more of an algorithmic knowledge which locates them in the same knowledge-acquisition stream as the engineer. There are some who deplore this irreparable loss of heuristic knowledge which, moreover, has not been without problems in the cement manufacturing industry. Beyond the debate on worker know-how, this change in know-how has modified the social equilibrium built up around the old production system.

The reconstruction of a single line of knowledge from the manual worker to the engineer makes it possible to rebuild a hierarchy around the same reference axis, which poses a problem for certain intermediate grades, in particular those in first-line shop-floor management. As has already been seen, one of the plants in the Group has even done away with the foreman's job. On the other hand, as Dadoy and Cohen-Hadria have pointed out even more forcibly, the restructuring of these commonplace jobs makes possible the development of a new function requiring an overall knowledge of the process, the raw materials and the equipment; a knowledge which can be progressively enriched and re-acquired by the operator. In this way, specific organisational practices permit a situation in which automation is not linked with a brutal reduction in job skills.

A point made by Cohen-Hadria, which is very important in obtaining a better understanding of the disruption caused by automation in the Group's plants, is that there has been a radical change in the rules of the game: the average level of qualification for newly hired workers is rising (technical baccalaureate, technical university diploma), to such a degree that there is a talk of a changeover from a professional qualification originating in the plant (semi-skilled trades, know-how) to a qualification of external origin (school and university training), and even a transfer of training responsibilities from the firm to external establishments. This also signifies that, within the plant itself, age is not only no longer a criterion around which the hierarchy is built up, but may be a factor of conflict or risk of older workers being set aside.

There are also other questions which appear around this new automation-dependent organisation of work. These relate, first, to the exclusion of certain categories of cement workers from the new promotion streams. This applies, in particular, to workers doing jobs upstream and downstream from the actual production process, i.e. quarrying and dispatching. Some factories have attempted to establish a sector system which allows a high degree of independence and flexible manning in the quarry and dispatch teams. Nevertheless, these workers are still, to a large extent, excluded from the promotion system as, in fact, they have always been.

These quarry and dispatch workers are the same ones who, still today, have the hardest physical working conditions, even though mechanisation and automation have made it possible to eliminate the most arduous jobs. In particular, the quarry machine-driver's work station still requires the same type of improvement that has been made to the working environment in dispatch jobs.

There is no questioning the fact that, within the factory, automation has removed workers from the jobs with the highest exposure to harmful agents. On the other hand, the trade unions emphasise the isolation of workers doing the rounds, the large distances that have to be covered and the vigilance tasks that have replaced the traditional physical arduousness.

Another basic point is what some authors call the segmentation, and others the stabilisation-destabilisation of the workforce. Looking only at the number of workers in the Group, automation seems to have had a very negative effect on employment. However, it seems that automation has been accompanied by an important growth in subcontracting, both for the most highly skilled maintenance tasks on the most sophisticated machines and also for the least skilled jobs of cleaning and maintaining the premises. Flexible manning has been instituted to the detriment of specialisation and expertise, with the latter being subcontracted. It is difficult to assess the extent of subcontracting in the Group, and this makes it impossible to carry out any analysis of the overall change in employment levels due to automation.

The use of temporary workers and subcontracting presents the trade unions and the whole of the employer-employee relations system with delicate problems (in the case of another branch of industry, petrochemicals, Coriat points out that three workers out of four in this sector are not covered by the petroleum industry bilateral agreement). In firmly stabilising the hard core of maintenance-production workers, the cement firms have freed themselves from the constraints of these agreements by making use of external firms that use contract labour. This disruption of the labour force caused by automation is not specific to the cement industry, but it will certainly be one of the major problems of years to come.

The social partners and the majority of investigators, in spite of the major reservations that have been expressed, have a positive view of the technological and organisational change that has taken place over the last 20 years in the Group under Study. A remarkable increase in wage levels and a reduction in occupational accidents has occured, attributable both to automation and to the development of an active safety policy. Moreover, as pointed out by Dadoy, these changes in the occupational structure have not resulted in a massive reduction in worker skills, and in spite of the older age group of the workers, they have occurred without too much in the way of problems or reluctance. The experiences gained from work organisation, which vary depending on the establishment, have not remained isolated; as is normally the case there is no doubt that the implementation of such a policy in this way has been made possible by the existence of a vital and active employer-employee relations system.

A final point that deserves mention but which, although emphasised by the trade unions, is still not supported by any specific study, is the increase in mental workload. However, what is pointed out in a comparative study on two cement works in the Group[48] is that there are a number of hazardous situations in the official operation of the installation which require numerous unforeseen adjustments. However, surveys which have been carried out show that the workers make few complaints about the workload occasioned by these adjustments.

In concluding this study, it is perhaps possible to draw up an initial balance sheet. In the Group under study, automation has entailed the disappearance of older trades and has, according to Cohen-Hadria, changed the rules of the game in hiring and promotion; the losers in these new rules of the game are undoubtedly the older workers. The contracting out of a number of operations has made production units dependent on external entities and on a number of specialists in new technologies. This leads to the comment that what we are seeing today is not so much an overall reduction in the qualifications of the workers in the plant, but a reduction in the qualifications of the production unit which does not possess all the expertise it requires; this may also be seen as the development of a new form of division of labour.

For the future, personnel management problems will continue to be crucial. What promotion can be assured for those young workers whose initial level of qualifications is constantly rising? Some feel that their careers will be more and more similar to those of management and will involve significant geographical mobility. This also means that production units located in a rural zone will include fewer and fewer rural workers. However, will this mobility be easily accepted? The conflict within the Pechiney Group at Noguêres should make us wary as regards this hypothesis.

Even in the most modern factories in the Group, traditional workers and, in particular, kilners are still indispensable. Will the Group maintain factories at different technological levels in order to preserve this specific know-how?

Are the new trades appearing in cement works the new trades of the automation and technological system era described by Touraine, which are not specific to cement works and which, according to Dadoy, would justify the establishmnent of a new form of general education? Or will we see the development of a specific cement works "trade", as proposed by Cohen-Hadria, which will bring together the knowledge of the raw materials and process, as well as of the system of process operation, control and regulation and the daily maintenance of the system? This latter choice has the disadvantage that it would permanently tie the workers to the industry.

Finally, it seems that although worker know-how has been essential in developing the more sophisticated forms of automation, this automation has not really undermined the old Taylorist division of labour between planning and implementation. An experiment, financed partly by public money, is currently under way in one of the Group's units to allow the workers to have much greater involvement in the choice and design of equipment. It is not yet possible to judge the outcome, but the stakes at play are important and the outcome should bridge the dichotomy which still separates these two worlds.

Notes

[1] Sontages Etudes de Marché (SOFRES): Les ouvrièrs (unpublished public opinion survey, 1980).

[2] See in particular the work of the Institut National de Statistique et des Etudes Economiques (INSEE) and of the Commissariat Général du Plan.

[3] The data given in this section are drawn primarily from O. Pastre et al.: Automation, travail et emploi, Report drawn up for the Mission à l'Informatique du Ministère d'Industrie (Paris, Ministère d'Industrie, 1979). This important report also contains a detailed bibliography; O. Pastre and J. Toledano: "Automation et emploi: L'hétérogénéité des conséquences sociales du progrès technique", in O. Pastre et al.: Emploi et système productif, Collection Economie et Planification (Paris, Commissariat Général du Plan, 1979); Société d'Etudes pour le Développement Economique et Social (SEDES) and Bureau d'Informations et Prévisions Economiques (BIPE): Recherche technologique et indépendance industrielle: Les Automatismes, Study carried out for the Délégation Générale à la Recherche Scientifique et Technique (DGRST) and the Comité d'Organisation des Recherches Appliquées sur le Développement Economique et Social (CORDES); Centre d'Etudes et de Recherches sur l'Emploi et les Qualifications (CEREQ): L'evolution des systèmes de travail dans l'économie moderne: Conséquences sur l'emploi et la formation, Proceedings of national days of study by DGRST, CEREQ and the Centre National de la Recherche Scientifique et Technique (CNRS) (Paris, 1981); Paris, Ecole Polytechnique, Centre Recherche en Gestion (CRG): "L'automatisation du processus de production", in Les Chaiers de l'ANRT, No. 3, Dec. 1978.

[4] "Les perspectives d'evolution des automatismes dans les matériels de manutention", in Options, 1981.

[5] Pastre et al., op cit.; Pastre et Toledano, op. cit.

[6] Pastre et Toledano, op. cit.

[7] Commissariat Général du Plan: L'impact de la micro-électronique (Paris, 1981).

[8] In operation: production of lift brakes and construction site machine chassis; under construction: production of gearbox housings, motor vehicle prototype components, missile bodies, furniture parts, alternators.

[9] Carried out by Sondages Etudes du Marché (SOFRES) for the Ministry of Labour, 1980.

[10] See SEDES and BIPE: Recherche technologique, op. cit.

[11] Rapport de la Commission "Informatique et libertés" (Paris, La Documentation Française, 1975).

[12] Simon Nora and Alain Minc: L'informatisation de la Société (Paris, La Documentation Française, 1978).

[13] Actes du colloque international "Informatique et Société" (Paris, La Documentation Française, 1980).

[14] Les Années 1980: Faits et choix économiques (Paris, Presses Universitaires de France, 1981).

[15] A. Danzin et al.: La société française et la technologie (Paris, Commissariat Général du Plan, 1980).

[16] P. Bonelli and A. Fillion: L'impact de la microélectronique (Paris, Commissariat Général du Plan, 1981).

[17] The Information Technology Commission is responsible for promoting the application of data processing in the administration and the economy as a whole and has as its role:

- to help users to define their computerisation programme and ensure coherence between the equipment policy of the public sector and the industrial policy as laid down by the Government;

- to provide users with the means and methods of a greater mastery of data processing;

- to evaluate the impact of data processing on the economic and social environment and to propose any measure aimed at making its introduction more harmonious;

- to take part in defining and implementing an education and training policy for data processing.

[18] In particular, the DGRST and CNRS are carrying out a joint research project on advanced robotics.

[19] Study by the Institut de l'Entreprise.

[20] Statement of J. Fouroux, in Informatique et société, Vol. III, Ch. 2.

[21] "Emploi, salaires, negociation - L'UIMM s'explique", Interview with M. Guillem, in L'Usine Nouvelle, No. 20, 14 May 1981, pp. 112-116.

[22] Survey carried out by the Crédit d'Equipement aux PME (Equipment Credit Institute for Small and Medium-sized Firms).

[23] Le Peuple, No. 1103, 16-31 Mar. 1981; Options, 3 Mar. 1981.

[24] Confédération française Démocratique du Travail (CFDT): Les dégâts du progrès: Les travailleurs face au changement technique (Paris, Editions du Seuil, 1977).

[25] Syndicalisme, 25 Sep. 1980; cf. also Industrie et Technique, special automation issue of 31 Dec. 1980; interview with Mr. Y. Lesfargue, National Secretary of the CFDT.

[26] See FO Magazine (Paris), Nov. 1980.

[27] Confédération française de l'Encadrement (CFE): La novotique (Paris, 1980).

[28] Since this study was completed, the Second "Anroux Act" (1982) introduced important changes in trade union rights, allowing the setting up of trade unions in enterprises of any size, strengthening the position of trade union and personnel delegates and extending the works councils' advisory powers; works councils could now also be formed by groups of associated small enterprises with a total of 50 employees. The Act also provided for the setting up of group councils including workers' representatives from subsidiary companies (eds.).

[29] The Third "Anroux Act" (1982) revises the Labour Code as it relates to collective agreements, with the aim of stimulating sectoral and enterprise-level collective bargaining and of ensuring wide coverage; negotiation is now compulsory, but enterprises with fewer than 11 employees may form optional joint occupational or inter-occupational committees (eds.).

[30] For a description of the production process, see in particular: G. Biezunski: "Les cimenteries", series of articles in Mesures, Apr. 1972-Sep. 1973; CFDT, Comité National des Salariés des Chaux et Ciments and Fédération Nationale des Salariés de la Construction et du Bois: Regards sur l'industrie cimentière (Paris, 1980); Ciments Lafarge: Rapport à l'Assemblée Générale ordinaire, 16 June 1981; and Syndicat National des Fabricants de Ciments et de Chaux: Rapport présenté à l'Assemblée Générale, 12 May 1981.

[31] Syndicat National des Fabricants de Ciments et de Chaux: L'industrie cimentière française (unpublished internal document).

[32] Lafarge Conseils et Etudes: Une philosophie nouvelle en contrôle commande de processus (unpublished internal document). The description of the operation of a cement works is taken from this study.

[33] See Yves Cottet: "Tendances actuelles des techniques et des technologies en matière d'automatisation en cimenterie", in CRG and AFCET, Aspects humains de l'automatisation: Le cas de l'industrie cimentière (Paris, 1978).

[34] Luttrin: "Historique de l'automatisation en cimenterie", ibid.

[35] The following description draws on Yves Cohen-Hadria: Automatisation et organisation du travail: l'exemple des cimenteries (Paris, CRG, 1979).

[36] ibid.

[37] ibid.

[38] Lebas: "Essai sur les formes nouvelles d'automatisation: l'exemple d'une industrie de procédé, la cimenterie", Annex 2 in Y. Bouchut et al.: Formes nouvelles de l'automatisation (unpublished research report; mimeographed, no date).

[39] Yves Cottet: "La conception moderne de Lafarge Conseils et Etudes dans l'informatique industrielle numérique: Installation à l'usine du Teil de ciments Lafarge France", in Ciments Bretons Plâtres Chaux, No. 1, 1980.

[40] Lebas, op. cit.

[41] Cottet: "La conception moderne ...", op. cit.

[42] See, in particular Association pour la Recherche et le Développement des Etudes Sociales et Economiques: Crises et pespectives de l'industrie cimentière (Paris, 1980); CFDT: "L'automatisation des cimenteries", in CFDT Aujourd'hui, No. 49, May-June 1981; CFDT, Comité National des Salariés des Chaux et Ciments and Fédération Nationale des Salariés de la Construction et du Bois: Regards sur l'industrie cimentière, op. cit.; R. Chico: "L'organisation du travail dans une cimenterie automatisée", in CRG and AFCET: Aspects humains ..., op. cit.; M. Dadoy: "La polyvalence ouvrière et sa rénumération", in Claude Durand (ed.): La Division du travail (Paris, Galilée, 1978); and Inter 325, published by the Confédération Générale du Travail, from Mar. 1973 to June 1981.

[43] M. Dadoy: Politiques de gestion du personnel dans l'industrie cimentière (Paris, AFCET and CRG, 1978).

[44] Convention collective et accords d'entreprise, 1972: Ouvriers de production centralisée et contremaître de production centralisée (unpublished internal documents).

[45] Cohen-Hadria, op. cit.

[46] B. Coriat: "Ouvriers et automates: Procès de travail, économie de temps et théorie de la segmentation de la force de travail", in Usines et ouvriers: Figures d'un nouvel order productif (Paris, 1980).

[47] Lebas, op. cit.

[48] de Terssac: "Repercussion de l'automatisation sur les modalités de réalisation de la tache: Etudes de cas", in CRG and AFCET: Aspects humains ..., op. cit.

CHAPTER VIII

AUTOMATION, WORKERS' NEEDS AND WORK CONTENT IN HUNGARY

(by Csaba Makó)

1. Some general characteristics of human needs and working conditions

In the past decade, both academic researchers and practical experts have paid great attention to the co-ordination of the structure of work demands and work expectations on the different levels of technological development of production. These specialists have generally come to the same or similar conclusions on the factors underlying the social and human aspects of performing a job.

Among the changes of a social nature which have taken place, emphasis has been placed first and foremost on the importance of changes in the internal composition of the structure of the labour force. In Hungary, for instance, the socialist industrialisation after the Second World War resulted in a basic change in the working class as a whole and in its composition. The number of industrial workers compared to all active wage earners rose from 37 per cent to 58 per cent between 1949 and 1973.[1] Important changes have taken place not only in the number of industrial workers but also in their distribution across industries. Table 31 clearly shows the changes in the composition of wage earners according to the different branches of the national economy.

The level of general and professional education of workers has also greatly increased. In the late forties, for instance, four-fifths of unskilled workers had not completed primary school, but this figure had decreased by half by the first half of the seventies.[2] Positive changes have also taken place as regards the qualifications of workers. The number of skilled workers and those who are in direct control of production has shown the most dramatical increase.

The overall changes taking place in the internal structure of manpower have had a significant influence on the economic activity of the sexes. While in the late forties, for instance, the ratio of male to female wage earners was 70:30, it today stands at 56:44. This means that the tension between work expectations and work demands, and the problems of co-ordinating them, have affected both male and female workers. As the level of education of each successive generation increases, so do human requirements. The increased desire to continue studying, which greatly influences people's ways of thinking and their needs, is clearly expressed in table 32. The number of younger and better educated workers has caused significant changes in work expectations: the number of workers under 30 exceeded 1 million for the first time in the early seventies.

The growth of human expectations of work obviously cannot be explained exclusively by increased general and professional education. Increases in the standard of living and in employment also make their influence felt. Most significantly, national income per head between 1950 and 1973 more than tripled, rising from US$300 to US$1,000.[3] Besides an increase in national income, the improvement and stabilisation of the standard of living of working people have also been furthered by the achievement of full employment, general social insurance and a free health service.

Table 31: Hungary: composition of wage earners by economic sector, 1949 and 1981

Economic sector	1949		1981[1]	
	No. of wage earners ('000s)	%	No. of wage earners ('000s)	%
Agriculture	2 100	53.1	1 030	20.3
Industry and building	856	21.7	2 090	41.2
Services	952	25.0	1 880	37.0

[1] Forecast.

Source: Gyorgy Pogany: Munkaero es munkaero gazdalkodas a szocializmusbanj [Manpower and manpower policy in socialism] (Budapest, Kossuth Konyvkiado, 1980), pp. 202-203.

Table 32: Hungary: educational level by age group (percentages)

Type of schooling	Age group		
	Born in 1940	Born in 1950	Born in 1960
Did not finish general (primary) school	34.0	13.0	9.0
Completed secondary school	24.0	39.0	40.0
Completed unfinished secondary school[1]	18.0	32.0	46.0
Did not continue studies	24.0	16.0	5.0

[1] The unfinished secondary school is a kind of vocational secondary school which does not give a final school certificate.

Source: Tamás Kozma: Bevezetés az oktatásügyi szervezetelméletbe [Introduction to the theory of educational organisation] (Budapest, Jegyzet a Pedagógustovábbképző intézet számára, 1980), p. 416.

In spite of these significant socio-political improvements, new social tensions have also developed in several areas, notably in the divergence between the health service as guaranteed free of charge by law and the actual conditions prevailing, and in the contradictions between the theoretical existence of full employment and so-called unemployment behind the factory gates. In is undeniable that these factors have an important and constant influence on the structure of human needs and expectations from work. We take it for granted that full employment does not reduce the need for safety at work, since in socialist industrial conditions this is guaranteed to every employee.

Having considered these factors influencing human expectations of work, the following question arises: How can the structure of work tasks and the structure of work demands meet such expectations? Unfortunately, the results of the division of labour and specialisation characteristic of Hungarian industrial organisations, such as the structure of work tasks, the levels of job hierarchy, and the decision-making processes, do not meet the needs of some groups of workers today, and this divergence can be expected to increase further in the future.

The internal development and functioning of Hungarian co-operatives and companies, as well as their organisational structure, are rather uniform. Large companies tend to dominate Hungarian industry, which means that negative effects are not directly apparent, but are manifested through excessive specialisation, formalisation and standardisation. These organisational characteristics impede development, making it more difficult to satisfy the varied human expectations of work and thus preventing the effective use of human resources. The structure of industrial work organisation makes it less possible (and often unnecessary) to make use of the individual and collective initiative of an educated workforce. This situation is well illustrated by the divergence between general and vocational education on the one hand, and the demand for work skills on the other. In 1977, for instance, among the 689,000 people with a final secondary school certificate, 83,000 were employed as skilled workers and 41,000 as semi-skilled workers.

2. Humanisation of working conditions: technological determinism and belief in organisational design

Work tasks and their structure, which play a decisive role in meeting human expectations in relation to work, have been formulated in Hungary according to Taylorist principles. In this field, the views of most academics and practical specialists are characterised by technological determinism. Work is divided into small discrete components which minimise the opportunity for individual or collective initiative. In their view, this can only be eliminated by automation and the spread of industrial robots.[4] However, we contend that the idea that the conditions and circumstances of industrial work can be improved simply by automation is an idealistic and unrealistic view.

In spite of the fact that in the past decade most of the material means for industrial investment were employed in the mechanisation of productive processes, the number of people working in automated, even in mechanised, enterprises remains relatively low in Hungarian industry. It is also characteristic that the mechanisation of basic industries is higher and that manual work dominates in the other sectors of production. Table 33, which includes data selected among economic branches, clearly reflects the relatively low level of mechanisation of industrial work and the decisive role of manual work.[5]

Not only is the overall level of mechanisation of production variable, but considerable divergences may also be found between branches of industry. Besides the general prevalence of manual work, for instance, important differences are to be found in transfer line production or in the ratio of people working on machines. Most people working on transfer lines, with the exception of workers in light industry, are engaged in manual work. With mechanised activities, machine work is characteristic, although - especially in the food industry - a high proportion of workers do manual work even when operating machines.

Besides these technological conditions of industrial production, we attach importance to those concepts which state that a change in the structure of work activities and the humanisation of work will not be brought about by the overall spread of automation and robotisation. These are the views of sociologists and psychologists who question the concept of leadership and management which considers the co-ordination of human needs and the demands of work organisation only through the adaptation of such needs to the conditions of work. The proponents of these views criticise the view that the machines and equipment, or the technology representing the level of production, can be used economically only in a certain type of work demand or work organisation. In other words, they reject the mechanical and deterministic approach according to which only a particular job structure or organisation can be optimum in industrial practice. Admitting the important role of the technological conditions of production, they call attention to the importance of socio-organisation factors and to the relative autonomy of production.[6] Technological determinism is often refuted in practice by work organisation reforms which co-ordinate human needs and work demands without ignoring the consideration of productivity.

The modernisation of organisations which neglects economic aspects, however attractive and ambitious it may seem, is bound to fail sooner or later. In order to meet human needs at a higher level - generally by improving QWL - the criteria, characteristics and structure of the changes considered must be taken into account. Although in a socialist society efficiency does not always play a decisive role in programmes for the humanisation of work, the initiatives of company management to maintain the efficiency and the marketability of the company are largely influenced by such programmes. We would like to emphasise this, despite the fact that there is evidently no direct relationships between the economic efficiency of the company and the satisfaction of workers' needs. In the case of socialist industrial relations, even if the different indicators of the efficiency of the company are deteriorating, the needs of significant groups of workers may still be realised.

Table 33: Hungary: composition of manual workers according to mechanisation of work in principal branches of industry (percentages)

Branch of industry	Non-mechanised activities			Transfer line production		Other mechanised production[1]		Machine monitoring[1]	Other	Total
	Supervision, inspection, quality control	Unskilled manual work	Skilled manual work	Manual work	Machine operation	Manual work	Machine operation			
	(1)	(2)	(3)	(4)	(5)	(6)	(7)	(8)	(9)	
Machine industry	7.9	19.8	22.5	7.5	2.6	10.9	19.4	2.4	7.0	100
Heavy industry	7.2	19.7	22.2	5.3	2.0	12.5	19.1	4.3	7.7	100
Light industry	6.5	19.9	11.0	7.8	17.1	11.9	19.1	1.9	5.8	100
Food industry	5.9	30.0	11.2	9.4	2.9	11.4	14.2	4.5	10.5	100
Total	7.0	20.8	18.3	6.1	4.1	12.5	18.5	3.9	8.1	100

[1] The essential distinction between manual work and machine operation is that in manual work the dominant tasks are loading or discharging machines rather than directing or controlling the machine itself. This can take place either on a transfer line or in the case of isolated machines.

Source: Kozponti Statisztikai Hivatal: Ipar munkaügyi adatfelvētel [Representative data of industrial labour] (Budapest, 1979).

Before examining the opportunities for and limits to the satisfaction of human needs at the workplace, we briefly describe the main features of the socialist industrial relations system.

3. Industrial relations at company level: the organisational triad

The difference between socialist and capitalist industrial relations at company level lies in the fact that the traditional organisational structure of management and trade union is supplemented by the role of the political party (the Communist Party). However, we have relatively poor systematic knowledge of the connections formed between the political party and the management and trade union in the company. This insufficient knowledge is partially due to the fact that the majority of social scientists consider the study of the relations between management and trade union as sufficient to understand the social phenomena in the socialist company. It is beyond the scope of the present paper to describe the structure and dynamism of industrial relations in Hungary, but it is necessary to mention the most important events of the last 15 years.

The new economic reform introduced in Hungary at the end of the sixties made basic changes in the roles of the social partners in industrial relations - economic management, trade union and Communist Party). The tendency of these changes may be briefly characterised as the expansion of the sphere of decision making of the trade unions in the field of working and living conditions for the workers, parallel with growing autonomy of company management. The leading role played by the Communist Party in the economy also underwent a change (i.e instead of concrete intervention at production level in the activities of company management, the indirect and more general elements of control were increasingly stressed). This new development in the role of the political party was based on the fact that the social conditions and consequences of the company's economic life are so complex that they can only be effectively dealt with by the mutual efforts of the economic and political entities and by the trade union. None of the partners in socialist industrial relations has the monopoly of knowledge and power concerning the social dynamics within the company. For instance, the structure and functions of the trade unions are more adapted than those of the political party to representing the interests of the workers concerning working conditions, wages, promotion, etc.

The world economic crisis in the mid-seventies made it necessary to go further in the direction of economic change. In the export-oriented Hungarian economy, improvements in the adaptability and flexibility of companies received high priority. With this new economic orientation it became obvious that company management could not cope along with the mobilisation of human resources. The only solution was to increase workers' participation through the active support of the trade unions and the Communist Party.[7]

In the late seventies the institutions of industrial democracy were expanded at company level. In spring 1977, the common decree of the Government and the National Council of Hungarian Trade Unions conferred an important power on workers' representatives (shop stewards) in industry. According to this decree, shop stewards elected directly by their workmates have the right to make decisions concerning such questions as collective agreements, wages, social funds, working conditions, etc.[8] Beyond these traditional areas of trade union rights, shop stewards also participated in the evaluation of the annual activities of company management. In practice, shop stewards experienced the greatest difficulty in controlling management's annual activities. The Communist Party organisations at central and company level had to intervene repreatedly in favour of the trade unions.

The changes in industrial relations within companies may be briefly summarised as follows. As a result of the new economic reform introduced in 1968 in Hungary, not only did companies' autonomy increase a great deal, but the trade unions also gained important powers in representing workers' interest. Parallel with the increasing influence of the trade unions, indirect and non-operative control methods were given more importance in the activity of the political party in company life. The democratisation of industrial relations shows that economic reforms cannot be realised without initiating political reforms. Along with political and social changes, the division of labour among the social partners of Hungarian industrial relations was essentially modified, and principally in the direction of a more balanced power distribution.

4. Case study - Automation in a motor vehicle plant: workers' needs and work content

It is rewarding to study the possibilities for and limits to the co-ordination of human expectations of work and work demands. How far can company management go in a particular direction towards satisfying workers' needs without making the efficiency and marketability of the company suffer? This question is particularly important under the conditions of socialist industrial relations where the efficiency of the productive activities of the company is not regarded as an absolute aim in itself, but the demands connected with it should be fulfilled by satisfying human expectations of work on a higher level. To co-ordinate workers' needs and work demands, however, one has first to analyse their structure and content.

In the following case study, the characteristics of the content and structure of workers' needs will be presented in order to determine the relative importance of working conditions and work demands to workers. Besides workers' needs, we shall deal in detail with such dimensions of work activities as the level of automation (mechanisation), the method of production organisation and the types of job, all which determine the structure of work demands and the possible combinations.[9]

4.1 Background to the company

The company was founded at the end of the nineteenth century, at that time with about 1,200 workers, almost exclusively producing railway carriages. In the course of development, production became more and more diversified, and by the late thirties the proportion of railway carriage production had greatly decreased. As a new venture, motor industry products were introduced and took over an increasing part of production. The number of personnel in the company quadrupled. Serious damage was caused during the Second World War, but in spite of this the company quickly recovered and, following the need for reconstruction, it once again predominantly produced railway carriages. Apart from several organisational changes in the sixties, a most significant change was brought about by production development in the early seventies, which was also combined with an important technological development. According to the conception of company management, the mass production of new or adapted designs began using manufacturing system which had already proved economical on an international level. In order to ensure this strategy of production, the company bought a complete plant and manufacturing systems, mainly from western European firms. Table 34 clearly shows the change in the structure of products of the company.[10]

The reason for the changes in the composition of products was the drastically decreasing market for the production of passenger coaches in the sixties and the low level of the technology employed. On the other hand, the COMECON countries represented an almost unlimited market for the heavy-duty diesel engines and rear axles for buses and lorries and large agricultural tractors. Along with technological development (for instance, the company was the first motor vehicle factory in Hungary to set up an automated transfer line), the composition of products was changed. The production strategy of company management gave priority to the vertical development of the production system (i.e. the overwhelming majority of the components of products came from its own plants). The development of the production system of the company closely followed that of the Renault motor company.[11]

At the time of carrying out its project on automation and industrial workers in 1974, the company employed some 10,000 manual workers. Of the total production an insignificant proportion comprised the manufacture of railway vehicles, the main products being rear axles used for road vehicles, engines, heavy lorries and tractors. For the purpose of the investigation, we chose the mass production engine plant and rear-axle plant, and the single-unit productive tool plant. We paid attention to the effects of automation on work content, working conditions and workers' attitudes.

4.2 The structure and role of workers' needs

The questions we considered important in the project were as follows: the workers' relationship to new machines and new equipment; and how far they were satisfied with the types of industrial work which were introduced due to technological development (automation). In the course of studying workers' attitudes to work activities, we paid attention to the examination

Table 34: Motor vehicle plant: breakdown of total sales by product, 1965-77 (percentages)

Product	1965	1974	1977
Railway vehicles/passenger coaches	38.7	7.7	3.3
Road vehicles, of which:			
Rear axles	24.0	35.1	35.7
Diesel engines	.	34.6	32.6
Lorries	-	5.4	9.7
Agricultural tractors	-	5.9	10.9
Subtotal	24.0	81.0	89.2
Other products	37.3	11.3	7.5
Total	100.0	100.0	100.0

- = nil or negligible.

of their needs. The analysis of the structure and content of workers' needs was important for us from several aspects. First, it constituted a general orientation in considering the relative importance of the questions studied. Our aim was not to examine and assess the effects of automation on work content and working conditions, but to understand the concomitant circumstances and consequences of development through the features which are most important to workers. To do this, the best basis of comparison was the structure of workers' needs and demands. Thus we may discover which of them are of secondary importance, or even negligible.

The relative importance of the questions studied was based on the people concerned (i.e. on the judgement of the workers). We are aware of the fact that the needs formulated by the workers are not necessarily the same as their real needs. Our methods of investigation included first of all interviews with workers, to discover their views as they formulated them; 540 workers were questioned in this way. The distorting role of consciousness on observation was also considered.[12] Bearing in mind what has been said about the examination of workers' needs, this is an appropriate method on the one hand for assessing the effects of the socio-economic environment outside the workplace, and on the other for determining their orientation towards work.

Table 35 sets out the structure of workers' needs as discovered in our investigation. The importance to the workers of the following components of work activities were examined: work content; relationships with co-workers and superiors; pay; possibilities for promotion; physical working conditions; and work strain. We found that pay is considered to be the most important, followed by work content and relationships with co-workers and superiors. Physical working conditions and possibilities for promotion are at the end of the scale.

The introduction of new machines, equipment and technological processes thus appears to workers as a question of pay and requires a solution on the part of the company. In certain respects, the results of our investigations (e.g. regarding work satisfaction) are certainly determined by the fact that all the three plants examined belong to a productive field where wages are highest, and the company too is an establishment where wages are the highest in Hungary.

Apart from the workers' assessments of their own needs, we considered it was important to find out the opinions of their superiors on the needs of the workers they controlled. In this way we were able to carry out cross-checks on the structure of workers' needs. Interviews were held with 30 workshop supervisors (i.e foremen) and 25 trade union representatives (i.e. shop stewards). Table 36 gives the results of this inquiry.

An overwhelming majority of the foremen (80 per cent) rated the financial aspect of greatest importance among workers' needs. In contrast to this, appropriate pay was considered most important only by approximately half of the workers, which means that their supervisors think the financial aspect of performing a job more important than do the workers themselves. Good work content was mentioned by both supervisors and workers as of secondary importance. Foremen and workers also shared the opinion as to the importance of opportunities for promotion within the organisation, which was mentioned as of least importance by nearly the same proportion of both groups. Divergences are noticeable, on the other hand, in judging the importance of relationships with one's superior and of physical working conditions: a good supervisor was mentioned in the third or fourth place by the workers, in contrast to the foremen's estimation, which ranked the importance of their activities and roles in fifth place. In judging the role of physical environment, the opinions of the supervisors differed to a large extent; one-third of them placed it in second place and another third in sixth (and last) place. Almost half of the workers interviewed mentioned the necessity of an appropriate physical environment in fifth place.

The evaluation of the trade union representatives in connection with workers' needs generally coincided with the workers' own judgements. One can, however, find a divergence in the estimation of needs in connection with work and pay. According to the shop stewards, appropriate work played a more important part in workers' needs than appropriate pay, whereas the workers interviewed rated appropriate pay as more important

Table 35: Motor vehicle plant: the structure of workers' needs

Factor	Percentage of workers who ranked the factor					
	1st	2nd	3rd	4th	5th	6th
Work content	26.7	28.2	13.6	20.8	8.3	2.4
Relationships with co-workers	10.5	13.8	26.2	29.2	15.1	5.2
Relationships with superiors	7.2	16.0	23.4	26.3	21.0	6.1
Pay	44.7	26.5	14.2	7.2	3.0	4.6
Possibilities for promotion	3.1	5.9	6.1	2.8	9.9	72.2
Work strain and physical working conditions	7.7	9.6	14.5	15.7	49.9	9.6

Table 36: Motor vehicle plant: superiors' opinions about the structure of workers' needs[1]

Ranking of workers' needs	Foremen	Shop stewards
1st	Pay	Work content
2nd	Work content	Pay
3rd	Physical working conditions	Relationships with superiors
4th	Relationships with co-workers	Relationships with co-workers
5th	Relationships with superiors	Physical working conditions
6th	Promotion	Promotion

[1] The superiors questioned carried out activities that were exclusively managerial in relation to the workers involved in this survey.

than appropriate work. Otherwise, the judgements of shop stewards and workers generally coincided as to the importance of suitable co-workers and supervisors, physical working environment and opportunities for promotion.

The opinions of trade union representatives and workshop supervisors differ in judging the importance of appropriate pay and appropriate work for workers. The foremen consider their workers more money-minded than do the shop stewards. According to the overwhelming majority of foremen, appropriate pay is of greatest importance among workers' needs. In contrast to this, shop stewards considered appropriate work to be most important for workers, although they did not deny the importance of appropriate pay among workers' needs.

Our examination of expectations in performing a job shows the same tendency from the point of view of the workers as from that of the workshop supervisors and trade union representatives. Although in some areas (e.g. the importance of appropriate work or appropriate pay) there is a slight divergence, the expectations in job performance are overwhelmingly judged in the same way by both superiors and workers. On the basis of this, one can assume that the structure of workers' needs is a suitable means of orientation to determine the importance and role of attitudes to work. This implies that, in studying workers' attitudes to work, we pay more attention to their opinions about pay and work content than, for instance, promotion within the organisation.

4.3 Technology, work organisation and work

After this investigation of workers' needs, we now analyse the nature of technical and technological development in the company, and the effects of automation on work and working conditions.

The productive units of the company examined clearly showed a development towards mass production and automation. We attempted to analyse them in this mixed stage of development, rejecting the possibility of abruptly removing transfer lines from their natural surroundings and comparing them to traditional machine tools. Instead, we tried to assess the state of technological development in the productive units as it existed in reality, and thus make it possible to examine the effects of automation on work. A technological scale with three dimensions[13] was used to describe the productive units. This distinguishes those aspects of technological development that are closely linked but are also separable. The three dimensions are as follows:

- level of mechanisation and automation of production: this shows the level of mechanisation of different types of production in individual jobs;

- increase in volume of production: this shows the aspect of technological development which is characterised by a development from batch production of individual orders to manufacture of standardised products in serial or mass production;

- development from single-unit organisation of production to flow production: the basis of traditional single-unit production is that the operations are uniform, whereas in flow production the emphasis lies on the uniformity of the product or products.

This approach attempted to analyse the three angles of technological development described above, which are partially interdependent, and the importance they may have for work and working conditions. We wished to go beyond the narrow approach which considers the consequences of technological development exclusively as a concomitant of the mechanisation and automation of production. In our approach we drew on the sociological theory which argues that the impact of technological development on work can only be understood in the context of the entire development of the production system. In the present study we deal with the consequences of the mechanisation (automation) of production on work organisation, work content and working conditions.

We were not able to make comparisons at the level of productive units due to the fact that most of them combined automated, semi-automated and traditional machines, so that the character of production was mixed. We therefore compared instead groups of automated with non-automated jobs. The analysis is based on observations at work, questionnaires given to workers, interviews with management, company data and other sources of information.

4.3.1 Comparison of work performed on automated and traditional machines

Regarding the impact of automation on skills, one can find at least three different schools of thought. According to the first school, which judges the impact of automation optimistically, both a vertical and horizontal increase in skills may be noticed. In the case of a horizontal increase, there is no change in overall level of skill, but rather a significant enlargement of the job content or sphere of activity, which results in quantitatively more skills than before and therefore implies an increase in skill level. The investigations carried out in socialist countries show this kind of process first of all.[14] The observation of automated transfer lines in Czechoslovak and Soviet plants showed a 15-40 per cent increase in skill for machine operators, machine adjusters and locksmith maintenance jobs.[15]

The second school is pessimistic. In the course of studying the production processes of the motor industry and the metal industry in general at company level, several researchers were surprised to find that the demand for expertise and ability on the part of manpower was decreasing. In important posts where long training and experience were necessary until the introduction of automation, the work demands became reduced to easily and quickly acquired activities. According to those who view the impact of automation in a pessimistic way, the need for manual skills is decreasing rather than increasing. The tasks formerly performed by machine operators and adjusters now belong to the spheres of activity of tool designers, technicians, engineers, etc. The relationship of worker to machine has changed considerably, the skilled machine operator being promoted to machine supervisor.[16]

Finally, in the view of the third school, automation does not exercise a significant effect on qualifications, or rather its impact is not homogeneous. In the opinion of the supporters of this view, important changes do not occur in the level of expertise, not even with those who perform maintenance jobs. Some also take a slightly more differentiated view according to which automation increases the demand for expertise in some cases and decreases it in others (polarisation). Automation introduced in the same branch of industry (e.g. motor vehicles) may also have a heterogeneous impact on expertise. For instance, the investigation dealing with the impact of automation at Renault showed an 11 per cent increase in skilled jobs, while a decrease in expertise was experienced in the remaining skilled jobs. At the same time, the need for expertise in maintenance jobs had considerably increased owing to the installation of automated machines and equipment.[17]

All the jobs included in our comparative analysis belong to mass production of flow organisation. They were performed on transfer lines or on automated machines of other types, on semi-automated machines, or on traditional machines linked closely or loosely with automated machines.

The <u>demographic characteristics</u> of the workers in the productive units of rear-axle manufacturing are summarised in table 37; 48 operators of automated machines were compared with 115 operators of traditional machines in flow mass production.

The demographic composition of the workers, whether they worked on transfer lines or revolving turning lathes, showed very slight differences. The operators of transfer machinery and automated machines were superior neither in education nor in professional qualifications to the workers on traditional machines. One way in which the operators of automated machines differ from the others was that most of them were male workers aged between 20 and 30. This is also borne out by the time they had spent at the company, nearly half of them having worked less than seven years there. In contrast to this, four-fifths of the workers in the productive units of rear-axle manufacturing equipped with traditional machines had length of service of more than six years with the company. This is primarily due to

Table 37: Motor vehicle plant: demographic characteristics of workforce in rear-axle manufacturing (flow mass production) (percentages)

Demographic characteristics	Operators of automated machines	Operators of traditional machines
Sex:		
Male	91.8	86.1
Female	8.2	13.9
Workers with elementary school certificiates and skilled workers' diplomas	30.6	38.3
Skilled workers	56.2	57.4
Technicians	8.4	-
Seniority (more than five years in the job)	14.3	59.2
Age (below 30 years)	69.4	44.3
Married workers	67.3	70.4
Workers with own flat	46.9	65.2

Source: Lajos Hēthy and Csaba Makó: Az automatizacio es a munkastudat [Automation and workers' perception] (Budapest, Hungarian Academy of Science, Institute of Sociology, 1975).

the attempt of company management to employ workers who are open to the new machines and equipment.

One may find a similar personnel policy regarding the organisation of the so-called autonomous work groups representing a new structure of job tasks and work demands. The measure of autonomy of action and responsibility for the workers has considerably increased, and job content has changed in these work groups as compared to former types of work organisation. According to experience in workshops, people are reluctant to change the jobs they have had for decades or have grown accustomed to, even for a more interesting job which carries more responsibility. This increased responsibility is a characteristic feature of autonomous work groups. In the development of posts characteristic of the new type of work organisation, company management tries to choose workers from the following three groups: firstly, from the young, who are only slightly tied down by routine and habit; then from the newly employed; and lastly from those who volunteer.

This phenomenon can be explained not only by the fact that the groups mentioned would be intellectually or psychologically more open than their co-workers who have been in the workshop for a longer period of time. The "openness" is certainly a common characteristic of the young, of the newly employed and of volunteers. However, these groups are also characterised by rather underdeveloped social relationships among themselves, and this is the reason why they are ready to accept initiatives coming from the company. In contrast to this, if company management can rely only on workers of long standing, where social connections are already developed and common interests have matured, then it is more difficult for the company and requires long negotiations to have a more advanced productive technology, or different working methods accepted.

The operators of automated and traditional machines, despite the fact that they worked in the same productive unit and their working conditions were therefore identical, had differing views on the development of their employment conditions (table 38). Workers were asked the question: How characteristic of your job are the following opportunities in connection with your work? The operators of automated machines had a more favourable opinion about both pay and opportunities for improvement of their professional knowledge and general education than their co-workers on traditional machines. Furthermore, a larger proportion of them expressed a favourable opinion about the possibilities of promotion - which is generally considered rare - and about employment safety.

Before describing the workers' opinions, we must emphasise the fact that those on automated, semi-automated and traditional machines work on the same transfer line. They put the same workpieces into their machines and they move or carry away the finished pieces with the same subsidiary equipment, such as moving stores and fork-lifts, roller lines and upper conveyors, etc. The equipment for removing turnings from machines is also the same. Furthermore, the recurrence and frequency of the work tasks is identical, and since there is flow production, workers depend on each other in the tempo of their work as well. It is not therefore by chance that the two types of operator notice relatively slight difference when comparing their jobs.

It is not by chance that the operators of automated machines and equipment spoke more favourably about their wages, since they earn more than the average. This is due to company policy which, by offering favourable wages, aimed to prevent any problems that might cause stoppages on the part of operators of machines and equipment with a large output and with a key role in production. The extent of loss due to a stoppage of this machinery would be far greater than the damage occurring from stoppages of traditional machines.

Workers on both automated and traditional machines were paid on a piece-rate system, the basic wage on machines with large output being supplemented by considerable bonuses, according to the quantity produced. The most common wage system for operators of automated machines was based on group piece-work (this applied to two-thirds of workers on automated machines). For operators of non-automated machinery, the typical wage system was based on individual piece-work, although even here one-third of

Table 38: Motor vehicle plant: views on employment conditions of workers on automated and traditional machines (flow mass production)

Item	Workers answering positively (%)[1]	
	Operators of automated machines	Operators of traditional machines
Good pay	53.0	28.7
Guaranteed employment	81.6	65.2
Opportunities for promotion[2]	10.2	3.5
Chances for improving professional knowledge	46.9	27.8
Opportunities for acquiring more general education	42.9	25.2

[1] Percentages of workers in whose opinion the item is characteristic of their jobs to a large extent (points 7-9 on a nine-point scale).

[2] i.e. appointment as group leader or foreman.

Source: Héthy and Makó. op. cit.

the workers were paid according to group piece-work. One may also find individual piece-work wages paid to workers on automated machines, although to a lesser extent than workers on traditional machines. For instance, among the automated productive units in the motor industry, the basis of wage payment was individual performance, rather than group achievement, as might have been expected. The use of this kind of wage system is, of course, provisional. The reason for its introduction was to maximise the possibilities of individual machinery and posts.[18]

In the long run, the use of piece-work wages may have several unfavourable effects. The individual competition between workers may increase in such a way that it may hinder the total output of transfer lines, when co-operation and mutual assistance is needed to ensure undisturbed production. That is the explanation for the fact that wages based on individual performance were paid to only 10 per cent of operators of automated machines.

Regarding mental and physical work strain, job descriptions show slight differences between, for instance, the structure and method of implementing information and the structure of work activities, etc., of operators working on automated and traditional machines, but one cannot conclude that there is an unambiguous tendency. Working conditions can be classified in several ways. From the point of view of our study, we aim to show the changes that occur in the major physical and mental demands of industrial work. Due to their importance, it is necessary to mention the development of physical and mental work strain, expertise and responsibility. Although these are analysed separately in most cases, we are aware of the fact that there is interaction between them. In reality they do not occur in isolation but jointly, and an individual description can only be useful for the purpose of analysis.

In the field of physical and mental work demands, similar tendencies have been found by researchers in different countries. In the development of physical strain, there is an inverse relationship between the different levels of automation and increasing physical demands. The physical strain demanded of workers at low levels of automation are high. In contrast to this, physical strain and fatigue do not occur at all in posts using automated equipment.

The situation regarding mental strain and automation is not so simple: one can find instances of both decreasing and increasing mental work strain in research studies. The contradictory relations between mental work strain and automation are apparent. In the strict sense of the term, mental strain which is connected with the supervision of the functioning of machines has decreased. However, one can agree with those who state that automated jobs involve repetitive tasks and require close attention from the worker, but lack any form of stimulation.[19]

As an illustration, we would like to present our data on observations at work concerning mental strain. In the course of our observations on the information processing activities necessary to perform the job in individual posts, we distinguished the three following stages:

(1) Collection and recording of information is the simplest form of information processing. In this field we did not find any important differences between the work on automated and traditional machines. One-third of working hours are spent on these kinds of activities.

(2) Classification, systematisation and combination of information require greater independence and more differentiated qualifications. These types of activities are more frequent with operators of automated than of traditional machines. The former spent 21 per cent of their working hours and the latter only 12 per cent on this kind of work.

(3) Evaluation and application of information requires the greatest creative abilities. Such activites are more characteristic of those working on traditional rather than on automated machines. The creative processing of information (e.g. determination of new parameters and their use) is not only possible in the case of traditional machine tools with a less rigid job structure, but is also a basic work demand. Workers in such jobs must undertake the greater part of evaluation and use of information, besides its registration and arrangement in connection with work processes.

Table 39 summarises the proportion of information processes at different levels of mechanisation.

Table 39: Motor vehicle plant: level of mechanisation and stages of information processing

Level of mechanisation	Occurrence of information processing activities (%)		
	Collection and registration	Classification and combination	Evaluation, comparison and application
Automated and semi-automated machines	36	21	29
Traditional machines	31	12	48

Workers' opinions about work strain and physical working conditions (table 40) follow the same tendency as job descriptions. Workers on automated and traditional machines who were asked to describe their working conditions had more or less similar opinions about the measure of being overloaded with tasks, the extent of physical work strain and the probability of occupational accidents and diseases.[20] The measure of mental strain and the time of starting and finishing work is more favourable to operators of traditional machines, though there is a relatively slight divergence. Workers on automated machines had a more favourable opinion only about the circumstances of physical work (e.g. lighting, vibration, etc.). Here, too, the difference was minimal, because there were many complaints about such factors as the pollution of the atmosphere, noise level, temperature and cleanliness of workshops, and crowdedness, even among operators of automated machinery. In short, the main finding was that there were only slight differences between workers on automated and traditional machines.

Regarding work content itself, workers on automated machines saw slightly more opportunities to improve their abilities and considered it more necessary to acquire new skills, which is understandable since they were using new machines and had to learn how to operate them. They considered their work more varied and more interesting. Their independence, on the other hand, was seen to be less, which is probably due to the fact that automated machines perform jobs of especially high quality. The attention of management and of those workers whose jobs depend on them is focused upon operators of automated machines. Responsibility was considered to be very great, but workers saw no possibilities for developing new methods of work. Indeed, company management preferred technological discipline and exact observation of the processes prescribed to initiatives by workers. It is a generally held opinion, among both workers and management, that the processes of production on automated machines are worked out in such a way that improving them is not possible and that an arbitrary change might make them much worse.

Table 40: Motor vehicle plant: views on work strain and physical working conditions of workers on automated and traditional machines (flow mass production)

Item	Workers answering positively (%)	
	Operators of automated machines	Operators of traditional machines
Overloaded with tasks	30.6	29.6
Physically tiring tasks	42.8	40.9
Mentally tiring tasks	16.3	9.6
Probability of accidents and diseases in connection with work	24.4	24.4
Adequate working hours (time of starting and finishing work)	65.3	81.7
Good physical circumstances (lighting, temperature, cleanliness, low noise level, etc.)	22.5	34.6

Concerning the impossibility of developing new methods of work, the operators of automated machines made the following statements:

> Automated equipment is totally different from traditional machines. There is a strict technological programme. Here the worker must meet the demands and must attend to the machine; there is no other task of possibility. Our knowledge is not enough to carry out new methods of work.
>
> . . .
>
> This equipment is an accurately designed unit which functions according to a strict technology. There is virtually no possibility for change or innovation.
>
> . . .
>
> The technology itself can't be changed very much, but there are unsolved problems. For instance, a greater precision of sizes could be used when forging, so that it would not be necessary to chip half the workpiece away.

The fact that the operators of automated machines do not have the power to intervene in work processes was confirmed by data deriving from the descriptions of individual jobs as well. Data gained from observation of the structures of job activities indicate the measure in which the worker performing his job can depart from instructions and rules; the measure of freedom of action for workers is determined by the particular characteristics of work, the method of execution and other properties of the work process. The degree to which the work is rigidly specified by the machine is a key element of the job content.[21]

According to the data derived from observation, in the overwhelming majority (91-98 per cent) of jobs operating automated and semi-automated machines and equipment, work activities are very exactly defined, in other words programmed. In contrast to this, in the majority (68 per cent) of machine jobs on traditional machines, there is a considerable opportunity to deviate from the previously recorded course of work activities. Thus, in the latter case, the workers have a greater possibility of action during the course of job performance.

Regarding the amount of human contacts during work, one can say on the basis of workers' opinions that they are not greatly influenced by automation. In the content of such contacts, however, there are more significant differences: people working on automated machines show more interest not only in work matters but also in one another's personal concerns than their co-workers on traditional machines.

Besides contacts with co-workers during job performance, one must consider the relationships which develop between workers and their immediate superiors (foremen or supervisors). This has been measured by the following five factors regarding supervisors:

- provides workers with appropriate instructions;
- does not "molest" workers without a reason;
- if workers have problems, he or she listens to them;
- arranges bonuses according to merit; and
- possesses the necessary expertise to perform the job.

The differing levels of mechanisation do not have an immediate influence on such areas of supervision as helping or listening to workers, or giving just recognition to their work according to merit. In spite of this fact, it is worth mentioning that workers, particularly those operating traditional machines, have condemned the methods of assessment employed by their immediate superiors. This aspect of management was stressed above others, which ties up with the fact that pay took first place in the structure of workers' requirements. Meeting this need depends to a great extent on the development of incentives as practised by the immediate superior in the workshop.

Workers on traditional machines were more satisfied than those on automated machines with the expertise of their immediate superiors and the instructions given to them during work performance. This divergence not only reflects the differences in professional skills and supervisory abilities between foremen, but indirectly it also draws attention to the fact that with the use of traditional machines, the reliability of machines and equipment, the quality of the product, the pace of job performance, etc., largely depend on the level of technical knowledge, skill and expertise of the immediate superiors. In contrast to this, with automated machines and equipment - as a result of precise regulation of production and a highly programmed work performance - the instruction and professional help of supervisors is in general less necessary.

Besides the divergences among automated, semi-automated and traditional jobs, there are much more important ones found within individual groups of other dimensions of technology, for instance, among traditional machine jobs. This means that in judging such characteristics of work content as variety, independence, responsibility, the possibility of making use of knowledge, etc., one does not necessarily find important divergences in every case between the opinions of workers on automated and traditional machines. Thus, the fact that a job is performed on automated machines does not necessarily imply that it is more independent and responsible, and that it develops abilities in contrast to a job performed on traditional machines. At the same time, regarding certain work characteristics, the activities performed on automated machines are more positive, for instance in respect to the interest of work or the necessity for acquiring new knowledge.[22]

However, in our research we have not experienced as great an effect of automation as we originally presumed.

The experiences described in this paper do not seem to confirm either the extreme optimistic or pessimistic views of how technical and technological advance transforms work content. This in itself is a very important scientific finding, since it means that our current theories are inadequate and that more comprehensive research is necessary. Thus, for the purposes of analysis and description, technical conceptions embracing the whole of the system of production are needed, in particular those which pay close attention to the characteristics of job organisation as well as to the level of mechanisation of production. In our view, the conditions for co-ordinating workers' needs and work demands cannot be fulfilled merely by examining one dimension of technology, the level of mechanisation. Rather, examination must be extended to the system of production as a whole, and to the technical, technological and social concerns of job organisation.

5. Co-ordination of workers' needs and work demands: opportunities and limitations

The results of the research dealing with the social effects of automation draw attention first and foremost to the differentiation of human needs. Workers claim varied needs in such aspects of work performance as work content, level of pay, contacts with co-workers and superiors, physical working conditions and opportunities for promotion. In the light of the experience of our own investigation and of other national research projects on workers' needs, one can say that workers consider pay as most important,[23] followed by need for meaningful work content. The need for satisfactory contacts with co-workers and immediate superiors come in third and fourth places. This is followed by the need for good physical working conditions, and finally by opportunities for promotion within the job organisation. Of course the structure of human needs as revealed by research may alter considerably, not only according to differing groups of workers, but in time as well; in spite of the fact that the overwhelming majority of workers place their need for pay in first place, there are others for whom work content and contacts with co-workers are of paramount importance.

Our aim in examining workers' needs in one or more companies was not only to reveal the exact structure and importance of human needs, but to clarify the assumption that workers had not only one, but several needs in their work. Thus, the view of "economic man", where the most important factor is the satisfaction of material needs, or that of "psychological man", where non-material needs are emphasised, are both erroneous, although these concepts can often be found in the practice of industrial relations. If company management does not take into account the differentiation of workers' needs, then a situation is produced in which workers, instead of varied needs at work, set up one single but rather strong claim. One can witness this phenomenon when "in exchange for" poor contacts with co-workers or immediate superiors, workers prefer to gain compensation in the form of higher pay and to compensate for their unsatisfied needs at work with outside activities.[24]

We also found a good example of a fortunate coincidence between workers' needs and work demands at the company examined. The female workers in the workshop, who performed simple semi-skilled jobs producing binding elements in the engine plant, were more satisfied with their work than the skilled male workers in the tool plant, who performed more complicated and exacting tasks. Of the workers in the tool plant, 13.1 per cent said that they were very satisfied, 44.5 per cent were fairly satisfied and 39.4 per cent only partly satisfied with their jobs. The respective proportions of the semi-skilled female workers in the above-mentioned workshop were 12.7 per cent, 55.5 per cent and 30.9 per cent. Management's attention should be drawn to the fact that it is worthwhile following workers' needs systematically. By doing this it is easier to fulfil the principle of placing the appropriate worker in the appropriate place.

Company management has certain possibilities for action in singling out appropriate workers (e.g. through examination of workers' needs, aptitude examination, etc. Furthermore, experiences abroad and the results of our own investigations alike showed that the "appropriate place" - the structure of job tasks - need not be taken as an unquestionable given, as an unchangeable factor. Within certain limits it is possible to change the structure of work stations and posts, for instance by choosing the method of organisation of production, or by combining jobs of differing character.[25] We also encountered possibilities for enriching the content of jobs in the workshops of the motor vehicle plant we examined. In the engine plant, on the other hand, this was made impossible by separating the jobs of macine operators and machine adjusters, and by filling them with workers who were specialised in these jobs separately. This shows that the impoverishment of a large number of machine operators' jobs was by no means a consequence of technology, but rather a result of the ignorance and inexperience of managerial decisions resulting from an underdeveloped state of awareness regarding workers' needs in connection with work content. As these conditions change, the unification of machine operators' and machine adjusters' tasks will obviously become possible, thus changing the content of semi-skilled jobs prevailing in the field of engine production today.

By means of similar solutions, one can counterbalance the harmful effects of organisational methods of production which, due to their constraint and rigidity, ensure only minimal possibilities of action for workers. Jobs performed within a flexible production organisation objectively ensure more favourable conditions of work so that workers can make use of their knowledge to develop their individual abilities and to formulate new work methods. In contrast to this, the running of automated machines within a rigid production organisation has a negative influence on the particularities of work content described above. Company management can influence the job, which is considerably simplified by this type of production organisation, in both positive and negative directions. One may speak about a positive organisational change if management enlarges the activities of the workers operating machines on transfer lines with rigid organisation of production by introducing adjusting, regulating or perhaps maintenance activities.

This once again concerns the unification of the work of machine operators and machine adjusters. The unfavourable effects of the rigid organisation of production is increased by management if the task of adjusting and regulating machines is not allocated to machine operators, but to workers specialised in this one particular job (i.e. machine adjusters), even where their professional qualifications and practical experiences are appropriate.

Management also has the means to compensate for certain negative effects of automation without introducing organisational changes. Monotonous jobs or other unfavourable working conditions can be made more attractive to workers by offering higher pay. In connection with this, we draw attention once again to the limited trade-off and constant changes in human needs. Thus, the solution of offering higher pay mentioned above will relatively quickly (generally within five years) lose any stimulating and motivating effects. Once higher pay becomes an accepted part of the job and of the workers' existence, any former incentive effect will come to an end. Management is therefore faced with the need for a further rise in wages or for an improvement in working conditions.[26]

Apart from the possibility of co-ordinating workers' needs and work demands presented above, one should also mention those obstacles which limit the possibility of action by management. Firstly, the question of available and economically applicable technology arises. Among the conditions of socialist industrial relations, the relatively limited material resources available to companies, and the possibility of obtaining central material resources basically determine the technical, technological and qualitative level of machines and equipment to be introduced, the measure of planned capacity, etc.

Secondly, besides the economic and technical aspect, one must pay more attention today, and especially in the future, to the sociological characteristics of manpower. The appropriate use of decreasing manpower resources will make it necessary to acquire a more thorough knowledge of workers' needs as regards work than we have today. The successful co-ordination of workers' needs and work demands can only be possible through a simultaneous consideration of the technical, economic and social aspects of production.

Notes

[1] Sândor Benedek: "Vâltozâsok a magyar munkâsosztâly belsõ strukturâjâban" [Changes in the inner structure of the Hungarian working class], in Társadalomtudományi Közlemények, 1977, No. 4.

[2] Központi Statisztikai Hivatal: Munkaügyi adatgyüjtemény a szocialista iparban foglalkoztatottakról [Labour data on workers employed in socialist industry] (Budapest, 1974), pp. 20-21.

[3] Hungary, Central Office of Statistics: Nemzetközi Statisztikai Evköny, 1970 [International Bulletin of Statistics, 1970] (Budapest); idem: Statisztikai Evköny, 1973 [Statistical Bulletin, 1973] (Budapest). National income per head was calculated at 1968 US$ values.

[4] Andor Berei (ed.): A szocialkizmus politikai gazdasagtana [The political economy of socialism] (Budapest, Kossuth Könyvkiadó, 1976).

[5] In this chapter, a distinction is made between mechanisation (or a low level of automation) and automation proper, since much of the discussion is centred on the former category (editors).

[6] L.E. Davis and J.C. Taylor: Technology, organisation and job structure", in R. Dubin (ed.): Handbook of work, organisation and society (Chicago, Rand McNally, 1976), pp. 379-419.

[7] See the following paper on the conditions of workers' direct participation: Csaba Makó and Lajos Héthy: "Worker participation and the socialist enterprise: A Hungarian case study", in Cary L. Cooper and Enid Mumford (eds.): The quality of working life in Western and Eastern Europe (London, Associated Business Press, 1979), pp. 296-326.

[8] Lajos Héthy: "Trade unions, shop stewards and participation in Hungary", in International Labour Review, July-Aug. 1981, pp. 491-503.

[9] In the course of presenting this case study of one company, we rely mainly on the results of the international project in Hungary entitled "Automation and industrial workers". A comprehensive account is contained in Lajos Héthy and Csaba Makó: Technika, munkaszervezet, ipari munka [Technology, job organisation and industrial work] (Budapest, Közgazdasági - és Jogi Könyvkiadó, 1981).

[10] Károly Fazekas: "Teljesitményhiány" és teljesitménybérezés a vállalati gazdálkodasban [Lack of performance and the piece-rate system in company management] (Budapest, Marx Károly Közgazdasagtudomanyi Egyetem, 1980; mimeographed doctoral dissertation), pp. 21-22.

[11] Patrick Fridenson: Histoire des usines Renault: Naissance de la grande entreprise, 1898-1939 (Paris, Editions du Seuil, 1972).

[12] We should like to mention the following example to illustrate the relatively independent movement of consciousness (distortion of observations). In the course of an American survey on working conditions in 1973, the researchers had several unexpected results as compared with a survey carried out using the same methods in 1969. The presumption of the researchers in 1973 was that the people concerned would have been less satisfied with their incomes. This was based on the fact that, due to considerable inflation since 1969, a 5 per cent decrease in the purchasing power of incomes had taken place. In spite of this fact, the number of people dissatisfied with their incomes did not increase. Contrary to this, another phenomenon was found in which people were completely unaware of any favourable changes in their living standards when they had actually improved.

In the course of the 1969 survey, only 8 per cent of the women gave accounts of discrimination at work. In contrast, in 1973, 13% per cent of women interviewed spoke of discrimination. The proportion of women who complained about discrimination increased in spite of the fact that several measures were introduced to improve the equality of the sexes between the two periods. In this case we may witness a phenomenon in which, due to the rapid development of women's consciousness, they reacted more sensitively to how they were being treated at work. See G. Roustang: "Enquêtes sur la satisfaction au travail: Analyse directe des conditions de travail", in Revue international du travail, May-June 1977, p. 297.

[13] For the theoretical basis of the technological scale, see Héthy and Makó: Technika, munkaszervezet, ipari munka, op. cit.

[14] L. Rzsiga: "Vlijányije naucsno-technicseszkovo progressza na roszt kvalifikácii rabocsevo klassza" [Influence of scientific and technical progress and increasing qualification levels within the working class], in Insztyitut Filoszofii i Szociologii: Naucsno-technicseszkovo revoljucija i rabocsij klassz v uszlovijah szocializma [The working class and the scientific-technical revolution in socialism] (Prague, CsSzAn, 1979), pp. 28-30; and V.V. Krevnevics: Avtomatizácija v szocialisztyicseszkom obsesztve kak uszlovije provüsenyija szogyerzsatyelnosztyi truda i udovletvorennosztyi trudom [Automation as a tool for enrichment of the content of industrial work in Socialist Society], ibid., p. 54.

[15] J. Auerhan: Technika, kvalificace, vzdelani [Technology, qualifications and education] (Prague, Czechoslovak Academy of Sciences, 1965). p. 147.

[16] C.R. Walker: "Changing character of human work under the impact of the technological change", in United States, National Commission on Technology, Automation and Economic Progress: Technology and the American Economy, Vol. II (Washington, 1966), pp. 293-315; H. Braverman: Labor and monopoly capitalism: The degradation of work in the twentieth century (New York, Monthly Review Press, 1974).

[17] M. Freyssenet: Le processus de déqualification-surqualification de la force de travail: Elements pour une problématique de l'évolution des rapports sociaux (Paris, Centre de Sociologie Urbane, 1974).

[18] In connection with this, attention must be drawn to the fact that the character of production activities and job organisation created objectively a co-operative type of work relationship in the motor industry. Following the logic of this kind of organisation, the jobs performed here would have required a wage system based on group work or time. In contrast to this, the need to draw on effectual and organisation reserves - at least temporarily - led to wages based on individual work performance. This phenomenon also shows that company management may select from several alternatives when choosing or introducing a wage system.

[19] L. Levin: "Effects of technological changes on the content of semi-skilled machine operators' jobs", in OECD: Manpower aspects of automation and technical change (Paris, 1966), p. 48.

[20] In connection with this, we should like to mention the widely held view that with the development of technology (e.g. appearance of automated transfer lines in automotive industry), healthy and safe working conditions will "automatically" increase. According to experts dealing with automation and the protection of the worker, there is no positive relationship between them. Admitting the fact that automation decreases certain kinds of physical strain (e.g. lifting, carrying, etc.), at the same time it increases other types of work strain (e.g. higher level of attention and greater proportion of routine jobs).

The new technological processes have an important impact on occupational diseases as well. In the past ten years in Hungary, the proportion of such occupational diseases as skin disorders, lead or mercury poisoning has decreased. On the other hand, the proportion of people falling ill due to harmful vibration and noise, and those suffering from occupational infections, has greatly increased. For a more detailed analysis see National Council of Trade Unions, Labour Safety Department: A munkavédelmi helyzet alakulása a számok és tények tükrében, 1966-77 között [Development of the protection of the worker as reflected in statistics between 1966 and 1977] (Budapest, 1977); and Gyorgy Horváth: Automatizácio és munkavédelem [Automation and labour safety] (Budapest, National Council of Trade Unions, Scientific Research Institute of Labour Safety, 1975; mimeographed).

[21] One must, however, also realise that the character of "compulsion" of work activity cannot be exclusively explained by the nature and peculiarities of the work process. It can develop as a result of factors which are outside work activities, such as the irregular supply of

materials and tools, methods of leadership and other disturbances (e.g. lack of time because of stoppages of machines caused by irregular and poor quality maintenance).

[22] F. Charvat et al.: The worker in the progress of changing technology, Czechoslovak National Report (Prague, Institute of Philosophy and Sociology, 1977), p. 17; F. Adler et al.: Scientific and technological progress and social activity, National Research Report (Berlin, Academy of Social Sciences, Institute of Sociology, 1976), p. 52.

[23] In connection with this, we would like to mention that according to the findings of sociological surveys dealing with managers' needs, pay is also at the top of the list. See István Kemeny and Ottilia Solt: A vezetói magatartások szociológiai vizsgálata [A sociological survey of managers' attitudes] (Budapest, Gazdaságkutató Intézet, 1973); (mimeographed); and Terēz Laky: Erdekviszonyok a vállalati döntésekben [Relations of interests in company decisions] (Budapest, Közgazdasagi és Jogi Könyvkiado, 1977), p. 98.

[24] The so-called "compensational" and "segmentational" interpretations are worth mentioning in connection with job performance and activities besides work. According to the supporters of the compensational view, workers who are discontented in the course of job performance try to find compensation in outside activities - for instance, they take on a greater role in family life and political activities. In support of this, the representatives of the segmentational view emphasise that the different fields of human activities - job, family life, political activities, etc. - result in different and autonomous attitudes, which are more or less independent from one another.

[25] Csaba Makó: "Une experience d'enrichessement du travail dans l'industrie hongroise", in Sociologie du travail, 1980, No. 4, pp. 390-407.

[26] In the course of enrichment and enlargement of work tasks, the so-called "abrasion" effect appears. The increased opportunities for making decisions and the positive effects of responsibility given to workers in order to increase efficiency "become worn out" in a few years. A further decentralisation of responsibility is necessary to maintain the efficiency that company management considers desirable. See J.R. Maher (ed.): New perspectives in job enrichment (New York, Van Nostrand-Reinhold, 1971).

CHAPTER IX

AUTOMATION, INDUSTRIAL DEVELOPMENT AND INDUSTRIAL WORK IN ITALY

(by Federico Butera)

1. Distinctive features of Italian industrial development

Italy is a highly industrialised country whose economic indicators are close to those of the most industrialised nations of the world. However, some characteristics make the Italian industrial system quite different from others. Its development has had three main phases.

1.1 The industrialisation phase (1945-60)

The process of intensive industrialisation is relatively recent; apart from the heavy mechanical industries, most of the industries were established or grew on a large scale only after the Second World War.

In the period between 1945 and 1965, industrialisation involved mainly the creation of big companies and large plants, almost exclusively located in the so-called "industrial triangle" of northern Italy, bounded by the cities of Turin, Milan and Genoa. In the other northern and central regions, traditional small-scale industries (food, glass, clothing, leather and shoes, etc.) grew slowly, while the southern (and insular) part of Italy remained basically rural, with the exception of a few huge, isolated process plants (so-called "cathedrals of the desert"). At the same time, in the industrial triangle heavy and light mechanical industries, including motor vehicles, textiles, chemicals, printing and engineering, were established or developed with advanced technologies and rational organisation patterns basically derived from the traditions of the United States of America.

The structure of the ownership of industrial companies is another Italian peculiarity. In addition to large and small private companies, a substantial number of basic industries were state owned through two main agencies: IRI, which controlled manufacturing industries like steel, motor vehicles, shipbuilding, heavy motors, aircraft, etc., and ENI, which controlled the oil industry. Lately another agency (EFIM) joined the two.

The process of rapid industrialisation had an important impact on society and on the life of the Italians: growth of large cities; impressive development of transportation facilities like motorways; availability of durable consumption goods; and changes in the educational system. Significant changes also took place on the institutional side: private and public industrial management acquired great influence in political affairs with the substantial agreement of the traditional rural and commercial ruling classes of the south. We may state that the growth of large enterprises was the leading force in economic and social affairs in those years. This industrial growth reinforced the tendency to political stability; the Christian-Democratic Party ruled the country along with small moderate parties up to 1962, when the Socialist Party was also brought into the Government. The Italian Communist Party, the largest among the Western countries, which was part of the Government immediately after the Second World War, remained permanently on the opposite side after 1947, in spite of its large base among industrial workers.

Alongside this political polarisation, the trade unions were divided in 1948 into three main confederations: CGIL (Italian General Confederation of Labour), which remains the largest including communist- and socialist-oriented workers; CISL (Italian Confederation of Workers' Unions), at that time basically christian-democratic; and, later on, UIL (Italian Workers' Union), including republicans, social-democrats and socialists.

Only in the seventies were attempts made to overcome this division, which still formally exists, even with a different internal composition and with an operational unification at policy and organisation level in the name of trade union unity. In the fifties, as now, each main confederation was divided into trade federations (metals, chemicals, textiles, transportation, etc.). For instance, the union-affiliated workers in the mechanical industries belonged to FIOM (metalworkers' union, part of CGIL), to FIM (part of CISL), or to UILM (part of UIL); nowadays all three federations are also part of a unified trade union, FLM (Federation of Metalworkers' Unions).

A high level of politicisation and division among the trade unions was characteristic of the first two phases of industrial growth in Italy.

1.2 The rationalisation phase (1960-68)

The sixties were the years of the most explosive phase of industrialisation in quantitative and qualitative terms. Some of the distinguishing features of those years were as follows:

- the reinforcement of organisation rationalisation in the large companies based upon job fragmentation, formalisation of control procedures, increase in bureaucratic rules, strengthening of hierarchical structures, etc.; in short, all that represented "the organisation" in industrial culture, as opposed to traditional organisation and the craftsman job structure;

- the expansion of mechanisation and - in process industries - of the first phase of automation. In contrast to the previous period, the technical solutions were widely studied and implemented by national technicians proud of their technical ingenuity and overall rationality. Technical progress tended to become a value in itself;

- the immense emigration flow of millions of southern peasants, fishermen, craftsmen, etc., to the fragmented but well-paid industrial jobs in the large cities of the north;

- the increased availability of consumer goods to the working class, reinforcing an instrumental attitude towards work and the acceptance of the importance of rationalisation;

- the level of wages, which were lower than elsewhere in Europe, and the relative weakness of industrial workers on the labour market, unemployment still being high;

- the high level of profitability and competitiveness of Italian companies;

- the beginning of industrialisation of central Italy; and

- persisting political stability along with the division of the labour market.

1.3 The crisis phase (1969-)

At the end of the sixties, some of the above-mentioned influences began to change:

- the saturation of the labour market in northern Italy, the improved strength of its industrial workers and the rise in salaries close to the European level;

- the permanent settlement of the immigrant workers and the rise in expectations related to their working conditions;

- the increasing complexity of Italian society and the decreasing economic and political influence of the large companies; and

- a turbulent international conjuncture, both on the economic and political front, and the beginning of the Italian economic crisis.

2. Workers' protest and trade union action: the rise of work organisation as a social issue

In the autumn of 1968 an unprecedented series of struggles were initiated by the industrial workers in the big northern factories. In addition to their extensiveness, those struggles possessed many unusual features in that they were conducted mainly by unskilled blue-collar and highly skilled white-collar workers in a spontaneous way and had as their main object such aspects of work organisation as piece-work, production rhythms, qualifications and hierarchy. They also displayed a certain amount of ideological radicalism. These events came as a surprise to everybody, but mainly to the trade unions which, however, were soon willing to patronise these struggles and to enable profound transformations in the workers' representation system by introducing directly elected Consigli di Fabbrica (factory workers' councils) and by fostering the unification process among the three trade union confederations. Subsequently the unions, which had by now acquired greater power, decided to tackle the issue of changes in work organisation as one of the main areas of demands and bargaining with the employers.

A poor working environment, monotonous and fragmented jobs, lack of qualifications and career opportunities and lack of workers' influence on production decisions came to be strongly criticised in the union literature. Demands for changes in the prevailing pattern of division of labour within the factories were furthered by means of more or less traditional union claims (more grades; rules for internal mobility; rejection of the principle of payment for harmful or dangerous working conditions; elimination of incentive systems; changes in administrative systems of work regulation, etc.). Work organisation became a social issue and the subject of industrial relations.

Industrial management, according to the tradition of the type of industrialisation process described, first tried to import from other countries remedies presented elsewhere as able to counteract workers' dissatisfaction and union demands: job enlargement; job enrichment; job rotation; and work groups seen as ready-made formulae. Visits to Volvo, lectures from foreign experts and simplistic experiments were frequent in 1971-73, but none achieved the immediate improvement expected in motivation and industrial relations.

A very limited number of companies undertook more serious and long-range planning in order to modify their work organisation in view of improvements in QWL as well as of productivity increases. They had realised that structural social and economic reasons necessitated a review of the prevailing patterns of work organisation. Many others made changes primarily within the perspective of responding to union demands: work organisation reform was first and foremost a cost issue and, secondly, one concerning industrial relations.

Management-union negotiations regarding grades, training, mobility, careers and work environment was intense during these years; however, a general agreement, as in Sweden, did not emerge. Starting in 1974, union interest in work reform seemed to become more ritual and verbal, and no procedure was generally reached on principles or on ways for implementing the changes.

3. The changed features of the industrial system[1]

An apparently surprising phenomenon was the increasing number of innovative changes in work organisation, in spite of less stringent union action at plant level. Workforce rigidity, increased cost of raw materials, inflation, reduction in exports, investments, etc., were pushing many companies towards more innovative, more flexible and less bureaucratic forms of organisation and a better utilisation of human resources; i.e. organisational change as an adaptation to more severe constraints or to uncertainties. The reduction in manpower was often counteracted by the trade unions, while fragmentation and acceleration of work were rejected both by the workers and by the need for cheaper and more flexible co-ordination and maintenance.

There were two types of adaptive responses to pressures: planned changes and unlabelled changes. On the one hand, the management of a selected number of important companies, mainly in state-owned process industries, promoted an integrated series of organisation development programmes including innovative job design and workers' participation in it: training programmes, research, organisation studies, action research, experimentation, etc.[2] Trade unions were urged to participate in those programmes, but responded negatively.

The small number of planned changes did have a main function, but they did not produce any general management-union agreement or any government involvement. However, there were other profound changes, as follows:

(1) A large but unknown number of small unlabelled changes in job design in large companies reflecting new principles of increased autonomy and responsibility at work (longer work cycles; jobs including production, maintenance and quality control; job rotation; group-based organisation; continuous careers, etc.) were introduced in an unplanned way for reasons concerning production and/or for negotiations.

(2) An unprecedented development took place in small and medium companies, partly as a result of the decentralisation of production cycles away from larger enterprises; most of these were autonomous sound small enterprises. These companies displayed a relatively high technological level and an "organic organisation".

(3) A profound modification of the labour market occurred: out of more than two million unemployed, it was estimated that only half a million (mainly located in the South) were willing to take any available job, while one million were seeking work "at achievable conditions" (regarding content, working time, location and career opportunities) and half a million only on very strict conditions. As a result, the amount of clandestine home-based work has greatly increased, while a totally new phenomenon is appearing: i.e. the immigration of around half a million people from Africa, Turkey and other developing countries to take up heavy and poorly paid jobs.

(4) An acceleration in technological development took place both in large and small enterprises. This may be characterised as follows:

(a) the whole range of substitution of transformation technologies is experienced, from more efficient small and large mechanical machine tools, to new chemical processes, etc. NC machines, machine centres and flexible manufacturing systems have been widely adopted, even in small units. Robots in assembly, welding and painting are appearing at an increased rate, and computerised control is the norm in process industries. Most of these technological innovations, unlike those in countries like the Federal Republic of Germany or Japan, are introduced piecemeal on a day-to-day basis. Most of them are labour saving, and reduce the transformation skills required of blue-collar workers. Because of the prevailing adaptive pattern, technological innovation is often developed in co-operation between the builder and the user;

(b) such areas of technological innovation as transportation technologies, materials handling, buffer systems, etc. Ingenious new solutions are applied and, moreover, a general computer-assisted systems approach to the physical displacement of material is often developed in various industries. Savings in the work in progress and higher manufacturing flexibility are pursued. A strong impact is often felt on the relationships between human tasks, so that the structure of work relationships and, consequently, on the content of roles are modified;

(c) data processing technologies, in Italy as elsewhere, show the most striking changes. The proportion of central process units (CPU), small special-purpose computers, peripherals, network and software packages, etc., introduced in Italy has been very high in all areas of industrial applications like management administration, sales, manufacturing and R & D; office automation is now appearing. The large multinational corporations dominate these developments in three respects: first, the existence of national computer manufacturers; secondly, the establishment in Italy of large subsidiaries performing autonomous R & D and manufacturing; and thirdly, the strong propensity to users not to buy turnkey equipment but to develop a technical-organisational philosophy suited to their needs. In fact, the boundaries of research into EDP philosophies and applications often overlap those of research into organisational philosophies and projects. The main issues are centralisation versus decentralisation, concentration versus diffusion of control on information systems, open- versus closed-loop feedback on processes, and design of jobs encompassing understanding of the control cycles of events versus design of jobs ancillary to EDP, etc. Issues like participation in EDP design and joint design of EDP and organisation are also becoming widely discussed;

(d) finally, innovations in what we call "systems technologies": complex and homogeneous systems of machines, information systems methodologies, control systems, integrated business functions through a total system design. Many managerial approaches (e.g. the management decision systems) are largely computer based. Approaches that are less technocratic, more participative and more oriented towards the social system are being developed, as in the case study presented below.

(5) The final change which has occurred in recent years is a cultural one, namely the development of a shared set of new principles and parameters on work and organisation design very different from the simplistic formulae of the early seventies. The realisation of a large number of interdisciplinary, action-oriented research projects on work and organisation, initiated by professionals on the one hand and by managers and trade union leaders on the other, have contributed to the wide diffusion in the professional and popular press of results of both successful and unsuccessful experiments. A new network of institutions and individuals has been informally established, crossing the barriers between different disciplines and conflicting institutions and providing for the emergence of a new industrial culture in the country. To conclude,

we may say that Italy does not have concerted programmes, government actions and established institutions directly confronting issues concerned with QWL, humanisation of work and of technologies, and so on. Nevertheless, there is a profound process of modification of industrial systems, where management, unions and government cope definitively, by various methods and in various contexts, with serious economic and social problems, where a large amount of experimentation and learning processes on work organisation are taking place, and where an emerging informal network of professionals and concerned students of work redesign is becoming active and innovative.

4. Case study - Joint design of technology and organisation in a new automated rolling mill

4.1 The NTM system: an outlook

NTM, which stands for nuovo treno medio, refers to a new seamless pipe mill that Dalmine, one of the largest steel companies in Italy installed in one of its plants in 1977-78. Several technical and economic features of the system are innovative, among them the large size, complexity and expensiveness of the huge "contrivance", which covers as much as 65,000 sq m.

Other characteristics of the pipe mill are equally impressive. Tubes are produced in a quasi-continuous flow at a rate of 35-50 seconds each, resembling a chemical process rather than a mechanical system. The 13 motors that drive the elements ("stands") in which the white-hot pipe is brought down to the desired length and thickness are machines as high as a three-storey building arranged sideways in an enormous ditch surrounding the main unit. The cabinets which contain the equipment to control the power output cover some hundreds of linear metres, thus giving a visual image of the large amount of energy required and the complexity of its control, even to a layman. The manufacturing and handling processes are controlled by a computer concentrated in the central control room and 12 processors allocated to the peripheral monitoring stations.

Among all this advanced technological equipment, very few workers can be met. People can be seen in the control rooms and offices, and a good many workers only appear near the machines when these stop or have to be serviced. Here, the mythology of the "automated factory", as a large self-regulating machine with a few workers whose tasks are restricted to supervision or maintenance jobs, may seem applicable. The technical complexity and an investment expenditure in excess of US$150 million also elicit the image of a huge technical and financial effort in the engineering and construction of the plant, unhampered by organisational considerations or social concerns.

What happened, however, is widely different. Behind and beyond the spectacular technical and economic complexity, a far greater socio-organisational complexity shows through, which appears to have been thoroughly studied in advance and managed in many ways by everyone in the plant.

This case study will, first of all, show that the nature of the plant design and implementation process has been mainly an organisational and social one, in two senses: both in that the most dedicated effort was made to organise people to develop and engineer a socio-technical system of an unprecedented complexity in the short run; and in that such effort was aimed at integrating technical and production/organisation aspects with the social system by projecting problems and effects which might arise in the ten years following the machinery start-up. The planned process of the combined technology and organisation design will be one of the focuses of this case.

The second focus will be the content of human work in handling the plant. It will become apparent that the high level of automation has reduced the amount of total work, as well as its craft content. It will also emerge, however, that the work was deliberately designed in such a way as to secure control and regulation, self-coordination, and innovation, for almost any role, quite unlike the automated factory stereotype, which can do without people. The case study also indicates the possibility of combining fully fledge automation and work rich in productive process control.

The combined design of technology, organisation and social system, and the adoption of advanced QWL criteria in the configuration of work content were implemented by using methods and innovative principles that may lead to some generalisation. What is, however, more susceptible to generalisation is a good diagnosis of the phenomena involved, the core of the entire experience. This was a diagnosis according to which a new processing technology and a huge investment made such technical and economic demands that it was essential to consider social and organisational effects in advance.

4.2 The subject matter

The present study is thus intended to illustrate and discuss the process of analysis, design and implementation of a new processing unit making up an entirely unknown socio-technical system, since the organisational unit, the scale of production, the primary and secondary technologies, the production organisation and the roles of workers were completely new both inside and outside the company.

The pages that follow are also intended to provide some contribution to the following issues and questions debated in the scientific literature concerning problems of alternative work and organisation models in highly automated environments:

(1) What is the degree of freedom in organisation choice? Here, the author's assumptions will be explored: work organisation evolves by successive (logical) stages, and at each stage the degree of freedom of organisational choice gradually narrows, while each stage is affected by sets of different internal and external variables. These stages are -

(a) total aggregated work, or the amount of work which is complementary to technology, and primarily influenced by the availability and cost of base technology, as well as the cost of labour;

(b) overall control system, influenced by the architecture of the production system as planned against the market;

(c) task content and inter-relation, mainly influenced by secondary technologies (information systems, instrumentation, machine tools, safety systems, etc.), which are, in turn, influenced by the performance system, and by the image of the social system;

(d) formal organisation structure (organisational units, jobs, procedures, etc.) influenced by economic criteria of boundary definition (technology, time, territory, results), social criteria (labour force composition and power distribution), and ideological and cultural criteria (organisational theories, authority concepts, etc.);

(e) system of task allocation to people, influenced by such variables as the internal labour market, occupational training and the stratification model;

(f) system of personnel management (working hours, pay, skill grades, punishment and reward scale, etc.), mainly influenced by industrial relations; and

(g) system of actual roles, which is affected by values, skills, institutions, goals, etc., of the external social system that people bring with them into the subsystem.

Here, a brief discussion will be devoted to the assumption that the most important, although not exclusive, stage in determining job design is stage (c), and, along with it, that the variables to be designed in advance and intentionally if a desirable job design is to be obtained are the secondary technologies. A more general discussion will be devoted to the relationship between contextual (external) variables and structural (internal) variables as the configuration factors within the scope of which the organisational choices can be made.

(2) Another issue discussed here will be the nature of the operation of distinguishing between internal and external variables, consisting in defining the boundaries of subsystems. Such an operation is formally one of the major manifestations of organisational choice, which falls well within the scope of managerial duties. A debate is being pursued in the literature as to whether this is an arbitrary operation, one that is guided by abstract technical and economic criteria (e.g. the

rules of technology, time and territory) or else influenced by economic or social corporate variables (e.g. product market, cost of labour force, labour market).

(3) Which are the variables (in terms of volume, production, process continuity, plant safety or degree of complexity of the production variables) that require automated process control systems? Which are the variables (of a technical, system implementation, cultural, social or other nature) that may or may not suggest a human control system to accompany or support the automated system? Finally, what positive or negative consequences are to be expected from entirely automated or combined systems?

(4) The likely characteristics of a system of human control in a highly automated process will be explored, firstly the object of such human control, followed by the configuration of the information system among the employees, the design of operating and supervisory roles, the task allocation system, and the kind of skills and abilities to be developed.

(5) The study will also cover the problems of the consistency and mutual support of the human control system and the production process, on the one hand, and the effort to design good working conditions (ergonomic design, work environment design, etc.) along with good management systems (recruitment, formal training, careers, working hours, etc.), on the other. In particular, questions will be explored concerning the harmonisation between personnel management systems that can be adopted in a new or modified socio-technical subsystem, and the working conditions and personnel management systems adopted throughout a company that has not undergone similar changes.

(6) A description will be given of problems related to the design of macrostructures, particularly the maintenance, quality control and engineering functions; and the problems of consistency between the macrostructures in the new part of the firm and the macrostructure of the company as a whole will be discussed.

(7) Finally, some attention will be given to the problem of defining criteria and indicators of cost-effectiveness and QWL.

4.3 The company and its external environment

Dalmine is one of the largest steel companies in Italy, a government-controlled enterprise of the IRI Holding Group, manufacturing pipes of all descriptions in seven processing plants in various parts of Italy. Its main distinctive product is the seamless pipe, in the manufacture of which the company also has a leading international position. The plant that accommodates the new processing system which forms the object of this case study is the largest in the company, located in a self-contained company town in the Italian countryside.

The company was first incorporated in 1906, using the Mannesman rolling process and highly rational organisational patterns on the German model. Both the technological process and the organisational structure were managed without problems for many years, through employing young, well-qualified engineers who were earmarked to become highly skilled managers, and hard-working peasants from the surrounding countryside, a few of whom were to become capable process or maintenance operators, and many more industrial labourers for heavy, unskilled tasks. After the Second World War, the company already had a sound economic standing and a high technical reputation. It had even become a social institution in the area, with its company town and comparatively plentiful social services.

In 1970, the company produced 820,000 tons of pipes, and achieved sales of 137 billion Italian lira (Lit), employing 12,400 workers. However, the equipment, as well as the technology and processes used, had become obsolete.

Material and labour costs were rising steeply. The simple organisational structure resting on a clear-cut functional differentiation of services, and a strict division of labour, tended to create rigidity, no longer mitigated by a spirit of co-operation which was now marred by growing labour conflicts. The workers' demands and the trade unions' increased bargaining power were producing changes in the work environment and organisation, as well as increasing costs. The competition from foreign manufacturers had become keener and keener, with major threats resulting from often lower unit prices, a greater adherence to market variations, the supply of "promised quality" (neither better nor worse than the quality required by customers), readier deliveries, and a more comprehensive marketing network.

A technological and organisational renewal, then, became a central issue for the company throughout the seventies. Many attempts at a localised improvement of equipment, with automated information systems, and organisational planning and marketing structures had already been made. It was in the first half of the seventies, however, that management became gradually oriented to an overall plan of organisational development, to be followed by a plan for all-embracing plant and equipment renovation.

This was the starting point for a co-ordinated set of organising and training activities with full support from top management. The earliest such actions impinged upon such macrostructures as the marketing, planning or maintenance divisions. Later on, projects were developed within production units, making it easier to connect the design of the "top" structures with work organisation. As a consequence, a large number of task forces, research and action research projects, and experiments co-ordinated by a specifically created organisational development department were established.

Various goals were set for these efforts: a planned change in the organisation; management training; a change in the work organisation and environment (as a result of the unions' demands); changed industrial relations rules (e.g. the formulation of work grades); and, last but not least, the development of a corporate culture, analytical tools and assumptions relating to further organisational changes to be used when introducing the new process facilities.

The second part of the overall plan was the construction of new facilities at the Dalmine plant, including a new steel mill and a new rolling mill (the NTM), both of an entirely new concept. The NTM project, involving a capital expenditure in excess of Lit150 billion, was passed in agreement with the top executives of the holding group in order to achieve a better quality of production, reduced product unit costs and expanded sales. The engineering of the primary facility was jointly conducted by Dalmine and by another company of the holding group, which took on the construction of the processing unit.

The work for the new steel mill was started in 1975 and that for the NTM in 1977. NTM was started up in 1982. The production figures for 1980, immediately before the full NTM start-up show that the preceding ten years were used by the company to reshape both its organisation and its know-how, and to qualify its production, by suppressing those products that exhibited the lowest added value. Although the gross annual production volume was not substantially increased over the period (886,000 tons), the product mix was improved in favour of products with a higher added value. The labour force remained basically unchanged (13,000 employees), but its composition changed in favour of higher qualification grades. Sales in monetary value went up from Lit137 to Lit598 billion.

The NTM product line, which was added to the older facilities (while replacing some products) will substantially increase throughput, although it poses severe problems of return on investment. It can be readily seen that the two strategic interconnected variables for the company in the next ten years will be: (a) the ability to market the increased output at a competitive price; and (b) the ability fully to use the productive potential of the new pipe mills. It is top management's awareness of the significance of both variables, rather than an abstract methodological preference, which explains the choice to design and start the new facility with methods, as described below, quite unusual in the Italian industrial environment, in view of the unprecedented degree of participation and integration of the various aspects under investigation.

4.4 The conventional system and its related organisation and quality of working life problems

The conventional processing line, now superseded by the NTM pipe mill, is a pilger mill. It is made up of a set of facilities which change a tubular billet into a pipe through a number of stations which process the work in a sequence. The billet is first made white hot in a furnace. A press then changes it into a piped billet (or "cup"), which is a white-hot billet with a hollow top in the direction of the length.

This work is fed to a raking set of rolls (OBO), which pierce the "cup" through its length. The pilger rolls then bring the piped billet down to the desired length and thickness. This is done by introducing a mandrel into the pierced white-hot piece and "aiming" it through a pair of cam-shaped rolls which move continuously. Under a number of heavy pounding beats by the jaws in the narrower section of each roll, the pierced tube is made to advance, thus becoming longer and thinner until it is turned into a pipe.

This process incorporates some continuity, controlled by the withdrawal directions given by the OBO operator. The core of the process is the operation of the pilger rolls, exhibiting a quasi-craftsmanlike stage, in which the operator must first select the right time to insert the hollow piece into the opening of the rotating jaws, then adjust the advance rate and the jaw gap. This operation is the time when variances, as caused or detected upstream, can be removed from the workpiece or when variances can be created which may affect the following stages in the process.

An action research study on this plant, conducted by the present writer in 1975-76, revealed the large number of equipment and process variances (and the strong interconnection between them) that must be kept under control to obtain results. The operators were found to be the only people able to control most of these variances in a timely fashion. However, they did so only in a random and inadequate way, for various reasons: their training background; the design of their roles bearing little or no responsibility content and oriented to the control of only certain parts of equipment; their physical dispersion along the line without any information system connecting their various positions; the existence of jobs strongly differentiated in their content, physical load and responsibility; and overlapping responsibilities between supervisors and operators. Finally, there was a critical integration between manufacturing, maintenance, quality control and planning. The research first introduced in the firm the principle of designing a socio-organisational system oriented to the control of variances, based on closed control loops, conscious co-operation and the definition of process boundaries in which partial results could be visible. The result was expected to create enlarged, multiple and result-oriented work roles, which were rich in operating, diagnostic and relational skills. The outcome of the research study on the conventional plant will not be discussed here. It is enough to say that the following problems were to be avoided in the new system:

- highly specialised and isolated jobs, responsible for machine operation rather than for the production process, as in the older system; and

- operators' roles converted into mere monitoring roles requiring no skills or knowledge, and which were monotonous and deprived of any contribution to the handling of both the equipment and the process, like in most automated factories.

4.5 The new production unit

The new production unit (NTM) is made up of a number of tools working on a continuous material flow to process almost always square-shaped bars - weighing anything between 360 and 2,500 kg and measuring 2 m long - into seamless pipes, 3.85 mm to over 25 mm in wall thickness and 26-28 m in lenth. The innovations over the conventional production units were threefold, as follows:

- production volume and continuity;
- processing and handling technology; and
- automation of process control.

4.5.1 Production volume and continuity

The production capacity is around 450,000 tons per year. The continuity of the process is high: a workpiece is passed through the main mill every 35-50 seconds, and approximately 400 pieces are moving through at any one moment.

The most critical management problem associated with such volumes and such continuous process is the handling of a very high product mix. The processed pipes vary in length, size, wall thickness, type of starting material, coating, machining and other factors. In other words, it is an almost continuous flow within which several well-identified products are processed.

Each workpiece in progress is associated with a specific rolling schedule. The distinctive features of each incoming workpiece and the related processing schedule must be identified at each station. Each piece is identified as soon as the billet leaves the steel works. Marks, weights, measures, photoelectric checks of through steps, or space discontinuities - all contribute towards giving the piece "a name and surname" at each station, to check whether it matches the pre-set rolling schedule. A "work tracking" automatic system is currently being completed to anticipate and follow the work flow in real time, and this is used as a data base for the scheduling system.

4.5.2 Processing and handling methods

The square billets from continuous casting are stored in a storage area outside the rolling mill. A few days before starting any specific rolling schedule, the bars in the required amount and type for that schedule are moved to another storage space within the rolling mill, cut to length by a saw or a hot working tool, mark-checked for the starting cast and the final intended pipe, stacked and, at the appropriate time, weighed and sent to the primary furnace. The handling of the bars within the internal storage area is secured by overhead cranes equipped with magnetic plates, beds and roll back-plates.

The bars are fed to a rotary hearth furnace, 46 m in diameter and 160 tons/hr capacity, and heated up to a temperature of 1,290°C. Each billet, as discharged from the furnace, is sent through a descaler, where it is subjected to a high-pressure (180 atm) water jet for the removal of the oxidation scales that have built up on the heated piece.

A rotary sizer is then used to bring the billet cross-section to size. The billet is then passed on to a press-piercing mill (PPM), which is one of the most innovative features in process technology throughout the production unit. The PPM simultaneously makes a hole by means of a fast forward and backward punch and carries out a rolling operation in which the billet cross-section is changed from a square to a round shape. The stand is a large steel box, more than 10 m high, which can be removed by an overhead crane and set up on the floor at each work change. The stands are duplicated to avoid any downtime, and in order rapidly to accommodate changing dimensions of the work. No piece elongation takes place at this stage.

An elongator mill then elongates the work as much as two or three times its original length and the hollow bloom is sent to the most powerful and important unit in the plant. This is a retained mandrel mill (MPM), which consists of the most significant innovation in the process technology.

The MPM performs the same function as the former pilger mill, but its size, technology and operation modes are entirely different. It is an eight-stand mill, which means that eight pairs of rolls are driven by 13 motors with 20,000 kW overall power output, gradually reducing the tube wall thickness and elongating the pierced body by up to five or six times. The advance of the hollow bloom is ensured by a cooled-down lubricated mandrel, which is inserted into the pierced body and advanced at a controlled speed, thus making for a constant advance of the tube, which is otherwise unaffected by the acceleration that the rotating rolls may tend to produce. The tube comes out of the last stand at a speed of approximately 4 or 5 m/sec (or 15 km/hr). The unit can discharge two tubes 28 m in length each minute.

The pipe has now been completed. It will later be rounded up on an extracting mill and sent, if necessary, through a reheating furnace and a stretch mill to be reduced in diameter and wall thickness.

The hot process ends here. The pipe is sent over to two finishing stations and one adjusting station, and to machining and hot-treatment areas, where the pipe is cut to the intended length, ground, inspected, heat-treated or painted, etc., depending on its end use (the cold process area).

4.5.3 Automation

Most of the scheduling, handling, processing and power input control operations are handled by a computer. There are two main systems, as follows:

(1) An automated management control system controls the entire cycle from production scheduling to material handling in progress. It is divided into two subsystems -

(a) the scheduling subsystem, handled by a computer with company-developed software, is connected upstream with the commercial offices through a number of terminals, and downstream with the factory, to which it supplies the production schedule down to such details as keying in individual rolling schedules;

(b) the material handling subsystem controls any information about the pieces being handled in the storage areas and in process (the tracking system mentioned above) by means of two computers.

(2) A technical automation system, which is made of three main subsystems -

(a) computerised in-plant control of the mechanical and electro-technical operation variables in the line, such as power input values, fluid control, machine operations, thermo-mechanical occurrences, and other variables. This subsystem is, in fact, made up of a number of highly integrated localised mechanical, electrical, pneumatic, hydraulic and electronic automatic controls, which are supported by a central computerised system (a "supervisor");

(b) this incorporates the rolling schedule control, under which the required rolling parameters are keyed in, the features of each workpiece in progress are checked and compared with the standards (i.e. temperature, size, weight and thickness), along with the plant performance rate (positioning rates, power input, fluid flow, etc.), and the relevant process variables are adjusted;

(c) this governs the physical handling of the workpieces through the processing facilities. The square billet, once it has been positioned on the working bed by the overhead crane, is automatically guided in its run through the various stages at which it is processed into a pipe. Provisions have also been made for manual operation. Due to the high advance rate, however, manual operation is largely restricted to test or maintenance runs.

The choice of the level and type of automation was made chiefly during the joint technology and organisation design stage that is the object of section 4.10 below.

4.6 Contextual variables, entrepreneurial choices and initial stages of the organisation formation

This and the following sections show how, given the base technology so far described, opportunities were used to select non-conventional approaches to secondary technologies and organisation and, above all, methods to design them.

First, however, it is necessary to summarise the most obvious motives and consequences of the strategic choices. Let us first assume that the work and organisation configuration is formed by logical stages, each of which influences (but without determining) the following stage, and is influenced by specific external (contextual) and internal variables.

4.6.1 Investment choice

Financial motives appear to be predominant in the investment choice, particularly in circumstances in which foreign competitors might have lower product unit costs, due to higher technological levels (processing equipment with a higher unit output and of a better quality), lower wages or higher manpower content in direct and indirect labour. Any partial cost-reduction plans had been unsuccessful, and the mature technology afforded no major improvements or useful partial renovations. The existing wages and manpower force could be not reduced because of the policies of a strong trade union. Further, any organisational improvements likely to influence efficiency and costs of commercial, administrative and technical labour were delayed by a set of long-established interests and cultural resistance.

As a second motive, in view of a world-wide crisis in the steel industry, the growing demand for financial funding by government agencies could only be met on the basis of sound investment projects rather than mere management considerations.

It thus becomes clear that, although the choice to invest in an entirely new plant was principally influenced by considerations of competition as well as by the need to increase the company's worth, it also had intervening variables in terms of the objective and subjective limitations to any improvement in the existing technology, the rigidity of manpower costs, the rate of bureaucratisation, and the degree to which organisation development plans could be implemented - all or most of these being socio-organisational dimensions. We may therefore put forward the view that any new investment project is chosen to obtain more competitive product unit costs in the actual or estimated absence of any possibility of influencing the costs of the production factors in the present configuration of the socio-technical system. In other words, the investment choice - in this as well as in other instances - is a combination of an economic assessment of the market/product and costs and a diagnosis of the internal change factors in any socio-technical system.

The investment choice, then, is simultaneously influenced by economic/technological and socio-organisational factors, both of which are partially external and partially internal. These considerations are estimated by management's fact-finding machinery, with a stricter appraisal of the economic/technological factors and a more subjective judgement of the socio-organisational factors. With respect to time, it would also be incorrect to imply socio-organisational consequences of technological choices because - explicitly here and implicitly in other instances - there is a time cycle which takes place as follows:

- organisational action, in an attempt to optimise the organisation with an unchanged technology, generating a diagnosis on the margins of changeability of the organisation;

- technological action, tending to attain unprecedented performance and organisation changes; and

- organisational action, to re-establish a "stylistic unity" of management control and optimal use of resources.

4.6.2 Appropriate technology

Within these premises, how can the most appropriate technology be selected? The image of a process of acquiring the most cost-effective and safe technology on the market of advanced technologies would be incorrect. This stage of the choice process has also been: (a) the result of a choice and combined optimisation of cost, technical and social parameters; and (b) the object of extensive designing - not only purchasing - activity by the company.

Thus, two main goals were first identified for the new production unit, i.e. a high physical plant productivity and constant quality, under the constraints of -

- operating costs (equipment, maintenance and manpower) such as to afford a lower product unit cost;

- technical-organisational consistency of the new unit with the rest of the company, within which it should be accommodated as another component; and

- preservation of the existing overall employment levels in the company (taking into account the fact that the NTM - the production volume being equal - employs only 50 per cent of the older system's personnel).

Secondly, management resolved to design the new production unit by entrusting another company of the same group with the design and construction of the equipment, but it also reserved to itself - although in a continuous interaction with the outside engineering company - the design of the processes, of production scheduling and control, and of the organisation. In summary, the company did not delegate the implicit redesign of either its product know-how or its organisation to its technology suppliers.

The base technology as such (the processing tools described so far) only directly affect the quantity of the labour force employed, some operation, handling, maintenance or other tasks being either suppressed or performed by machines. This is the first stage in the formation of the overall work. As will be seen, other elements of the organisation and work are only affected in an indirect fashion. It would then be arbitrary to expect any correlation between the technology (misrepresented as a base technology) and the organisation (extensively understood as inclusive of jobs, management control systems, the social role system, etc.).

A stage of overall work formation which logically follows is one in which the features of the control system are first outlined. In this case, it changes (according to Woodward's scheme)[3] from personal and impersonal/administrative to impersonal/technical. This means that the control of the production process and human resources takes place mainly through technical procedures handled by a computer This occurs less because the base technology has changed than because the production system has changed.

The two fundamental goals of the system were said to be a constant quality and a high productivity at decreasing product unit costs. The processing facilities are well equipped to meet these goals, as long as they are not halted, and the control system remains consistent with the characteristics of the production system as a whole. The production system to meet these goals is not only identified by its operating complexity and speed of the individual tools but also by the fact that these tools are arranged in a sequence, so that the work flows from one to the next in a continuous manner and what occurs at any one stage can be either corrected or aggravated at a subsequent stage. This amounts to saying that the most relevant features of the production system are its continuity and integration. The operating complexity and speed of the individual tools, and the continuity/integration of the process require a very large number of strongly correlated parameters to be controlled in very narrow time units with the utmost reliability.[4] The high control operating speed and reliability requirements can only be met by an automated data processing system (i.e. a computer-aided control system).

The system of control of human resources (operators, supervisors and technicians) becomes dominated by the impersonal/technical aspect of the automation system. However, no base technology <u>per se</u> can dictate any automated control system. Theoretically, the facilities described here could well be controlled by a large number of people performing information collection, processing, comparison and transmission tasks. What influences the choice is rather the chain of relations between performance goals (constant quality and high productivity), connotation of the production system (continuous-flow integrated process), complexity of the control of the individual machines and the system, the need for speed and reliability, and lastly, the cost of the labour force.

4.6.3 System of work relations

The third stage of the overall work configuration is the one at which the content of human tasks and their connection (the system of work relations according to Herbst) are defined, these being the building blocks of the organisation.

The base technology and the computerised configuration of the control system alone cannot determine the structure of work relations, which is formed only when the system of secondary technologies is chosen, involving the selection of automated information systems, instrumentation, environmental protection systems, etc. The most relevant aspect in our case study is the level and type of automation. Its selection obviously depends not only on the choices made as to the selection of base technology and computerised control configuration, but also - and above all - on a variable that becomes all important at this stage: the actual image of human behaviour in the production processes, and the roles that people can and must assume under such automated control system.

This stage cannot usually be distinguished from the previous one, either logically or chronologically. That is to say that, unintentionally, the social system is designed at the same time as the production system and secondary technologies. This risk was avoided in our case, when timely assessments were made of the social system and its relationship with the technical system, in a specifically planned phase of the project, which will form the object of a separate discussion. Suffice it to say here that the relevant analysis concluded that both the operators and the technicians could and had to participate in the control of the production process, by using, integrating and completing the computerised system.

The reasons for such conclusions were as follows:

(1) The number of variables to be kept under control is very large. Further, the knowledge of the laws of the system is not all embracing, and there will be a high number of variances (i.e. variables falling outside the range expected under the present knowledge of the system). To control these variances and to develop more accurate system laws, the utmost commitment of the social system is required.

(2) The plant is very vulnerable, and even the least mistake can have very serious consequences: it is most appropriate for the operators to have and to exert maximum knowledge and responsibility on what is taking place in and along the production line.

(3) In past experience, the personnel assigned to work on the new plant has been shown to be able positively to control the production processes and to develop greater degrees of skill: the main content of career development plans could be the ability to control variances.

(4) The new plant is the greatest technical/organisational effort the company has ever made, and it requires maximum participation; such an effort will last at least ten years, and participation must be developed within everyday work.

The result has been a control system characterised as follows:

- closed loop on the operator (the computer ensures the control, and the operator is aware of its logic, establishes its set values, evaluates its process, integrates it, takes action by exception and, in very specific instances, replaces it); and

- partly fixed, partly variable memory (the operator is able to help in fixing the standards and parameters that govern the computerised system).

The formation of overall work and of the organisation is summarised in table 41. So far we have considered only the first three stages of the process, but we shall return to this scheme again when all the elements have been collected to complete the picture.

4.7 Human tasks and task relationships

In section 4.10, below, we present the long planned process by which the base technology was designed, along with the configuration of the production system and the automated information system, in order to show how considerations about the social system and the organisation were systematically taken into account even at the earliest stages, when the "bone structure" of the organisation first became formed as aggregated work, control system and task structure.

This section will briefly discuss the task types present in the production unit in order to clarify the way in which the tasks were grouped together to form organisation units, jobs and roles (stage 4 in the formation of overall work), to allocate work to people (stage 5), to design management systems (working hours, qualifications, training, career structure, etc. - stage 6), and to outline the social system (stage 7). In describing the tasks, let us first review the disaggregated work, without specifying who does what, and how. The tasks may be grouped as follows:

1. Conversion tasks:

   (a) direct conversion tasks. These are unmanned in the hot process phase. In adjustment and finishing there are, however, a variety of human tasks supported by machine tools (e.g. grinding, cutting, coating, etc.);

   (b) handling tasks. These are unmanned, except for a few cases in which pieces must be hooked. Occasional human tasks may occur (e.g. when moving pieces must be set free following accidental locking);

   (c) machine set-up tasks. Unlike continuous-flow chemical plants, the NTM involves a large number of time-consuming set-up tasks. Some are lengthy and heavy manual operations (e.g. manoeuvring, assembling or disassembling, or welding), while others are very

Table 41: The formation of overall work and of the organisation (influencing stages and internal and external variables)

Stages	Contextual (external) variables	Internal variables	Choices as to goals and parameters	Choices as to the technology and the production system	Stages of formation of overall work and of the organisation
1. Investment choice	Competitors; funding problems; cost of direct and indirect labour	Estimated difficulty in improving cost/performance only through organisational development	Reduced product unit costs	Base technology (more powerful and reliable equipment)	Amount of necessary work
2. Appropriate technology	Same, plus predicted market	Same, plus base technology as selected	Same, plus performance system (constant quality; high productivity; low production costs)	Continuous, integrated-flow production system	Computerised (impersonal/ technical) control system
3. System of work relations	Same, plus EDP technologies available	Same, plus continuous-flow process; productive and social role image	Same, plus control reliability and control readiness	Secondary technology system (level of automation)	Content and relationship of human tasks; human task system (structure of work relations)
4. Formation of overall work	Same, plus theory of business in conditions of turbulence of external environment	Same, plus organisational cutlure; socio-occupational traits of manpower	Same, plus intrinsic readiness, adaptation and competence	Organisational mechanisms (management control subsystems)	Formal organisation (organisation unit, roles)
5. Allocation of work to people	Same, plus labour market	Same, plus industrial relations	Same, plus flexibility of use of manpower	-	Task assignment (planned job rotation
6. Design of management systems	Same, plus labour cost; industrial relations policies and general system	-	Motivation and co-operation; control of wage flow	-	Personnel management system (high, differentiated qualification, training)
7. Outline of the social system	Social roles	Informal organisation; real organisation	-	-	Social system

- = not applicable.

Note: The variables exert their influence on the other variables both vertically and horizontally.

fine and delicate manual operations (e.g. the actual setting up, or assembling and disassembling with the aid of precision instruments). In almost every instance, the objects to be handled are large-size parts;

(d) auxiliary tasks. In any large plant, there is a great variety of planned or unplanned auxiliary tasks, such as technological cleaning, routine lubrication, etc.;

(e) data collection and transmission tasks. These include many visual and perceptual checks, instrument readings to be taken, collection and recording of written or keyed-in data, that must be made on primary and auxiliary facilities, work in progress or documents, and conveyed to other people. This involves an enormous amount of data to be collected and transferred outside the automated system;

(f) data comparison and processing tasks. The existence of an automated control system does not dispense with the need for operators to compare a number of data against the standards formally set by the system or through one's personal experience, and to process this data, whether formally or informally. All of this will be in the form of appraisals, calculations, consultations, comparisons, assumptions, etc., concerning widely differing aspects of either the process or the facilities;

(g) direct regulation tasks. Several devices or tools are not regulated by automatic instruments. For example, an overhead crane, mandrel turning and boring machines, finishing tools, etc. In other words, there is a large amount of work which is similar to the operations of a conventional workshop;

(h) human control tasks connected with the automated system. The choice as to the configuration of the automated system implies the existence of the following human tasks:

  (i) tasks of definition of the parameters within which the processing system must operate. Depending on the characteristics of the tube under rolling, a set of mean-value rules are established to control the system variables, which take into account the existing knowledge of the relations between equipment variables and product variables;

  (ii) equipment setting tasks. The above rules are passed in the form of a rolling schedule to those workers who have the task of plant setting through the appropriate console instruments (e.g. roll gaps, number of rolls to be lifted, mandrel advance rates, etc.). These rules can be slightly altered by the operator on the factory floor, depending on special occurrences known only to him, such as mandrel wear, conditions of the equipment, upstream variances, etc.;

(iii) monitoring tasks. In the event of unusual, exceptionally serious occurrences, a number of warning signals would lead to associated emergency actions, even including total plant stoppage;

(iv) checking tasks. The automated system, through terminals, print-outs and instrumentation, supplies the various work stations with data concerning -

- rolling mill and process variables (e.g. power input, motor r.p.m., advance rates, cutting and thrust forces, furnace temperatures, etc.): a terminal conveys this data in real time, through both a display and print-outs intelligible to workers;

- product variables (e.g. length, wall thickness, weight, external surfaces, temperatures, etc., of the work being processed): many of these variables are visually checked following processing;

- flow variables (e.g. number of pieces, frequency, piece identification, etc.);

(v) comparison tasks. Two types should be distinguished:

- instant comparison. The parameters detected do not, in themselves, give any meaningful result unless compared against some standard. Any deviation from the standard may often immediately indicate a specific malfunction, and may sometimes require a diagnosis whereby the conditions of either the mill or the process can be inferred from the specific indication. For example, any slow-down in the mandrel advance speed may result from defective motors, mechanical problems, a setting mistake, faulty lubrication nozzles, or other defects. The computer has a limited range of diagnostic outlines; other diagnoses must be made by people, who can only do so if they are thoroughly aware of the system variables and laws. This is the core of what we have called "variance control";

- historical comparison. The recorded data allows for a comparison of plant operation variables, rolling variables and product variables (results). This can usually be done only after the event; for example, in the 35 or 50 seconds during which a tube flows through the MPM, it cannot be said whether or not the rolling process is attaining the required diameter, wall thickness, length, ovality, straightness, etc. Both the time and methodology of the task are such that this type of comparison can be effected even in places other than the process location, and by people other than mill operators;

(vi) decision-making and adjustment tasks. The comparison made as described above may suggest corrective actions (adjustments). These may be -

- automatic start-up: the computer visually displays which conditions are missing for automatic operation, so that human operators can take corrective actions, and then start the plant;

- transmission of information, more or less fully processed, from bare data to diagnosis, to those having authority to take action (e.g. maintenance, engineering department, etc.);

- adjustment of running variables (i.e. correction of pre-set values such as speed, roll gaps, etc.);

- input to change the standards;

- process stoppage in emergency;

- manual plant operation: this is only possible in a regular manner in the first stages of the process; it is restricted to test and maintenance runs on primary equipment, for reasons of speed and continuity.

Conversion tasks may therefore be summarised as follows:

- there are a large number of manual tooling and set-up tasks, most of them requiring skills and knowledge;

- there are a certain amount of conventional, machine operation tasks;

- there is a share of controls not covered by the automated system, particularly those which poorly influence the process continuity;

- the automatic control system is in a loop closed on human tasks. In other words, human tasks have a number of possible interactions with the control system, and those performing these tasks must then have a knowledge of the system laws and be in an information loop with other people. Moreover, the system relies partly on variable memory (i.e. people can change computer storage containing the standards, in real or delayed time modes, under certain circumstances);

- most of these tasks are interlinked in time relations (sequentiality, interdependence), and/or operational relations (joint execution), and/or information relations (two-way information flow), and/or by objectives (interdependent objectives), and/or by variance transmission/absorption (high-level variance transmission from one task to another, also exhibiting considerable chances of suppressing or minimising the imported variance).

We may therefore conclude that task composition by content exhibits a picture of high differentiation but also an average high level of required skills and knowledge (Blauner); and that inter-relations between tasks show a high level of technically required or permitted co-operation (Meissner).

2. Maintenance tasks. These are tasks devoted to the restoration (and innovation) of the technical, organisational and social systems (Susman), which can be classified as follows:

   (a) Processing plant maintenance tasks. These comprise a substantial share of the human tasks in the plant, due to its complexity. Unlike other similar plants, they are characterised by -

      (i) a high craft specialisation (e.g. mechanical, hydro-pneumatic, electrical, electronic, building, thermal, and other maintenance tasks);

      (ii) rationalisation of maintenance actions (e.g. accessibility, ad hoc instrumentation, procedurisation, etc.);

      (iii) a stronger and more organic differentiation of first-line action and planned maintenance tasks;

      (iv) a greater signifcance of inspection and diagnosis, due to the newness of the plant; process control cycles are initiated by these tasks.

   In summary, these are tasks contributing to specialisation at the operating stage, and to "socialisation" at the inspection and diagnosis stages;

   (b) rolling standard and operating practice maintenance tasks. These are tasks to work out and update process laws and operating rules. The automatic information handling system collects and processes a large amount of data. The tasks comprise -

      (i) identification of data to be acquired;

      (ii) acquisition of non-programmed data;

      (iii) setting of data processing modes;

      (iv) reading of data;

      (v) formulation of "process laws";

      (vi) setting of "operating rules".

   Except for (ii) and (iv), these are highly specialised tasks requiring great abstraction and overall insight;

(c) information system maintenance tasks. These are the tasks traditionally making up the roles of EDP personnel (system architects, system engineers, programmers, etc.). In view of the very close interconnection of electronic technology with the process, some phases of these tasks are necessarily shared with the users;

(d) logistic system maintenance tasks. These include the preservation and development of the scheduling, work progress and storage plans. The logistic system in this production unit is not the "traffic regulation" superimposed on the process, but rather an integral part of the production system. The joint optimisation of the various subsystems is the driving force for a strong integration;

(e) socio-organisational system maintenance tasks, including maintenance and revision of the qualification, working time and pay systems. These are largely performed under trade union bargaining conditions.

In summary, maintenance tasks contain a high level of specialisation, and the boundaries between maintenance and innovation are continually crossed due to the novel nature of the processing unit. They require strong integration - or even overlapping - between them and with conversion tasks. Moreover, the integration of maintenance and innovation tasks is the material base that has required an integrated design methodology, as discussed below.

3. Co-ordination tasks. These are intended to interconnect human tasks between them and with the machines, information systems, products and flows (Thompson, Susman). They comprise the following:

(a) technical integration tasks. The strong integration of machines, operations, products and flows has been seen to be largely built into the system. However, it must be also managed. This takes place through management procedures or tasks aimed at the resolution of conflicting objectives or the absorption of variances crossing over the boundaries of any one subsystem. For example, the decision to impose a high withdrawal force in the last MPM stand, where the mandrel has the highest differential speed in relation to the tube, would result in a better rolling effect but higher mandrel wear. Such a decision is an act of technical integration, specifically of tube-oriented and mandrel-oriented tasks. On the other hand, in the face of an expected fault (a variance affecting both maintenance and operating tasks), the decision whether to stop the plant immediately or to wait for conditions less disturbing for production is a decision integrating maintenance and operating tasks;

(b) organisational integration tasks. The object of these tasks is the use of resources. They consist of working out and deciding what to do before or after a process, who does what, what should be done in isolation, what should be done co-operatively, etc.

To summarise, the complexity of the co-ordination tasks in the production unit is very high. It may be expressed by the formula

$$I = f\ (U, N, C) \quad \text{(Galbraith)}$$

where

I = the amount of information to be handled to secure effective co-ordination

U = the uncertainty as to task requirements (necessary resources, variances, time required, etc.)

N = the number of elements to be interconnected

C = the amount of connection between the elements

As indicated, the value of I is very high. The conditions then prevail for the simultaneous establishment of very known form of co-ordination (Thompson), by programmes, by supervisors and by mutual adaptation.

4.7.1 Conclusions on the nature and interconnection of tasks

The above listing of tasks has been made because it illustrates a number of functional requirements that human tasks posed to the organisation in terms of the base technology, production system and secondary technology system design: open roles; orientation to variance control; closed-loop information flow; high co-operation; multiple co-ordination; and flexibility of use. It will become clear in the next section how the organisation design is less the result of abstract job design criteria than of functional needs and an image of the social system, which are both already built into the task structure, largely through appropriate timely choices made on secondary technologies.

4.8 Work station and work environment configuration

The material working conditions were considered before plant construction was started, covering such physical aspects as noise, fumes, vapours, glare, etc., in terms of protection against fatigue and accidents, or such psycho-physiological conditions as perceptional load control, stress protection and others.

We know little about the degree to which these aspects were considered by the designers of the base layout. However, we do know that this took place while important aspects of secondary technologies were designed, such as interior architecture, instrumentation location and configuration, EDP peripheral configuration, work environment control devices, tools and facilities to perform manual tasks, etc. Considerations of working conditions were interlinked with considerations on the optimisation of human control of the process, and together they have affected the configuration of the final solutions along with other technical and economic factors.

The most significant of these solutions appeared to be -

- the installation of control rooms, in which the operators remain most of the time. In the hot-process area there are four such control rooms: billets and furnaces; PPM; MPM; and sizing mill and discharge. These are large, elevated, glass-walled rooms, from which it is possible visually to control the assigned process area. The entire control and checking instrumentation and the peripherals of the EDP system are concentrated there. The premises are air conditioned and sound proofed, which affords protection from noise (at a very high level in the vicinity of cutting tools) and airborne lubricant vapours;

- control and adjustment instrumentation. Here the most significant unit comprises the consoles containing instruments and controls. Changes were made to the layout originally presented by the manufacturer, which optimised electrical cable routes and connections but made the perceptional load too high. The consoles are now located under the windows and exhibit well-separated areas of emergency control, plant setting (divided into equipment setting and rolling setting facilities), manual control (motors, pushers or roll-ways), and digital indicators. Closed-circuit television sets are mounted on the consoles to allow the operator to have an all-round view of the plant, with visual display units showing sequence values. Minicomputers complete with printers and monitors for appraisal of automatic process control are located just behind the consoles. An intercom system and telephones connect the various control rooms with each other and with the rest of the factory;

- fume venting systems, protection shields, safety devices, etc., were mainly installed from the beginning;

- hoisting and conveying systems for work in progress were designed mainly to avoid the operators working in the vicinity of, or under, moving pieces. One cannot say, however, whether safety or flow-speed considerations prevailed;

- specific equipment and procedures were designed to facilitate fast and rational tool changes (e.g. rapid stand-change, automatic mandrel change, etc.) and for maintenance purposes.

In addition, in the operators' view, the physical working conditions have been significantly improved in the hot-process area over the former layout. Nevertheless, there are still problems of noise, dust and vapour in the cold-process area. The impression is gained that any improvements may have been dictated by synergism, as well as by process safety and the need for speed.

Any stress problems typical of automated factories appear to be moderate. This is due, however, to the work organisation as well as to the environment and hardware features, as will be illustrated in the next section.

4.9 The organisation design

A lengthy agreement on the NTM organisation was executed by the company's management and the trade unions in March 1980. This was the final act in an effort to compose human tasks into an organisation. Overall work (as completed in stages 4-6 shown in table 41) was almost entirely brought under configuration.

4.9.1 Production sub-areas

The elemental unit of the production organisation structure is the sub-area, i.e. an organisational unit tending towards homogeneity, including a number of workers enjoying some operating autonomy to develop a set of result-oriented, integrated activities. No such terms as production group, semi-autonomous team, or others, are used because these units are different from each other, although they share the same principles.

The basic idea is to design the organisation units by defining the "minimum critical specifications" of their socio-technical components (i.e. boundaries, number of employees, objectives/results, variance control cycles, tasks, work stations, roles, operating mechanisms, decision-making and co-operation systems and multi-skilling), through adopting the same principles but differentiating the areas depending on technology and task characteristics (Davis). These components are described as follows:

(1) The sub-area boundaries are defined according to result criteria (each sub-area must have a meaningful measurable result) that can be obtained through integrated activities and effective control cycles among those working in it; technological homogeneity criteria (task components, however differentiated they may be, must have some homogeneity in technical content); territory and process contiguity criteria (human and technical resources must be able to act promptly and effectively); and skill criteria (the task components must require a significant share of operating skills and knowledge for the control of production events). Thus, seven sub-areas have been identified in production -

(a) Billet Sub-Area, from billet collection from the steel mill to billet sequencing in front of the furnace;

(b) Furnace-PPM-OBO Sub-Area, from furnace inlet to the withdrawal of the hollow bloom from the OBO;

(c) MPM and Mandrel Mill Sub-Area, until the rolled tube leaves the extracting mill;

(d) Sizing and Storage Sub-Area, from the rolled tube to the sized, inspected and stored tube before it is sent through the adjustment and finishing stages;

(e) Heat-Treatment Sub-Area;

(f) Adjustment and Finishing Sub-Area; and

(g) Tool Restoring Sub-Area, particularly for rolls.

(2) The number of workers in the units varies between 11 and 19 persons per shift.

(3) Objectives/results. Each sub-area has one priority goal (e.g. stable quality, handling efficiency or quality level) and a set of micro-objectives as variances to be controlled.

(4) Variance control. A list of variances to be kept under control has been identified for each sub-area. These variances can be non-standard values of either the product, the process, the plant or the organisation (e.g., in the MPM Sub-Area, incorrect tube temperature, mill setting deficiencies, tube surface or dimensional anomalies, etc.). For each variance, the task that can control it and the task that can regulate it are identified, together with its origin and consequences. As a result, new tasks or new content of existing tasks may emerge, these being dictated less by the amount of time devoted to them than by the productive significance and the knowledge about the relations between the occurrences in the plant. Since the workers who control are other than those who regulate, the need for further functional communication and co-operation may arise.

(5) Each sub-area includes an identified set of differing tasks to be performed (see section 4.7, above). The decision has been made to include conversion but not maintenance and co-ordination tasks, except for surveillance and diagnosis and tasks for data input into co-ordination and maintenance subsystems.

(6) Work stations have been identified in each sub-area. They do not correspond to jobs but more simply to task groupings to be performed in a given physical place, with the aim of preventing other tasks belonging to a different station from being performed at the same time.

(7) Interactions and communications have been identified within each sub-area and with the outside, and these must be carried out in practice.

(8) Result orientation, the stress on variance control and interactions are such that workers do not have jobs but roles (i.e. sets of tasks which are looked at in relationship with others' tasks as well as with regard to their functional meaning - Parsons). The principles

underlying role design may be defined as specialisation without parcelling, functional co-operation and knowing the purpose served by what is done.

(9) Only one role is identified for each sub-area. Theoretically, there should be total multi-skilling. This is planned in the sense that each operator learns how to perform every task in the sub-area, as a partially pre-set sequence, within two-and-a-half years. Both the actual performance of tasks and the carrying out of the role ("actual ability to operate in a complete autonomous manner in all the activities of the sub-area") are checked at different fixed times. At each time, a higher qualification grade is assigned, up to level 5 (the highest ranking qualification for factory workers) in sub-areas having a high technological content, and up to level 4 in any other sub-area.

With respect to the original design, a significant exception has been recently introduced in the present design. This is a role which is only present in the PPM and MTM sub-areas, and only assigned to one member of each sub-area, who is employed at level 6: the chief operator-preparer. In addition to tasks and responsibilities in common with the other team members, he has specific goals, and more substantial work authority in matters of -

(i) control of variables affecting efficiency, productivity and quality (tooling and parameter checking); and

(ii) changing the "technical memory" (updating operating practices).

(10) Training courses are planned for every worker who becomes a member of the system, centred on the operations required by the tasks, as well as on the NTM technology and organisation.

(11) Technological manning is laid down in trade union agreements. In the NTM manufacturing process, there are about 380 workers and 45 supervisors and technicians. Approximately 140 and 20, respectively, are employed in maintenance.

In summary, the resulting organisation pattern is generally consistent with the task characteristics (structure of work relations and nature of tasks), and with the employees' aspirations for more responsibility and higher skills. This pattern overcomes certain classical principles of work organisation (e.g. jobs as a summation of tasks, fixed task assignments, orientation by prescription, training as merely operating learning, promotion by steps, etc.). Instead, new principles have emerged: roles; job rotation; result orientation; training; career by growth in the role; information possessed and used; and certain self-regulation of skills.

However, certain elements still in force in the rest of the company do survive and are restated here: differentiation by formal skills and grades among the workers; and the placement of the teams making up each sub-area within a hierarchical structure. Nevertheless, the innovations described above will change both the social system and QWL, as discussed below.

4.9.2 The structure of production supervision

The first-level supervisor is a shift supervisor, who holds hierarchical responsibility. However, the prevailing trend is less towards control actions within the individual sub-areas than supplying the sub-areas with the required resources and information; handling the boundary relations between sub-areas and production services; and handling any variances beyond the scope of the sub-areas.

There are three types of shift supervisors. One is largely oriented to rolling operations, the second to work flows and the third to the cold-process area. The role is a very delicate one, which will probably be formulated in practice during the life of the unit. To cover this role, technically skilled workers have been promoted, former foremen retained and high-school graduates newly hired.

Three department supervisors (rolling, adjustment/finishing and tool restoration), oriented for the short or medium term, are responsible for the processing pattern, working schedules and the reconciliation of any conflicting objectives. They take action by exception and report to a manufacturing manager, oriented to the medium/long term who is responsible for the technical and economic results of production or improvement plans.

In summary, a potential role emerges for the foreman as a boundary manager and a traditional role remains for the other positions. A review of the actual functioning of these structures may provide evidence of new meaningful characteristics, particularly in the management/innovation relationship and in the fast transition from control styles of the personal/administrative type to control styles of the impersonal/technical type. However, as yet it is too early to tell.

4.9.3 Direct production services

The manager of the NTM unit is the hierarchical superior of two service structures in addition to manufacturing: the Planning Department and the Manufacturing Techniques Engineering Department.

The Planning Department sets forth the manufacturing and equipment procurement plans and looks after their implementation. Due to the high computerisation level of the system, the technological content of this service is also high.

The Manufacturing Techniques Engineering Department (the NTM Engineering Department) has played and still plays a major role in defining the "system laws" and in formulating rules and practices. This department is the centralised collector of any data related to technical and production parameters, and it issues proposals concerning both the equipment and the production cycles, as well as carrying out studies and experiments, which initiate technological proposals. It also collects observations and suggestions about operating rules and practices, and formalises and issues them to the manufacturing units in the form of operating rules and practices, servicing and training, for the purpose of "favouring the acquisition of autonomy in approaching regulation of the various machines, and particularly in control and action capabilities to secure a better management of the manufacturing process", according to a corporate document. This unit accommodates the centre of the "technical memory of the system", designed in an open two-way fashion vis-à-vis the "local memories" (manufacturing units, roles) (De Maio).

The consistency of the system design of the production units closest to the NTM, and this centralised department made up of highly skilled technical personnel, appears to have a structural reason which has been already mentioned. In any such complex technical system, the knowledge of the system laws is not given at the start. Rather, it is probably the result of ten years' research effort requiring the mobilisation of the knowledge and abilities of the largest possible number of people. The image of a huge automated machine with a fixed memory that cannot be changed (or possibly changed only by a few "magicians") appears to be far from reality in this as well as in many other instances.

4.9.4 Production services outside the NTM organisational unit

The foregoing has demonstrated the technical integration between the state of the plant and the plant operation, and between quality control and the definition/management of the rolling parameters.

During the organisation design work, assumptions were made as to an organisational integration - at a macro- and micro-structural level - of manufacturing with maintenance and quality control, at least in terms of emergency maintenance and ordinary process control. The final choice was, however, to leave the various structures separated. NTM maintenance and quality control are now handled by individual departments having jurisdiction over the entire plant, and only current process inspection is performed by the Manufacturing Department.

Various considerations may have dictated this choice. There is a powerful organisational tradition at a macrostructural level in the steel-working industry, originating from a principle that can be termed "competitive" and stated as follows: maintenance, quality and operation which pursue different goals in competition will generate the best possible break-even point, and thus optimum management efficiency. This tradition has consolidated managerial skills, procedures and approaches which tend to reject more integration of organisational forms. This same tradition, together with a strong division of labour and skills between shop-floor workers (poorly qualified in the past) and maintenance workers (craftsmen), has gradually developed different expectations, interests and cultural orientations, and even occasional mutual hostility, within these two groups of workers. Work recomposition has often posed serious problems for both the management and the trade unions.

Another factor in this decision might have been that some maintenance activities are of a production plant operation type rather than actual production services, as, for example, in electricity, oil and water power plants. The number of workers employed (over 140) could very well have been an added reason. Finally - the most important contingency, in our view - maintenance and quality integration in the most important area of the entire plant could have created a factory within a factory, involving delicate problems of balance of influence and homogeneity in the formal organisational pyramid.

Many technical and procedural improvements have been introduced in both departments, the communication system has been strengthened, and the unit composition has been reviewed and developed for greater efficiency. However, no significant change in the organisational pattern over that prevailing in most steel-working companies can be observed.

The maintenance function, in its two subdivisions of planned and shift maintenance, is arranged into sub-areas, the boundaries of which are defined on the basis of the type of action required (shift, planned and central surveillance actions), equipment technology, maintenance workers' type of trade, and territory. Nevertheless, company management believes that the boundaries, content and roles in this area may undergo some changes following an increasing knowledge of maintenance problems and depending both on the settlement of the organisation structure and on the actual manifestations of planned roles.

4.10 The socio-technical system design process

How have the described technical/organisational choices been reached? Has the type of design process had any impact on the nature of the solutions adopted? These questions may have often occurred to the reader. We have, however, chosen to describe first the problems and the nature of the structural variables, and secondly the solutions adopted through the design process, before approaching the design process itself. As we show below, the process model selected is of relevance to those operating within other contexts, particularly if related to the problems and the object to which it has been applied.

In summary, the initial problems which suggested that a planned process should be conducted in the ways described below were the following:

- a high capital expenditure, extremely challenging for the company's future, and thus the need for high certainty;

- a production unit having a very high level of integration of commercial, technical, organisational and social aspects, requiring an overall design to keep any interdependence under strict control;

- an overall system involving a strong time interdependence of the design choices as arranged both in sequence and in parallel, with the need to control designing times;

- an entirely new system requiring an extensive study into every aspect, and adequate training of those who were, sooner or later, to look after it;

- a system likely to change radically the conditions in which people were accustomed to work, and thus the need for a better appreciation of the features, requirements and evolution opportunities of the social system in order to design an appropriate technical-organisational system;

- a system likely to modify employment levels, working conditions and personnel management systems, requiring solutions which were acceptable to both management and trade unions in the earliest stages to avoid pre-set conditions that might later restrict bargaining margins;

- a system the design and management of which required the contribution of jurisdictions and authorities that were widely separated and communicated poorly at that time, with the need for joint efforts and the breaking down of cultural or functional barriers within the company; and

- an extremely "technology-dominated" system, with the need to balance the influence of pure technologists.

In the early seventies, a team of expert engineers conducted commercial and technical investigations to explore the opportunities to start a new manufacturing unit featuring the above characteristics. Once the decision had been made, the same people worked with the construction company in plant engineering. This was the Plant Project. Another formally constituted team was asked to conduct market surveys, and to design the economic control system (the Commercial Project). The Organisation Development Department, which had promoted organisation projects for some years, was asked, in 1975, to set forth some assumptions as to the organisation to be built around the NTM, and the labour force structure.

These efforts clearly demonstrated that no organisation design and implementation work could be carried out effectively either after defining any technical aspects containing organisational constraints (what we have termed "secondary technologies" in this chapter) or by separate functions operating with a "service outlook".

Top management fully accepted this diagnosis, and, in 1976, established an NTM Management Project, a temporary department reporting to the plant manager. The object of the team was to investigate and analytically design any major management control subsystems. In practice, this was what we have called the overall work and organisation of the NTM. About 40 managers and technicians were assigned to the Management Project, most of them full time. They were to have a matric placement between the Project and their original functions. The team was given complete freedom to study and to seek advice. Final decisions were made within a smaller committee chaired by the top executive.

Nine subprojects were established, each of them devoted to one management control subsystem: Management Cost Control; Technical-Economic Management of Equipment; Warehouse and Transportation; Work Organisation and Personnel; Production Planning; Quality; EDP System; and Technology. These corresponded to organisational activities as well as to organisation units reporting to the major corporate functions. How could the "service outlook" be avoided? Five steps were taken, as follows:

(1) People were assigned to each subproject from different corporate functions. Each problem was dealt with in existing functional terminology, and this reduced the "service outlook". The technical and managerial level of the participants was high. Experiments previously made in the company with vertical groups which also included workers or trade union representatives were not repeated in this instance. Their participation was invited in other ways, as illustrated below.

(2) It was decided to start with analytical organisational elements and actual problems rather than with functions. A large number of such dimensions were identified for design. Some work items were added as checks for new technical-organisational approaches, goals, malfunctions to be avoided or knowledge to acquire. The result was a list of more than a hundred "activities", which were program-coded and incorporated in a PERT program handled by the central processing unit. These included, for example, such items as operating practices, control system, safety and environment, management control variable inter-relations, maintenance work organisation, storing criteria, training needs, organisational trends for the NTM system, line goals and constraints, etc.

For each activity item, a reference sheet was prepared to include questions or constraints posed by people foreign to the corporate functions. More team-work and research work was carried out than in other similar instances. The connection of each activity with respect to other activities was identified in an input/output relationship in the form of a 200 x 200-cell matrix to show the degree of correlation. Relations between activities and problems were thus illustrated. It also emerged in people's consciousness, among other things, that personnel goals and problems had to be approached in designing such technical dimensions as the EDP system, the layout, the operating practices, etc. Correlations with the work of the Plant Project were identified, thus leading to active relations with base technologists.

(3) The next step was to set up a working time-frame to explore every interconnection without time extensions. The result was a subproject work plan which included as goals for the technical-organisational design QWL parameters such as safety, good working environment, the development of human capabilities, social integration, clear-cut rules, variety, autonomy, complexity, control ability in role designing, co-operation, learning opportunities, etc. These parameters, far from being systematically developed, were derived from in-company surveys, needs expressed by the workers on previous occasions, trade union demands, the knowledge of the team members or literature.

An unprecedented circumstance was that here, for the first time, these parameters were expressly, though not systematically, used to design the more technical dimensions of both the technology and the organisation rather than, say, a personnel management system or a job description (i.e. not the automated system but microprocessor software, not an organisational pattern but order processing). An unusual communication experience between technical and personnel people was brought into play.

It was at this stage and by this approach that the automated system with the features already described (i.e. variance control cycles, input/output system of the terminals, the console layout and overhead, cut-off systems on moving equipment for maintenance workers' protection, magnetic hooking systems, replacement of manual methods, sound-proofed roll-ways, etc.) was designed. Even the organisational design proper benefited from such a full and integrated approach. Such organisational innovations as the sub-areas, role design, etc., appeared justified even in technical and economic terms, everyone being persuaded that a job design that took QWL into account was not merely the result of trade union demands or a sociological fashion.

(4) The groups' work proceeded satisfactorily, with extensive, multi-disciplinary investigations, feasible operating suggestions and timely reports, in a climate of co-operation and enthusiasm. Yet, how was one to escape the danger that these groups might become islands of different technocrats doing different things? It was essential to link the subprojects with the company's ordinary processes so that individual subprojects could elicit questions, joint tasks and experiments with elements outside the project.

The first preventive action against this danger was the huge rolling mill project looming ahead, which attracted the whole attention and concern of everyone in the company. A second was top management's strenuous support, and a third was the trade unions cautious, though not hostile attitude. In other words, these groups of people were facing a huge challenge, which was under the whole company's social, as well as organisation, control.

Various steps ensured an operating relationship between the NTM Management Project and the rest of the organisation. On the management side, a far-reaching formalisation of the EDP and PERT output kept both executives and supervisors aware of the state of the art. A formal questionnaire system was introduced, with questions that top management posed to the group, which was expected to answer them in writing. The question-and-answer records were widely circulated. Any decision about major issues often and, even animatedly led to this procedure.

On the trade union side, the unions had been receiving what were for them the most important guarantees since the very beginning - that employment would be not reduced, despite the higher productivity, because of pre-established development plans. The company's management had informed the unions in full detail about all the commercial, economic and technical aspects of the investment project, thus fully meeting its "obligation to inform" as laid down in the collective labour agreement.

Such long-pursued union goals as professionalità (occupational skills, qualities and status) or an improved working environment were accepted by management as among the goals of the new rolling unit. Detailed information was provided throughout the course of the project. Basically, the three main reasons for the unions' hostility to automation had been removed: a dramatic reduction in employment levels without any compensation in substantial development plans; uncontrolled impoverishment of professionalità; and a one-sided modification of the content of industrial relations. These were the premises that led to the agreement mentioned above, which was a complete innovation in Italy, as it covered virtually all aspects of the organisation in addition to conventional industrial relations issues (manning, grades, mobility, etc.) and met a need for organisational changes requiring ongoing consent and understanding.

(5) The final step was to ensure connection of the Project with the implementation stage. Indeed, the most significant relationship of the Project group with the field took place at the stage of plant installation and start-up. This is described in the next section.

4.11 Workers' inclusion and participation in plant installation and start-up: formation of the social system

The erection of the new rolling mill unit began in 1977. Simultaneously, the workers to be assigned to the line were being recruited. The Work Organisation and Personnel Subproject looked after recruitment, training and placement. This stage appeared to be an integral part of the preceding process.

4.11.1 Recruitment

To make up the NTM manpower, one of two methods could be selected: (a) a representative sample of the plant's workforce; or (b) only the best workers. The second approach was chosen, at the expense of any future reproduction of the NTM organisation throughout the workforce, and to the advantage of the workers' active contribution to a prompt start-up and ongoing growth of the system.

A number of criteria were adopted for selection. The first was objective (younger age group, high educational level), the second aptitudinal (abstract and concrete intelligence, tool-handling ability, resistance to perceptional load), the third motivational (mainly volunteers) and the fourth organisational (acceptance by proposed supervisors). Workers to be assigned to the hot-process area came largely from the existing plant (approximately 70 per cent). Those to be assigned to the cold-process area were largely newly hired. The first core of 40 or 50 people were recruited three years before the new mill was started, and 3 or 4 months before work began. The others were recruited as the plant trial runs progressed.

4.11.2 Training

A training course of 7-15 days was planned, and it was to become a standard pattern. In addition to technical training, it presented the organisational model, its individual components and underlying concepts, environmental problems and the personnnel rules system. The first courses were much longer (three months) because the trainee groups were, in fact, discussion groups to provide the company's management with feed-back on the proposed organisational model.

A total 10,000 work-hours of training had been administered at the time of writing (to factory and office workers), as well as on-the-job training.

4.11.3 Placement

The first workers were assigned to plant erection jobs at the end of each training course. This was judged as a valuable technical and organisational experience. Shift supervisors were appointed later. In practice, the functional structure was established prior to the hierarchical structure, which was conducive to team-work and accountability.

Plant trial runs went on for several months. The rank-and-file worked, learnt and made suggestions and comments that were later used for many changes, mainly to mechanical and assembling details. There was ample participation in this set-up process, but it was not given the same formal structure as that conferred on the new steel mill, in which the workers were partners to the project team. Manning was complete at the time of writing, with approximately 470 shop-floor workers and 45 clercial workers (technicians and supervisors) in manufacturing, and approximately 140 shop-floor workers and 20 clerical workers in maintenance.

4.11.4 The formation of the social system

It is in this final stage that the formal organisation developed on paper must be applied to real life, showing all its values and limitations. The management of the control cycles, the discovery and use of the system laws, conscious co-operation and the acquisition of open roles only become operational (unlike any prescribed grade, rank or job) when people are called in for implementation. People will even build, change, enrich or reduce these elements in the implementation process, depending on their abilities and motives but, above all, on the objective conditions of the technology and management systems, and the multiform social roles that each worker brings in from the outside world. An experimentation process takes place, in a constructive, as well as fulfilling atmosphere.

The conventional organisation is based on the principle of uniform and fixed work roles, while pretending that social roles can be ignored. However, a crisis has occurred throughout Western countries because of the intrusion of social roles in the conventional scheme, often with destructive effects. At other times, the same tendency might have been constructive, when a real organisation had been produced on the basis of new, more socially and productively effective principles.

New organisational principles had been already worked out in actual terms by workers, technicians and supervisors within the same company, as evidenced by a case study carried out by the Institute of Action Research on Organisation (RSO).[5] These principles were then largely included in a formal way within a large new structure. Their success will be demonstrated at a later date, as to -

- whether the designed role system is effective (suitable, flexible, controllable, etc.) in the face of the changing requirements of this specific technical system and of a social system made up of the specific individuals recruited; and

- whether the designed role system can be reproduced; i.e. whether it contains such operating traits and features that permit its use for new organisation units (service and administrative work departments, R & D, etc.) which are different in their nature and applicable to different people, more representative of the company's total workforce.

The main feature, in our view, of the take-off phase has been the continuous growth of the role system and the learning process it involves, or the continuous generation of the actual organisation. It is too early as yet to tell whether the role system will continue to grow.

It is now possible to envisage the complete outline of the formation of overall work (table 41) which was only partially presented above. We now see its complexity and, in this specific instance -

- the grounds for an organisation designed in advance of its "invisible" stage (i.e. stages 1-3, those of the technologicial design); and

- the grounds for a planned design process which (though different from the present one in stages, subjects and content) could also constitute an effort to investigate and design the socio-technical system in an integrated, comprehensive and concrete way.

4.12 Some indications for appraisal

A few evaluations have been already made. It should now be added that it is very difficult to supply any accurate measurements or appraisals, as actual start-up began only in January 1980, and no specific follow-up study has yet been conducted.

Results are judged to be good in production terms. Production rates have been 85 per cent and plant efficiency 80 per cent of the standard, the planned product mix has been completed, qualitative results are already better than in the older mill, and equipment wear is to standard on mandrels (representing 6 per cent of production costs).

In terms of QWL, the opinions of the workers interviewed in the hot-process area have been favourable as to the environment, work content, advancement and learning opportunities, and social climate, and these opinions are shared by the trade union representatives interviewed. Common findings point primarily to the satisfaction of contributing towards and controlling such a complex system, the feeling of playing roles that can also be extended outside the work community, and a common-sense display of the many things that have been learnt. None of the interviewees would ever choose to revert to their older jobs.

Some environmental complaints do exist in the cold-process area (high noise level and vapours in some places). The work is certainly less rich in content. However, a climate of substantial growth opportunities exists even among this very young group of workers, of whom very few have reached the expected qualification ceiling.

The above opinions must be weighted for the pioneering climate which, however, will in all likelihood survive for as many years as required to set up and develop the system to completion; and for the particular composition of the labour force, the mode of the age curve being 23-24, workers' educational level being no lower than junior high school, with 10 per cent of high-school graduates, and clerical employees' education level including only high-school and university graduates.

A few facts appear to substantiate the above judgements and impressions, the accident rate being 6 per cent lower than the average for the entire plant, absenteeism 33 per cent lower and no hour having been spent on strike for issues specifically affecting the NTM. The labour turnover has been nil.

4.13 Summary and conclusions

The main conclusion of this case study is that the negative effects of automation on jobs cannot be eliminated, in a new, large-scale, automated factory, once it has been built. Indeed, we have seen the far-reaching complexity and integration of the system, whose effects cannot be acted on after the system has been set up. It is only possible to design human work in the earliest stages while the technical system is being designed.

A model illustrating the various dimensions of overall work has been presented (table 41). It is formed by successive logical stages, which also become time-steps in planned design. The first three stages are the "invisible stages" of work design. Here, no mention is yet made of jobs, job rotation, qualifications or people. These stages are, in fact, used to predetermine most of the elements leading to any one type of job design and social system. They are also used specifically to fix the amount of necessary work, the production system and the system of human tasks or the "work relationship structure" (Herbst).

The last item, under which task content and inter-relations are defined, is the "bone structure" of work organisation or - as we prefer to call it - the overall work. This is, in particular, the stage at which action must be taken if the job and organisation designs are to meet specific requirements. As shown, the design of the human task system takes place indirectly, when secondary technologies, and the EDP system and instrumentation are designed. This can be done in the wrong way, as is usually the case when the image of the roles and the social system one has in mind is left implied (Davis). In the NTM unit, however, these aspects were made explicit through well-defined parameters, dimensions and indicators, and accurately planned design processes.

The case study has also shown that, although secondary technologies (and the task system with them) can be planned with social organisational parameters in mind, they cannot be planned in an arbitrary fashion according to abstract models, as this kind of planning is affected by a number of economic, technical and social variables. These are societal variables, the company's internal variables and strategic objectives, which have been thoroughly reviewed in the study. On the one hand, this substantiates the findings of authors who have asserted the unique nature of the organisation (Davis, Cherns) and the contingent character of any organisational approach (Woodward, Burns and Stalker, Lawrence and Lorsch). On the other hand, our results have indicated the very real possibility of identifying laws of relation between the variables that can help meet QWL requirements without jeopardising production effectiveness and efficiency (Trist, Emery, Davis). Finally, the study has confirmed that overall work is the "minimum scale" of the work design in the event of substantial technological innovations.

The case study has offered the example of a feasible model of far-reaching, closed-loop and variable-memory automation, in which a highly sophisticated approach to the automation of the information system can easily coexist with a system of human control of the process - this human control system being oriented both to the checking and integration of the automated system, and to the modification of the memory of this same system.

It has been found that the existing task mix has a further definite influence on job design. Successful results have been obtained in this case because there was a whole range of manual and non-manual, highly interconnected tasks with a good technical content, in addition to the control tasks.

Secondary technologies have a powerful impact on the task and work-relation system but they do not determine it in full. The system has also been the object of a specific design in the case study. A substantial effort has been made in designing the system of information flow to and between human tasks, and the variance control cycles.

The organisational model that was worked out has a twofold interest. It was intentionally "pre-set" at an early stage, and it contains novel principles, including -

- groups with boundaries defined according to results;

- operators' open roles;

- planned job rotation;

- the training system;

- an ongoing and growing career process based on growth of skills and expertise rather than on automatic promotion;

- the supervisor's role as a boundary manager; and

- the configuration of the engineering department as an "open memory", etc.

General organisation design problems have been discussed, such as the high degree of innovation in work organisation (particularly in the functional work system) which, because of its localised nature, has the only constraint of full consistency with the specific socio-technical system, as compared with the macrostructural system and the personnel management entities, which have problems of consistency with the overall organisation system, because of their general nature. Another problem discussed was the reproducibility of the organisational model in other units with different people. The projecting role (i.e. not only one of resistance) of the social system in the organisation was also considered.

Finally, the planned joint technology and organisation design process used in this case was illustrated and discussed. It exhibits a high possibility of generalisation, less in its method, content, organisation structure or subjects, which would be different in individual instances, than in such central issues as -

- the all-encompassing approach that has covered almost the entire scale of overall work;

- the detailed concrete analysis of even the most minute aspects of overall work;

- the structural diagnosis as to the relationship of external and internal variables, the interdependence of organisation elements with the design method (overcoming both the abstract methodology of organisation development or dogmatic job design), the interdisciplinarity and interaction of functions;

- the use of factual problems as a starting-point;

- the amount of research and debate generated;

- the stress on activity as a generalised learning process;

- the attention paid to, and connection with, the company's ordinary processes;

- top management's support;
- the presence of every subject that counts in the process for whatever reason;
- the use of the implementation stage as an extension of the design process; and
- the ongoing character of the organisation design.

Notes

[1] For further reading, see F. Butera (ed.): Le ricerche per la trasformazione del lavoro industriale in Italia, 1969-79 (Milan, Franco Angeli, 1981), in particular the essays of Fantoli, Marri, Mollica, Oddone and Zara; idem (ed.): Il cambiamento dell'organizzazione negli anni '80 (Milan, Franco Angeli, 1983); and G. Della Rocca: Il ruolo delle parti nella progettazione e applicazione di nuove forme di organizzazione del lavoro (Dublin, European Foundation, 1983).

[2] F. Butera: "Crisi, dibattito, trasformazioni nell'organizzazione del lavoro", in Politica ed Economia, No. 6, 1978.

[3] J. Woodward: Industrial organisation: Behaviour and control (London, Oxford University Press, 1970).

[4] The parameters to be controlled are: machine mechanical operation parameters; machine electrical operation parameters; machine hydro-pneumatic operation parameters; machine operating mode parameters (speed, r.p.m., cylinder gaps, etc.); product parameters (weight, length, wall thickness, temperature, surface appearance, etc.); work progress parameters; and others.

[5] F. Butera: L'organizzazione del lavoro nel laminatoio "Pelligrino" della Dalmine (Milan, RSO, 1977; mimeographed).

CHAPTER X

AUTOMATION AND WORKING CONDITIONS AND ENVIRONMENT IN JAPANESE INDUSTRY

(by Haime Saito)

1. National background

1.1 Major features of automation in Japanese manufacturing industries

In Japan, the process of technical innovation leading to widespread mechanisation and automation in production has developed since the late fifties. This development, linked with the rapid economic growth of the country throughout the sixties and seventies, has brought with it a marked influence not only on production management but also on the working life of production workers.

The most prominent types of automated production processes adopted have been process automation and mechanical automation. Process automation has involved almost all kinds of continuously operated processes as in oil refineries, synthetic chemical plants, power plants and iron and steel industries. Typically, operation of the controlled processes is centralised in the central control room, where the operators are mainly engaged in vigilance work monitoring dials and panel displays which indicate changes in the processing systems. This central control room operation is usually supported by miscellaneous kinds of patrolling and other subsidiary work. Mechanical automation has evolved in the manufacture of electrical and other machinery, automatising to various degrees the operation of specific machines. The use of transfer machines and the more widespread use of NC systems is now seen in fabricating and assembly lines, often connected with conveyor lines, as a typical kind of mass production. Mechanical automation similarly brings with it a large variety of simple and repetitive tasks that cannot be immediately automated.

This process of automation has progressed in conjunction with economic growth, but it has also continued through the period of low economic growth in the wake of the oil crisis. Government statistics show a steady increase in the workforces of manufacturing industries amounting to a little over 40 per cent of each sex during the ten years from 1961 to 1970. The increase for the succeeding three years was 5.9 per cent for men and 3.6 per cent for women. In the following two years until 1975, the recession period, the number of workers experienced a decrease that was more evident for women, who were mostly engaged in simple tasks and more liable to personnel cuts. Such trends might have been favourable for continued introduction of automation, including industrial robots. However, in the following period of low economic growth, there was a slight increase in the number of female workers in manufacturing industries, mainly due to the employment of part-timers.

An emerging trend is the increasing use of micro-electronics, with wide-ranging effects on industrial society. There is a steadily increasing number of NC machines and robots which make use of an electronic controlling device connected with transforming parts, the sequence and the speed being programmed in advance. A survey by the Ministry of Labour in 1980 revealed that NC machines, machining centres and transfer machines were installed in 47.1 per cent of 4,897 undertakings employing 30 or more workers taken as a sample of general machine industry.

In large undertakings employing 1,000 or more workers, these machines were found in 93.7 per cent of cases. Further, according to the United States Robots Associations which collected data in March 1979, Japan led other industrialised countries in the number of industrial robots, with 14,000 as compared with 3,225 in the United States, 850 in the Federal Republic of Germany, 570 in Sweden, 360 in Poland, 185 in the United Kingdom and 110 in Finland.

As a factor which has promoted the rapid adoption of new technologies in industry, R & D by the management side of private enterprises in Japan may be mentioned. This has played a role with the guidance of, but without the direct intervention of, the Ministry of International Trade and Industry and other governmental agencies. Several banking institutions, including extra-governmental bodies, have played a role, too, by supplying funds for adopting new production systems. The importance of such funds may be indicated by the fact that, during the period of high economic growth, the ratio of debt of enterprises to their own internally available funds was markedly higher in Japan than in other industrialised countries of Europe and North America.

Thus the automation of production has had multiple aspects. It is not merely a problem of replacing part of human work by automatic control-aided or computer-aided systems. We should examine not only the change in quantity and quality of work of automatic machine operators, but also changes in the working life of all those affected by the progress of automation. Monotonous work, for example, and associated alienation problems are equally serious for operators dealing with automatic machines and other workers doing peripheral jobs.

1.2 Changes in quality of work brought about by automation

Automated production processes have brought about drastic changes in the required knowledge and skills of workers. The energy expenditure of new jobs has markedly decreased, and the transition from traditional muscular work to information processing and monitoring has demanded a complete reorganisation of workers' skills. As new workplaces are usually confined to a centralised operating space, the working environment has generally improved. Further, these changes have accelerated standardisation of work and higher productivity under a more and more strictly organised division of labour.

On the other hand, this extreme division of labour, an increased rate of work and simplified work demands have created new problems of monotonous, repetitive tasks, while increased production has also meant a larger number of workers performing such tasks. Many perform such tasks at the periphery of, or between, automated machines, with the result that a "gear-like" feeling, as if the workers themselves had become part of the mechanical processes, has spread among both automated machinery operators and repetitive task workers.

These changes have brought about significant effects on the physiology and psychology of the workers. We should note that the effects have multiple facets. They comprise elevated nervous strain in processing a large amount of information, locally concentrated load by repetitive manipulations, postural strain by restricted working movements, eye strain, and night and shift work in a greater number of workplaces. The psychological effects of monotony and boredom at work, and of alienated working life are becoming increasingly serious. These effects are integrated in a worker.

Not all these effects can be explained by the problem of monotonous work alone. In fact, there are contrasting differences between vigilance tasks of process automation operators involving mental strain and repetitive tasks of conveyor flow workers. For example, as we shall see in the case studies, the most serious problem for the former is shift work, and for the latter is boredom. Rather, the inhuman nature of work as a whole should be the focus of our attention.

This paper is an attempt to discuss the implications of these two major types of jobs created by automation in relation to working life.

1.3 Aspects of industrial relations in automation of production in Japanese industries[1]

After the Second World War, management-labour relations in Japan were newly regulated by the Labour Union Law (1949), the Labour Regulations Law (1946) and the Labour Standards Law (1946). The Employment Security Law (1947) and Unemployment Insurance Law (1947) were also relevant. These were the basis of new industrial relations in post-war Japan.

More recently, after the high economic growth supported by technical innovation, some laws were newly enacted to meet the change in industrial circumstances. This time national policy laid its emphasis on securing a labour force which was suitable to changing high-level technologies: the Employment Countermeasures Law (1966) and the Vocational Training Law (1969) are such examples. The Juvenile Workers Welfare Law (1970) and the Women Workers' Welfare Law (1972) were also focused on promoting the welfare of young and female workers. Further, the Labour Safety and Occupational Health Law was established in 1972. As the employment situation deteriorated in the wake of the oil crisis, 1974 saw a revision of the law concerning vocational training facilities and in 1976 of the law concerning promotion of employment for older workers. These legislative measures have in general strengthened traditional management-labour relations in Japan.

The most characteristic feature of industrial relations in Japan consists of the intra-enterprise organisation of trade unions, linked with the lifetime employment system and seniority-based wages. This has to some extent created favourable conditions for introducing automation, as decisions could be agreed upon through negotiations between management and the representatives of a trade union covering all employees. This has usually proceeded in accordance with enterprise efforts to minimise the effects of automation on the employment status of the workers concerned, in order to appeal to their feeling of belonging to the enterprise. It would be too simplistic, however, to think that this enterprise-oriented system of industrial relations has generally and unconditionally favoured the spread of automation. On the contrary, the system initially appeared to hamper the widespread use of automation which was then considered to lead to dissolution of lifetime employment and of seniority-based wages. Retraining of skilled and older workers was also considered to be a problem, as they would resist the acquisition of new skills which might damage their seniority within the enterprise. Under these circumstances, most enterprises tried to retain traditional industrial relations in spite of accelerated automation, by modifying the content and organisation of new jobs.

Thus, in Japan, in an attempt to comply with the lifetime employment system, management was usually committed to the retraining and redeployment of older workers. This has helped incorporate into automated and related jobs some modifications to allow organised group work and to make room for seniority. Perhaps another factor which favoured retraining and redeployment of workers inside a large enterprise was the retirement system at age 55, so that blue-collar workers were still adaptable enough to benefit from retraining. Moreover, many newly established plants at the time of high economic growth were flexible enough to absorb workers relocated from other older plants of the same enterprise.

Thus, it may be said that automation in Japan has developed to the extent which could be attained by retraining and redeployment of workers within the framework of lifetime employment and seniority-based wages, though the latter underwent certain changes by extensively introduced, complicated job allowances.

The trade unions have also contributed to alleviating the impact of automation on employment and on general working conditions. They have tried, since as early as the late fifties, to mitigate the disadvantages due to automation, either by nominally opposing its introduction or by negotiating a system of prior consultation whereby the enterprise is obliged to negotiate with the unions prior to the introduction of automation. Employment security, the reduction of hours of work towards 40 a week, technical education and retraining, as well as the prevention of work intensification and industrial accidents, were among the major demands of the unions. The slowing down of the pace of personnel redeployment and guaranteed wages were also common demands of large trade unions in both private and public sectors. Moreover, we can see some of the influence of these union policies in the organisation of automated jobs incorporating older workers. The basic trends of automation and, in particular, the technical content of newly created jobs, however, have been affected to a limited extent.

As of 1977, when the Government studied systems of labour-management consultation, 51 per cent of private enterprises had such consultation systems, including 69 per cent of large-scale enterprises employing 1,000 or more workers and 19 per cent of small enterprises employing fewer than 30 workers.

As a result, the effects of automation may have been particularly noticeable in sectors of production where it was applied mainly to young, female or unorganised workers, or where it gradually penetrated into such areas of mass production as conveyor flow systems. These effects are discussed in this report.

1.4 Some future aspects of the use of micro-electronics in production

At the present time, the most pressing question related to the industrial use of micro-electronics is apparently its effects on employment, although the above-mentioned survey by the Ministry of Labour contends that the effects are not yet serious.

A comparative study carried out in 1975 and in 1979 on plants with and without NC machines disclosed that the number of regularly employed workers had more or less decreased from 1975 to 1979 in both cases. Rather the introduction of these machines had resulted in the maintenance of employment levels and in curbing the decline of employment through expanding production capacity and increased demand. Until now, the use of NC machines has had the two-sided effect of labour-saving and job-creating, the latter being more prevalent among small industries. The labour-saving effect in large-scale industries seems to be offset by the transfer to new jobs of excess labour.

Countermeasures taken by management when NC machines were introduced comprised, according to the survey, the transfer of workers to new jobs with new skills (65 per cent), reduction of the personnel concerned (29 per cent) and recruitment of new personnel needed (24 per cent). Of the 29 per cent which reduced personnel, only 4 per cent resorted to immediate personnel cuts, while the remaining 25 per cent transferred workers to other jobs inside the same enterprise. Further, vocational training was given by sending trainees to machine manufacturers (67 per cent), training inside the plants (50 per cent) or training in public vocational training schemes (2 per cent).

This government survey was chiefly concerned with the issues of employment and skills training. Little is known yet about working conditions as affected by new NC machines or robots. It is suspected that progressive use of micro-electronics, and in particular of robots in production, will have both positive and negative effects on skills development, conditions of work, safety and health of workers, and industrial relations.

It cannot be denied that labour-saving robots serve to eliminate hazardous and monotonous work. This is especially true when robots replace simple and repetitive tasks in hazardous workplaces or in the assembly of parts. Welding robots, for example, contribute to eliminating the adverse effects of both repetitive tasks and welding fumes. Thus, in a car manufacturing plant, robots are used in one-third of welding workshops, while management says that they will soon find wider use in other units.

A negative effect of using robots appears at intersections between automated machinery in which a human role remains to perform unmechanised tasks. The pace of work at these intersections may be determined by that of robots. The problems that we have seen so far involving mechanical automation will continue to exist also in the case of robot technologies.

Moreover, it is feared that a a production system supported by robots will further alienate blue-collar workers. Engineers or technicians may increase their role in controlling and maintaining robots, but workers performing tasks on the periphery or in the neighbourhood of robots are certainly deprived of this technical know-how. In addition, they are likely to feel as if they are regarded by management as similar to robots, having no thoughts or emotions. Besides, the introduction of robots might stimulate management to adopt production at night. New shift systems might come to be aplied not only to workers dealing directly with robots, but also to those working beside robots.

Under such circumstances, a careful examination of the problem of robot technology is urgent. In this chapter, we first present the results of a survey on monotonous work and its countermeasures, and then examine three case studies on the wide-ranging effects of robot technology.

1.5 Results of a survey on monotonous work and its countermeasures[2]

This survey was carried out using a questionnaire form and some interviews, in order to examine practicable countermeasures against monotonous work. Two kinds of monotonous work were involved, i.e. repetitive assembly work incorporated into mechanically automated plants (conveyor flow work) and sustained vigilance work in process automation plants (gauge-watching work). The survey comprised 25 factories (belonging to 16 companies) in which 25,884 workers were doing conveyor flow work in industries involving electrical machines, motor vehicles and food products, and 11,232 were doing gauge-watching work in electric power plants, iron and steel works and petrochemical plants.

If the workers in each company are divided into those doing monotonous work and those doing non-monotonous work, then the distribution is variable between different industries. The results are shown in table 42. Those who were engaged in monotonous work, according to the judgement of the companies, accounted for 69.2 per cent of all the conveyor flow workers. The rate was the highest in the motor vehicle industry, with 72.4 per cent. The corresponding rate was only 16.8 per cent among gauge-watching workers in electric power plants, iron and steel works and petrochemical plants. The iron and steel works showed the lowest rate of 10.8 per cent. This result reflects the general thinking that gauge-watching work is not necessarily monotonous. This must be taken into account in interpreting the results described below.

Of the 25 factories, questionnaire replies were collected from 956 workers. Of these, 683 were conveyor flow workers from 18 factories

Table 42: Japan: workers doing monotonous and non-monotonous work by type of work and by industry

Type of work	Industry	Total no. of workers	Percentage doing		Total
			Monotonous work	Non-monotonous work	
Conveyor flow	Electrical machinery	7 067	67.0	33.0	100
	Motor vehicle	17 226	72.4	27.6	100
	Food I (confectionery)	1 323	41.0	59.0	100
	Food II (soft drink)	268	64.5	35.5	100
Total		25 884	69.2	30.8	100
Gauge watching	Electric power plant	534	41.3	58.7	100
	Iron and steel	8 564	10.8	89.2	100
	Petrochemicals	2 125	34.7	65.3	100
Total		11 232	16.8	83.2	100

of ten companies; 196 from electrical machine plants, 139 from motor vehicle plants, 158 from confectionery plants (Food I) and 190 from soft drink plants (Food II). The remaining 273 were gauge-watching workers from seven factories of six companies; 195 from power plants, 40 from iron and steel works and 38 from petrochemical plants. the majority had been working at the plant for more than five years. In the following subsections we summarise the results for these 956 workers.

1.5.1 Working time

Hours of work, shift systems and rest breaks, and weekly rest-days of the factories studied are shown in table 43. Shift systems were being applied in all the industries involved. Two-shift systems of 7 hr 30 min-8 hr a shift were seen in electrical machinery, motor vehicle and food industries, whereas three-shift systems were seen in motor vehicle, electricity, iron and steel and petrochemical industries. Among the three-shift systems, hours of work differed between industries: 8 hr in the motor vehicle industry; 12 hr per night shift in power plants; 7 hr 45 min,

Table 43: Japan: working hours, shift systems, rest periods and working week

Type of work	Industry	Working hours on day duty (hr/min)		Working hours in shift system (door-to-door—actual; hr/min)	Rest periods (min)					5-day working week (%)			
		Door-to-door	Actual		Meal-time	1st break	2nd break	3rd break	Total break	Always	3 times a month	2 times a month	Others
	Electrical machinery	9.00	8.00	2 shifts: day 8.45 → 7.30	60	10	8	8	86	80	-	-	20
				2 shifts: night 9.00 → 7.30	60	5	5	-	70				
					45	10	10	-	65				
		8.50	8.00		45	15	15	-	75				
Conveyor flow	Motor vehicle			3 shifts by 4 teams 9.00 → 8.00	60	10	10	-	80	100	-	-	-
				2 shifts: early 8.48 → 8.00; night 8.24 → 7.48	60	-	-	-	60				
				2 shifts: day 9.00 → 8.00	40	-	5	-	45				
				2 shifts: late 8.41 → 7.56	50	7	7	-	64				
	Food 1	8.20	7.20	2 shifts: early 8.15 → 7.30	40	10	10	-	60	-	50	50	-
				2 shifts: late 8.15 → 7.30	45	5	5	-	55				
		8.14	7.14		45	-	-	-	45				
	Food II	8.30	7.30		60	10	10	-	80	-	-	-	every other week 100
	Electric power	.	.	3 shifts by 4 teams: day 8.00 → 7.00, 6.00 → 5.00; evening 6.30 → 5.30, 5.00 → 4.00; night 10.30 → 9.30, 13.00 → 12.00	60 min divided at workers' discretion				60	-	-	-	100
Gauge watching	Iron and steel	.	.	3 shifts by 4 teams (other types found: 3 shifts by 3 teams and 2 shifts by 2 teams): day 8.30 → 7.45; evening 8.00 → 7.15; night 7.30 → 6.45	One rest break of 45 min				45	100	-	-	-
	Petro-chemicals	.	.	3 shifts by 4 teams: day 8.00 → 7.00; evening 8.00 → 7.00; night 8.00 → 7.00	60 min divided into 3-4 breaks at discretion				60	-	100	-	-

- = nil; . = not applicable.

7 hr 15 min and 6 hr 45 min in day, evening and night shifts, respectively, in the iron and steel works; and 7 hr only in petrochemical plants. It should be noted that, for conveyor flow work, there was no weekend work, but that for process automation plants, operation was fully continuous including weekends.

As for rest breaks, conveyor flow workers usually took a break in the morning and another in the afternoon, each of 5-10 min, in addition to a lunch break of between 40 and 60 min. In the case of process automation plants, breaks were usually taken alternatively.

The five-day working week was being adopted by all the plants, though in some cases six days were worked every other week or once a month.

Overtime work was quite common in all the plants (table 44). Those who worked overtime every day accounted for 68 per cent of motor vehicle workers, but 33 per cent of all conveyor flow workers. The rate was much lower for gauge-watching workers (7 per cent). However, overtime was also worked by the latter from time to time.

As table 44 further shows, the percentage of those who could take almost all their paid holidays was only 21 per cent in the motor vehicle industry. This percentage was higher for other conveyor flow workers, the average for all conveyor workers reaching 38 per cent. The percentage was much higher for process automation workers (80 per cent on average).

1.5.2 Feelings of monotony and extent of job rotation

As table 45 shows, 35 per cent of conveyor flow workers said that they always experienced feelings of monotony at work, whereas the corresponding rate was only 5 per cent in the case of gauge-watching workers. If we include those who sometimes experienced monotony, the rate was 76 per cent for the former and 49 per cent for the latter.

The percentage of workers having rotation of jobs among themselves differed among industries. There was a tendency to rotate jobs at relatively short intervals of within a day or daily among conveyor workers and among iron and steel workers. Rotation of jobs between the morning and the afternoon work periods was reported by 6 per cent of electrical machinery manufacturing workers, 9 per cent of motor vehicle workers, 8 per cent of Food I workers and 3 per cent of Food II workers, whereas the rate was 15 per cent in iron and steel works. No such rotation was made in power plants and petrochemical plants. Rotation day by day was made by 7 per cent of conveyor workers and 5 per cent of iron and steel workers, while the rate was 8 per cent in power plants and 11 per cent in petrochemical plants. Rotation at much longer intervals of a few months or at irregular intervals was seen for both groups of workers, involving 10 per cent of conveyor flow workers and 19 per cent of gauge-watching workers.

Rotation of jobs at short intervals may be regarded as a measure against monotony. This may be the reason why it was applied by some

Table 44: Japan: overtime work and annual paid holidays taken (percentages)

Type of work/industry	Extent of overtime work				Annual paid holidays taken up				
	Scarcely any	Some-times	Almost every day	Total	Scarcely any	About one-third	About one-half	Almost all	Total
Conveyor flow									
Electrical machinery	27	34	39	100	19	22	26	33	100
Motor vehicle	12	20	68	100	40	35	12	12	100
Food I	35	60	5	100	13	14	16	57	100
Food II	2	74	25	100	6	14	35	45	100
Total	19	47	33	100	18	20	24	38	100
Gauge watching									
Power plant	19	69	12	100	1	1	15	83	100
Iron and steel	10	87	3	100	3	8	25	64	100
Petrochemicals	-	100	-	100	-	-	5	95	100
Total	15	78	7	100	1	2	17	80	100

- = nil or negligible.

Table 45: Japan: workers experiencing feelings of monotony during work, and extent of job rotation (percentages)

Item	Conveyor flow					Gauge watching			
	Electrical machinery	Motor vehicle	Food I	Food II	Total	Power plant	Iron and steel	Petro-chemicals	Total
Feeling of monotony									
Always	29	28	23	55	35	4	5	8	5
Sometimes	45	40	49	32	41	41	28	63	44
Never	26	32	28	13	24	55	67	29	51
Total	100	100	100	100	100	100	100	100	100
Job rotation									
None	74	73	70	87	77	74	67	45	71
Within a day	6	9	8	3	6	-	15	-	2
Day by day	7	6	13	3	7	8	5	11	8
Every few months	4	4	4	-	3	8	10	33	11
Irregular[1]	9	8	5	7	7	10	3	11	8
Total	100	100	100	100	100	100	100	100	100

[1] Every few months, every six months or every year.

conveyor flow workshops. Rotation at longer intervals, on the other hand, was presumably aimed at having workers acquire multiple skills during their career.

1.5.3 Attitude towards the job

As shown in table 46, an interest in work and willingness to continue to work in the present job were fairly variable between industries. The proportion of those having an interest in work seemed to be somewhat larger for process plants (29 per cent) than for conveyor work (17 per cent). The majority of the workers in both groups said they had neither interest nor no interest in work, but were indifferent to their jobs.

Concerning willingess to work, 23 per cent of conveyor flow workers and 31 per cent of process plant operators showed it to a large degree and about half to some degree. It was rather unexpected that, even among process automation operators, only a minority had positive attitude towards continuing their jobs.

Those who complained of nervous strain due to work accounted for 23 per cent of conveyor workers and 34 per cent of gauge-watching operators. The motor vehicle workers had a relatively high rate of 28 per cent. If those who had sometimes experienced nervous strain due to work were included, the rate was 74 per cent for conveyor workers and 92 per cent for gauge-watching workers. The conditions causing such nervous strain would be different for the two groups: for conveyor workers, this would be related to being kept busy, while for process plant operators, the strain would be based on complex stress associated with their responsibility and uncertainty about work.

As the last part of table 46 shows, the percentage of those who were very satisfied with their working life was small, 12 per cent among conveyor flow workers and 17 per cent among process plant operators. This feeling of satisfaction, if those who felt it to some degree were added, was shared by 44 per cent of the former and 61 per cent of the latter. This difference could be linked to whether the work required complex skills in performing the job.

1.5.4 Assessment of working environment

Generally speaking, technical innovation has made working environment clean in appearance and often air conditioned. On the other hand, however, large-scale mechanisation with high speed of production have brought about new environmental problems, in particular with respect to noise, vibration and hazardous chemicals.

Table 47 gives percentages of workers who mentioned environmental conditions as a problem. Conditions mentioned comprised noise, inappropriate air-conditioning, dust and other factors in the case of conveyor workers. The main problems for gauge-watching operators were noise, dust and

Table 46: Japan: attitudes towards work, nervous strain and satisfaction with working life (Percentages)

Item	Conveyor flow					Gauge watching			
	Electrical machinery	Motor vehicle	Food I	Food II	Total	Power plant	Iron and steel	Petro-chemicals	Total
Interest in work									
Have interest	19	20	24	7	17	28	43	3	29
No interest	10	13	7	17	12	4	2	11	4
Indifferent	62	55	59	56	58	62	55	81	63
Forced to work	9	12	10	19	13	6	-	5	4
Total	100	100	100	100	100	100	100	100	100
Willingness to work									
To a large degree	23	28	31	13	23	34	48	5	31
To some degree	47	47	50	36	44	50	50	61	53
Not in particular	22	18	11	32	22	11	2	29	12
Scarcely any	8	7	8	19	11	5	-	5	4
Total	100	100	100	100	100	100	100	100	100
Nervous strain at work									
Feel it very much	21	28	20	19	23	37	33	25	34
Feel it sometimes	54	48	61	47	51	55	59	69	58
Feel it hardly at all	21	17	16	27	22	8	8	3	8
No strain	4	7	3	7	5	-	-	3	0
Total	100	100	100	100	100	100	100	100	100
Satisfaction with working life									
Feel it very much	12	14	15	6	12	17	28	-	17
Feel it some degree	37	36	36	20	32	47	44	26	44
Feel it hardly at all	41	42	43	61	47	32	18	58	32
Hard to reply	10	8	6	13	9	4	10	16	7
Total	100	100	100	100	100	100	100	100	100

- = nil or negligible.

Table 47: Japan: environmental conditions mentioned by workers as problems (percentages)

Environmental conditions	Conveyor flow					Gauge watching			
	Electrical machinery	Motor vehicle	Food I	Food II	Total	Power plant	Iron and steel	Petro-chemicals	Total
Illumination	13	13	4	6	9	8	15	3	8
Noise	41	55	63	71	57	55	58	82	61
Vibration	2	5	2	5	3	8	8	24	13
Heat	14	15	11	19	15	9	13	3	9
Cold	30	36	30	44	34	15	28	3	15
Humidity	4	6	11	22	11	3	3	3	3
Odour	16	15	10	4	11	1	-	47	11
Dust	18	21	29	29	24	12	35	3	15
Ventilation	20	14	10	11	14	7	5	3	5
Others	9	5	4	3	5	24	13	16	20

- = nil or negligible.

vibration. Especially for the former, ear-plugs or muffs seemed to be useful to a certain extent.

It should be noted that many of the workers in both groups felt that they were not well protected against these environmental conditions in spite of advanced technologies.

1.5.5 Fatigue due to work

As shown in table 48, the percentage of those who were usually very tired due to work was as high as 38 per cent for motor vehicle workers and 27 per cent on average for conveyor flow workers. The proportion was smaller for gauge-watching operators. Concerning recovery from fatigue, day-to-day accumulation of fatigue was reported by 73 per cent of motor vehicle workers, 64 per cent on average for conveyor flow workers, and 53 per cent of gauge-watching workers.

With regard to parts of the body fatigued, many conveyor workers mentioned mainly the eyes, shoulders, legs, and arms and hands, apparently related to repetitive operations and working postures. This contrasted remarkably with the parts of the body fatigued in the case of gauge-watching operators, who complained mainly of fatigue in the eyes, head and legs, due obviously to nervous strain and moving around.

The worst result as to fatigue was thus seen in the case of motor vehicle workers. This would be connected with the fact that this industry ranked similarly as lowest in terms of overtime work, annual paid leave and rest. As those workers had high frequencies of complaints in the shoulders, neck, arms and hands, their fatigue must be linked to the working method characteristic of conveyor systems.

As the main means of recovery from fatigue, the workers of both groups unanimously mentioned sleep and rest in first place. Sports came in second place except for motor vehicle workers, who mentioned drinking instead. Surprisingly, few workers mentioned cultural or leisure activities as the usual means of recovering from fatigue due to work.

1.5.6 Job assessment by workers and desire to change

In general, many Japanese traditionally think that their work is of value. But in recent years, this sense of value of industrial jobs has deteriorated. As seen in table 49, those who desired to change their job accounted for 24 per cent of conveyor flow workers and 30 per cent of process plant operators. Those who wished to continue the job were less frequently seen in the former (34 per cent) than in the latter (45 per cent).

The reasons for wanting a change was strikingly contrasted between the two groups of workers. The major reason among conveyor workers was "monotonous work", while that among process plant operators was "shift work".

Table 48: Japan: fatigue due to work (percentages)

Item	Conveyor flow					Gauge watching			
	Electrical machinery	Motor vehicle	Food I	Food II	Total	Power plant	Iron and steel	Petro-chemicals	Total
Daily fatigue due to work									
Very tired	28	38	20	24	27	17	20	13	17
Tired to some degree	61	56	60	58	59	60	67	63	62
Not tired	11	6	20	18	14	22	13	14	21
Total	100	100	100	100	100	100	100	100	100
Parts of body fatigued									
Whole body	19	24	32	22	24	21	13	37	22
Head	16	9	11	9	11	26	30	37	28
Eyes	47	27	41	36	39	40	58	47	44
Neck	24	13	15	25	2	14	15	18	15
Shoulders	46	36	33	24	35	19	23	11	18
Arms and hands	27	35	21	10	22	7	8	21	9
Back/low back	21	19	22	11	18	4	10	3	5
Legs	28	48	40	27	35	20	40	42	26
Others	3	7	3	8	5	10	3	8	9
Fatigue recovery									
Chronic fatigue	12	18	10	18	15	8	5	-	6
By holidays	53	55	48	41	49	44	50	57	47
By sleep at night	29	25	38	32	31	41	40	38	41
Scarcely tired	6	2	4	9	5	7	5	5	6
Total	100	100	100	100	100	100	100	100	100

- = nil or negligible.

Table 49: Japan: job assessment and desire to change (percentages)

Item	Conveyor flow					Gauge watching			
	Electrical machinery	Motor vehicle	Food I	Food II	Total	Power plant	Iron and steel	Petro-chemicals	Total
Willingness to continue the present job									
Willingness to continue the present job (A)	32	31	59	18	34	47	67	11	45
Willing to find another job, if possible (B)	22	34	11	32	24	29	5	58	30
Neither A nor B	46	35	30	50	42	24	28	32	25
Total	100	100	100	100	100	100	100	100	100
Reason for desiring to change[1]									
Low wages	10	15	23	24	18	9	9	25	11
Long hours of work	1	8	2	11	6	1	-	-	1
Shift work	-	13	6	1	5	56	9	100	56
Work too fast	14	18	-	5	10	-	-	-	-
Monotonous work	33	66	27	45	44	5	9	8	6
Uninteresting work	10	10	4	23	12	1	-	25	2
Unable to display one's own ability	17	13	15	18	16	3	-	17	4
Unable to acquire skills	15	13	17	23	17	5	-	25	6
Unable to have hopeful future	21	8	6	19	16	4	9	5	7
Human relations	6	9	10	16	11	1	9	-	1
Others	6	13	13	2	8	3	-	17	3

[1] Those who wished to find another job = 100.

- = nil or negligible.

This difference between the two groups was apparently related to their perception of uninteresting work and few possibilities of acquiring technical skill development and promotion, especially among the conveyor flow workers. It should be noted that 66 per cent of motor vehicle workers mentioned "monotonous work" as a reason for desiring to leave the job.

1.5.7 Monotonous work and absence rates

As mentioned above, each enterprise was requested to classify production jobs as monotonous or not. When the absence rate was compared between the two groups of workers, it was higher for those doing monotonous work than for those doing non-monotonous work in all of the five industries which supplied such data (table 50). The average absence rate for the former was 114 per cent of that for the latter. This ratio was higher for the conveyor workers (116 per cent) than for the process plant operators (110 per cent).

It is remarkable that not only fragmentary conveyor flow work but also gauge-watching work in process plants could result in higher absenteeism than non-monotonous traditional work.

1.5.8 Actual measures taken in each industry

The chief measures which each industry had taken to cope with the monotonous work were as follows. In parentheses are shown the effects of such measures which were pointed out by management.

(1) Electrical machinery industry:

(a) improvement in workers' participation in planning of work system (raised morale of workers);

(b) adjustment of the work speed to a reasonable level (reduction of workload);

(c) introduction of unmanned production systems using robots in assembly lines of very small parts (decrease in very repetitive tasks);

(d) putting in practice the total five-day working week (decrease in labour turnover and absence rate, effective use of leisure);

(e) improvement of hot working environment (increased work efficiency and reduction in absence rate);

(f) introduction of background music (some effects in preventing sleepiness and irritability of workers);

Table 50: Japan: absence rates in relation to monotonous work (percentages)

Type of work/industry	Sex	Those doing monotonous work (A)	Those doing non-monotonous work (B)	A as % of B
Conveyor flow				
Electrical machinery	Male	4.60	3.82	120
	Female	7.46	6.46	115
Motor vehicle	Male	3.33	3.05	109
Food II	Male	4.63	3.87	120
			Mean	116
Gauge watching				
Power plant	Male	4.70	3.87	121
Iron and steel	Male	8.04	8.16	99
			Mean	110

(g) gymnastics and other forms of physical training in workshop (promotion of workers' physical fitness and change in mood).

(2) Motor vehicle industry:

(a) workers' participation in planning of work system; improvement in the average number of proposals per worker per year reached 36 in a certain factory, the proposals including improved work-rest schedules and conveyor flow conditions (workers' opinions were activated so that they felt themselves in control of work);

(b) change to small-scale work groups and workers' control of work (workers felt the workplace to be their own);

(c) adoption of job rotation systems (decrease in labour turnover);

(d) guidance and job training by a one-to-one system (promotion of human relations);

(e) introduction of unmanned production systems (redistribution of personnel to those jobs in which workers could display their abilities and inventiveness);

(f) gymnastics and physical training in workshops (mood change and preventive effect against low back pain).

(3) Food I (confectionery):

(a) introduction of new job rotation system in goods inspection so as to rotate every 30 min between three jobs (alleviation of nervous strain and higher detection rate of defective products);

(b) change to small-scale work groups (sign of workers' increased consciousness of participation);

(c) improvement of air-conditioning and its wider application to workshops (more comfortable environment and raised working efficiency);

(d) adoption of background music.

(4) Food II (soft drinks):

(a) adoption of job rotation in bottle inspection tasks so as to rotate every 15 min or 20 min between two kinds of jobs, i.e. empty bottle inspection and finished goods inspection (alleviation of feeling of monotony);

(b) change to small-scale work groups (workers' mood change and improvement in awareness of responsibility).

(5) Electric power plants:

(a) job enlargement in two workshops (enlarged worker knowledge and skills and positive attitude towards jobs);

(b) workers' participation in improvement of working systems through the proposal system; every change in work systems is usually to be carried out with participation of workers (positive attitude towards jobs);

(c) introduction of unmanned production systems (release from monotonous work and increased motivation of workers).

(6) Iron and steel industry:

(a) adoption of job rotation systems (workers' mood change and alleviation of mental fatigue);

(b) workers' participation in improvement of working systems (improved morale of workers).

(7) Petrochemical industry:

(a) job enlargement by expanding the sphere of patrolling (mood change through preventing falling into work routine);

(b) application of gymnastics and physical training (reduced nervous strain, mood change and promotion of recovery from fatigue).

1.6 Conclusions and recommendations

Automation has been brought forth as an innovative method of mass production. It has led, on the one hand, to high speed of production and, on the other, to automatic controlling instead of skilled labour. The speedy production has necessitated more advanced division of work, whereby the role of human operators is merely to intervene in intersections which remain to be automatised. In mechanical automation in particular, workers assigned to such intersections are forced to perform simple work with a great number of repetitive motions. In centrally controlled systems, characteristically seen in process automation, operators are forced to keep vigilance on dials, gauges and alarms which should tell them what to do in an unanticipated moment. The central control room operators may be physically underloaded, but they must keep constant watch, sometimes transacting a large quantity of information at a time.

A common feature in both types of operation is increased occupational stress resulting from lack of autonomy of the pace of work. Repetition of simple operations in intersections of automated assembly lines or of process plants easily leads to "acceleration stress". The essential problem here is that the operators are to fight against the effects of monotony and yet to keep pace with the line speed.

Another important feature linked with automation is the increase of work at abnormal hours, either in the form of shift systems or excessive overtime work.

Studies seem necessary, first of all, in order to find a reasonable speed of work assigned to intervening operators of automated systems. This speed should allow them to have variations in pace of work, change working postures, chat with co-workers and have rest pauses, even if short.

The rotation of jobs has proved effective towards reducing adverse effects of fragmentary tasks, but it should be based on our knowledge about the kinds of occupational stress of the jobs to be rotated. Rotation should enable workers to "rotate" their working postures, their way of using parts of the body and the mode of work as a whole. The cycle period of rotation should be long enough to ensure such effects.

The participation of workers in planning improvement for better work systems is essential, as this improvement should release workers from the rigid control of work. We should pay attention to the fact that release from acceleration stress requires a way of thinking different from the usual management policy of first raising production, while reducing workers' spare time as much as possible.

Concerning the problem of participation of workers in planning new technological choices, there seems to be a general perception outside Japan that the Japanese systems of participation are traditionally fairly strong.

However, in the author's view, this does not hold entirely true. Rather, we may say that the technological decision-making processes inside a private Japanese enterprise should be characterised by the strong position of management. Such decisions by management would rely much on internal consultations with its own managerial staff, as well as on negotiations with the enterprise-based trade union leaders. This system carries obvious limitations as to participation of workers themselves in determining the content of their work. As discussed earlier, the main purposes of these management-labour negotiations used to be to secure employment and retraining and to moderate redeployment of personnel within the enterprise.

While top management in Japanese firms see the long-term merit of constantly renewing production technologies, workers are more likely to be concerned about the prosperity of their company, in which they are to work on a lifetime basis under the seniority system. The central concerns of most trade unions have so far been the increase in wages, rather than new production systems and their consequences such as monotony, de-skilling or health hazards. The more popular form of employee participation in production technologies has been small group activities organised by management, such as quality control circles and zero-defect groups. Needless to say, these group activities have been in line with management policy for higher productivity, though they often included improvement of work systems. As these activities are usually led by foremen, suggestions by workers have little to do with work content improvement, the final purpose being betterment of the quality of products that would be compatible with the given productivity.

Under the circumstances, we should make efforts to study and promote organisational forms of worker participation which are best adapted to the Japanese situation. Some experiments carried out in Japan on new forms of work organisation also reveal that particular emphasis on co-determination by the workers is needed. In this respect, institutional activities for the design of more humanised work as part of the design of new technologies are still generally weak in Japan. Some private bodies such as the Japanese Union of Scientists and Engineers, the Japan HR Society and the Japan Proposal Activity Society have contributed, with the help of industries concerned, to the spread of quality control circle acitivities, but it must be noted that these bodies have not aimed directly at improving QWL. More important has been research and information dissemination by individual researchers and professional research units like the Institute for Science of Labour. This institute, which from 1921 represented scientific criticism of Taylorist theories, has been promoting more humanised work design based on its research results. Institutional support in this direction, with resources from govenrment and industry and collaboration with international organisations, seems imperative.

Particular attention must be paid to the problem of night and shift work. In view of its adverse effects on health and daily life, shift work must be restricted in a reasonable way. In mechanical automation plants or conveyor systems, shift work must be avoided as much as possible, along with limitation of overtime work. In process automation plants, shift systems must be improved so as to prevent accumulation of fatigue and ill-health as far as possible. We should recall the fact that the principal reason for desiring to leave a gauge-watching job in these plants was the condition of shift work.

As for the increased use of microprocessors and robots in industry for the purpose of economising on personnel or solving a personnel shortage, we should be vigilant about their accompanying effects. It is certain that these developments carry the means to solve problems of hazardous or monotonous work. Perhaps, first of all, a national policy should be formulated in order to avoid any increase in unemployment. At the same time, however, precautions are necessary in designing work at intersections of automated processes and in operation-maintenance of robots or program-controlled machines. These new types of work must be ergonomically designed to minimise unnecessary acceleration stress and machine-pacing. In this respect, we should fully assess our findings from fragmented jobs in mechanical automation and vigilance tasks in process automation.

In conclusion, the following measures are recommended concerning jobs in automated production processes:

- to set up a practicable limit of work speed or work tempo, in particular in strictly machine-paced operations, as in the case of conveyor work at intersections of automated machines. Criteria for the limit should include, among others, variety in work pace, free change of working postures, freedom to chat, chances of spontaneous pauses and reduced feeling of monotony;

- to adopt a job rotation system that enables each worker to perform different kinds of work in different postures and using different parts of the body. This rotation must be organised so as to reduce feelings of monotony due to work. The selection of jobs to be rotated for a worker and the cycle period of rotation must be determined with the participation of the workers concerned;

- where possible, to attempt to increase freedom by workers to control their work or take rest breaks by formulating small groups of workers. They should be relatively free from the high tempo of key automatic machines which principally regulates the line speed. This measure would enrich the skills of workers by avoiding repetition of very simple manual or inspection tasks. The use of "buffers" seems essential in many cases;

- to design work stations from an ergonomic point of view. Elementary ergonomic considerations of working height, chairs, working motions and inspection methods, as well as of environmental conditions, will greatly ease the workload. Noise, lighting, vibration and air-conditioning should be carefully reviewed;

- to limit the length of a work period and to insert short breaks as frequently as possible. An appropriate length may range from 10 min up to about 60 min depending on kinds of work and on the chances of taking free pauses. Good resting facilities, including couches, should be provided;

- to restrict shift work and overtime work, and to improve shift systems if this is inevitable. Long night shifts should be avoided, and the cycle period of rotation of shifts should be short. A five-day working week without habitual overtime work must be a general rule. Measures to support the taking of the full complement of annual leave are necessary;

- to attempt to open the way to promotion for every worker through enlarging the range of skills to be acquired and raising the skill level in accordance with innnovative technologies. There should always be opportunities open to workers to display their individual abilities. Perhaps it would not immediately be possible optimally to design all jobs within an automated plant. Rather, it is vital to assure flexibility in planning new jobs, taking fully into account the expectation of workers towards a better working life.

2. Case study 1 - Changes of workload in an automated power plant

The type of work and workload before and after introduction of automation were compared in a thermal power plant in the Tokyo district. The study, undertaken in 1956 and 1961, gave a typical account of drastic changes in job characteristics by a shift to automated operations.

The methods used included time study, measurement of energy expenditure, assessment of fatigue by functional tests, and subjective fatigue and environmental measurements. The study also covered night work and accident records.

Before introducing automation, the working environment was generally hot (often 35°C or more), noisy (73-95 dB(A)) and dusty, except for a switchboard room of generator operators. With automation, the environmental conditions of operators improved considerably and the muscular load was reduced a great deal. However, increased mental workload seemed to give rise to new problems as well.

2.1 Time study results

As shown in table 51, the work of operators at the older plant included a variety of physical tasks such as repairing, heavy materials handling, walking and carrying, the rate of actual working time being relatively low. The energy expenditure at that time corresponded to physically "moderate" or "lower-grade heavy" work. After automation of the power plant, operators also worked at a variety of tasks other than pure watch-keeping and inspecting, but they stayed mainly inside the control room and spent more time than before on vigilance tasks and work discussions, including telephone conversations. Thus the energy expenditure was reduced markedly to about 700-800 kcal per shift, a level corresponding to physically "light" work.

Table 51: Automated power plant: comparison of results of time study between operators of an old and a new plant (percentages spent in various activities)

Activity	Old plant operators			New plant operators		
	Boiler	Turbine	Generator	Boiler	Turbine	Generator
Arranging	7	14	12	8	19	9
Inspecting with operations	17	34	18	24	18	24
Recording	3	16	36	8	18	9
Carrying and walking	39	5	5	5	5	2
Going up/down-stairs	5	2	1	-	1	1
Work discussions	-	-	-	15	21	34
Others	4	-	-	5	6	3
Waiting and resting	25	13	22	30	8	7
Energy expenditure per 420 min (kcal)	1 262	-	-	744	741	706

- = nil or negligible.

2.2 Work fatigue

The difference in workload described above was reflected in results of fatigue tests. Before automation, the level of sweating was fairly high and physical fatigue was predominant. Sweat volume during a shift was on the average 2.34, 2.65 and 1.81 litres for boiler, turbine and generator operators, respectively. As for maintenance workers, sweat volume per shift was 4.53, 2.33 and 2.36 litres in boiler, turbine and generator sections, respectively. Thus the mean body weight loss on a day shift was about 1-2 kg. The critical flicker fusion frequency (an indicator of the activity level of the cerebral cortex - CFF) decreased on average by 3-6 per cent from the beginning to the end of a day shift, 4-10 per cent during a night shift. After automation, the sweat volume per shift was reduced to less than 1 litre. However, variation of the CFF was also significant among the new plant operators, the mean decrease being 4-7 per cent after a day shift and 3-9 per cent after a night shift.

More interesting was the difference in average rates of fatigue complaints between the two periods. Table 52 gives mean numbers of complaints after work by workers in the old and new power plants for each of three groups of ten-item fatigue feelings. The old plant operators complained of 2.14 items out of ten physical symptoms, but only 0.78 items out of ten mental symptoms and 0.74 items out of ten neuro-sensory symptoms. These values were similar to those of maintenance workers, implying that the fatigue of the old plant operators was similar to that of manual workers such as maintenance workers. On the other hand, the mean number of physical symptons complained of by the new plant operators was 2.23 items (i.e. as many as that of the old plant operators), but the relative weight of mental and neuro-sensory symptons increased, with 1.56 items for mental symptons and 1.31 items for neuro-sensory symptons. After a night shift, not only physical symptons but also mental and neuro-sensory symptons notably increased.

2.3 Nervous strain

In the new power plant, the main tasks of operators turned out to be of a conceptual nature, relying mainly on watch-keeping of dials and gauges based on technical knowledge. The pressure from the vigilance period increased while the frequency of manipulation became far less than before, a manipulation requiring more precautions. This new situation seemd to cause particular nervous strain among the operators. Many of the new operators felt irritable, uneasy about work and unrelieved from work. As table 53 shows, the proportion of operators feeling irritable on account of work accounted for 36 per cent in the modernised power plant, as compared with 18 per cent in an oil refinery and 9 per cent among a sample of manual workers. The percentage of workers who felt neither irritable nor uneasy was only 36 per cent in the power plant, as against 46 per cent for oil refinery operators and 69 per cent for manual workers. This was not due to inexperience on the part of the power plant operators, because in this power plant the percentage of those feeling irritable did not differ between experienced, skilled operators and those with incomplete skills.

The increase in irritability among the new operators had obviously resulted from increased job strain, larger responsibility and uncertainties about what they would do in the immediate future. The operators, though they were freed from sequential manipulative operations, were not sure about what they would do next. Thus, a reduction in muscular load resulted in the loss of a stable working pattern.

The study indicated that further automation would not free the operators from vigilance tasks and accompanying nervous strain. It was suspected that the simplification of the operators' roles would aggravate rather than improve such problems of nervous strain.

Table 52: Automated power plant: mean rates of complaints of fatigue among operators in an old and a new plant after a day or a night shift[1]

Item	Old plant (day shift)		New plant	
	Operators (34)[2]	Maintenance workers (26)[2]	Day shift (48)[2]	Night shift (60)[2]
Physical symptoms	2.14	2.11	2.23	3.04
Mental symptoms	0.78	0.91	1.56	2.31
Neuro-sensory symptoms	0.74	0.95	1.31	2.15

[1] Number of items mentioned out of three groups of ten items each.

[2] Figures in brackets give the number of workers questioned.

Table 53: Irritability and uneasiness at work among operators at a thermal power plant and an oil refinery, and a sample of manual workers (percentages)

Item	Watch-keeping operators		Manual workers (91)[1]
	Thermal power plants (78)[1]	Oil refinery (50)[1]	
Often irritated by work	36	18	9
Often uneasy about work	28	36	22
Neither irritated nor uneasy about work	36	46	69

[1] Figures in brackets give the number of workers questioned.

3. Case study 2 - Work incorporated into automatic production of fluorescent lamps

The human role of production with partially automated lines is typically seen in assembly plants. A study was conducted in 1975 in a fluorescent lamp assembly plant which was using various machines able to process and assemble lamps automatically. Human work incorporated into the assembly lines mainly accounted for feeding, inspecting and packing, where simple and repetitive work was prevalent.

3.1 Production system and employees

An example of how an assembly line was composed is shown in table 54. Typically, almost all the workers along this line were performing simple and repetitive tasks of feeding or inspecting in a fixed working posture of either standing or sitting. Work space per worker was commonly narrow amid many large automatic machines. The work was strictly paced by the speed of machines or conveyors, allowing little room for variation in cycle time of work. The noise at working sites ranged between 71-87 dB(A), reaching 80 dB(A) at many sites.

In this and other assembly lines studied, all the work was being done by young female workers, the majority of whom lived in a company dormitory. They worked a two-shift system applied between 6.00 and 22.00 hrs (night work between 22.00 and 5.00 is prohibited in Japan for women). The shift change took place at 14.00, and a meal break of 45 min and 10 min rest break were taken in each shift. The duration of employment was mostly short; less than one year for 25 per cent of workers, from one year to less than three years for 58 per cent, and more than six years for only 10 per cent.

3.2 Cycle time and rotation of jobs

The pace of work was quite rapid. As table 55 shows, cycle time for dealing with a valve was usually only around 4 sec in three of the four lines (N, C and S). Cycle time at the fourth line (H) of the table was the longest, but only about 11 sec. Perhaps due to this short cycle time, the duration of a work spell, in which each worker continued to work at the same job, was limited to 30 min in all of these lines. The number of valves processed in this 30 min spell ranged between 150 and 1,400, usually 400, 430-500, 300-500 and 160-180 for the four lines, respectively. The number of valves handled by a worker during a whole working day amounted to about 5,000-7,000 in the first three lines of the table and over 2,000 in the fourth line.

Rotation of workers was commonly practised, with an interval of 30 min. Some examples of rotation systems are given in figure 18. The rotation usually took place without a break. Actual work spells were thus for the morning shift: 120 min-140 min-75 min-90 min; and for the afternoon shift: 150 min-105 min-80 min-80 min.

Table 54: Composition of fluorescent lamp assembly line using automatic machines and ten workers

Production process	Automatic or manual	Number of workers	Outline of work
Supply of valves	Manual	1	Take a bundle of valves from a hoist and supply them to a bucket
Valve washing	Automatic		
Spraying fluorescent paint inside valves	Automatic		
Fixing fluorescent membrane	Automatic		
Inspecting the paint	Manual	1	Inspect appearance of valves while rotating about ten at a time and set them on a wagon
Fixing and sealing a cathode	Automatic	2	Feed a valve and a cathode, then another cathode to other end
Enclosing	Automatic	1	Feed a valve to the vacuum machine
Capping both ends	Automatic		
Inspecting (4.8 sec/ valve)	Manual	1	Inspect moving valves and remove defective ones from a conveyor
Fixing a base	Automatic		
Soldering wires	Automatic		
Ageing[1]	Automatic		
Detecting defective valves	Automatic		
Final inspection	Manual	2	Inspect wires, caps, marks and outside view, and remove defective valves
Packing	Automatic		
Inspecting packages	Manual	1	Arrange in case of need
Filling in cardboard cases	Automatic		
Loading cases	Manual	1	Put finished goods on a pallet one by one

[1] Heat treatment to increase hardness, strength and electrical conductivity.

Table 55: Cycle time, work duration and daily output of four different fluorescent lamp assembly lines

Assembly line	Process	No. of workers	Posture[1]	Cycle time/ valve (sec)	Duration of a spell (min)	Output
N	Valve supply	1	st	non-paced	30	29 000/day
	Inspecting paint	1	st	1.3	30	1 400/30 min
	Cathode sealing	2	st	4.7	30	400/30 min
	Enclosing	1	st	4.8	30	400/30 min
	Inspecting wires	1	si	4.8	30	400/30 min
	Final inspection	2	si	1.5	30	-
C	Inspecting paint	1	si	3.5	30	500/30 min
	Cathode sealing	1	si	4.2	30	430/30 min
	Basing: sealing	2	si)			
	screwing	2	si)	8.4	30	215/30 min
	wire fixing	2	si)			
	Ageing	1	si	4.2	30	430/30 min
	Final inspection	1	si	4.2	30	430/30 min
S	Valve supply	1	st	1 bucket/ min	30	9 500/day
	Spraying paint	1	st	6.7	30	270/30 min
	Fixing membrane	1	st	4.0	-	-
	Cathode sealing	2	si	4.8	30	380/30 min
	Enclosing	1	st	4.8	30	380/30 min
	Basing	2	si	4.5	30	400/30 min
	Ageing	1	st	3.5	30	500/30 min
	Final inspection	1	st	-	-	-
H	Valve supply	1	st	10.0	30	180/30 min
	Spraying paint	1	st	10.5	30	170/30 min
	Sealing	2	st	11.5	30	160/30 min
	Enclosing	3	st	12.0	30	150/30 min
	Wiring	2	si	-	30	160/30 min
	Silicon membrane	1	st	11.0	30	160/30 min
	Basing and ageing	1	st	11.0	30	160/30 min
	Final inspection	2	st	-	-	-

[1] st: standing; si: sitting; - = nil.

Figure 18: Fluorescent lamp assembly: rotation systems

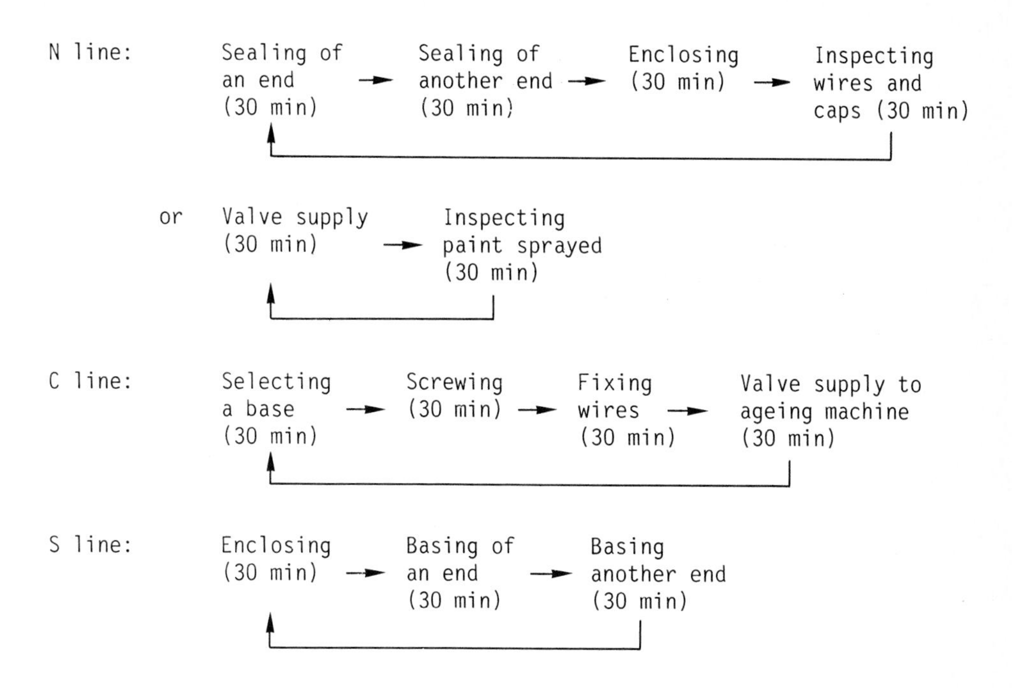

3.3 Feelings of fatigue and monotony

The results of recording the feelings of workers by means of a questionnaire are shown in table 56. A total of 68-95 workers completed the questionnaire. The percentage was higher for feelings related to the monotonous character of simple and repetitive work, such as often feeling tired of work, thinking of matters unreleated to work, feeling time elapsing too slowly and becoming sleepy. About 30 per cent of the workers claimed that they had such feelings during work. In general, the percentage of those feeling that they were kept busy with work was also high, indicating a shortage of rest pauses. The proportion reached 45 per cent in the latter half of the afternoon shift. The roration systems seemed to have prevented higher percentages from experiencing feelings of monotony, however.

Table 57 gives percentages of workers complaining of fatigue at various parts of the body, the great majority experiencing fatigue at one or more parts. Those who said that their whole badoy was tired were seen more frequently among standing workers (47 per cent at the last period of the morning shift) than among sitting workers (18 per cent). Standing workers felt fatigue especially at the neck and shoulders and the legs and feet, while sitting workers also felt fatigue at various parts, in particular at the eyes, the neck and shoulds and the back.

Those who felt fatigue at the neck and shoulders at the end of the morning shift accounted for 62 per cent of standing workers and 88 per cent

Table 56: Fluorescent lamp assembly: workers complaining of feelings related to fatigue and monotony during working hours (percentages)

Item	Morning shift			Afternoon shift	
	6.00-8.30	9.15-12.20	12.30-14.00	14.00-18.30	19.15-22.00
Often feel tired of work	24	27	33	32	32
Become weary	17	15	13	16	15
Often think of matters unrelated to work	29	28	32	32	39
Kept busy with work	31	34	32	43	45
Feel time elapsing slowly	28	33	38	31	24
Become sleepy	30	38	30	24	27
Feel heavy in the head	22	17	18	14	14
Feel irritable	12	16	9	24	22
Aware of noise	12	16	10	13	11

Table 57: Fluorescent lamp assembly: workers complaining of fatigue in various parts of the body due to work during the periods 6.00-8.30 and 12.30-14.00 (percentages)

Item	Standing workers		Sitting workers	
	6.00-8.30	12.30-14.00	6.00-8.30	12.30-14.00
Complaining of fatigue	84	83	92	94
Whole body tired	42	47	15	18
Body parts fatigued				
Eyes	15	22	26	39
Hands and fingers	6	5	2	2
Wrists and arms	16	24	15	20
Neck and shoulders	49	62	83	88
Middle back	21	23	17	39
Lower back	11	21	17	33
Legs and feet	57	81	2	6

of sitting workers, while those who felt fatigue in the legs and feet accounted for 81 per cent of the former but only 6 per cent of the latter. Fatigue at the lower back was more frequent among sitting workers (33 per cent) than among standing workers (21 per cent), apparently resulting from forced working postures. On the other hand, fatigue in the eyes, apparently related to the mode of work, was significantly higher among those sitting (39 per cent) than among those standing (22 per cent).

Judging by these results, rotation between standing and sitting jobs could contribute to reduction of bodily fatigue. The results also point to the shortage of rest pauses during work and the need to reconsider the length of a work spell to be followed by a break.

3.4 Changes in heart rate, subsidiary movements and errors

The types of work in these assembly lines were not physically heavy, but heart rate monitored during a spell of work supplying flare pipes maintained a relatively high level of between 100 and 110 beats/min (figure 19). The mean level of about 105 beats/min was about 50 per cent higher than the resting level, falling to about 90 beats/min only while the worker turned to watching for a while. Similar results were also obtained from other workers. Of 16 workers doing ten different tasks, six exceeded the mean level of 100 beats/min for a work spell. This was relevant to the high pace of work, demanding locally concentrated use of the upper extremities.

Various kinds of subsidiary behaviour were directly observed and recorded. This was done as a means of studying the effects of monotony and nervous strain on workers, considering that keeping to a prescribed task performance for a long period would lead to "compensatory" motions irrelevant to the task itself. These included looking aside, chatting, yawning, appearing drowsy, resetting the buttocks on a chair, crossing and uncrossing legs, stretching the body or the legs, changing foot positions and such trifling motions as touching the hair or shoulders with a hand. The increasingly frequent appearance of these could mean progressive fatigue.

Observation was carried out on standing and sitting workers in four subsequent work sessions during both morning and afternoon shifts. There was an increasing tendency over time in the frequency of movements in the lower part of the body by both standing and sitting workers, the movements being significantly more frequent in the latter half of a shift. Changes in working posture showed no increase, but subsidiary movements in the upper part of the body increased in the final period (12.00-14.00 in the morning shift; 20.30-22.00 in the afternoon shift) in the case of standing workers. Restrictions in free change of posture and long continued work periods would account for the higher rate of compensatory movements.

The relevance of a shortage of rest pauses is clearly illustrated by the fact that the frequency of chatting and spontaneous pauses progressively increased, especially among standing workers. Chatting was normally seen

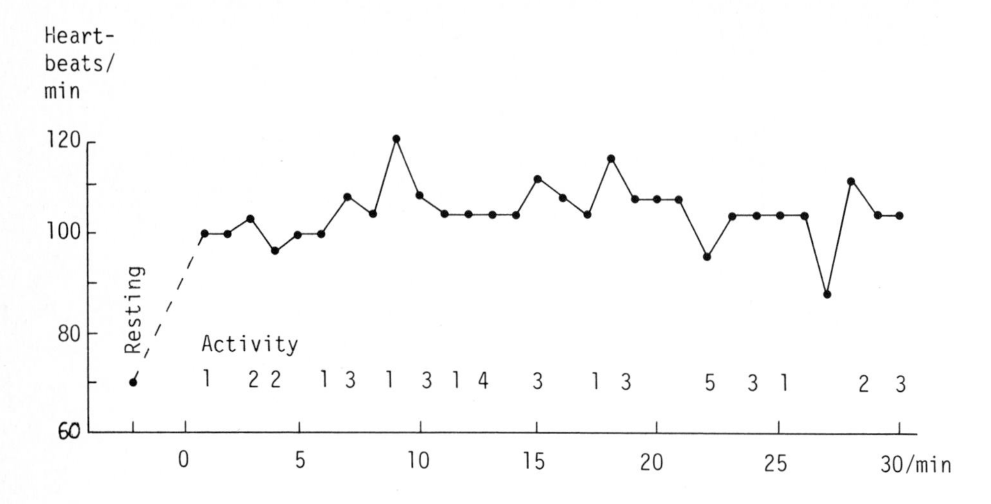

Figure 19: Fluorescent lamp assembly: change of heart rate of a worker during a spell of work supplying flare pipes to a parts-assembling machine (measurement between 9.30 and 10.00 hrs)

Key: 1: checking differences of flare pipes; 2: cleaning the machine; 3: supplying flare pipes to the machine; 4: opening a package of flare pipes; 5: watching.

at a rate reaching 20/100 min among both standing and sitting workers, but it reached an average level of 60/100 min in the last period for the standing group. Spontaneous pauses were also frequent from the start of a shift, beginning around 1/5 min and increasing to a range of about 1/2 min at the end of a shift.

As figure 20 shows, the rate of incorrectly arranged filaments was as frequent between 5-10 per cent in the latter half of a shift, pointing to a progressive deterioration in the quality of work.

3.5 Recommendations

Taking into account all these results, some countermeasures may be considered necessary to reduce fatigue and monotony on the part of the assembly workers working with automatic machines. The following recommendations would be relevant also to other kinds of assembly line work of a repetitive nature resulting from high-efficiency automatic machines:

- alteration of standing with sitting tasks as much as possible. Rotation of workers within a shift period should take this into account;

Figure 20: Fluorescent lamp assembly: change in the rate of work errors while arranging filaments

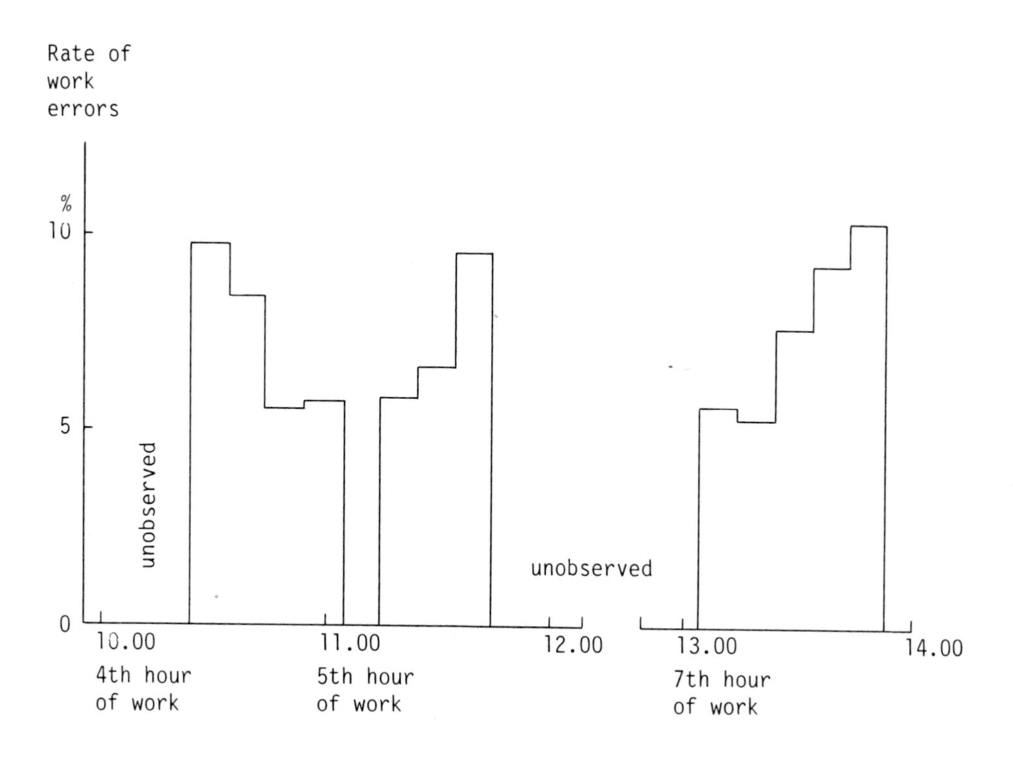

- rotation of tasks for a worker involving tasks with different characteristics, such as manual handling versus inspection. This is important even if the workplaces concerned are separated;

- a break after a maximum of 60 min continuous work. Rotation without such a break may not be desirable; in any case, a break should come after 60 min of continuous work. Each break should last for at least 5-10 min. An increase in breaks would also have positive effects on the efficiency of work;

- trials to shorten the interval of rotation, possibly to 20 min;

- ergonomic improvements in work methods and work space. Good chairs for sitting workers and couches for standing workers should also be provided.

4. Case study 3 - Inspection work in soft drink bottling plants

This study was undertaken to ascertain the effects of mechanical automation on workers carrying out mainly inspection work. Technical innovation in bottling plants has greatly changed the methods of work, using a series of automatic empty bottle washers, metal-plate conveyors and automatic filling machines. Ingredient mixers and feeding equipment have also been extensively automatised. As a result, the major tasks of the bottling line workers are focused on inspection of bottles.

The rate of work along a bottling line is primarily determined by the capacity of the main automatic washing machine. The rate is so high that a worker has to deal with a few to several bottles a second. This rapid pace of work seems to present a particular problem for workers.

4.1 Pace of work

The capacity of each automatic washer of the bottling plants studied ranged between 150 and 320 bottlers per minute (bpm). Each washing unit was connected with 2-4 conveyor lines. Thus the oldest type had a capacity of 300 bpm (150 bpm x 2 lines) and the most modern type 1,280 bpm (320 bpm x 4 lines). This made it impossible for workers to deal manually with each bottle and forced them to work entirely at the pace of machine, mainly in the form of inspection. As a result, the work was characteristically very simple and repetitive, the task being to detect defective bottles. Only a small number of workers were also engaged in operating machines as an additional task.

There were three kinds of bottle inspection: "primary inspection" for detecting rough defects before bottles were washed; "empty bottle inspection" for detecting fine defects and dirt after washing; and "finished goods inspection" for checking the filled volume of the liquid and detecting any slight defects in the finished product. When a defective bottle was found, it had to be removed quickly. Even when a fault-detecting machine was installed, the confirmation of the automated inspection by the eyes of an inspector was required. Inspection workers were inevitably overloaded with rapid eye movements while keeping a fixed sitting posture and frequently stretching the arm to reject defective bottles. This contrasted with machine operation, which consisted mainly of keeping watch while standing and slowly walking around.

4.2 Psycho-physiological changes in bottle inspectors

The bottle inspection was found to be very monotonous. This monotony of continuing inspection was well reflected in the fall of the cerebral activity level of inspectors and the occurrence of various kinds of subsidiary behaviour.

Changes in the CFF and the rate of chatting among bottle inspectors working at the rate of 200 bpm were compared with those of manual workers doing three different kinds of repetitive tasks. The CFF continuously declined during a working day for all these groups, but the decline was the most marked for bottle inspectors. The bottle inspectors had scarcely any time for chatting, while experiencing drowsiness. They were seen to fall, from time to time, into a short-duration doze.

When the rejection rate and the frequency of subsidiary movements were compared among different speeds of bottle inspection, there seemed to exist an optimal speed. Figure 21 shows results for six empty bottle inspectors working with four different speeds of 150, 200, 250 and 300 bpm. The work-rest system was the same for all the four speeds. The rejection rate was the largest for 200 bpm, while the percentage of workers whose colour-naming time after work was prolonged by more than 10 per cent was lowest at 200 bpm. The frequency of heaviness in the head, as well as that of subsidiary movements, was the lowest at this speed, too. A speed slower or faster than 200 bpm resulted in less efficiency and more fatigue. Despite existence of an optimal speed, however, the bottle inspection work seemed to yield a farily large number of defective bottles overlooked by the inspectors at any speed.

4.3 Effect of the length of a work spell of inspection

Next we carried out an experimental study by changing the length of a spell of empty bottle inspection between four periods of 10, 15, 20 and 30 min. Over each period, the bottle speed was kept constant at 200 bpm. After each spell of bottle inspection, the inspectors did miscellaneous tasks such as cleaning their workplace for the same period of time as the inspection spell. Thus the total period of inspecting bottles was the same for all groups.

In the early hours of work, the rejection performance was better the longer the period of inspection, but this relationship was soon reversed. In the late afternoon, the inspectors' performance was highest during 10 min spells and lowest during 30 min spells. The decline of eye accommodation capacity was the least for 10 min spells. Further, the decreasing tendency of the CFF was similar for all four groups, but the 10 min spells seemed to delay the start of such a decline. These results established that longer work spells had more adverse effects on inspectors. We should note that inspection work relying on rapid and complex eye movements could be delicately affected by the work rate and the spell length. A short spell is therefore recommended.

A comparison was also made of changes in the near-point threshold (the smallest distance from the eye at which a small object can be seen without blurring) of six bottle inspectors between one of the existing rotation systems (10 min empty bottle inspection-5 min miscellaneous work) and a test system (10 min empty bottle inspection-10 min finished goods inspection-10 min miscellaneous work). In the former system, the near-point threshold gradually increased after initial shortening, but the prolongation was inhibited by the test system. As the empty bottle inspection was the most

Figure 21: Empty bottle inspection: performance, fatigue feeling and frequency of subsidiary movements in relation to four speeds determined by the bottle-washing machine

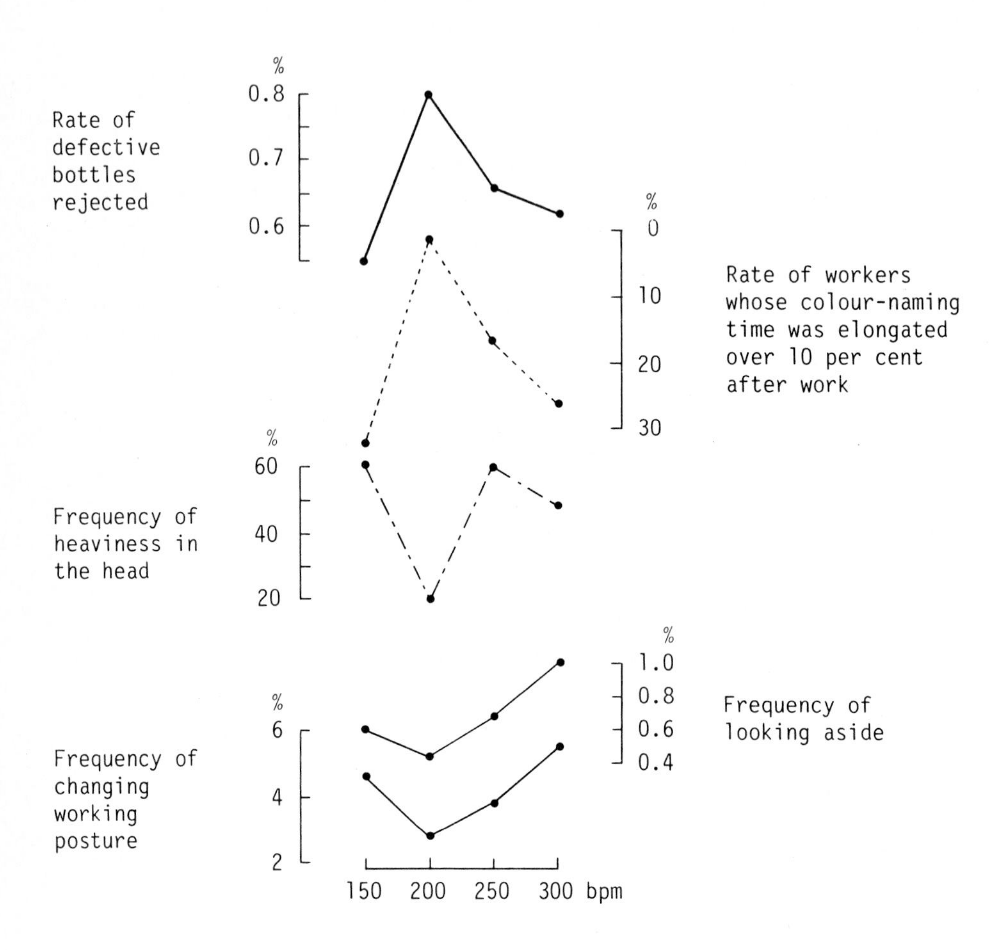

fatiguing, the rotation of workers involving both empty bottle inspection and finished goods inspection did have favourable effects. A long cycle period of rotation consisting of a greater number of different tasks seems desirable.

4.4 Relationship between work organisation and absenteeism

The absenteeism data were compared between various work organisation systems applied in 30 bottling plants of 14 companies which were producing the same brand of soft drink. The data were collected from January to March 1970, together with information about technical conditions and conveyor speeds. In the 30 plants, bottle inspection tasks were combined in 22 different configurations related to sequence, combination and time spent in various tasks. The tasks were broadly grouped into four work systems, as shown in figure 22. The bottle inspectors were mostly young women, although a small number of young men were included.

Figure 22 shows the absence rates for both sexes for different line speeds, different rotation systems and different proportions of the time spent in empty bottle inspection. For either sex, the absence rate was higher the greater the number of bottles inspected per minute. The differences were very significant. Further, the absence rate was higher when the empty bottle inspection (E) was alternated only with waiting and cleaning (W) or with machine operation (O), and not with other forms of inspection (F, P). Accordingly, a larger proportion of empty bottle inspection in work time also seemed to raise the absence rate, as shown in the bottom diagram of the figure.

These results were suggestive of the importance of both the work rate and the rotation scheme. While the high bottle speed accounted primarily for raised absenteeism rates of bottle inspectors, short rotation cycles resulting in a high proportion of empty bottle inspection, the most fatiguing task, also had an unfavourable effect.

The results confirmed our previous findings that both cycle time of unit operation and rotation scheme affected the absence rate of assembly workers. In a cosmetics processing plant employing many women, we had found that the mean number of days of absence per month was 2.4 for 56 belt conveyor workers without rotation of jobs, 1.7 for 24 belt conveyor workers with one or two rotations a day, and 1.4 for 46 table system workers. In a transistor radio parts assembly plant employing young women, the mean number of days of absence per month was 1.44 for 76 workers with a cycle time of 57 sec, 1.30 for 56 workers with a cycle time of 70 sec and 1.20 for 78 workers with a cycle time of about 90 sec.

The fact that young workers in these mechanised plants could not have any foreseeable prospect for future promotion by skill acquisition would have influenced the results obtained. It is striking, nevertheless, that the work rate and the rotation plant further enhanced dissatisfaction of the workers suffering from monotonous work and thus raised actual absence rates.

Figure 22: Empty bottle inspection: absence rates in relation to working speed and work rotation systems

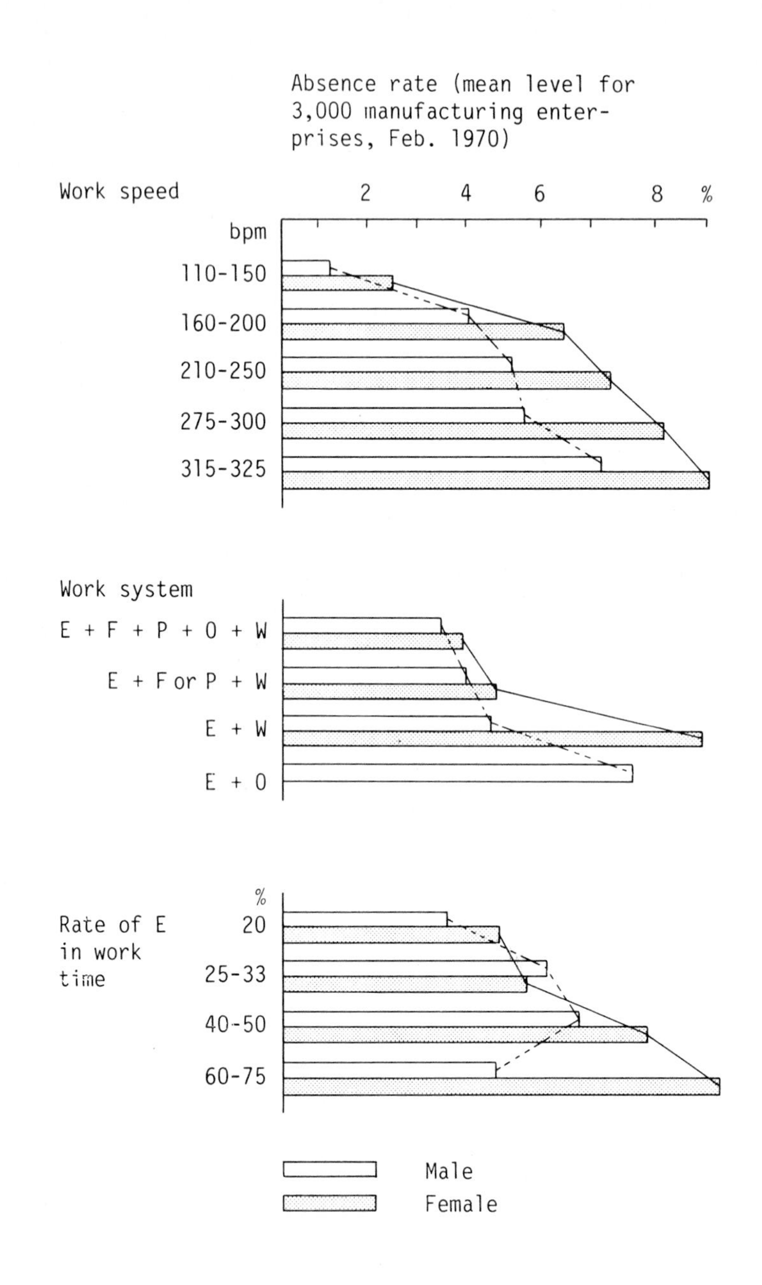

Key: E: empty bottle inspection; F: finished goods inspection; P: primary inspection; O: machine operation; W: waiting, including cleaning of work stations.

Two findings from a previous survey carried out by the Committee of Experts on Monotonous Work organised by the Ministry of Labour, in which the author also took part, confirm a close relationship between the possibility to display skills on the job and job satisfaction or willingness to continue the job.[4] For example, among workers employed in the iron and steel industry on work classified by enterprises as not monotonous, about 45 per cent said they could display their ability on the job, almost as many expressed satisfaction with the job and 60 per cent wanted to continue in their present job. On the other hand, among workers in precision instrument manufacturing engaged in work classified as monotonous, only 17 per cent said they could display their ability on the job, while a mere 15 per cent felt satisfied or wished to carry on in the same job.

4.5 Recommendations[4]

Based on these findings, the following recommendations concerning bottle inspection work would be valid also for other forms of simple, repetitive work:

(1) The conveyor speed should be adjusted to an optimum one as far as possible, so that not only work efficiency but also the workers' fatigue may improve. At the very least, a conveyor flows so fast as to suppress opportunities to chat with co-workers or to make subsidiary movements should not be adopted.

(2) Attempts must be made to find the most desirable combination of a spell of work and a rest, taking into account both work performance and fatigue. While this combination would depend on kinds or speeds of work, a work spell of as short as 10 min should be considered for certain particularly fatiguing methods of work such as empty bottle inspection.

(3) The cycle time should be made as long as possible for simple and repetitive tasks.

(4) Rotation schemes should include a variety of tasks, not, for example, only two simple tasks. As far as strictly paced work is concerned, any rotation cycle must include tasks in which the worker may be independent of the machine pace.

Notes

[1] For further reading, see K. Ishikawa: The Japanese system of quality control (in Japanese) (Tokyo, Nikkagiren Shuppansha, 1981); and Rōmu Gakkai Nihon: New views on the Japanese system of labour management by the Japanese Society of Labour Management (in Japanese) (Tokyo, Chuo Keizai Sha, 1978).

[2] Japan, Ministry of Labour, Labour Standard Bureau: Report of a survey on conditions of monotonous work (in Japanese) (Tokyo, 1968); see also idem: On monotonous work in recent times (in Japanese) (Tokyo, 1979); and idem: Modes of monotonous work and their effects on health of workers (in Japanese) (Tokyo, 1979). The author took part in the first survey as a member of the Committee of Experts on Monotonous Work. The other two reports were written by two different survey teams, in which the author took part as chief investigator.

For further reading, see H. Saito et al.: "Studies on monotonous work: Comparison of different work control systems", in Journal of Science of Labour (Rōdō Kagaku) (Kanagawa), Dec. 1969 (Part I), pp. 679-739; idem: "Studies on monotonous work (II): Comparison of different work control systems", ibid., May 1971, pp. 243-286; H. Saito: "Overcoming monotonous work and alienation" (in Japanese), in Rōdō no kagaku, Vol. 28, No. 8, 1973, pp. 19-26; idem: "Technical innovation and working life: Weakness of self-realisation through automated work and feelings of worth" (in Japanese), ibid., Vol. 30, No. 3, 1975, pp. 4-8.

[3] See H. Saito et al.; "Studies on bottle inspection tasks (I): Comparison of different work control systems", in Journal of Science of Labour, Aug. 1972 (Part II), pp. 475-525; idem: "Studies on monotonous work (III): On the optimum conveyor line speed in empty bottle inspection", ibid., May 1972, pp. 239-260; and idem: "Studies on monotonous work (IV): On the reasonable work spell of empty bottle inspection", ibid., June 1972, pp. 337-350.

[4] Japan, Ministry of Labour, Labour Standard Bureau: Report on a survey on conditions of monotonous work, op. cit.

CHAPTER XI

JOB DESIGN AND AUTOMATION IN THE NETHERLANDS

(by C.L. Ekkers)

1. Introduction

Decisions about the automation of administrative or production processes usually take place at the level of individual enterprises, chiefly with the objective of a better control of the production process and/or higher productivity. However, the effect of these decisions are not limited to the enterprises where the decisions are taken, but have a profound influence on other aspects of society, e.g. the labour market (in a quantitative and qualitative sense), industrial relations, the educational system and other social and cultural aspects.

These more general effects are to a large extent determined by the way in which applications of modern technology lead to changes in existing jobs, the appearance of new jobs and the disappearance of other jobs. It is therefore important to study the overall changes in jobs and the job structure as a consequence of automation in order to gain insight into possible more general effects of automation in society. Furthermore, the design of jobs and work organisation in enterprises where new technologies are being applied is a significant influence contributing to the more general effects of automation in society. Job design therefore, is a crucial factor that must be taken into account when automation takes place with respect both to effects at the level of the enterprise itself and to the more general and long-term impact on society.

Unfortunately, the effects of automation on work content and work organisation have long been accepted as relatively unavoidable consequences of the technology in question (technological determinism). This is understandable, because in the earlier stages automation led in many cases to improvement of working conditions (with regard to process control tasks, for instance, control rooms became separated from the production processes with consequent less exposure of the workers to noise, dust, toxic substances, etc.). Positive effects have also been reported on work content, because automation led to the centralisation and integration of certain types of production processes, resulting in more complete and interesting jobs.[1]

More recently, however, attention has also been focused on possible negative effects on job content and work organisations as a consequence of automation. In the Netherlands two developments must be maintained in connection with this shift of attention. The first of these is a growing concern about possible negative effects on society of the application of new technologies. The challenges and dangers of the application of micro-electronics are important topics for discussion, especially with respect to possible effects on employment and the quality of work. Second, there is an increasing interest in the Netherlands in activities aimed at the improvement of work and working conditions, in other words the humanisation of work.

We will discuss these developments in more detail in later sections of this chapter. At this stage it is sufficient to note that it is becoming increasingly clear that the negative effects of automation on job content are not unavoidable, but that technological and organisational alternatives are available and that ways must be explored to use new technology for the humanisation of work.

2. Extent of automation

Automation is a very broad term, indicating a process in which work formerly performed by people is taken over by machines. Often a distinction is made between mechanisation and automation, where automation is seen as a stage following that of mechanisation. In the mechanisation stage the work is performed by machines, but people are needed to operate the machines and to perform the parts of work that are not (yet) mechanised. The workers are relieved of much of the physical effort accompanying manual work, but at the same time in many cases fewer qualifications are needed and an organisational structure is introduced that allows less room for individual autonomy and leads to monotonous, repetitive and short-cycle work, according to the principles of so-called scientific management. The best-known example of this type of work is that of the assembly line.

In the automation stage, machines need no longer be operated by people, but the introduction of more or less advanced control systems makes self-regulation of (parts of) processes or systems possible. Of course, there is no sharp boundary between these two stages: automation includes mechanisation, and many production processes are a mixture of manual work, mechanisation and automation. Nevertheless, elements of automation are introduced in more and more types of production systems. There is also the tendency for partly automated production systems to become automated to a higher degree. At the same time, we see that the better control over production processes reached by automation facilitates enlargements in scale and a growing complexity of production systems. In this way, higher production can be reached with fewer workers, whose work consists of monitoring and occasionally adjusting the process.

When we look at the question why production processes are automated, a number of reasons can be mentioned:[2]

- reduction of labour costs;
- better control of information streams;
- better control of production processes;
- improvement of internal organisation;
- rise in production;
- improvement of working conditions;
- limited availability of workers with certain specific qualifications.

We may therefore conclude that the decision to introduce automation is made mainly for economic reasons. The degree of automation chosen, however, is dependent on the technical possibilities. It is the rapid development of these technical possibilities, especially in the field of micro-electronics, combined with the rapidly diminishing costs of electronic components, that makes automation economically attractive for an increasing number of applications. All kinds of production processes, for which automation was formerly not possible for technical or economic reasons, will be influenced by automation in the near future. When we restrict ourselves to the present situation, however, two main types of applications can be distinguished, namely automation of production and transportation systems on the one hand, and automation of administrative systems on the other.

2.1 Production and transportation systems

Applications are found in -

- oil refineries: a very high degree of automation is possible here (computer-controlled systems);
- chemical industry: this is traditionally strong in the Netherlands. The degree of automation is very dependent on the type and scale of process;
- electrical power plants: a very high degree of automation is already possible, and the trend is towards still larger and computer-controlled systems;
- transportation systems: rail, air and sea traffic control systems are gradually becoming more centralised and more automated. The degree of automation is, however, still relatively low;
- metal industry: applications in this field consist of blast furnaces, steel factories, NC machines and, recently, industrial robots and computer-assisted design and manufacturing (CAD-CAM).

It is estimated that up to 100,000 workers, almost exclusively male, are carrying out more or less automated tasks in these fields.

2.2 Administrative systems

Quantitative data are available about automation in the administrative sector thanks to a large national survey in 600 enterprises.[3]

Automation here is rapidly increasing. The total amount yearly spent on administrative automation in 1977 may be estimated at about 3.0 billion guilders amounting to about 1.5 per cent of GNP. The number of people active in administrative automation is between 50,000 and 70,000, about 1.5 per cent of the total working population of the Netherlands. About

45 per cent of these people work in development and programming, and 55 per cent in data processing. Women, who form 29 per cent of the total, are underrepresented in the higher jobs; for instance, 98 per cent of data entry is performed by women. Automation attracts mainly young people. In development and programming, 60 per cent are aged less than 34, and for data processing this proportion is almost 70 per cent. Automation occurs in all types of enterprises, but a very high rate is seen at present in banking and insurance companies.

3. Effects of automation

For many years the effects of automation on the labour market, both in a quantitative and in a qualitative sense, have been under discussion. The expectation is that further automation will lead to a continuing loss of jobs, both in the production and in the administrative sector. These effects are particularly noticeable at times of economic stagnation, as has been the case during the last few years. This leads to rationalisation processes in many enterprises without compensation by growth of other enterprises. During the last few years, therefore, we have seen a sharp rise in the unemployment rate in the Netherlands, which stood at over 400,000 out of a working population of less than 5 million at the end of 1981. In the near future the most dramatic effects of automation are expected in the administrative sector, where the development of microelectronics offers economical possibilities for making human labour superfluous. The restoration of economic growth and the redistribution of work among the population (e.g. by shorter working hours) are seen as possible ways to reduce the effects of these developments.

With respect to qualitative changes, much attention has been directed at shifts in the qualification structure of the labour market as a consequence of the application of modern production technologies. Official studies in the Netherlands[4] expect a need for higher qualifications ("upgrading"), although much of the literature points in the direction of downgrading[5] or polarisation (a decrease in the middle category of jobs.[6] Recent research in the Netherlands[7] shows downgrading tendencies among blue-collar workers and polarisation tendencies among white-collar workers for the period between 1960 and 1971). During the same period, however, the educational level of the population has risen. This leads to so-called qualitative discrepancies on the labour market, implying that there are still a considerable number of unskilled jobs for which it is difficult to find personnel. It is therefore not surprising that in 1971 more people held jobs below their level of education than in 1960, according to the results of the study.

Besides this global approach aimed at assessing the effects on the qualification structure of the labour market, it is possible to look more specifically at changes in the content of work (e.g. in terms of autonomy and possibilities for personal development) and in working conditions (e.g. in terms of exposure to noise and toxic substances), and at the effects of work and working conditions on the workers (e.g. in terms of stress, health complaints, job satisfaction, sickness absenteeism, etc.). This way of studying the effects on the quality of work is mainly limited to certain branches of industry or to the impact of certain specific types of new technology. The scope therefore is narrower, but the information it gives is more detailed and better applicable to problems of jobs design. The method used is that of case description or a comparative analysis of a number of cases. In the Netherlands, a few studies of this type have been carried out during the last few years. These provide evidence that automation does not necessarily have a positive effect on the quality of work and can even have adverse effects. A study of the quality of work and labour market problems in three seaport regions shows, for instance, that automation leads in a number of cases to more rigid working procedures and loss of autonomy of the workers.[8] More shift work also occurs, which must be judged negatively because of its effects on health and social life. As a positive aspect, however, the improvement in physical working conditions is mentioned.

Another study, the results of which will be presented in the second part of this chapter, directs itself at a specific type of function in relationship with automation, i.e. operating functions in different industries.[9] Within these functions, operators have to control more or less automated technical systems, such as chemical processes, electrical power plants, transportation systems and computer systems. The work takes place in control rooms where information from the production process is presented on panels or VDUs and from where the operator(s) can control the process (so-called human control tasks). From this study it appears that the quality of work in terms of work content (autonomy, variety and degree of active involvement in the production process) is lowest at higher levels of automation. The regulating activities, which can make these tasks interesting for the operators, are taken over more and more by automatic equipment. Passive monitoring becomes the task of the operator in highly automated systems, with monotony and boredom as a consequence.

NC machines (e.g. lathes) are another form of automation evaluated in a study in the Netherlands.[10] Skilled work is here replaced by simple routine work. Programming the machines is work of a higher level, but the number of these jobs is relatively small.

With respect to the effects of automation in the administrative sector, few research results are available in the Netherlands. A study based on statistical data of bank personnel[11] showed that the number of jobs in the banking sector had tripled during the last 20 years. The growth was concentrated, however, on jobs of a relatively low level, often performed by women. The role of technological developments cannot be specified on the basis of this material. Older studies on the tasks of clerical workers (e.g. Green report) a tendency towards increased routine, less responsibility and more rigid working procedures as a consequence of automation. On the basis of these and other studies, several authors[12] conclude that recent technological developments in the Netherlands have not led to a higher quality (and indeed in many cases even to a lower quality) of the remaining work in terms of work content and qualification levels.

At the same time, however, it is claimed that this is not a direct effect of technological development as such, but rather a consequence of the way new technologies are applied in work situations, which reflects the goals, interests and strategies of management more than the characteristics of the new technologies. When new technology is introduced in combination with traditional organisation methods based on scientific management (horizontal and vertical division of labour, fractionisation and simplification of tasks, etc.), automation often leads to work degradation and a diminishing influence of the workers on the production process.

Such a process is shown in a study of the so-called HAGA group (Restructuring, Automation and Consequences for Employment at the University of Utrecht.[13] In an electro-technical industry, an existing organisation structure based on job rotation and semi-autonomous groups was broken up by the application of new automated equipment. The new situation consisted of individual jobs with far less degree of freedom for the workers. Automation here was used by the management to increase its control over the production process at the cost of control of the workers over their own work situation. This reasoning implies, however, that in principle it is also possible to use new technologies for an improvement in the quality of work, or in other words for the humanisation of work. This will be discussed in the next section.

4. Humanisation of work

During the last few years there has been an increased interest in the Netherlands in the improvement of work and working conditions. Improvement is seen in terms of "human" criteria, such as the absence of factors injurious to health and the presence of possibilities for personal development and influence on the work situation. Therefore, the term humanisation of work is often used. This term covers a broad spectrum of activities, directed at, for instance -

- physical and chemical working conditions (e.g. noise, heat, vibrations, toxic substances, etc.);
- ergonomic aspects of the work situation (e.g. design of tools, control panels, worker machine interaction, etc.);
- work content (e.g. qualification level, autonomy, variety, physical and mental workload, etc.);
- conditions of employment (e.g. wages, working time); and
- industrial relations (e.g. democratisation and participation).

Because this paper is devoted to job design and automation, we will pay most attention to approaches concerning work content or quality of work, although it becomes increasingly clear that for the humanisation of work an integrated approach is needed, taking all aspects into account.

There are a number of reasons for this increased interest in work and working conditions. In many enterprises there are problems of rising sickness absenteeism, low productivity, lack of motivation, and the difficulty of finding people for certain kinds of work. This indicates that there is a discrepancy between what people are expecting from their work situation and what they find there. As a consequence of, among other things, the rising educational level in society and the democratisation processes of the last few years, people are demanding more from their jobs in terms of work content, autonomy and possibilities for individual development. As discussed in the previous section, developments regarding work content of the available jobs in the Netherlands have not been in accordance with these higher demands.

A second problem is the large number of people who withdraw permanently from the labour market for medical reasons. This number is growing each year, amounting to a total of over 400,000 persons receiving the WAO payment (invalidity benefit) in 1981. About 40 per cent of these people are declared medically unfit for work for psychological or psychosomatic reasons. It is assumed poor work content (stress reactions as a consequence of too high a mental workload, lack of variety, etc.) is one of the possible factors that influence these figures.

A third factor stimulating activities relating to the humanisation of work consists of technological developments in the field of microelectronics that will enhance automation processes in the industrial and the administrative sectors. It is clear that these developments will influence the quality of work, and there is much discussion about the question of how new technologies can be used to improve the quality of work instead of leading to further work degradation. This interest in job design in relationship with automation is, however, fairly recent. Although there is a long tradition of research into the quality of work in the Netherlands,[14] relatively little attention has been directed until now at automated work situations. Instead, most efforts have been invested in work structuring experiments in mechanised industry (e.g. short-cycle assembling tasks) and in participation and democratisation experiments.

An interest in the problems of automated work situations came first from the field of ergonomics. Traditionally, the accent in this approach is not so much on the quality of work but more on problems of effective task performance in worker-machine interaction. More recently, however, there has also been a shift of attention towards the problems of health and well-being of operators.[15] Under the title "systems ergonomics", a way of designing worker-machine systems is indicated that makes use of the criteria for job design developed in the socio-technical systems approach adopted by the Tavistock Institute of Human Relations in London.[16]

Concern about work and organisation design in relationship with automation, especially in the administrative sector, is also growing in the field of organisational psychology and sociology. Projects have been started up recently at some universities, often in collaboration with social scientists in large administrative organisations. Here, too, a socio-technical systems approach is generally being followed. Reports of completed design projects are still rare, however.[17]

4.1 Government activities

The effectiveness of activities aimed at humanisation of work, certainly in the long term, depends for a great deal on the macro-conditions, in terms of government policy and industrial relations, under which these activities take place. On the government side, a growing interest in improving QWL may be perceived. Humanisation of work is regarded as one of the ways in which qualitative frictions on the labour market can be solved. A positive effect is also expected on the level of sickness absenteeism and permanent unfitness for work.

Of course, there is a long history of legislation in the Netherlands with respect to health and safety at work. Very recently, however, legislation has been extended to the domain of work content and well-being of workers. The new Act on Working Conditions, the implementation of which will begin in 1982, is directed at the promotion of health, safety and well-being of the workers. Well-being must be promoted by avoiding short-cycle work and by giving possibilities for personal growth and development, autonomy, contact with co-workers and knowledge and results of the work. This relates very clearly to the content of work. The law will gradually be implemented during the eight years to come, and it is generally expected that it will have a stimulating effect on research and other activities directed at the improvement of work and working conditions.

Furthermore, the Government financially supports the improvement of workplaces by allowing the employer a subsidy of 50 per cent of the costs of projects aimed at the improvement of poor working conditions. Most projects are concerned with physical working conditions, but job design projects can also be supported by this arrangement. Financial support is also granted to projects at the Socio-Economic Council (SER), in which the Government, employers' organisations and trade unions are represented. An example is a long-term programme, started in 1973, consisting of scientifically guided experiments directed at participation and democratisation in enterprises.

With respect to technological developments related to micro-electronics and automation, the interest of the Government is directed at the challenges and dangers for the national economy and the societal effects of the introduction of the new technologies. A special governmental advisory group reported on these matters recently.[18] The advisory group expects an increase of 170,000 unemployed in 1990 on account of the rapid implementation of micro-electronics, under the assumption that no compensatory measures are taken. A slower implementation, however, will lead to about the same loss of jobs, due to less competitive products and prices. The advisory group therefore recommends a strong innovative policy making use of micro-electronics combined with compensatory measures to reduce the effects on unemployment.

With respect to qualitative changes, the group expects a greater demand for skilled workers and a decrease in the demand for unskilled workers. This expectation is in contradication with most of the literature on qualitative effects of automation, as discussed in section 3. The advisory group recommends further studies on these qualitative changes, also taking into account changes in policy with respect to education and training.

The years to come will show whether the Government will succeed in developing an effective technology policy on the basis of these recommendations.

4.2 Activities by employers and trade unions

Employers and trade unions in the Netherlands do not usually include aspects of work content in their negotiations about conditions of employment and working conditions. The responsibility for the content of work is left primarily to the employer, and the unions do not normally try to exert a direct influence on the quality of work in particular enterprises or branches. Indirectly, however, they have an influence on work legislation, e.g. via the Socio-Economic Council. This has had its effects with respect to problems of safety, health and democratisation, but work content has been beyond the scope of legislation until very recently.

With respect to technological developments, the policy of the trade unions is also directed particularly at the quantitative effects on the labour market and the impact on work classification and wages. The interest in the effects on the quality of work is growing, but this has not resulted in a clear influence on trade union policy until now.

This situation led in the past to a rather reluctant attitude of the unions towards job design experiments. Often there was a certain amount of mistrust, because the experiments were seen as serving rationalisation objectives on the part of management. On the other hand, an important impulse for management to develop a job design policy was lacking, because there was no pressure in that direction from the side of the trade unions. Although it is argued[19] that a socio-technical job design is in the interest of the employees and of the employers, because of the resulting greater flexibility of the organisation, many employers prefer to stay with the classic forms of work organisation. Too much participation and autonomy for the employees is seen as a threat to control over the production process and the organisation. This is probably an important reason why many work structuring experiments remain at the periphery of the organisation and why many employers seem to prefer personnel problems, such as high sickness absenteeism and high turnover, to increased democracy. Another reason is that the financial consequences of poor work content and unfavourable working conditions in terms of costs of sickness absenteeism and unfitness for work are not at the charge of the individual enterprise but of the social security system as a whole, thus reducing the urge for measures at enterprise level.

These considerations explain why, in most cases of introduction of new technologies in the work situation, the old forms of work organisation are maintained. Problems arising from low acceptance by the employees of the new technologies are handled by allowing a certain amount of "user participation". This implies that in the last phase of the process of automation the employees participate in decisions about how to work with the new equipment and the design of the workplace. The possible amount of influence on the part of the employees is rather limited in such cases, because the more fundamental decisions about the use of technology and the work organisation are taken at earlier stages of the process of automation.

The fact that job design is not a central issue in the system of industrial relations in the Netherlands is one of the reasons that most job design activities, in the past have had a rather isolated and temporary character and have not led to more systematic approaches on a larger scale. This does not seem to provide very fertile ground for the development of job design activities in relationship with automation in the future. However, this situation may soon change, partly as a consequence of an increased interest on the part of the trade unions, and partly under the influence of the Act on Working Conditions, which prescribes participation of employees at enterprise level regarding safety, health and well-being at the workplace.

5. Case study - Results of a project on control tasks

After these introductory sections, we now move on to our detailed case study on the quality of work in relationship to a specific type of automation. We have chosen the study of so-called control tasks.

In these task situations, one or more human operators supervise and control a complex, more or less automated technical system. This is done from behind a control panel (interface) in a central control room or computer room. The panel displays information from the technical system (via VDUs, dials, polygraphs, visual and acoustic alarm signals, etc.), as well as containing controls, so that the operator(s) can bring about changes in the technical system. Examples of these tasks can be found in the chemical industry (oil refineries, ethylene plants, electricity generation (nuclear power, conventional electrical power plants), steel industry (blast furnaces, hot strip mills), traffic control systems (air traffic control, sea traffic control) and computerised administrative systems (banking and insurance).

In the Netherlands, many thousands of people are carrying out work of this type, and the expectation is that this number will increase in the years to come. As a consequence of automation, more production systems will acquire a "process-like" character, which makes centralised control possible. The jobs of an ever-increasing number of people will therefore consists of monitoring and controlling automated technical systems. Furthermore, these jobs themselves are subject to continuous change under the influence of new technological developments, increasing automation, growing complexity and scale enlargement of technical systems. It is therefore important to know how the quality of this work is being influenced by these developments.

Early research[20] gives a favourable picture of this type of work situation in comparison with work in mass-production systems (assembly line). The work is less fragmented and short-cycle, the workers have an influence on the complete production process instead of only on a small part, the work is therefore more interesting and challenging and offers greater possibilities for social contact and integration of the worker in the organisation.[21] Other authors, however,[22] doubt if this positive picture will be maintained at the ever higher levels of automation that are made possible by modern technology.

A point often mentioned is that increasing automation reduces the interesting and challenging aspects of the job. In advanced stages of automation, the tasks of operators can be restricted to a single start-up of the technical system and subsequently to passive monitoring (i.e. awaiting the occurrence of alarm signals). Thus only incidentally is an appeal made to the skills of the operator, while for the greater part of working time appropriate tasks are lacking. It is to be expected that these situations will have a negative effect on the employees involved in the long run.

A further investigation of these and other problems in relationship to increasing automation was the aim of a project carried out by the Netherlands Institute of Preventive Health Care/TNO and subsidised by the Socio-Economic Council.[23] The project consisted of a comparative analysis of 24 work situations in different enterprises with varying degrees of automation. In each of these situations, data were gathered with respect to the technical aspects (type and degree of automation), work content and organisational aspects. The operators were interviewed regarding their job satisfaction and health complaints. Data on sickness absenteeism were also collected. In this way a picture was obtained about the actual state of affairs with respect to job design and this specific type of automation in the Netherlands.

In the following section, we will first pay attention to some developments related to this type of automation. We will also present a brief outline of the general characteristics of the work content in control tasks.

5.1 General characteristics of control tasks

Technological and economic development in recent decades have affected in a number of ways the characteristics of industrial production systems, transport and shipment systems, and information processing systems. Product diversification and the necessity for a firm control of product costs have led to production on an enlarged scale. At the same time, there has been a growth in complexity of the technical systems involved in the development of improved control devices and automated information processing equipment. Extra dimensions have been added to these developments by the introduction of the computer (i.e. fast processing of huge amounts of data and automatic control of complex, technical systems).

This process of growth is generally referred to as "automation". As the examples indicate, applications can be found in a variety of settings. Due to the general characteristics of automation, the systems referred to have the following common features:

- system and/or subsystem operations are to a lesser or greater extent automatically executed;

- information presentation with regard to the state and time history of the process and system components is centralised and displayed by means of an interface;

- this interface in most cases also contains control devices;

- the human operator can either adjust the controlled variables directly, or has the possibility to change set points of the automatic control system.

Before discussing possible problems for the human operator in these work situations, we will briefly examine a model of the human control task and the activities of the operators in these systems.

A general outline of the human control task is presented in figure 23. The task of the human operator is to control the complex, more or less automated system within specific limit values or to modify the state of the system if required. Therefore the operator compares information from the technical system (1), presented via an interface, with desired values (2). Control activities can be executed via the interface (3 and 4). Depending on the degree of automation, a smaller or greater part of the control activities are carried out without intervention of the operator (1, 5 and 6).

Control of the system is effected either directly (3) - in some cases by field operators - or is by means of set-point control (4). One should realise that such a worker-machine system is embedded in an organisational structure. This in turn affects the operators (e.g. when management requires changes in product quality and/or product composition). Furthermore, the technical system itself can be influenced by the environment (e.g. weather conditions affecting air-traffic control). A technical system can also be connected to other technical systems (e.g. electrical power plants which are mutually connected by means of the national grid system).

5.1.1 Activities of the operators

Given this general outline, we will now first look into the various activities the operator has to perform. In most automated technical systems, a considerable amount of time is spent by the operator on *monitoring* the system. The operator carries out an active observation of the displays, or in a more passive way reacts to alarm signals. In general, less time is spent on *control* activities (dependent on the degree of automation). These control activities may be related to adjustments of the production process, as well as to reactions with regard to disturbances in the system. Often, however, a large amount of time is spent on *communication* activities. We can make a distinction here between consultation with co-operators or the shift leader on the one hand, and indirect control of the system by means of orders to the field operators on the other. Finally, there are a number of activities which are aimed at the *future effectiveness* of the system, such as administration and operations concerned with system maintenance.

To clarify this rather abstract presentation of the general characteristics of control tasks, a concrete example of a work situation in a pharmaceutical factory is given.

Figure 23: Model of the human control task

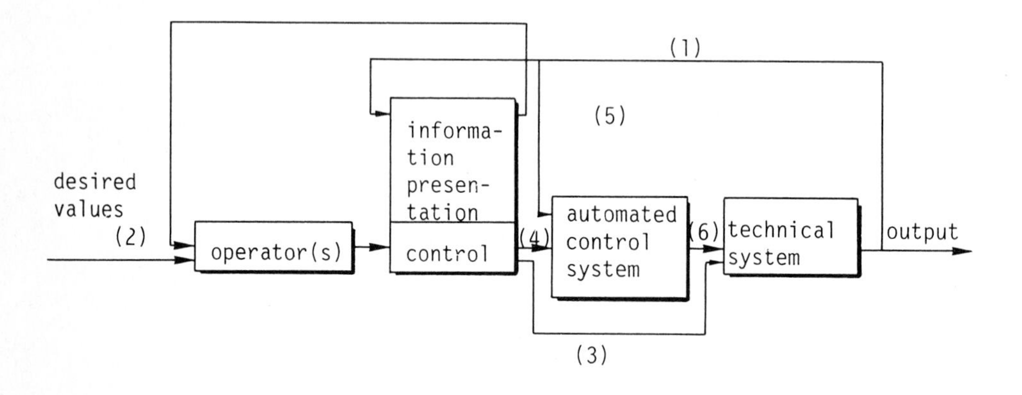

5.1.2 Work situation in a pharmaceutical factory

In this factory an intermediate product is made that is used in other plants for the production of antibiotics. The production process consists of a large number of mixing processes, chemical reactions, centrifugations, crystallisations, etc. The production takes place in batches with a cyle time of 4-8 hours within which, using a certain quantity of raw materials, a complete production process is finished, after which the product is carried away. Then the next production cycle is started. (Batch production must be distinguished from continuous production, where the production process runs continuously in the same direction and there is a continuous input and output of materials.)

This production process is controlled from a central control room by two panel operators, each responsible for certain sections of the process. To do this they have at their disposal a fairly large information and control panel. They also receive information about the process from the plant operators, who monitor the process in the plant itself. These plant operators can also take orders from the panel operators to carry out control actions that cannot be effectuated by means of controls on the panel. In the normal course of events, the panel operator and the plant operator together ensure that the conditions for a certain phase of the production process are fulfilled. This phase is then started by a simple switch controlled by the panel operator, after which the part of the process concerned takes place almost automatically. The task of the panel operator is then to monitor the successive steps in the production process very carefully so as to detect possible disturbances. Only when disturbances occur and in cases of start-ups or shut-downs does control take place manually instead of automatically.

In terms of the different levels of automation that we shall distinguish in section 5.2.2, this work situation can be classified at level 2b.

The operators work according to a four-shift system, each shift consisting of 13-14 who rotate between four different functions. The panel operator always gives guidance to the other operators in production and technical matters. Higher in the hierarchy is the assistant shift leader and the shift leader. The four shifts are under the supervision of a department manager and an assistant department manager.

In summary, we may say that we are dealing here with a relatively complex production process with a degree of automation that leaves the operators enough work to do and a real influence on the production process. The team of operators can function almost autonomously under normal circumstances, but intensive co-operation is needed between the operators in the team. From interviews with the operators, it appears that they are satisfied with their jobs and appreciate the high degree of craftsmanship that is necessary to fulfil the tasks. Another positive aspect is the high degree of autonomy over their work.

5.2 Design of the study

We selected 24 work situations, most of them in different companies. The situations had to meet the following criteria:

- the human operator must monitor and/or control a more or less automated technical system (panel operator);

- all information input and control output of the panel operator must be concentrated in a central control room or computer room.

The work situations selected can be classified with respect to the type of production process in the following groups: chemical industry; power generation plants; traffic control systems; administrative computer systems (i.e. the task of the console operator who controls the functioning of the computer system, starts up programs, etc.); and four other work situations that could not be classified according to these categories.

5.2.1 Data collection

In each of the work situations, several groups of data were collected. The first group was formed by data about characteristics of the technical system, the work content and the organisation. These data were gathered at the management level (i.e. design engineer, personnel manager). Secondly, data were obtained by means of structured interviews with the operators (± 10 per task situation). These data had to do with the perception and evaluation of the technical system, the work content and the organisation by the operators. They also filled out a questionnaire about job satisfaction and subjective health. Figures on sickness absenteeism were supplied by the personnel department. In all 223 operators participated in the research.

Audio-visual recordings were also made in 10 work situations to gain insight into the actual course of events during work.

5.2.2 Levels of automation

For our purpose it is necessary to classify the 24 worker-machine systems with respect to their level of automation. In the literature several scales for levels of automation have been presented. Although these scales can be very useful to gain insight into developments in specific activities,[24] or into levels of automation of different parts of the process,[25] they are too detailed to be used in this investigation. We therefore make a distinction between three levels of automation on the basis of three different configurations of control systems. The distinction is based on the relative contribution of the operator in the control of the technical system. These three levels are illustrated diagrammatically in figure 24.

Figure 24: Configurations of control systems

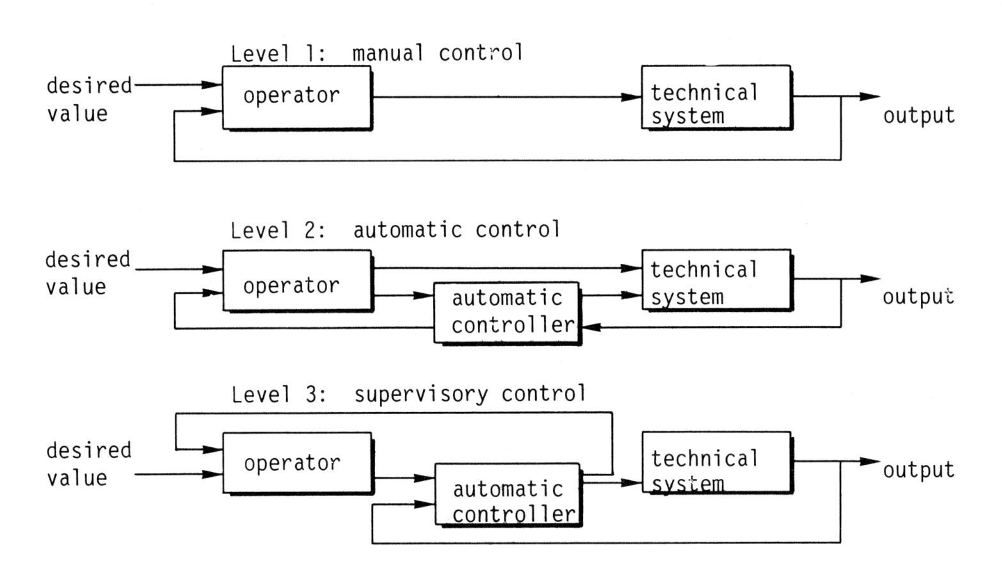

At level 1, the operator forms part of the closed loop of the control system for 100 per cent of the time. He or she receives information about the output of the technical system, compares this information with the desired value and performs control actions in order to minimise the deviation between the output and the desired value. In general, this form of control is called manual control. Each arrow in the diagrams denotes a vector of signals, so that a number of variables are controlled at the same time.

An example of this level of automation is the work situation of the operator on a hopper dredge ship. Hopper dredge ships have to ensure sufficient depth in some particular fairways. For this purpose these ships are equipped with two suction pipes, each manually controlled by an operator.

At level 2, the operator as well as the automatic controller can execute actions to minimise deviations between the output of the system and the desired value. The operator can adjust the set points of the automatic controller as a means to optimise the control of the system. In general, the automatic controller assists the operator in pre-processing the complex information of the system. This form of control is often called automatic control.

A number of examples of this type of configuration can be found in industrial situations. Sometimes the major part of the control actions are executed by the operator, and the automatic controller is merely assisting the operator (level 2a). In other situations, however, the major part of the control actions are performed by the automatic controller (level 2b). One should realise that the automatic controller can be an analogue electronic or pneumatic device as well as a digital computer. A further example of this level of automation is the work situation in the pharmaceutical factory that was explained in more detail above.

At level 3, the technical system is controlled by the automatic controller and the task of the operator is mainly a monitoring one. If necessary, the operator can intervene in the control actions of the automatic controller by adjusting the set points. As can be seen from the diagram, the operator forms no part of the closed loop of the control system, but merely supervises this system. In some cases the task of the operator is limited only to intervening in emergency situations, when the automatic controller is unable to solve the problem. In general, this form of control is called supervisory control.

Examples of level 3 automation are work situations in highly automated power plants (conventional and nuclear) and oil refineries, where most of the time the process is self-supporting and the operators have only occasionally to intervene in the process.

At the present time, all these levels of automation can be found in industry. Higher levels of automation are most often of a more recent date, however, because modern technology is necessary to reach these levels. Lower levels of automation will also remain, however, as is shown schematically in figure 25.

5.3 Some results

From the many results obtained from our study,[26] we shall discuss the most important consequences of higher levels of automation. The main results are summarised in table 58.

Figure 25: Different levels of automation during the last decades

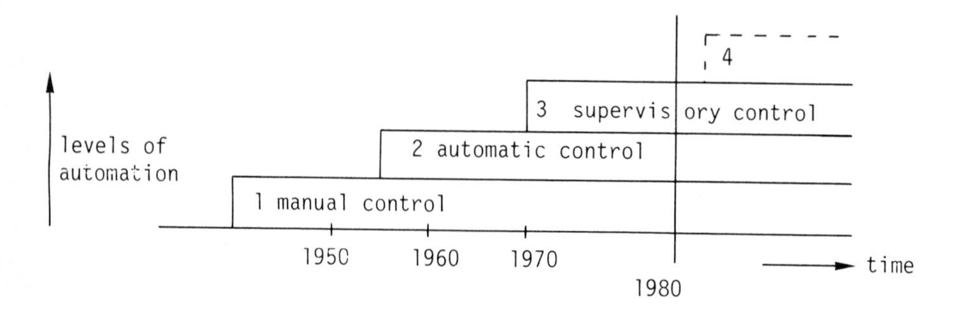

Table 58: Control tasks. comparative overview of 24 work situations at different levels of automation

Component of the work situation	Item	Automated production processes and transport systems[1]				Computer systems[2]
		I	IIa	IIb	III	IIa
Aspects of the technical system	Degree of automation	I	IIa	IIb	III	IIa
	Complexity	--	-	+	++	+
	Dimensions	--	-	+	++	-
Aspects of work content	Time spent in control (%)	21.9	6.2	3.8	2.6	15.6
	Time spent in monitoring (%)	69.7	43.3	48.6	44.6	42.9
	Time spent in administration, maintenance etc. (%)	0.3	41.2	36.3	31.8	35.7
	Time with no activities (%)	8.0	9.1	11.5	21.1	5.8
	Degree of routinisation	+	=	-	-	++
	Degree of procedures	+	=	-	-	++
	Quantity of information to be processed	+	+	-	-	++
	Complexity of information to be processed	-	-	+	+	-
	Feeling of having the system under control	++	-	-	=	-
	Uncertainty in control	--	+	-	-	+
	Frequency of disturbances	=	=	=	-	+
Work organisation	Number of team numbers	-	-	+	++	+
	Co-operation between team members	-	-	+	++	++
Subjective experience of the work situation	Total score for job satisfaction	+	+	-	--	--
	Feelings of achievement	+	+	-	-	--
	Amount of variety	+	+	--	-	--
	Subjective workload	+	+	-	-	+
	Satisfaction with co-operation	=	-	+	+	+
	Autonomy	++	+	+	-	--
	Attitude towards shift work	++	=	+	--	+
	Subjective well-being	+	=	+	-	=
	Subjective health (psychosomatic complaints)	-	+	+	-	-
	Sickness absenteeism (frequency)	-	+	+	++	++

[1] No. of work situations = 19.

[2] No. of work situations = 4.

Computer systems are presented separately because of a different pattern of results. ++ = very high score; + = high; = = average; - = low; - - = very low.

5.3.1 Work content

With respect to work content, it appears that at higher levels of automation less time is generally spent on control activities and that the percentage of time during which no work is available strongly increases. Furthermore, we see, that control activities are more of a routine character at lower levels of automation. At the same time, the monitoring is of a less passive character (more actively gathering the information from the production process) than at higher levels of automation (more passively waiting for disturbances). These findings are in accordance with the observation that more advanced automation leaves mainly the non-routine activities for the operator, which leads however to long periods of inactivity, at least in the work situations studied here.

As regards the method of working, we see that at higher levels of automation less use is made of fixed working procedures. This is to be expected because here the activities consist mainly of dealing with disturbances, for which it is difficult to indicate general procedures. At higher levels of automation, the information from the production process that the operator needs to make decisions about control actions has a more complex character than at lower levels of automation. This has to do with the fact that higher automated processes are generally also more complex. In a number of cases, however, this leads to uncertainty for the operator about the correctness of the decisions taken. The feeling of really exerting control over the process is in many cases lacking at higher levels of automation.

The results found in the work situations consisting of controlling computer systems are presented separately in table 58. This was done, because these show to an extreme degree a number of characteristics that result from the relatively low levels of automation used for controlling computer systems (the job of the so-called console operator): much time is spent on control activities with a strong routine character and rigid working procedures, resulting in a high level of workload for the operator.

5.3.2 Work organisation

With respect to some aspects of the organisation, it was found that higher automated processes are generally operated by a large number of people. The work teams consist of more operators (sometimes 10-15 versus 1-2 in many cases with a low degree of automation, and within the teams intensive co-operation is necessary. This is due to the fact that the higher automated processes are normally also more complex and extensive than the processes at lower levels of automation). Nevertheless, there is not much difference in organisational design between the different work situations. The operators are supervised by a shift leader, the next step in the hierarchy being a functionary who exercises supervision over the different shifts. Then follows the department manager, production manager or plant manager. In none of the work situations investigated were experiments being carried out with other than the traditional hierarchical forms of organisation. Nor were any efforts under way to improve the work content by job structuring (e.g. job enrichment) activities. In some cases, however, a system of job rotation existed within the team of operators.

This lack of attention to new forms of work organisation in an ordinary sample of work situations in the Netherlands illustrates what was concluded earlier in this chapter, namely the temporary and isolated character of job design activities in the Netherlands.

5.3.3 Subjective experience of the work situation

The highly automated work situations are generally more negatively experienced by the operators than the work situations with lower levels of automation, with the exception of the work situations implying the control of computer systems. These have a low degree of automation but are judged very negatively by the employees. The negative experience of operators in highly automated work situations and on computer systems are concentrated on the content of the work they have to carry out.

In highly automated work situations, operators complained about monotony, boredom, lack of activity, no feelings of achievement in the work, etc. In the work situations involving the control of computer systems, the complaints referred to the routine character of the work, the lack of variety and the superficial knowledge gained about the functioning of the computer system. The consequence of the latter is that the work has to be carried out according to rigid procedures, and also that promotion possibilities to other functions are very limited.

Consistent with these findings is the fact that on average the frequency of sickness absenteeism is higher in work situations with relatively high levels of automation and in work situations involving the control of computer systems. There are also lower scores on a questionnaire about subjective well-being and higher scores on a questionnaire about psychosomatic complaints. These results make clear that negative effects of badly designed work situations are not limited to feelings of dissatisfaction but have a more profound effect on the functioning of the employees, even outside the work situation.

To illustrate these findings, a more detailed description is given of a work situation in an electrical power plant. In contrast with the situation in a pharmaceutical factory described above, this work situation has a very high level of automation and shows a number of the problems which accompany such application of modern technology.

5.3.4 Work situation in an electrical power plant

For generating electricity in the Netherlands, about 36 electrical power plants are in use. Most of these are conventional plants using gas, oil or coal, as fuel. Two plants use nuclear energy.

As a first step in the generation of electricity, steam is produced by heating water by means of gas, oil, coal or nuclear energy. The steam is led to a turbine by the aid of which electricity is produced in a generator. The electricity is supplied to the national grid system which transports it to local transformer stations and from there to the users of the electricity. The steam production part of a power plant consists of a large boiler fired with oil, gas or coal by means of which controlled quantities of steam can be produced at specified temperature and pressure. Connected to the boiler is a complicated system for the supply of fuel, air and water, the discharge of combustion gases and the transport of steam to the turbine. The work situation described here consists of controlling this part of the production system.

This task is assigned to a shift of six operators. These operators spend most of their time in a central control room from where the control of the system takes place. For this purpose a large panel is installed in the control room with a great number of information-displaying and control instruments, together with alarm instruments. Under normal circumstances the system is controlled entirely automatically; only during start-up and shut-down procedures and during disturbances does the operator switch to manual control. In terms of the different levels of automation distinguished above, this work situation can be classified at level 3. The task of the operators consists therefore of monitoring the process and waiting for irregularities, although they occasionally make rounds in the plant to carry out inspections and control actions that cannot be performed from the control room. The large number of operators per shift is necessary, in case of a severe disturbance, to keep the very large and complicated system in close co-operation under (manual) control. Most of the time, however, the operators do not have meaningful work to do.

The operators work in a continuous five-shift system. Within the shift, a so-called first operator has a co-ordinating task with respect to the other operators. Higher in the hierarchy is a chief operator who supervises more shifts in other parts of the plant.

The results of our study in this work situation can be summarised as follows. The very large and complex production system with a very high degree of automation is monitored by a team of operators. The first operator is permanently in the control room, while the others also make rounds in the plant itself. Under normal circumstances it is necessary to wait for irregularities, which means that for most of the time no meaningful work is available. The operators have fairly high qualifications in order to be able to control the system in case of disturbances. As a consequence of the high degree of automation, however, they can seldom apply their knowledge. Therefore they do not experience the feeling of having the system thoroughly under control and of influencing the production results. Working in a team is positively experienced, but because of the low frequency of disturbances there is not much opportunity for real co-operation. In principle, the shift would be able to function as an autonomous working group but in practice there is close supervision by higher technical supervisors, especially during the rare moments, when something interesting is going on, such as a starting up or shut-down procedure or a disturbance. This practice is followed in such situations, because under manual control the system is very vulnerable to incorrect control actions. It is not surprising that we found in this work situation a low degree of job satisfaction and subjective well-being. The figures for sickness absenteeism were also relatively high.

5.4 Discussion

The starting point of our study was the question whether increasing automation of production systems, as it has taken place in the Netherlands during the last decades, has led to an increase or a decrease in the quality of work. The results of our study give reason to assume that the latter is the case; they are therefore in contradiction with the more optimistic views expressed in publications of 15-20 years ago. An explanation for this is that during the most recent years, technological developments have proceeded rapidly. Scale enlargement, increasing complexity, centralisation and computerisation have resulted in work situations quite different from those that were discussed some 20 years ago.

One of the most important differences is that ever more tasks have been taken over by equipment designed for automatic control of processes. In highly automated production systems, therefore, the task of the operator(s) often consists of switching on the technical system and thereafter monitoring the process and waiting for disturbances. The technical design is aimed at as low a frequency of irregularities as possible. This leads to a poorly balanced job structure in which long periods of inactivity and boredom alternate with short periods of very intense activity, when dealing with disturbances. The latter is often a stressful event, because determining the cause of an irregularity is not always easy and has to take place under pressure of time. In many cases there are risks to the safety of the environment, which implies that technical and human errors can have severe consequences. As an illustration, we may point to recent accidents involving chemical plants and nuclear power plants. This means that human operators carry a heavy responsibility for the proper functioning of the process. At the same time, however, they see their tasks being increasingly taken over by electronic equipment.

As a consequence of this, a discrepancy arises between the responsibilities of the operator and the actual degree of involvement in controlling the technical system. This can lead to feelings of uncertainty, as we have seen in our study, and in some cases to less adequate reactions in emergency situations.

To summarise, we may say that the relatively low degree of job involvement and job satisfaction we found in highly automated work situations can be attributed largely to a less favourable division of tasks between the human operator(s) and electronic equipment. Seen from the perspective of the organisation, there seem to be possibilities of meeting these problems, at least partly. Because at higher levels of automation technical systems are often also on a larger scale, controlling and monitoring the system has to be performed by larger teams. Therefore a certain degree of job involvement could be based on the functioning within the group. This appears to be true: at higher levels of automation, operators are more satisfied with the co-operation in the team. At the same time, however, there is a loss of autonomy of the individual operator: at lower levels of automation the operator generally functions more autonomously than at higher levels of automation, where control of the system is frequently taken over by higher levels in the organisation. This applies particularly to special situations such as disturbances, starting up the system, etc. Thus, even during the scarce interesting moments of the job, the operator often does not have the chance to gain full control over the system. The conclusion may be reached that organisational solutions (e.g. division of responsibilities) can reduce the negative effects of highly automated task situations. At the same time, we must conclude that these possibilities are used very seldom.

Another problem is created by the work situations in administrative computer systems. Here the console operators cannot complain about a lack of activity. However, the routine character of the work and the rigid procedures according to which it has to be carried out have a negative impact on the workers. The operators can influence the actual functioning of the system only to a limited degree. Furthermore, their jobs do not require a high qualification level, so that their understanding of the functioning of the system is merely superficial. Here also, it is a problem of division of tasks between the human operator and the electronic equipment. In this case, only routine work is left for the operator, which leads to a number of negative consequences.

5.5. Consequences for job design

We can conclude from our study that considerations of work content and organisation in many cases play only a small role in the design of automated production processes. The consequence is that operating jobs often consist of an arbitrary collection of tasks that have not (yet) been automated for technical or economic reasons. Especially at higher levels of automation, this implicit method of designing jobs appears to result in problems for the workers, at least according to the results of our study. This is not surprising, because increasing automation most often implies transferring the control over the production process from the human operator(s) to the electronic control devices. Therefore it becomes increasingly difficult to design a job for the operators that offers them a sufficient amount of autonomy and control over the process and enough meaningful activities to keep them involved in their jobs.

As already discussed, this is a problem also from a production and technical point of view, because a low involvement in the job can enhance errors in cases of disturbances. An additional factor is that for higher automated production systems, that are often also more complex, higher qualification levels are required of the operators. This means in many cases higher expectations with respect to autonomy and control over the work situation.

These consequences taken together make it increasingly important to undertake efforts to ensure that application of modern technology results in work situations that fit the possibilities of human operators and satisfy their needs. How can this be accomplished? In existing production systems, there are generally few possibilities to change the work content of the operators. This is caused by the fact that the way of interaction between the operators and the technical system is determined for the greater part by the technical design of the production process and its control system. Therefore major changes in work content can only be achieved by technical changes in the (control of the) production system.

In most cases such a solution will not be feasible because of the financial consequences. Changes in the work organisation (e.g. the division of tasks between operators) may, however, be possible in a number of cases, though even here the technical design can impose limitations. For example, a very centralised method of information presentation about the production process (e.g. on VDU) makes it difficult to divide the responsibility for the control of the process among more operators. This means that already from the very beginning, in the phase of the technical design of productions processes, one has to take into account the requirements from the point of view of human possibilities and needs. This is the only way of systematically preventing the negative effects reported in our study. The problem is, however, that the technical design of production systems in most enterprises is the task of technicians. The design takes place according to technical and economic criteria, and considerations with respect to the work situation of the operators play only a minor role.

To effect changes in this one-sided technical approach is one of the major problems to be solved before we can expect improvements in this area. We have already mentioned the alternative of a socio-technical systems approach, according to which technical system and work organisation are designed in close inter-relationship, making use of technical, economic and social criteria (section 4). One must realise, however, that it will be increasingly necessary to influence the technical decisions in the design process. In many examples of the socio-technical approach, the accent lies on the design of jobs and work organisation, given the fact that basic technological choices have been made. Nevertheless, increasing automation will more and more limit the degree of freedom for designing appropriate jobs.

It will therefore be necessary to use "human" criteria from the very beginning of the technical design process to avoid a situation in which one can only realise marginal improvements. The starting-point of this approach is that the products of modern technology have a high degree of flexibility, which implies that there are far more technical design alternatives than are usually taken into consideration. These different alternatives must be assessed systematically with respect to the question of how much degree of freedom is left for designing an appropriate work organisation. A central issue here is the amount of control over the process that is left to the human operator(s), or in other words the allocation of functions either to the operator(s) or to the electronic control equipment.

5.6. A schematic outline of a design process

To show how social criteria can be brought into a technical design process with the objective of finding the optimum allocation of functions between the operator(s) and the control equipment, we will now explore the design process further. Figure 26 gives a schematic outline of the design process of automated production systems. Although each design process is to some degree unique, in general five phases can be distinguished which are independent of the specific characteristics of the system to be designed.

During each phase of this design process, a number of alternatives must be generated and must be compared with each other on technical, economic and social criteria. Alternatives generated in subsequent phases will be partly based on the choices made in earlier phases of the design process. These feedback loops have not been included in the schematic outline.

5.6.1 Phase 1: the architecture of the production system

The beginning of all design processes is the formulation of the objectives of the production system. These objectives are the starting-points for the definition of the separate functional tasks to be performed by the system. The definition of relations between the tasks of the system and the relative overlap takes place in the next stage of this first phase. In large-scale systems this can already result at this stage in decisions about centralised or decentralised control which, for example, can have consequences in the next phase for the number of control rooms to be installed. From the points of view of the human operator, it should be emphasised that decentralisation should not result in isolated jobs.

5.6.2 Phase 2: the structure of the system

This phase is particularly important for the ultimate structure of the job of the operator, because it is at this moment that the level of automation is for the most part determined.[27] Here care should be taken that enough control possibilities are allocated to the operator. This can imply that certain activities which one would like to automate because of differences in skills between workers and machines[28] should nevertheless be allocated to the operator. Of course other considerations (costs, effectiveness, control of the process, safety, etc.) can oppose against such an allocation. The results of this study have shown, however, that decisions in this stage should take into consideration the fact that too high a level of automation, as well as too many routine tasks for the operator, will both lead to an undesirable situation in terms of human criteria.

Figure 26: Schematic outline of the design of an automated production system

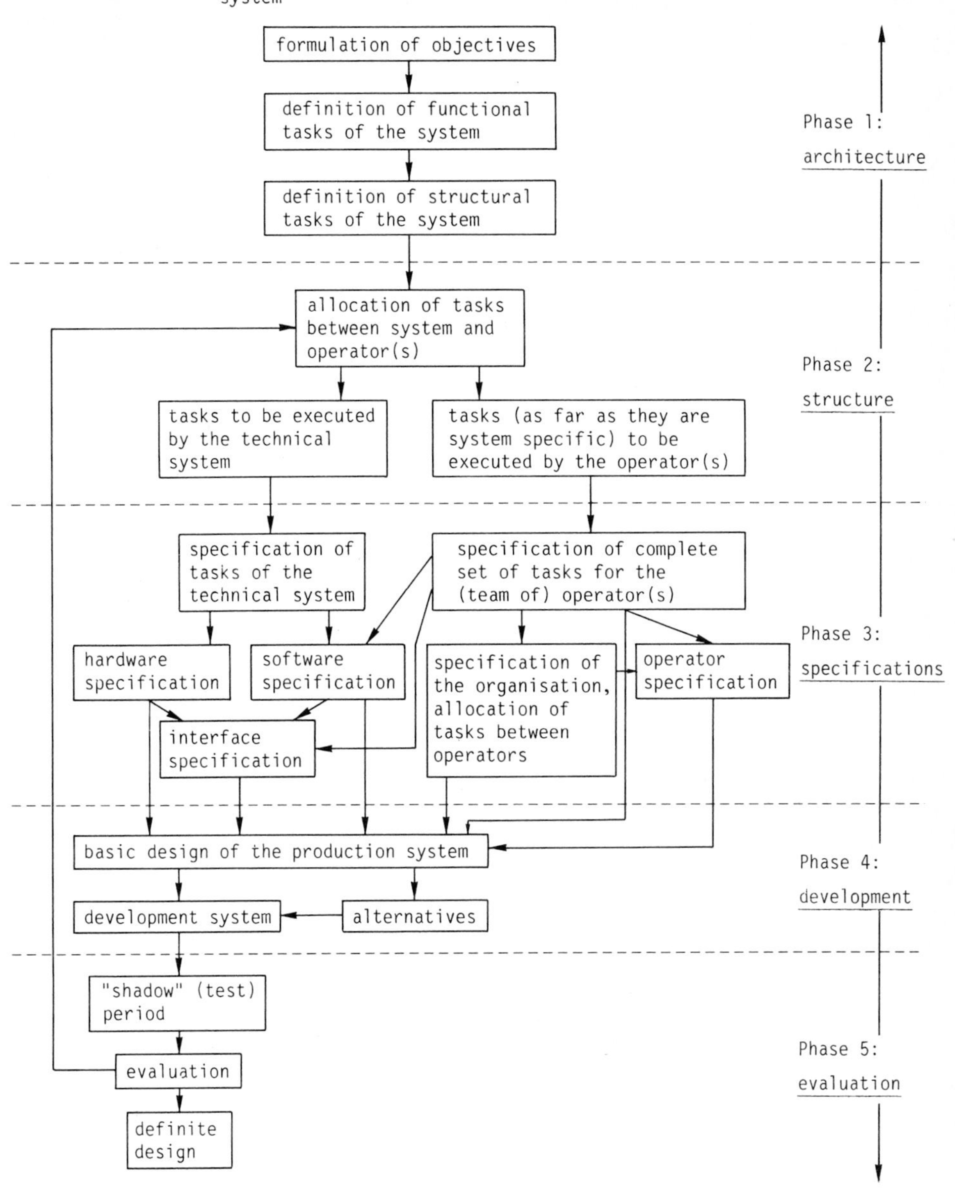
formulation of objectives
definition of functional tasks of the system
definition of structural tasks of the system
allocation of tasks between system and operator(s)
tasks to be executed by the technical system
tasks (as far as they are system specific) to be executed by the operator(s)
specification of tasks of the technical system
specification of complete set of tasks for the (team of) operator(s)
hardware specification
software specification
specification of the organisation, allocation of tasks between operators
operator specification
interface specification
basic design of the production system
development system
alternatives
"shadow" (test) period
evaluation
definite design
Phase 1:
architecture
Phase 2:
structure
Phase 3:
specifications
Phase 4:
development
Phase 5:
evaluation

5.6.3 Phase 3: the specificiations of the system

Here the work content of the human operator is ultimately defined. Given the allocation of tasks which has taken place in phase 2, it must be decided at this point how tasks will be divided between operators, and also if tasks will be enlarged with activities that are not directly related to the control of the technical system (e.g. planning, maintenance, work programming, etc.). At the same time, the competence of the individual operator as well as the team of operators with respect to the higher levels of management in the organisation (the degree of autonomy) must be drawn up. One should take care to establish operators' jobs in such a way that mutual co-operation is necessary without the need for a rigid hierarchy within the team of operators.

The worker-machine interface must be specified in this phase as well. Ergonomic criteria should be applied, but the way in which supervision and control will take place must be in accordance with the chosen organisation of tasks within the team.

5.6.4 Phase 4: development of the integral system

In this phase, the previous design specifications must be integrated in the total system. In order to fit the subsystems to each other, modifications will be available. These modifications should not only take place at the expense of the "human" criteria because these specifications are "soft" compared to criteria related to costs or quality of products. At this phase the flexibility of recent technology developments must be emphasised and brought into practice by the design team. Alternative solutions and designs must also be taken into account. A too rigid approach to this phase can lead to realisations which can no longer be adjusted or modified at a later date.

5.6.5 Phase 5: evaluation of the design

During evaluation of the final design, it is necessary to investigate how far the production system is in line with the objectives formulated in phase 1. It is now possible to examine whether the allocation of tasks between the operator(s) and the system was correct, or must possibly be modified.

Finally, the chosen structure of the organisation can be evaluated. In a practical situation this evaluation generally starts within a "shadow period", during which separate subsystems are tested and the first integral system tests take place. Small-scale simulations are executed and conceptual faults in the system can already be traced. This preliminary information about the system can be used to implement modifications before the final, and in general protracted tests are executed. Of course, the opinion of the operators who will work with the worker-machine system has to be taken into account during the evaluation phase.

A design process as indicated above can be applied to different types of automated production systems in different types of organisations. The schematic outline therefore has a rather general and abstract character. In most cases, one would have a multidisciplinary team in which technical and economic knowledge, as well as know-how from the behavioural sciences, is incorporated. The users, who will finally have to work in the production system, should also be represented, particularly in the case where they have experience of supervision and control of comparable technical systems. In cases where the trade unions play an active role with respect to this type of problem, they may be represented as well.

In the course of such a design process, many practical problems have to be dealt with. Incorporating social criteria into a technical design is not an easy matter. In the long run, however, it is the only way in which modern technology, which is seen by many people as a threat to QWL - and which it is indeed in many cases, given the present method of application - can be used for the improvement of work and working conditions.

Notes

1 F.C. Mann and L.R. Hoffman: "Individual and organisational correlates of automation", in Journal of Social Issues (Ann Arbor), Vol. 12, 1956, pp. 7-17; R. Blauner: Alienation and freedom: The factory worker and his industry (Chicago, University of Chicago Press, 1964).

2 Automatisering [Automation], Report of the Socio-Economic Council (SER), (The Hague, 1968).

3 A.J. Hamnink: Administratieve automatisering in Nederland, 1976-1980 [Administrative automation in the Netherlands, 1976-1980].

4 Automatisering, op. cit.; Maatschappelijke gevolgen van de micro-electronica [Social consequences of micro-electronics], Report of the Rathenan advisory group (The Hague, Staatsuitgeverij, 1980).

5 J.R. Bright: "Does automation raise skill requirements?", in Harvard Business Review, 1958, No. 4, pp. 85-98; H. Braverman: Labor and monopoly capital: The degradation of work in the twentieth century (New York, monthly Review Press, 1974).

6 H. Kern and M. Schumann: Industriearbeit und Arbeiterbenusstsein [Industrial employment and worker consciousness], Series of Studies of the Institute of Social Research (SOPI), Göttingen (Frankfurt, Europaische, Verlaganstalt, 1970).

[7] G.J.M. Cohen and F. Huijgen: "De Kwalitatieve structur van de werkgelegenheid in 1960 en 1971" [The quantitative structure of employment opportunities in 1960 and 1971], in Economisch-statistische Berichten, No. 65, 1980, pp. 480-87.

[8] Commissie Zeehavenoverleg: Kwalitatieve aspecten van de arbeidsmarktontwikkelingen in zeehaven-gebieden I [Qualitative aspects of the evolution of the labour market in maritime ports] (The Hague, 1980).

[9] C.L. Ekkers, et al.: Menselijke stuur-en regeltaken: Eindrapport [Human control and regulating functions: Final report] (Leiden, NIPG/TNO, 1980).

[10] B. van Leusden: Taakstructur en numerieke besturing [Structure of functions and numerical control] (Twente, Technical High School thesis, 1970).

[11] H. Wierama: Arbeidsverhoudingen in het bankwezen [Labour relations in the banking system] (Nijmegen, SUN. 1980).

[12] L.U. de Sitter: Op weg naar nieuwe fabrieken en kantoren [Towards new factories and offices] (Deventer, Kluwer, 1981); J.J. van Hoof: Technological development and structure of qualifications in the Netherlands, Paper for the European Research Seminar on the Relation between Technology, Capital and Employment, 1981.

[13] HAGA: Herstructurering, automatisering en gevolgen voor de arbeid [Restructuring, automation and consequences for employment] (Utrecht, University of Utrecht, 1981).

[14] H. van Beinum and R. v.d. Vlist: "QWL developments in Holland: An overview", in International Council for the Quality of Working Life: Working on the quality of working life (Boston, Martinus Nijhoff Publishing, 1979).

[15] K.S. Bibby et al.: Man's role in control systems, Plenary paper, Sixth Triennial IFAC Congress of the International Federation of Automatic Control, Washington, 1975.

[16] F.E. Emery and E. Thorsrud: Form and content in industrial democracy (London, Tavistock Publications, 1969).

[17] See, for example, J.T. Allegro and E. de Vries: "Project 'Humanisation and Participation' in Centraal Beheer", in International Council for the Quality of Working Life: Working on the quality of working life, op. cit.

[18] Maatschappelijke gevolvegen van de micro-electronica, op.cit.

[19] de Sitter, op. cit.

[20] Mann and Hoffman, op. cit.; Blauner, op. cit.; G. Brenninkmeyer: Werken in geautomatiseerde fabrieken [Working in automatic factories] (Amsterdam, Swets and Zeitlinger, 1964).

[21] M. Fullan: "Industrial technology and worker integration in the organisation", in American Sociological Review, Dec. 1970, pp. 1028-39; W.C. Vamplew: "Automated process operators: Work attitudes and behaviour", in British Journal of Industrial Relations, Nov. 1973, pp. 415-39.

[22] G.I. Susman: "Process design, automation and worker alienation", in Industrial Relations, Feb 1972, pp. 34-35; for an overview of the literature, see J.M. Shepard: "Technology, alienation and job satisfaction", in Annual Review of Sociology, 1977, No. 3, pp. 7-21.

[23] Ekkers et al., op. cit.

[24] Bright, op. cit.

[25] Kern and Schumann, op. cit.

[26] For a full report, see Ekkers et al., op. cit.

[27] J.E. Rijnsdorp: Levels of automation, a multidimensional approach, Report to the Fifth International Conference on Production Research, Amsterdam, 1976; Bibby, op. cit.

[28] P.N. Fitts et al. (eds.): Human engineering for an effective air navigation and traffic control system, (Washington, DC, National Research Council, 1951); R. Coburn: Human engineering guide to ship system development (San Diego, California, Naval Electronics Laboratory Center, 1973).

CHAPTER XII

AUTOMATION AND WORK DESIGN IN NORWAY[1]

1. Country overview

Both historically and geographically, Norway has always been on the periphery of Western Europe. Industrialisation already had a long history in other European countries when it reached Norway at the end of the last century. Before 1900, Norway's largely agrarian economy maintained a level of basic self-sufficiency, but the country was undeveloped. This began to change around the turn of the century with the development of Norway's extensive hydroelectric power resources which created cheap electricity for new industry.

Today, with the development of North Sea oil, Norway has become one of the most affluent industrialised countries in the world. But oil does not tell the whole story. As the Economist observed in its special issue on Norway:

> Even without oil, Norway in the twentieth century has been a major economic success story. It has moved from being a low productivity and almost subsistence economy, based largely on the exploitation of primary materials - fish, timber, iron ore - to one based on high technology ...[2]

For our purpose, neither technology nor economics is sufficient to explain this rapid growth. A major factor in Norway's industrial history is the relatively early development of a strong trade union movement with close ties to the Norwegian Labour Party (DNA). DNA has provided the country with a socialist government almost continuously since the mid-thirties, and both work and working conditions have been important topics for the Labour Government. In the early sixties, a government inquiry into industrial democracy resulted in the widely discussed Norwegian Industrial Democracy Programme.[3] This programme, perhaps the first systematic national effort in the world to enhance and democratise QWL, based explicitly on socio-technical theory[4] and on labour-management co-operation.

In the early seventies, partly as a result of experiences from the Industrial Democracy Programme, attention became focused on the effects of new technology, particularly as related to the interests of trade unions. A series of field studies completed in close collaboration with the Iron and Metal Workers Union[5] clarified trade union demands and policies concerning technological change. This development of trade union consciousness of the QWL consequences of automation, and the experiences of trade unions and employers with more democratic forms of work design from the Norwegian Industrial Democracy Programme, led to the pioneering "technology agreements" between the national labour market parties in 1975 and to the innovative Norwegian QWL law in 1977.[6] These developments in turn strengthened trade union participation in technological change and more humanistic and democratic forms of work. In short, after a belated start to industrialisation in the late nineteenth century, Norway, by the late 1980s, had become a leader in developing innovative and democratic approaches to the problems of automation and work design.

Although Norway may have entered the twentieth century as a relatively poor, pre-industrialised society on the periphery of Europe, it leaves it as a "post-industrial" society more in the centre of innovative industrial relations and QWL experiments than on the periphery. Norway is a particularly relevant case to examine concerning the issue of automation and work design, since this topic has been an important public issue for almost two decades. We will place these developments in their national context by first describing relevant socio-economic and technological developments. These influence and are influenced by Norway's industral relations system, especially its legal and political aspects. Finally, we describe the field of work organisation design in Norway and identify main research groups as a basis for introducing our case studies.

1.1 Socio-economic conditions

Norway is at once a small and a large country. It is small in terms of population (4 million), but larger in land area than the United Kingdom or the Federal Republic of Germany. The resulting low population density is exaggerated because the relatively small population is highly spread out in a country that is very long (more than 1,600 km) and narrow (in places well under 160 km), and laced with high mountains and deep valleys. In fact, over 60 per cent of the surface area consists of mountains, with the only significant flat area located around the Oslo fiord. With a very low population density and enormous barriers to transportation, Norway was one of Europe's least developed countries at the end of the last century, when the Norwegian economy was primarily based on farming, forestry and fishing.

Some industry could be found prior to the twentieth century. For example, wood products were a significant economic activity, even in the nineteenth century, in the production of cellulose and paper. These industries advanced significantly in the twentieth century with the development of Norway's hydroelectric resources. The growth of this energy resource significantly changed the Norwegian economy and altered the economic structure in fundamental ways. This growth promoted the development of the electro-chemical and electro-metallurgical industries in the first two decades of this century, and within these areas Norway had a leading position in the world.

This process of industrialisation during the last hundred years is reflected in statistics on the distribution of employment among various sectors of the Norwegian economy. Initially, as is typical, the largest proportion of employment was in the primary sector. As this became less dominant, it was replaced by growth in the secondary sector which in turn, as a society moves into what has been called a "post-industrial" phase, is replaced by the service sector as the major source of employment (table 59).

The single most striking economic fact about contemporary Norway is that the decade of the seventies was a period of continued economic growth. GNP increased at an average annual rate of 4.6 per cent (reduced to 3 per cent for the last half of the decade if oil revenues are excluded).[7] Thus, by most indicators Norway is in a much better economic condition

Table 59: Norway: percentage of employment in main economic sectors, 1865-1979

Economic sector	1865	1910	1950	1973	1979
Agriculture, forestry, fishing and hunting	59.8	39.0	25.9	11.4	8.6
Industry, mining, construction, electrical power	13.6	25.0	36.5	33.9	30.1
Public and private service	12.6	17.8	16.2	30.7	29.3
Commerce, transportation and other	14.0	18.2	21.4	21.0	31.9
Total	100.0	100.0	100.0	100.0	99.9

Source: N. Ramsøy et al.; The Norwegian society (Oslo, Universitetsforlaget, 1968); Statistical Yearbook, 1980.

than most other industrialised countries. This is because of the large oil deposits that have been discovered in the North Sea and are now being rapidly developed.

Oil was found in the North Sea at the end of the sixties, but it was not until the early seventies that these resources began to be developed. It is quite clear that these enormous oil reserves have for the time being placed Norway in a class of its own in relationship to other European countries. Since 1976 Norway has been the only European country to be a net exporter of oil. The importance of oil for the Norwegian economy has been particularly strong and increasing, especially since the large oil price increases after 1973. As a result, Norway is the only industrialised Western country that could aim at, and to a large degree, achieve a relatively high degree of growth and low unemployment in an international economy that suffers from instability and stagflation. Norway has financed this growth through large foreign loans and has provided selective support to certain industries that have been disproportionately affected by an adverse international economy.

At the beginning of the eighties, Norway had one of the highest rates of GNP per head in the entire world and an unemployment level of less than 2 per cent. Despite low levels of unemployment, Norway has never had culturally different foreign workers in any significant numbers. Today, in a labour force of 1,700,000, there are only 80,000 foreign workers (4.7 per cent) who are mostly from other Scandinavian countries, and this proportion has been relatively stable. The result is still a fairly tight labour market, especially in the area of skilled labour related to the oil sector. The main labour market problem in Norway is one of redeployment rather than unemployment. There is, however, relatively high unregistered unemployment among youg people and women, especially in rural areas. The increasing rate of technological change and automation seems likely to increase the problem.

Nevertheless, we can easily conclude on balance that the Norwegian economy has developed rapidly in this century from pre- to post-industrial and that Norway seems to be one of the Western countries least affected by the international economic crisis.

1.2 Main trends in automation

The picture is much less clear concerning the development and implementation of new technology in Norway. Despite a number of government reports on the issue of new EDP technology, there exists no survey of computers or other data processing equipment for Norway as a whole. Such a census would be difficult because significant amounts of new technology are imported. Even in those areas where Norway has traditionally developed new technology, such as process control and machine tooling, current, comprehensive and systematic survey data are lacking. Thus, while we can say that levels of automation appear to be high in certain sectors and increasing in others, we cannot make a conclusive analysis. We do, however, have some data on computer-based process control technologies, which are relevant to the case studies reported here.

The first computer control system in Norwegian industry (developed by researchers at the Foundation for Industrial and Technical Research - SINTEF - at the Norwegian Institute of Technology in Trondheim), was installed in 1966 in a cement factory. A number of similar systems were installed in 1967 and 1968 (e.g. in aluminium and paper production industries). Thereafter, only a slow development occurred until 1972, when the number of systems increased rapidly (see figure 27). In the first half of the seventies, process control computers were installed in a number of large companies within the chemical industry, in the paper and paper production industries, and in the production of cement and metals. This growth levelled off somewhat in the middle of the decade. From that point on, it became more common to install microprocessors in the production process, in order to achieve more direct operation control or to actualise simple feedback control systems. With the implementation of new technology, there was a tendency at first to let it perform the existing operations, but after a while it was used to increase the degree of automation.

Another important trend in automation in Norway was that the tendency towards the use of larger main frame computers ceased. Control was instead allocated to a larger number of smaller computers, which could be either gathered together in one location or decentralised. The purpose of this was to make the system less vulnerable to failure, and to reduce the transfer of large amounts of data over long-distance cables. Furthermore, the equipment was used more for analysis and for administrative routines.

The development of data processing in Norway has been described as a development in three phases.[8] In the first phase, a large computer is typically installed in the "computer department". Data is fed into the large computer, manipulated and stored in data banks. The computer department then sends the results to users. In the second phase, a common data

Figure 27: Trend in use of process control computers in Norwegian industries

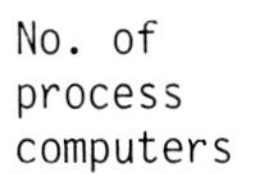

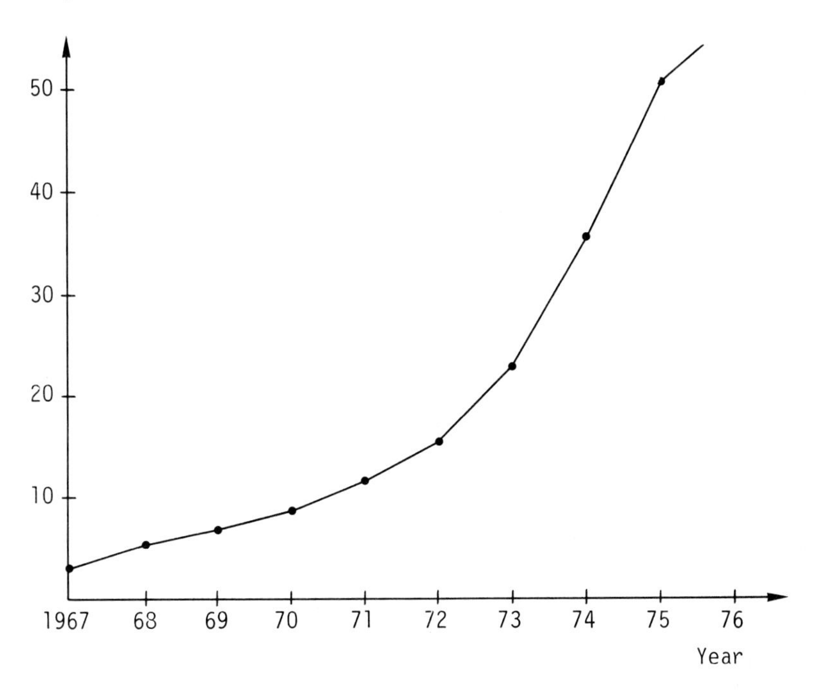

<u>Source</u>: Arthur B. Aune: <u>A survey of the use of process computers in Norway for the operation of production plants in the fields of chemical, petroleum, paper and process industry</u> (Trondheim, SINTEF, 1976).

base is established for groups of users. According to Nygaard and Fjalestad: "Now a common conception of the organisation, materialised in the data base, is enforced upon all users of the computer".[9] In the third phase, the walls between the computer department and the rest of the organisation are "penetrated by communication equipment". The organisation becomes a network of people, machines and information processing equipment.

We may say that the development increased between the mid-sixties and the mid-seventies, when about 50 process computers were in use. Parallel to their increasing number, computers are becoming larger and are covering more complex processes.

1.3 Industrial relations: the legal and political context[10]

One of the dominant characteristics of working life in Norway is the State's relatively deep involvement in industrial relations. This has its basis in Norway's experience under industrialisation, especially in the origins of the Norwegian trade union movement (LO) founded in 1899, the DNA and the organisation of employers into the Norwegian Employers' Federation (NAF) in 1900. Industrial relations in Norway are based on the relationships between LO and NAF which are highly institutionalised, centraised and comprehensive.

Today in the private sector LO and NAF are clearly the largest and most dominant organisations, so that agreements and relations between them often influence strongly the relations between other organisations representing employers and employees. In the private sector, at the beginning of 1979, NAF represented approximately 9,200 member companies, with a total of 379,000 employees, while LO had approximately 602,000 members (including a large proportion of public employees). In other words, both are fairly large organisations covering in practice almost all industrial work. Those organisations which are not formally members of LO or NAF tend to follow LO-NAF agreements.

Norway's first trade union (created among typographers in 1872) was founded some 15 years before the DNA was created. From the beginning, the DNA was the political spokesman of the working classes and labour movement. After the creation of the LO in 1899, the LO took over trade union representation, but maintained close contact with the DNA. This close connection between the labour movement and the DNA has been of major significance in the period since 1935, when the DNA, with only minor interruptions, has controlled the Government. This has given the Norwegian trade union movement a unique chance to realise its interests through the apparatus of the State.

With the signing of the first Basic Agreement between LO and NAF in 1935, the final barriers to freedom of organisation were removed. The Basic Agreement includes the principles of freedom to organise and the right to elect shop stewards and other representatives, and outlines the main structure within which negotiations occur. It is normally renewed every four years, at which time it can be revised. The signing of the first Basic Agreement signified a formal acceptance of LO as a legitimate partner in public policy-making and affirmed its permanent place in Norwegian society. The Agreement both presupposed and facilitated a transition from an industrial relations system based exclusively on confrontation to one characterised more by co-operation than by conflict.

Membership of LO grew rapidly, and in fact more than doubled during the five years prior to the Second World War (from 173,000 in 1934 to 360,000 in 1939). With its biannual revisions, the Basic Agreement has been substantially extended since the mid-thirties and this, together with the ideology of co-operation in industrial relations, has produced a relatively long period of industrial peace in Norway. Norway typically is one of the advanced industrialised countries with the fewest number of working days lost to strikes.

This broad consensus has dominated the post-Second World War period in Norway. According to some observers, it resulted in part from the personal relations established between labour leaders and employers in opposing the Nazi occupation and reinforced by the common task of social and economic rebuilding of the country. With the election of a non-socialist government in 1981 and the difficult economic issues posed by oil income (especially for non-oil related, labour-intensive industry), it remains to be seen how much of this consensus persists in the final two decades of this century. In part, however, the main structure of industrial relations and labour market policies are so well-institutionalised that few major changes are expected in the short run.

The relationships between employers' and employees' organisations are regulated by what has become a fairly comprehensive system of laws, regulations and agreements. For our purposes, the most important of these include the Basic Agreement (Hovedavtalen, including the data agreement, Dataavtalen) and laws regulating industrial conflict (Lov om arbeidstvister), shareholdings (Aksjeloven), quality of working environment (Arbeidsmiljøloven) and the national sickness benefit scheme.

The Lov om arbeidstvister came first (1915) and helped to lay the foundation for the industrial relations systems. This law gave collective agreements predominance over any local agreement, as well as making distinction between conflicts of interests and conflicts in law (rettstvister). It also stated that strikes would be illegal in the period between wage negotiations. Legal conflicts were to be adjudicated by the Labour Court whose decisions are final.

The law further requires that when wage negotiations cannot be successfully completed, the conflict must be put before the national arbitration service before there can be further action by either employees or employers. If the representatives of employers and employees reach agreement under their negotiations, the results of such agreements are sent out to be either rejected or accepted by their members. If the Government finds that a conflict may adversely affect national interests, it can then propose to Parliament a law which brings the conflict before compulsory arbitration. A decision reached by the compulsory arbitration board is final and cannot be appealed against. A dispute may also be taken to compulsory arbitration if the parties to the conflict wish to do so on their own.

In 1973, the shareholders' law was changed in a number of respects, and in addition, in companies with more than 200 employees, the employees were given the right to elect one-third of the representatives to a new body, the company assembly. The company assembly then elects the company's board of directors and has the final say in questions concerning major investments and changes which substantially affect the daily working life of employees. This in effect limits managerial prerogatives. Moreover, the company assembly functions as an advisory organ for the board of directors. Most important, employees were given the right to elect one-third of the representatives in the company's board of directors. (the employees must have at least two representatives).

The Basic Agreement also specifies a number of organs where the employees are represented. Works councils are to be established in companies with at least 100 employees, and in large companies we also find departmental works councils. These have an advisory function but no final decision-making prerogative.

The quality of working environment became subject to new regulations in 1977 with the passing of legislation entitled the Act relating to Worker Protection and Working Environment. One important aspect of this law is that it does not simply set minimum standards for negative factors which should be eliminated or reduced, but also legislates for the increase in positive factors concerning QWL. In addition, it places strong emphasis on the employees themselves taking an acting role in investigating and improving their own workplaces. Finally, a significant characteristic of this law is the attempt to regulate socio-psychological aspects of working life, aspects that often seem to be problematic in relation to new technology.[11]

To implement the new QWL law, the Government also made available considerable funds for training, company improvements (in the form of long-term, low-interest loans and matching grants) and research. New research funds to the amount of 10 million Norwegian Crowns (NKr) were divided approximately equally between the Ministry of Labour and Municipal Affairs and the Work Environment Committee, Royal Norwegian Council for Technical and Scientific Research. These new research funds, related to physical, medical and socio-psychological problems in working life, are only a part of the total public funding allocated to work research in Norway, which has been estimated at NKr65 million or more than US$10 million annually.

In short, QWL is important in Norway, and the industrial relations system is extensive, highly centralised and co-operative. In fact, there is centralised negotiation with decentralised and co-operative implementation. The challenge to central authorities in LO and NAF is to create conditions for local initiative and co-operation. It is in this context that we must understand Norway's innovative solution to the effects of new technology - the data agreement between LO and NAF.

1.4 The data agreement between LO and NAF

The first local data agreement was made in 1974, and immediately afterwards LO and NAF began negotiations on the question of technological development. The first national data agreement between LO and NAF was completed in 1975. This agreement gave employees the right to co-determination in the implementation and development of new technology. The agreement also created the possibility of electing special data processing shop stewards.

Legal support for the data agreements was given in the 1977 Act on quality of working environment, section 12 § 3:

> The employees and their elected union representatives shall be kept informed about the system employed for planning and effecting the work, and about planned changes in such systems. They shall be given the training necessary to enable them to learn these systems, and they shall take part in planning them.

Data agreements like the one between LO and NAF today cover most of the private sector and the public sector at both national and local level. The agreements affect computer-based systems which are used in planning and execution of work, as well as in systems for data storage and use of personal data. Any data-based system having consequences for workplaces and QWL is included and is to be evaluated as a total system. In other words, in addition to technical and economic criteria for evaluation, social criteria are also to be included. The agreement requires that information concerning the system be given in an easily understandable manner and that all necessary documentation be made available to data processing shop stewards and employees participating in concrete projects.

The most important provisions relating to participation are:

> If the employees of an enterprise so desire, they may elect, preferably among the existing shop stewards, a special representative to safeguard their interests and to co-operate with management within the scope of this agreement.
>
> If the size of the enterprise and the extent of the use of computer-based systems so require, the employees may elect, in agreement with management, more than one special representative. It is recommended that these representatives form a working group, and it is presumed that the time needed is made available.
>
> The two central organisations recommend that, in addition to the shop steward representative, employees who will be directly affected by projects covered by this agreement should to the greatest practical extent be involved in project work.

The task of the "data shop stewards" is described as follows:

> The representative(s) shall, on the basis of their particular qualifications, be at the disposal of the employees and the other shop stewards, e.g. in connection with their involvement in concrete projects. The representative(s) shall also contribute to the co-ordination of the employees' involvement in matters covered by this Agreement.

Establishing a formal agreement or enacting a law is not the same as making real changes in everyday affairs, but substantial implementation efforts have been made in training and in creating the various co-operative councils, joint working groups and representative boards. Gustavsen and Hunnius[12] provide a very comprehensive and current summary of the working environment law. A government task force is assessing the first four years of experience. Its report is planned to be presented to Parliament in 1982 as part of a legislative evaluation and possible revision of the law.

New electronic data technology is as much a public issue in Norway as in other European countries and has been the focus of reports from several government commissions.[13] Experience with the data agreements is still limited. Preliminary figures from the Ministry of Labour indicate that some 700 data processing shop stewards have been elected, and many of these have already completed some training. According to a recently published government report, over 150 local, company-level data agreements in eight of the largest and most politically significant trade unions have been established, and some 16 unions have passed resolutions on the issue.[14] In addition, this report finds that there was comparatively little activity at the local level from 1976 to 1979, shortly after the first national data agreement was signed, but there appeared to be a significant increase in 1980.[15] Some trade unionists have also suggested that companies be required to apply for permission to implement new technology that could negatively affect employment levels or work environment. Right of veto for employees has also been discussed.

Despite this wide and apparently increasing interest among trade unions in dealing with these issues and utilising the rights granted by these reforms, there is, as we shall see illustrated in the case studies, so far only limited progress in practical implementation. These reforms deal with issues of increasing importance to trade unions, but unions still have only limited experience and resources in dealing with such relatively non-limited conventional and complex issues as socio-psychological health factors and sophisticated computer technology. These are issues where Norwegian work researchers and enlightened technologists have made some important contributions and have relevant research currently under way.

1.5 Research approaches to data technology and quality of working environment

The correlation between automation and skill has occupied social scientists since the mid-fifties. Originally there were two main positions, one claiming the upgrading effects of new technology,[16] and the other down-grading effects.[17] (Several researchers provide a review of this literature.)[18] More recent research has given credit to both positions.[19] The "polarisation thesis" states that both de-skilling and increases in skills might occur at the same time, so that the result is a concentration of jobs at the extremes of the qualification scale.[20]

The other main problem is to what degree one is free to choose organisational solutions within a given technological context. On the one hand, we have technological determinism which posits technological development as the prime mover behind social change. This development is usually seen as passing through different stages (i.e. craft technology, mass production technology and, finally, automated process technology).[21] On the other hand, we have a "changing environment" approach[22] which leans heavily on factors external to the organisation to understand the organisation's inner life. Here we also find the socio-technical school which claims that different organisational solutions are possible, given the same kind of technology.

In Norway we find a double approach to research on work design - QWL on one hand and data technology from a trade union perspective on the other. Both these approaches, which have spread to a number of other countries, are at present centred in three different research institutes: the Work Research Institutes (AFI) and the Norwegian Computing Centre (NR) in Oslo and the Institute for Social Research in Industry (IFIM) in Trondheim. All three of these institutes have in common the basic assumption that human and social factors should weigh at least as much as technical and economic ones in the development of a high-quality working environment. Beyond this, they each have distinctive approaches to the humanisation and democratisation of work under conditions of technological change.

A full description of each institute exceeds our present limits. The AFI is best known for work on the Norwegian Industrial Democracy Programme and socio-technical systems[23] and the socio-psychological provisions of the Acron quality of work environment.[24] The Norwegian Computing Centre has concentrated on data technology and trade unions[25] and on research and development concerning data agreements and participation.[26]

These two approaches are considered by some to be mutually exclusive. Initially, the AFI approach focused on workplace organisational change planned in collaboration with a labour-management co-operation committee. Although grounded in socio-technical theory, AFI tended to emphasis co-operative forms of developing democratic organisation. NR initially worked exclusively for trade unions, explicitly concerning the effects of new technology on trade union interests, and tended to emphasise development of consciousness among trade unions (both at local and national levels) and the evolution of trade union policy demands concerning new technology. NR's pioneering work directly facilitated development of the first data agreement in 1974. More recent developments of both these approaches have shown that they are by no means mutually exclusive and aspects from each may be combined with advantage in concrete projects.[27]

In the third institute, IFIM, located at the Norwegian Technical University, which is Norway's main centre for basic technical research, has a tradition stemming from the Industrial Democracy Programme (which started there in the sixties) but more recently has developed a research programme on the workplace and organisational consequences of new technology. The first project completed in this programme was undertaken in collaboration with NR and provides the case study from the aluminium industry reported below.[28] The two other case studies are from projects currently in progress in the areas of worker participation in technological change and QWL in the process industry. IFIM currently allocates approximately half of its research capacity (over five work-years) annually to this research programme, and projects are under way in advanced machine tool systems, word processing systems and information systems design. The projects tend to be undertaken in collaboration with engineers and scientists in each of these areas.

The overall aim of the programme is both to produce findings about the consequences of different kinds of new technologies and to increase the awareness and understanding of technical researchers about the human consequences of their technical design and development work. IFIM researchers, for example, are at present working with design engineers on projects concerning robotics, information systems design and CAD-CAM.

1.6 The wood products industry in Norway

Since two of our three case studies concern the wood products industry, we will now take a closer look at this industry and its technological development.

Forestry has always been a significant part of the economic life of Norway. At least since around 1500, the export of timber has been an important activity, and in 1698 the first paper production started in Norway. The first cellulose factories were built in the 1880s. In the period after 1945, the wood products industry was one of the country's most important sources of foreign currency, and when the post-war reconstruction period was completed early in the fifties, it was the dominant industrial branch. This industry is still important, even though a large number of companies have disappeared. In fact, the production of cellulose and paper has increased. In other words, the development is clearly in the direction of fewer and larger, more productive companies. One important precondition for this development and for increasing productivity has been the extensive use of new technology and automation.

1.7 Automation in the paper-making industry

Fortunately, current data is available on the degree of automation that exists and that which is planned in the paper-making industry. This information was acquired as part of the industry's analysis of the present situation, and as a forecast. Data were also collected on what each company expected to achieve or had achieved in relationship to reduction in number of workers, increased product quality, increased productivity and improvement of working conditions. Table 60 summarises the main results of this investigation.

As we can see from table 60, exactly half (112 out of 223) of the process units investigated are or will shortly be computer controlled. There is, however, no way of analysing on the basis of this data how comprehensive a computer system is, so that even small simple systems are registered as computer control systems. Our conclusion must nevertheless be that computer-based production control processes have found a permanent and expanding place in the Norwegian paper-making industry, especially in connection with bleaching and paper machine control, where almost 100 per cent of the process is computer controlled. The main advantages up to now have primarily been improved product quality and working conditions, and only secondarily reduced labour costs and increased productivity.

1.8 A pre-case

As part of the earlier-mentioned Norwegian Industrial Democracy Programme in the sixties, we find an experiment that is of particular relevance to the case studies we shall describe in the next section. This experiment in more autonomous forms of work organisation occurred at Hunsfos pulp and paper mill in southern Norway in the mid-sixties. Since this

Table 60: Norway: automated processes in the paper-making industry

Steps in the paper production process	Total no. of processes	Automated processes	Automated processes planned
Log handling	19	1	3
Wood chips production and handling	14	0	2
Sulphite boiling	3	2	0
Sulphite, chemical and mechanical	7	1	2
Planing	17	1	5
TMP and refining	6	0	3
Bleaching	8	6	1
Hollandery	44	8	11
Paper and carton machines	60	24	16
Flat-pressing	3	1	0
Rolling	16	4	3
Page-cutting	8	5	0
Packing	18	8	5
Total	223	61	51

experiment is very well documented elsewhere, we shall only summarise the most relevant aspects here.

Hunsfos pulp and paper mill, located in a small town on Norway's southern tip, employed in the mid-sixties 900-1,000 persons, most of them with deep roots in the local community (it was not unusual for some of them to be third-generation workers in the factory). The production integrated the different operations of mechanical pulping, chemical pulping and paper-making. After discussions between the management, the unions and the researchers, it was decided to use the Chemical Pulp Department as the first experimental site.

The technical system there was very similar to the ones we will describe in the cases that follow, except that there was no chemicals recovery unit (as in case study 2) and initially no control room. The system included five conversion processes: boiling, screening, bleaching, boiling acid preparation and bleaching liquid preparation. The work was done by four shift teams plus one day-time worker, a total of 29 employees. The management hierarchy consisted of a shift foreman, a general foreman, a production engineer, the pulp mill manager and top management. There was a high degree of specialisation and highly formalised division of labour, with virtually no multi-skilling of the workers and a traditional foreman's role (i.e. a hierarchically structured work organisation following traditional "scientific management" principles such as "one worker-one job").

The experiment aimed at increasing local autonomy in daily work with respect to tasks and decision making. The work organisation was therefore redesigned to include more maintenance tasks, multi-skilling and group work (e.g. creation of an information centre - a rudimentary form of control room - and a group-based quality bonus). This required extensive training and a restructuring of the foreman's role. The result was the successful implementation of a non-hierarchical form of work organisation in which the work groups exercised more direct and effective control of the pulping process. In effect, the experiment demonstrated that work organisation could be substantially changed from hierarchically managed to self-managed (i.e. that a semi-autonomous work group could be established) without any substantial change in technology.

Even though there was no fundamental restructuring of the management hierarchy as a whole, this experiment in the mid-sixties was one of the first in the world to demonstrate the possibility of organisation choice within the same technology. In short, the Hunsfos case shows how an old (non-computer based) technology can be combined with new (non-hierarchically structured) work organisation.

2. Case study 1 - Two cellulose factories with different levels of automation

2.1 Introduction

The company we studied is one of the largest industrial workplaces in Norway. It lies in Norway's most industrialised area, where industrial activities started over one hundred years ago. Today the company's operations are concentrated in one site, containing multiple interdependent processing facilities and some 2,800 employees. We investigated two processing facilities, both producing cellulose.

Table 61 gives comparative data on the two facilities. Both are highly comparable in terms of technology, product and company policies. To increase comparability even more, we focused primarily on the bleaching department in each. The main difference between the two is age. The bleaching department in the older facility, which we will call Cellulose 1 (C 1), started in the mid-fifties, while the newer cellulose operation, Cellulose 2 (C 2), was established in the early seventies. The principles for control of the processes in the two factories are the same, but the more up-to-date equipment in C 2 gives a better basis for remote control. However, C 1 and C 2 are organisationally very different, illustrating quite separate organisational design choices. This difference is further emphasised by the quite striking development of C 2's organisation during the seventies.

Table 61: Comparative data on two cellulose factories with different levels of automation (C 1 and C 2)

Factory	Start-up	Production (tons)	No. of employees	Types of production job titles	No. of employees per shift	No. of bleaching employees per shift	Absenteeism rate (%)	Organisational levels
C 1	1954	90 000	100	21	9	2	9	3
C 2	1972	168 000	65	5	8	2	6.6	2

We therefore base our report on an analysis of C 2 using data from C 1 to make selected comparisons. To a certain extent we are able to describe developments in C 2 during the seventies. We should, however, make clear that, despite important changes in organisational design, there have been no significant technical changes since C 2 started up in 1972.

Perhaps the most striking contrast between the two factories is in the way the work is organised. In the C 2 factory, the work is clearly organised on the basis of semi-autonomous work groups. Workers have a broad range of competence and have in effect taken over the traditional foreman's role, so that they have actual responsibility for personnel decision making, quality control and training. The C 1 factory has a more traditional form of work organisation, with functionally specialised jobs that are co-ordinated and led by a foreman. Whether this very sharp contrast in work organisation is a necessary consequence of technology or more a product of mangerial philosophy is a subject to which we return later.

2.2 Methods and data

We can divide our data into two main types. The first type resulted from semi-structured interviews lasting approximately 60 minutes each, while the second type consists of specific information on the company's operations and policies. This information came from a number of meetings where representatives of management and the largest of the trade unions in this industrial complex met with researchers. The purpose of these labour-management meetings was to identify basic information of a more practical character which everybody could agree on.

We conducted a total of 40 interviews, of which eight were with management and the remainder (32) with operators. Our interview data has not yet been completely transcribed and verified so that this analysis must be considered preliminary and limited. However, we believe that the main pattern we describe here will be supported in our final data analysis.

2.3 Process control technology

C 1 is a relatively old factory, and shows it. There are sufficient operational difficulties for it to be closed down in 1982, and it is therefore somewhat neglected in terms of maintenance. The machinery is worn out, and unexpected production stoppages often occur. The bleaching department is an integrated part of a relatively long production sequence. In C 1, several different qualities are being produced, which implies changes in processing. In C 2, on the other hand, there are only two qualities, and hence a more stable production.

In the case of C 2, the bleaching department is to a much larger degree an independent unit. However, in the main, both factories make the same product using a similar chemical-based process. The main technical difference is that C 2 has a six-step bleaching process compared to four steps for C 1, the two extra steps being connected to a chemical process. C 2 also has a computer-based process control system that is more comprehensive than that of C 1.

Process control of the bleaching process in C 1 was, until 1978, carried out in part manually (i.e. by the physical manipulation of valves in the various machines out in the factory), and in part on the basis of simple analogue instruments on a display panel, also in the production hall. This analogue form of instrumentation was improved in 1978 with a new control system based on a computer. The new computer system controls only a small part of the production process, acting as a limited supplement to the analogue instruments that were directly connected to the machinery and controlling only the feeding of chemicals, the temperature and quality control for the first bleaching step (out of four). The change from one quality to another is done manually. Because of this, a significant part of production process control in the C 1 bleaching department is based on simple analogue instruments and on manual adjustments. The relatively frequent production stoppages amplify the tendency towards manual control.

In contrast, C 2's bleaching department has a control system that from the beginning was based on computer systems. In principle, the computer-based control systems in C 1 and C 2 are identical but the C 2 system covers the entire process. Process control is based on full analogue instrumentation, with a computer as the main control unit. All monitoring occurs in the control room. Process system control is based on models that are developed by the company's own engineering department. In such a situation, there are three possible levels of process control -

- the computer controls everything, but if it breaks down, the system reverts to -

- use of analogue instruments such as controllers in the instruments panel; or

- direct manual control of valves in the production process.

The computer probably makes 99 per cent of process control decisions, but process operators have the opportunity to choose one of the two other forms for process control if they find it necessary.

Taking both factories, we can see a definite development in the level of automation if we take our point of departure as the situation in C 1 before 1978. There we found a relatively large degree of direct manual process control. After 1978, there was an increased use of computers, but only as a supplement. The third step can be seen in the use of computers as a supervisory process control system in C 2. This change in the level of automation has had consequences for, among other things, work tasks, training, skill levels and task structure.

2.4 Work tasks

In C 1, there are two bleach workers on each shift. Each has a different job, and is separate from the other. One, the process operator, is responsible for operating the bleaching process, while the other is responsible for testing product quality and other laboratory work. The work of the process operator consists of monitoring many quite different kinds of production equipment, particularly the manual setting of valves, etc. There are approximately 600 different valves in the bleaching shop, many of which are quite difficult to manipulate, but need to be manually operated. The complexity of the bleaching process, the rather large area of work, and the variety of different kinds of machinery all require that the bleaching operator must have a long period of training and a high degree of practical experience. The job, classified as skilled labour, is a key position in the plant.

The other worker in the C 1 bleaching department tests product quality at different stages of the process (43 tests per shift). The laboratory work in connection with the tests also requires relatively long training and experience. This worker regulates the operation in part as a result of the tests made, in addition to monitoring a panel of about 30 different instruments and recording the status of a number of critical functions during the rounds. Another task is to help the co-worker with operating the process when necessary.

These tasks were not fundamentally changed after the implementation of the computer back-up system in the C 1 plant. In short, the computer in C 1 did not take over many process control tasks.

The bleaching operation in C 2 is also controlled by two operators, but they work closely together as a team, sharing the work equally. The operators communicate with the computer via a terminal so that necessary information on the current state of the shift operation or weekly reports can be generated on the basis of permanent programs. These reports contain data on the quantity of production, the consumption of supplementary materials, etc. In addition, the computer can produce an instant overview of a number of central process control parameters, in contrast to C 1, where the operators have to go to the production hall in order to obtain the corresponding information about the process. Furthermore, the two workers make a number of tests manually on the production floor. They analyse these according to present criteria, and the results are fed into the computer via a terminal. Finally, the operators also have control of production throughput before and after the screening process. Their process control work here is primarily manual: they adjust different valves on the actual production equipment.

2.4.1 Training

The staff in C 2 receives a thorough training to achieve the position of "process operator", which is the minimum required. This position requires -

- three years of experience (or two years of experience plus graduation from a technical school);
- demonstrated competence in three production jobs in the company, of which two must be in critical areas;
- completion of a basic course and an operations course (approximately 480 hours, including both theory and practice); and
- participation in an apprentice programme in the different departments within the company.

Training in C 1 is less comprehensive and systematic. It has traditionally occurred through an on-the-job training programme where an individual works together with an experienced operator. A certain amount of written material has also been available, but has not been the basis of systematic training. In recent years, new employees have received exactly the same training in C 1 and C 2 in anticipation of the closing down of C 1 in 1982 and transfer of the employees.

2.4.2 Skill level

The technological development described above is paralleled by a systematic change in skill requirements. The difference in skill requirements between C 1 and C 2 is relatively clear. In C 1 there are 21 different kinds of jobs. Parts of those jobs are also found in C 2, while the rest of the "old" jobs are incorporated in the technology. In addition, the new jobs in C 2 also consist of new tasks for the workers, such as the old foreman's tasks. The jobs in C 2 are what we may call "patchwork" jobs; they seem patched together from pieces of jobs left over after automation.

According to our interviews, the difference in C 2 compared to C 1 is an increased need for conceptual skills and a reduced need for motor skills. In addition, there is a new requirement for the ability to comprehend and interpret information quickly, an increased need for knowledge about the process as a whole, and the ability to make quick and correct diagnoses and decisions. Finally, skills are distributed more equally among the C 2 operators (everyone must be competent in at least three of the five jobs in the C 2 plant). The operator's knowledge is to a large degree passive - it consists of a repertoire of highly skilled but little-used responses, necessary primarily in critical situations but relatively unused when production is stable, as is typical of process technologies generally.

2.5 Work organisation

As we have stated, the technological and production differences between the two factories are not in themselves sufficient to explain the striking differences to be found in job design and work organisation. We describe these differences here and suggest why and how they arose. It appears that they arose in part because of circumstances, but mostly they seem to be the result of a consciously planned organisational development effort in C 2. In other words, the work design differences between C 1 and C 2 seem to be primarily (though not exclusively) the result of choice of organisation rather than choice of technology. Technological change appeared to make more autonomous forms of work design feasible, while other factors such as values, management philosophy and actual knowledge made them desirable and achievable.

2.5.1 The nature of the work

As is typical of the process industry, the work itself consists primarily of monitoring the process to keep it within critical parameters. As one of the workers said, "the company really makes money when we don't do anything". This is a typical statement which illustrates that when the production process operates continually without problems, then the work itself is primarily that of monitoring. Since conditions can be relatively stable and non-problematic for weeks at a time, the work itself can become highly routinised and tedious. This kind of routine can be broken in part by cleaning up in the production area and by taking samples.

Both functions are done on a daily basis. Ironically, the work is more routine in the somewhat more technologically advanced C 2 plant. One of the reasons for this is that there are more breakdowns in C 1. Another is that the C 1 plant produces many different kinds of cellulose, which implies several changes in processing requiring a high level of attention and concentration from the process operators, while the C 2 plant produces at maximum only two types of cellulose quality. However, there are at least two other factors which contribute to defining the character of the work. One is that the computer-based process control system in C 2 provides a more secure and stable production process. The other is that the equipment in C 2 plant is relatively new and maintained at a high level, so that breakdowns are quite seldom in comparison to C 1.

2.5.2 Different forms of work management

The traditional hierarchical structure which we find in C 1 places the foreman in the central position for the organisation and management of the shift. In addition, among the operators themselves, there is a system of ranking which is based primarily on seniority. This ranking system is not only a matter of social status, but also a result of the wage system. The C 2 factory, on the other hand, has a rather "flat" organisational structure among the operators: everyone has the status of "process operator", with the same wage. There is no foreman's position on the shift, so that promotion in the sense of hierarchical advancement is not possible. Each shift group is self-managing, and the workers themselves advance this as a positive characteristic.

Initially, during the plant's first years, there was a type of shift foreman's position. However, the workers who held these positions were engineers by training and therefore did not have the traditional foreman's background of having risen from the ranks. They became superfluous as the process operators grew increasingly competent in managing the production process. Therefore, the job of shift engineer became gradually less and less inspiring for these employees, and they began to apply for other jobs. Our interviews with workers, engineers and managers revealed a high degree of agreement in describing this development. When the last shift foreman left, there was a question of what to do.

Management today looks back at the situation as having presented three alterantives: first, they could employ new engineers; second, they could maintain the traditional foreman's position by promoting one of the process operators; third, they could remove the position completely and reorganise for self-management. Why was the third alternative chosen? One important reason lies in the ideals developed through the Norwegian Industrial Democracy Programme, started by Einar Thorsrud and colleagues in the mid-sixties.[30] The idea of autonomous work groups was in the air. In the early seventies, a labour-management redesign team from the plant participated in a workshop sequence over one year run by researchers from the AFI in Oslo.[31] Advanced skills training to increase the operators' process control competence was a key element in implementing the redesign. Another consequence of redesign is increased flexibility because most of the operators are able to perform more than one job.

Traditionally, the foreman has played a key role in the process industry as the person with the most in-depth knowledge and the most practical experience concerning adquate functioning of the process. Therefore, both process operators and managers are often very strongly dependent on the foreman's insight and competence. In the C 2 factory, however, additional training resulted in a transfer of knowledge from the foreman to the operators. Thus, the traditional dependency on the foreman's role was very much decreased and within a few years disappeared.

Perhaps the most fundamental reason for the decision not to hire foremen was that such a role had become unnecessary. The process operators were able to control the production process on their own. Clearly both the new computer system in C 2 and new organisational ideas contributed to this development. The computer control system improved the possibilities for continuous overview and monitoring of the process and more timely process control. The AFI workshop contributed new ideas about work organisation and provided practical organisation development assistance over a period of time.

However, these are not the only important conditions that supported the development of more autonomous forms of work organisation in C 2. The majority of the operators are relatively young (most are 30-35 years old) and interested in exercising more control over their own work. In addition, they all went through the training programme together when the factory started. With a common point of departure, support from management and researchers, an equality in skills training, and a shared interest in a more self-managed form of organisation, they developed equal task competencies and substantial autonomy.

The contrast to C 1 is striking. C 1's work organisation is based on segmented tasks which generate different levels of competence and lead to an internal status hierarchy. The foreman in C 1 has maintained his traditional role based on widespread process competence. Again, the plant and those who work in it are older and have not been through an organisation development process. We should make clear that we are not evaluating one form of organisation as superior to the other. Indeed, workers in each factory appear to evaluate their own organisation as best. Our most relevant finding is that similar technology in similar plants allows for quite different kinds of organisational designs. Here we find a more advanced form of computer control associated with a more self-managed form of work organisation. An older factory with a less advanced computer control system has a more hierarchical form of organisation. Furthermore, the difference is not limited only to workplace organisation. The organisation development project in C 2 seems also to have been important in reducing the management hierarchy and evolving new production planning and control structures consistent with increased worker self-management.

One consequence of the departure of the engineers in C 2 was that the process operators required new channels of communication and co-ordination with plant management. Therefore, a formal "co-ordination meeting" was established, and has endured as an effective solution to the problem. In this meeting, one contact person from each shift participates together with the management of C 2. All general personnel matters are decided at the contact meeting, and it is of special importance to note that everything concerning hiring, firing and holidays is the responsibility of the meeting. A similar reorganisation with contact meetings was attempted in the C 1 plant, but it was not successful. In the C 2 plant, this arrangement is a necessity, but in the C 1 plant, with its traditional hierarchical structure, this form of participative decision-making conflicted with the allocation of responsibility and authority in the formal hierarchy.

2.5.3 Control room: a consequence of control system centralisation

There are control rooms in both C 1 and C 2. In C 1 we find three small control rooms with one to three people in each. Each control room is physically placed near the particular process that it is to control. There have always been control rooms in C 2 so that we must look to the C 1 factory which at the beginning did not have control rooms to identify the consequences of establishing them.

In the beginning, as we have described, measurement and control instruments in C 1 were placed directly in the relevant equipment out in the production hall. To monitor the production process, operators had to walk around and attend to the different pieces of equipment. With the installation of control rooms, this kind of monitoring work could be done directly from the control room. Nevertheless, the tradition that operators should be out in the production area did not disappear from the C 1 plant. There was a norm that it was not quite right to "sit on your rear" in the control room. One reason for this norm is that much of the equipment in C 1 is in such poor condition that workers need to be out on the production floor in order to act quickly and decisively in the case of equipment failure.

In the C 2 plant, the control room has a much more central role in process control. In contrast to the computers in C 1, the computer in C 2 actually controls the entire process to a large degree. The process operators, therefore, naturally spend most of their time in the control room which has become a social centre as well. The design of the room itself has also taken this into consideration (coffee making, place to take breaks, etc.). In C 2 the way the technology was organised seems to have facilitated the group-based organisation design.

2.6 The creation of groups among the workers

There is a stronger tendency towards a sense of social collectivity and equality in the C 2 plant. An important basis for this is the wage system and the attitude of the workers towards equal pay. As one of them said: "Those who begin working here must be willing to complete the courses that are necessary to be a process operator, so that they can get the same pay as we get." It is also based on the attitudes workers have towards the shift's contact person. This person is elected for one year, and the role is rotated among all members of the work group. Such attitudes were not expressed in C 1. In our opinion, the fact that there is no one in a supervisory position (such as a foreman) contributes to a stronger sense of equality and integration among the process operators than is usual in more traditionally organised factories, such as C 1.[32]

In addition, the more integrated control room (covering most process functions) in the C 2 plant provides better possibilities for social interaction among the process operators. They tend to meet each other often as part of carrying out tasks. Furthermore, within the control room conversations and discussions occur without the interference of noise and other environmental distractions. This increased possibility for contact between workers, in our opinion, leads to stronger social integration. These different factors reinforce each other, as for example in C 2, where the control room is larger and more comfortable, so that people spend more time there. In addition, there is no social ranking, in contrast to C 1 where the shift foreman is often in the control room, which is also less physically adequate than the C 2 control room. These factors in the C 1 situation decrease the possibilities for social integrations among process operators. Nevertheless, we should note that C 1 workers today experience their foreman more as part of the shift work group than was the case a number of years ago. Our data do not allow us to assess whether the development of control rooms in C 1 (a technological change) resulted in decreased social distance between foreman and work group.

2.7 Changes in the physical aspect of quality of working life

Physical working conditions were improved simply by creating a control room which separated the process control work from the actual production process area. This in itself reduced gas, steam and noise levels to which workers were exposed. In addition, the C 2 plant has all the QWL advantages that a new plant would be expected to have in comparison to an old plant, such as C 1. For example, in the old C 1 factory, the chlorine tower is placed inside the factory building, which substantially increases the risk of chlorine gas poisoning. In the C 2 plant, the chlorine tower is built outside the factory building, thus reducing the danger. Furthermore, the computer control of the process in C 2 leads to a substantial reduction in physical effort, for example in the manual adjustment of valves. In the C 1 plant, these valves are often quite difficult to reach and manipulate, so that a design which reduces the need for manual control has clear advantages.

As we have said, computer control of the process in C 2 has led to a very stable system. This means that there are few production stoppages in the continuous process, which in turn results in fewer operations and adjustments on the process equipment and a reduction in the risk of injuries.

2.8 Trade union and worker influence

The local trade union looks at the developments we have described in two ways. On the one hand, it sees them as positive since no jobs have been lost and the physical working environment has been improved. On the other hand, it is afraid that jobs will be lost in the future, and a number of shop stewards are dissatisfied with the possibilities they have had to influence the selection and design of new technology.

The trade union referred to above had a relatively minor role in the design and development of the C 2 factory in 1972, its main engagement being to negotiate on the total number of jobs. It did not participate in the actual planning of the factory, purchasing of equipment on the original design of the organisation, although it did participate to some extent in the organisation development effort which in turn led to more workplace self-management of tasks and a higher degree of equality.

Since 1975, as we have described in section 1.4, new laws and national agreements have been created in Norway to increase workplace self-management and trade union influence in technological change. The company and the trade union followed up in the late seventies with the establishment of a work environment committee, a local data agreement and the election of a data processing shop steward.

The effects of this new system of participation and worker influence can be seen in the planning and development of a new drying department in the factory. In this case, labour-management work groups have been created in all relevant areas (e.g. the purchase of instrumentation, training, etc.). Process operators are represented in all of the work groups. In addition, the president of the local trade union is a member of the main project planning committee. In other words, the formal rights and structures for worker influence are in place, and workers are in fact participating in the design of the new drying department. Preliminary indications are that the design will go further in the same direction as the earlier organisation development project in C 2.

While this is a de facto positive evaluation by workers and managers of the C 2 philosophy of self-management, it does not mean that the trade union is completely satisfied with the actual practice of participation as established by law and formal agreement concerning data systems. The local trade union president complains that too little time is available to evaluate new technology, that the local trade union does not have sufficient resources to make such an evaluation, and that it does not know enough about data technology. The union has not set its own goals in relationship to technical change, nor does it have any overall strategy as a guide to tackling this type of problem. Other factors which limit the union's impact here are the small number of active members and the degree of co-operation between shop stewards and members, which does not function in a satisfactory way. In addition, the project groups which have been created are only advisory. All of the important decisions, according to the trade union, are made by higher-level management. For example, in the new drying department which is now under construction, the project group strongly recommended a control system similar to that used in C 2, but management decided instead to use another kind based on digital display.

According to our interview data, workers evaluate participation in project groups as positive in that they receive more information on what is going to happen. The more detailed design of workplaces, placement of equipment, etc., does include a larger degree of worker participation. However, the possibilities for choice here are limited by the earlier decisions concerning the choice of control system, type of equipment, etc. The workers feel they have not influenced these higher-order decisions. In short, our preliminary analysis of this case suggests that further developmental effort may be required to take full advantage of the right to participate that is formally established by law.

2.9 Preliminary conclusions

We have described the development of two factories which are technologically comparable but show quite different kinds of organisation. C 1, the older factory, maintained its hierarchical form of organisation even after a new computer control system was introduced in 1978. C 2 started with basically the same computer control system in 1972, but evolved a non-hierarchical and more self-managed form of organisation, in part as a result of an organisation development effort. The new drying department, which is now under construction, will apparently extend both the technological trend (by using a digital rather than an analogue-based computer display system) and the organisational trend (by emphasising extensive training, a competence-based wage sysem, and self-managing work groups).

Parellel to these developments we find a tendency to increased worker participation in technological change and the design of work. This reflects new national labour-management agreements (dataavtale) and laws (arbeidsmiljølov) that were established in the mid-seventies. However, we also find that these formal rights have not yet been fully realised in practice to the satisfaction of the trade union. Nevertheless, the trend is clear. In the early fifties, when C 1 started, the possibilities of participation were minimal. In the seventies, the employees participated in making C 2 increasingly self-managed, although not in the original design of the factory itself. The more extensive participation in work groups designing the new drying department may be an indication that worker participation in work organisation design and development may be more of a reality in the eighties.

Finally, we find limited confirmation that increased automation in process technology is associated with more self-managed forms of work organisation and other QWL benefits. Two factories with quite similar products and technologies have quite different work organisation. The older, slightly less automated factory is hierarchically managed. The new factory is more self-managed. However, technology alone is not sufficient to explain these organisational differences. At best it may have been a facilitating condition supporting the organisation development effort that aimed at increased self-management. As with our pre-case from Hunsfos, the main explanation appears to rest less on technological determinism and more on organisational choice.

A final, unambiguous conclusion is difficult. Our preliminary analysis reveals a chain of events in which ultimate causation is difficult to assess, given our limited data. The more comprehensive computer system in C 2 led to the creation of a more integrated control room where operators developed a better overview over and insight into the production process. This tended to modify the foreman's role. But C 2 was also a new organisation with younger workers and no long tradition of foreman as "boss". Furthermore, this tendency towards modified hierarchy in the direction of self-management was substantially enhanced by the organisation development project which aimed specifically at establishing a self-managing work group. Given this amalgam of technological change, a new organisation and planned organisational change, we can only conclude that technology does not appear to be the one determining element in work design.

3. Case study 2 - Computer-based process control in a cellulose factory

3.1 Company background

This company, located in a medium-sized Norwegian city, is part of a larger Norwegian corporation with total of 1,500 employees. The plant site we investigated has approximately 660 employees, divided among three integrated factories which produce sulphite-based cellulose and finished paper. All of these factories are in the same geographical area. Our study is limited to an examination of the cellulose pulp mill which recently went into production and which was at the time one of Europe's most highly automated pulp mills.

This mill's design was the result of years of uncertainty in the company about what to do with the cellulose production. Several different alternatives had been discussed, and the decision to build the new factory led to the closing down of an old factory in the same geographical area. At that time (1976), the normal procedure would have been to install, if not conventional (i.e. non-computer based) process control equipment, then at least a conventional (i.e. manual) back-up system. This plant's innovative characteristic was, then, that it relied totally on computer-based control of the process. The plant is also interesting for our purpose because, while advanced technologically, there were no special organisation development efforts (in contrast to the "pre-case" and C 2 in our first case study).

The firm experienced no problems in finding experienced operators for the new factory. Almost without exception, the operators who began working in the new factory had at least some experience of the conventional production of cellulose. For most of them, though, this experience was rather limited because they had only worked for some months in the old factory while waiting for the new one to open. The old factory was abandoned and the workforce dispersed, so that comparison with the new factory was difficult.

The situation in the old factory resembled the original C 1 factory in the first case study. The buildings and the equipment were old, maintenance work was superficial, and the physical work environment was poor. There were no control rooms. The process, which was to a large degree open and visible, was controlled manually in the factory. However, the number of persons per shift and the number of organisational levels were the same in the old and new factories in this study, and the basic type of production equipment was largely the same. The main difference between the two lay in increased production and productivity, in the physical enclosing of the process and in the way the process was controlled (i.e. new buildings, some new equipment and a new highly automated, computer-based process control system). In many ways the differences between the old and new plants paralleled the differences between C 1 and C 2 in our previous case, with the important exception that there was no organisation development effort.

Today approximately 100 individuals are employed in cellulose production in the new plant. Of these, 30 process operators and their foremen have the responsibility for process control. The factory, officially opened in spring 1979, produces 65,000 tons of cellulose per year, both from spruce and fir, and approximately 50,000 tons of the annual production is processed further in the parent company's own paper factory. The rest is sold to other paper mills which do not have their own cellulose production.

3.2 Data base

The study of this factory is part of a larger study of the consequences of automation in the process industry generally. Data were acquired by three researchers who visited the plant for a period of approximately one week, during which time the plant manager, operations manager, two shift foremen and six process operators on one of the five shifts were interviewed. Additional interviews were also obtained with the plant's business manager who had responsibility for personnel administration, the plant's training manager and representatives from the local trade union (the president and the data processing shop steward). We used relatively unstructured interviews, which lasted from half an hour to over five hours, although most were less than one hour. These interviews constitute the main data source. In addition, we acquired written materials such as descriptions of the production control system, the shift plan and production flow diagrams.

This case study is not yet completed, but is part of an ongoing study which will conclude with the publication of several case studies and a theoretical analysis.[33] It should be understood that our analysis is based on incomplete data and must be considered as preliminary to our final result.

On a general level, dried wood chips and chemicals go into the process and cellulose comes out. If we look more closely at the process, we can see that the dry chips are mixed with acid and boiled in the digester department. When the boiling is finished after approximately 12 hours, the acid is washed out and later reclaimed. The fibre mass is further cleansed and strained, after which the cellulose is ready for input into the paper production process. In this factory, the cellulose either goes directly through a pipeline into the corporations' paper production facilities, or is dried and sold to external customers.

The acid which has been used is first evaporated with steam to increase the proportion of acid in the mixture. Then the mixture is burned to get rid of the organic parts. In the last stage, fresh chemicals are added to obtain a correct strength of the acid before it is led back to the digesters again. This part of the process is new. Previously the used acid went into a nearby river, causing pollution.

3.3 The production process

The production sequence is presented diagrammatically in figure 28.

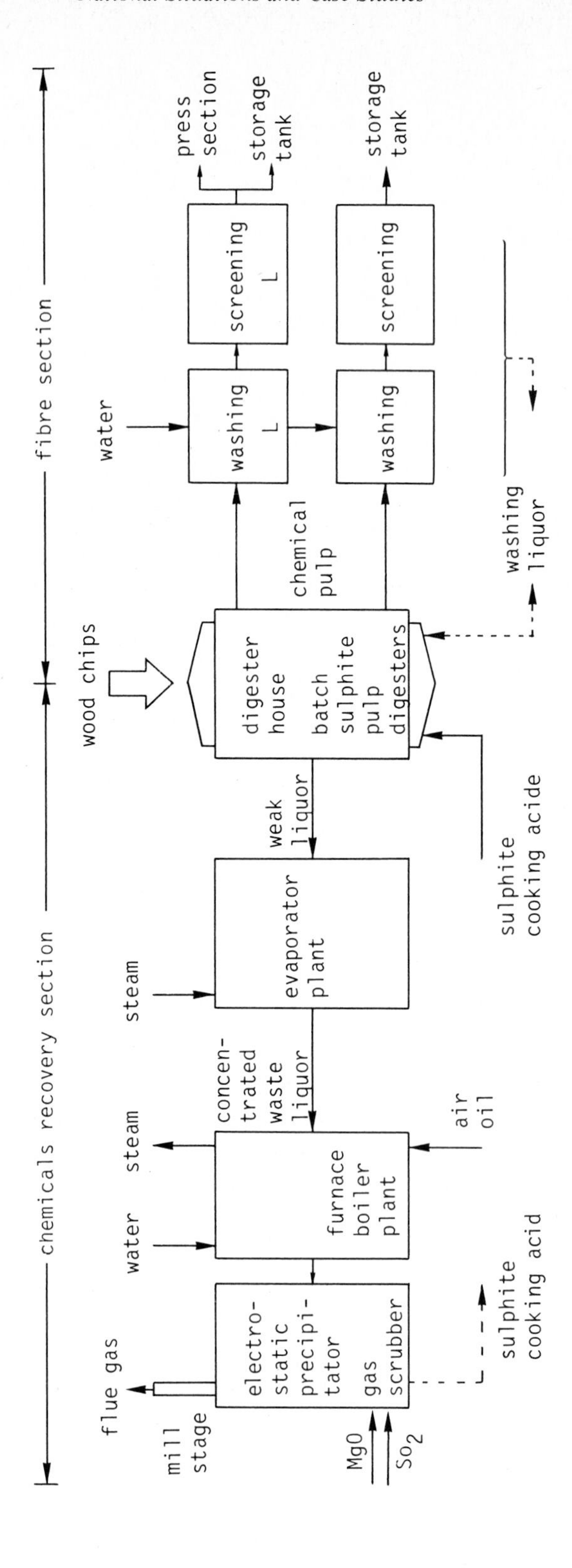

Figure 28: The principles of sulphite pulp production

3.4 The technology

3.4.1 Production technology

The production technology is of the conventional type found in most cellulose factories. The digesters, the core of any factory of this kind, are rather old; they comprise six in all, three from the old factory and three from another cellulose factory that has been closed down. What has changed is the way the process is controlled.

3.4.2 Computer-based process control

When cellulose production was transferred to the new factory, management decided to go all out for modern computer-based production process control and monitoring. In one of the corporation's other factories, simple instrument panels were replaced by multicoloured VDUs, and a simple terminal that permitted direct access into the process. The economic results were so positive that the company decided to base production control in the new factory entirely on computer systems. The most striking feature of this system is that it is designed with a computer back-up instead of the more usual direct manual back-up.

The control technology in the new factory is totally dependent on the computer system, work tasks being distributed to five computers. Process operators communicate with the system through two independent computers, each of which serve two process operators. The remaining computer is used for analysing data, compiling statistics and as back-up for the other two.

The main frame computer is used to administer these work tasks and to perform certain service functions. It is also connected to an operator work station with a colour VDU and a keyboard input/output device. Through use of the keyboard, the process operators can choose between a total of 21 different flow charts. A colour code indicates which valve is open, which motors are operating, and the level of different tanks and storage units. Moreover, valves and motors are started, stopped and closed off by using the same operating keyboard.

The system is basically duplicated. Each of the operators have available two independent and identical VDUs. If a mistake or problem should occur in one, the operator can use the other. For the process computer, if there is a mistake or a shut-down in one system, the reserve computer automatically takes over.

3.5 The control room

The control room is a rectangular room of approximately 35 square meters. A traditional control room for a pulp mill of comparable size would normally have some 10-20 meters of control panels, gauges and tables with instrument read-outs. In this new plant, however, the entire process is controlled basically through two pairs of data screens. The control room, though small compared to conventional control rooms, also has important social functions. It serves as a lunch room and partially as a lounge for other groups, especially on the day shift. It is possible to see part of the factory from the control room, but the operators sit with their backs to the interior windows. The reason for this is to avoid glare on VDUs from the exterior windows.

Figure 29 shows the layout of the control room. The operators of the two main lines are grouped together. The fibre line (1) is to the left and the reclamation line (2) to the right. The fifth terminal (3) is used for training purposes and is not connected to the process control.

3.5.1 Work tasks

Each shift consists of six process technicians and a foreman. The foreman has in addition responsibility for the cleaning department and the press department, which are located in an area outside the factory proper. Three of the process technicians have responsibility for the fibre line, while the other three are responsible for the chemical reclaiming line. On both lines, two of the three operators are on the job all the time, and the third is free. Originally it was intended that the operators rotate within "their" line, so that one operator should always be out in the factory. To operate the process properly, it was maintained, the operators must have a clear picture of what was going on in the factory. Furthermore, the only way to teach newcomers what was really happening and to maintain the more experienced operators' knowledge, was to send them out there. However, at the time of our interviews, few, if any, of the operators were seen out in the factory unless they were taking the necessary quality control samples.

On the fibre line we find the digester operator and the screener. The digester operator is responsible for cooking the wood-chips, which is the most manual job in the control room. By depressing the proper key on the keyboard, this worker takes care of chip feed, refilling of acid, heating and cooking, and decides when the mass is properly boiled by checking fibre samples. Then the boiler is emptied, and the sequence is repeated. The feeding of acid is determined as a function of the amount and quality of the chips, and can be calculated using a table. The crucial point in this job is to decide when the mass is finished. Stopping the cooking sequence too early causes the screener many problems and leads to a low-quality product. Stopping it too late may result in "black cooking", which means that the entire batch must be discarded.

The screener, located next to the digester operator in the control room, has to make sure that the mass is thoroughly screened and free from contamination, then washed before it goes further into paper production. This part of the process is highly automated, but is described as the most demanding of the steps in the process. The reason for this is the changing quality of the input. When the mass is free from contamination, the job is simply to monitor the process so that it proceeds normally. If there have been problems in cooking, however, then the screening can become clogged up and the mass will not flow as it should. The screening process

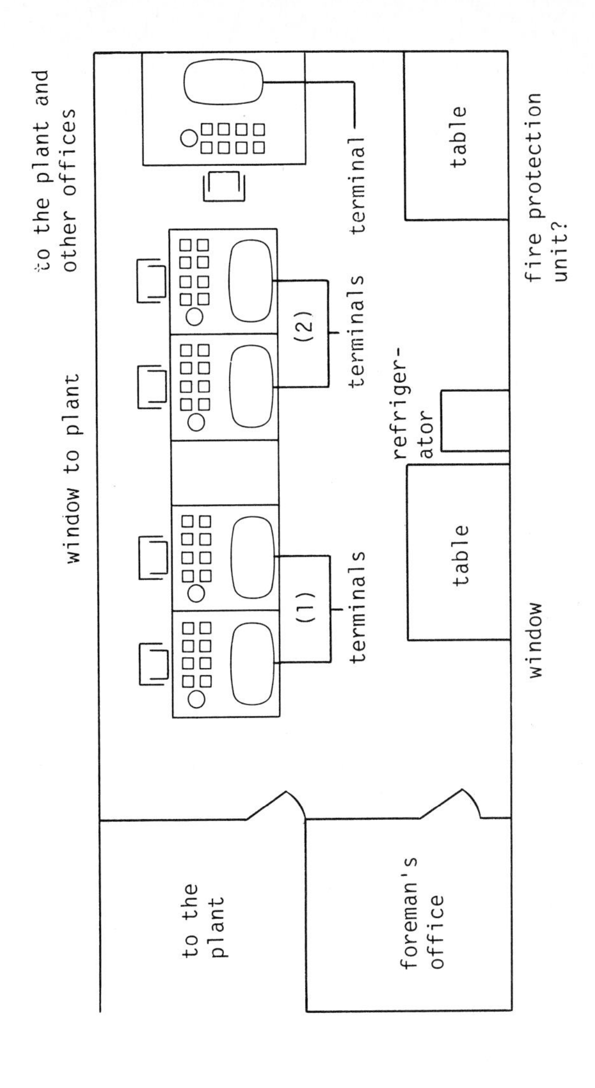

Figure 29: Computer-based process control in a cellulose factory: layout of the control room

operator must then intervene and control the process manually. This requires a thorough knowledge of how the subprocesses in this step actually function. The third worker on the fibre line has a somewhat unclear status, but usually acts as a reserve who takes over when one of the others are out in the factory making the necessary tests or taking a break.

On the recovery side, the process is continuous, stable and highly automated. Manual intervention from the process operators' side occurs very seldom, and work mostly consists of monitoring the process. The most stable of the tasks is recovery of the chemicals, and the person responsible for this part of the process takes over the responsibilities of the foreman when he is out of the control room.

The division of labour between the process operators on a shift used to be much more rigid than it is now. In the old factory, as in the Hunsfos pre-case, each worker had responsibility for a step in the production process, and was employed in a position entitled, for example, boiler (or possibly even first boiler, second boiler, etc.), but not process operator. Today it is management policy that everyone on the shift learn as many parts of the total process as possible, so that co-workers can change jobs with each other. There is no formally developed pattern for job rotation, but in practice the two who work on the fibre line and the two responsible for the chemical reclamation line overlap with each other. In order to operate the furnace and boiler, a special boiler's certificate is required, so that one worker on each shift has responsibility for this task.

Although this is the main pattern, there are variations between the different shifts, depending on what kind of custom has developed and on the level of knowledge of the different process operators. Today some of them have sufficient experience to operate the entire process, while others can only operate a small part of it. Nevertheless, the long-term goals of management are clear: extensive multi-skilling.

In continuous process production the role of the foreman is often problematic. We saw in the first case study (C 2) that the foreman's role was eliminated. The same would also have been possible here, although not without severe problems. The foreman does not perform the tasks which traditionally belong to the foreman's role (e.g. in batch production, like assignment of work, supervision of performance and quality control). Operators have their own jobs, but the individual operator's contribution to the finished product is hard to evaluate.

The foremen we talked to looked upon themselves as "errand-boys", as one of them put it. Besides tabulating the number of hours worked by each member of their shift, most of their day was filled with shift support activities, like keeping contact with the maintenance department, the paper factory and higher levels of management. In addition, they performed the internal co-ordination on the shift. This was necessary because the foreman also supervises a number of people outside the control room including, for example, those working in the handling of timber (input) and the drying and packing of the finished product for external sale (output). Thus both the foremen's relation to the process itself and their other responsibilities make the term "errand-boy" an accurate one.

3.6 Consequencs of the new control technology

When a new factory replaces an old one, many factors change - the technology is updated, new employees are hired and the building is often more functional than the old, worn-out one. Therefore we again have difficulties in isolating the consequences of technological change from consequences due to these other types of changes.

One obvious technological change is that the new computer system offers the process operators a different medium through which the process is controlled. The operators communicate with the process through a "tracker-ball" system located beside the keyboard. With this "ball", they can identify for the computer on the process flow diagram the exact parts of the process they want to manipulate. The actual manipulation is done through the keyboard. The production technology itself is, as we have mentioned earlier, more or less unchanged. The motors that have to be started through the computer in the new plant are equivalent to those valves and motors that were operated manually in the old plant. The main differences are that this is now done remote from the process and with the help of the computer system above described.

It is difficult to say what impact this has on the operators' possibility of learning the process properly, of the way they run the process, of their skill level and of their opportunities for taking proper action if anything unexpected occurs. We do not have adequate data because workers found it difficult to compare their present work situation to the one in the old abandoned factory on the relevant QWL dimensions. Indeed, many of them had had only a short period of work in the old factory. Thus, while many could express some opinions, their views seemed to be largely guesswork about what the long-term consequences might be.

The first problem that arose after the opening of the new factory was that many of the older workers did not transfer to the new system. Some did not even try, and others tried and failed. Today only three of four older, more experienced workers (out of about 20) are left in the control room. Some of these 20 have reached pensionable age, others have left the factory for other reasons, while others again are working with materials handling, either in dealing with the incoming timber or on the drying and packing of the finished product. The point is, however, that most, if not all of them had and - for some of them - still have problems with getting used to the new technology. We are not able to give a satisfactory explanation of why this is so. One explanation offered was that if one is used to operating a process in a certain way, and has done so for many years, then it is hard to change. One acquires habits and becomes familiar with the technology and the rhythm of the process. In the old factory, a worker who opened a valve, started a motor or increased the pressure could immediately hear and partly see what was happening.

The new control room offered a good physical work environment, comfortable chairs, silence and possibilities for social contacts with colleagues. But what was lacking was a feeling of what was really going on "out there". This explanation is exemplified by older workers who, having, for example, started a motor through the computer, actually went out into the factory to see if the motor was really running. Another consequence of this removal from direct engagement in the production technology is the way control of the system is performed. Newcomers often tend to make quick changes, resulting in very hard pressure on equipment, while the older, more experienced workers do it carefully and correctly, knowing from experience how high-pressure load change affects the equipment. This, we think, is a more sophisticated and probably correct explanation than saying that problems are due to the older workers' "irrational" resistance to any kind of change.

When it comes to the possibilities of learning the process, we were told that the new technology gave the operators better chances to learn the total process and to see how the different parts are connected. They do not, however, learn the process as it is, but a picture, or a theoretical model of it. The model they learn is quite abstract, very similar to, but nevertheless not the process itself. Sitting together in the control room also makes it easier to communicate with each other, and easier to see how poor results early in the process cause problems in later stages. this, together with the more accurate measures and information the system can give on any point in the process, makes it easier than before to operate the process nearer to an optimum point.

Whether this technology will increase or decrease the operators' abilities to take proper action if anything damaging should occur in the process, it is difficult to say. In this kind of production, this is not a critical factor in the same way as, for example, in the oil-producing industry and in nuclear power plants. If it is true that workers acquire a better understanding of the total process (which we find very likely), it would be less probable that they would make large and costly mistakes. If however, as some of them maintain, they know less and less about what is really going on out in the factory, the possibilities for errors and wrong actions might increase. In short, the ultimate consequences of automation appear to be mixed or at best undecided, and highly contingent on such non-technological factors as training, work organisation and informal working practices.

The removal from the physical process makes, as we have seen, a more abstract understanding of the process necessary. One does not any longer learn from the process itself, but from a picture of the process on a screen. One has to learn to understand symbols and diagrams and develop a basic trust in the technology. If a lamp or a blinking arrow indicates that a valve is open, one has to rely on that as a fact, as one cannot see the valve and hear the mass passing through it. The level of abstraction is clearly increased substantially.

In this new factory, management's intended policy was that everybody should learn as many jobs as possible, preferably all of them. This policy, however, was not followed by either monetary or other incentives, so that it is up to the operators themselves as to how much they in fact pursue multi-skilling. Workers have the possibility to increase their skills and some of them already can perform several jobs. More usually, however, they know the jobs on their line, and it is rare that they look along the table to see what the others do.

We also found that some "unlearning" is to a certain extent necessary for those used to the old way of monitoring the process. One no longer needs to know where the different valves and motors are located, how to stop minor leakages and to detect signs of breakdowns. This is today exclusively the job of the maintenance department. The manning on the shift is, nevertheless, arranged so that three persons are responsible for two screens, in order that one operator should always be in the factory in order to prevent the tendencies we have described. This does not function as it was planned, and control room operators do not willingly take a walk in the factory except for taking the necessary periodic product samples.

3.7 Evaluation of the changes

Representatives from both management and the trade union express a positive attitude towards the changes described. When the new factory was being planned, the decision to go all the way and rely totally on computer-based process control was a radical one. At the time, many thought that such an unconventional design was premature. Originally, management did not think such a control system would be feasible. However, cost-saving projections with the new computer technology convinced them otherwise. An independent analysis from an external consulting firm showed that not only would such a solution provide a secure and stable operation, but that it would also be less costly than a conventional system.

The trade union at this point in time did not have a strong opinion on either of the alternatives. Their main concern was that there would be no workforce reduction. Technological design solution was at that time, and to a large degree still is, not of central interest to the local trade union as long as the number of jobs and the wage levels are guaranteed. the national data agreement provides for a data processing shop steward at the local level, and one has been elected in this company. However, this is still a relatively new arrangement and has not yet been developed here in its final form. Both process technicians and the officers of the local trade union evaluate the new process control technology as positive, and the local union supports among other things the company's plans to increase data processing capacity by buying a much larger computer.

3.8 Summary

The change which we have described here was a radical one in many ways. The new factory involved a change not only in technology but also in the way of organising work and the job design of the operators. In addition, the operators' physical work environment was radically improved. Nevertheless, because so many other aspects changed at the same time as the technology, it is difficult to asses what is due to technology in the new situation, and what are the consequences of the other changes that have taken place. There is, however, no denying the fact that the skills needed to produce cellulose in this factory are different from those needed earlier. A worker has to learn to control the process from a distance, by means of an abstract theoretical model represented through symbols on a screen, rather than by manual adjustments of the process out in the factory.

The skills needed to run the process today may also be transitional. Even though the equipment is sophisticated, the digesting and cleansing of the mass is still manually performed with the help of the new equipment. Increased automation, which is on the drawing board, may involve further dramatic changes in the skills needed today.

In 1977, the company's decision to rely completely on a computer-based control system without conventional back-up was a radical one. However, modernisation of the wood products industry has increased rapidly in recent years, and today, more than four years later, we can find companies which have gone even further in automating process control. Continuous automated digesting is feasible, and the company under study has plans for what is called a sequence control of digesting. In the foreseeable future this might, if necessary steps are not taken, result in the disappearance of knowledge among process operators about how the process can be controlled manually. This condition of indirect process control, which we have described as one of the main consequences of the new technology for operators, is already in the process of becoming typical for this industry.

However, casual relations between technological change and work design are neither simple nor direct. This case study is useful because the change in work systems we have identified cannot be explained by any planned organisation development effort. It may therefore be useful in illustrating the kind of organisational change than can follow closely and naturally in the wake of this kind of technological change (i.e. computer-based control systems in a chemical batch process technology).

Significantly, although automation in this plant is much more advanced than that in our pre-case and even in case study 1 (C 2), and there seems to be some work organisation change in relation to the foremen's role and type of skills, work organisation is still more hierarchically managed than self managed. In short, this case study presents an example of new, highly advanced use of more automated production control technology which is associated with an organisational change in the direction of a modified form of hierarchy. We note that this QWL change evolved without any explicit organisation development effort of the kinds we found in the pre-case (old technology) and in case study 1 (new technology).

4. Case study 3 - Technological and work design change over 26 years in an aluminium plant

4.1 Introduction

The case study described here is part of a recently completed research project for the Norwegian Chemical Workers Union (NKIF).[34] The purpose of the project, consistent with the LO-NAF data agreement, was to provide NKIF with insight into the consequences of new technology in the chemical process industry and to help it to develop its own strategy for advancing its interests along with the introduction of new technology.

The main strategy question was how to give the local trade union a resource base so that its members could by themselves decide on what QWL demands they should make concerning new technology. This project aimed both at increased knowledge on long-term strategy for the national trade union and at developing methods and a knowledge base to be used locally to evaluate the local QWL consequences of proposed new technologies before they were implemented. In short, the project aimed at developing new knowledge to advance trade union interests at both local and national levels.

The following material is the result of a co-operative effort between the researchers and a local trade union which participated actively in the research project. The company studied produces aluminium in a small town on a fiord on the west coast of Norway. We base our report on the results of a study of working conditions that was completed primarily by a working committee from the local trade union, with the active support and advice of researchers from IFIM.[35] The researchers' role was to help formulate the design for the research and to act as a discussion partner during data analysis and report writing, but the workers carried out the actual research entirely on their own.

The local trade union used the company's own job descriptions for work in the smelting plant as the data base for assessing changes and working conditions. Three waves of technological change were identified during the company's almost 30-year history, and job descriptions for selected categories of jobs were analysed for each of these periods. In addition, the workers in the local research committee had long experience in the factory (several of them were hired as young men when the factory started up) and supplemented their main data with individual interviews and group discussions with other workers, who also had considerable experience in the company during these different periods. The criteria for evaluating these changes and the final evaluation of them was developed by the local trade union's research committee, partly in co-operation with the researchers. Since the main research work was carried out by the local trade union on its own, this case study is also an example of participatory research at work.[36]

4.2 The company's history

The first production of aluminium in Norway began in 1907. The raw material for aluminium production, aluminium oxide, is imported from other countries. This company imports its aluminium oxide from Jamaica. The easy access to hydroelectric power and its relatively low cost, as well as Norway's access to sea transport, made this country a particularly favourable place to produce aluminium. Further processing of Norwegian aluminium into other products occurs primarily in other countries, first and foremost in Scandinavia and Western Europe.

After the Second World War, the company came under Norwegian state control as it had been expanded by the Germans during the Occupation. The corporation today consists of a main headquarters in Oslo and three production facilities located in western Norway. Today the company is 75 per cent state owned, but operates as a normal profit-making enterprise.

The aluminium production plant on which this report is based was built immediately following a decision in the Norwegian Parliament in 1951. The production facility was placed in a small town where the main economic activities were agriculture and fishing. Development of the hydroelectrical potential of this region, and the construction and operation of the aluminium production facility created many new possibilities for employment. The factory is still the only large industrial company in the local town and has contributed to a large increase in population, since workers were recruited not only locally, but also from other parts of Norway. Today the company employs approximately 1,200 of the community's 10,000 inhabitants and almost all of its industrial workforce.

The local plant began with two large halls filled with smelting pots. In the mid-sixties the company built a third and more modern electrolysis hall. With the completion of this third hall and modernisation of the two others, production of aluminium doubled, even though the number of employees remained the same as before. We return to this later in the report.

The price of aluminium and the aluminium market has been increasing regularly, except for the first few years of the seventies. The rising price of electrical power and competition from other aluminium-producing countries has decreased the comparative value of Norway's natural advantage, cheap hydroelectric power, and has led to major efforts to increase efficiency in the aluminium industry.

In addition, restrictions on the further development of hydroelectric power and more demanding laws and regulations concerning natural resources have forced aluminium companies to invest in more energy-efficient production facilities and in facilities for reclaiming chemicals and other by-products that previously were considered to be wastage.

The local trade union was established in 1952, and within the first year had 150 members. Today 800 workers belong to the union (practically 100 per cent of the industrial workforce). Prior to the beginning of this research project, the union was not concerned with technological change, but rather with traditional industrial relations issues such as negotiations on wages and working time and, more recently, physical working conditions. The consequences of new technology were discussed only to the extent that it could affect the total number of jobs in the plant. Technological change had not resulted in a decrease of jobs, so this was not seen as a problem.

The local trade union was included in the project on its own initiative. It established a working committee of three members who participated in the research project. One of these was later elected to be the data processing shop steward, who now has as much paid time off as he needs to work on the consequences of data processing. The working committee began its investigations in 1978 and completed its report in 1980.

4.3 Production process and technology

In order to produce raw aluminium one begins with aluminium oxide which, with the help of an electrolysis process, is changed into pure aluminium. The smelting process occurs in relatively small ovens or pots (Søderberg type) where production of approximately 1,000 kilos during the course of a 24-hour period is highly energy consuming. Every ton of aluminium produced requires between 15,000 and 19,000 kilowatt-hours. This is about as much energy as an average Norwegian family will use in their house during the course of one year.

The smelting process takes place in three large electrolysis halls, each containing from 200-300 ovens or pots. The plant we studied produces 130,000 tons of aluminium per year. (The corporation as a whole is Europe's largest producer of aluminium.) As the process of electrolysis changes aluminium oxide into pure aluminium, each pot must receive additional quantities of aluminium oxide in order to maintain the smelting process at an optimum level. A large truck with aluminium oxide is driven between the rows of ovens and each oven is filled with the amount of aluminium oxide that is required at that particular stage for that particular oven. This is a permanent and regular task on each shift. Other chemicals must also be added on a regular basis. The pot is drained approximately once a day through the use of a vacuum pump. The molten metal is collected in batches and delivered to the foundry, where it is cast into different shapes as required by different customers. Further processing and shaping of the finished aluminium occurs only to a very limited degree in this plant.

The most critical point in the production of raw aluminium is the development of an oxide layer around the electrode which occurs only after a certain amount of time and a certain very high temperature ("bluss"). For optimum functioning of the oven, this critical point should occur in a regular and planned manner approximately once every 24 hours. When the layer of oxide is formed, it must be removed quickly because it considerably overloads the oven both in terms of energy consumption and in terms of creating a hot point. If not properly attended to, this hot point can burn a hole in the oven, which creates major problems for production, health and safety. The main tasks in maintaining optimum operation of an oven are to maintain a constant electrical charge, a constant quantity of material and a constant effect of the electrodes (raising and lowering them and replacing them when they are used up).

Traditionally all these tasks were manually performed. The temperatures during the actual conversion process of aluminium oxide to aluminium are so high, and the chemicals used are so corrosive that no direct measuring of the state of the transformation itself is possible. Regulating the oven in relationship to these tasks therefore requires a high degree of skill and experience with the particular ovens involved. As production technology has developed in recent years, many of the tasks have become mechanised and then automated. The technological developments in this plant can be conveniently divided into the following three periods:

- period 1 (1952-56): little mechanisation. Authoritarian foreman control;
- period 2 (1956-67): introduction of heavy mehcanical equipment. Less authoritarian work organisation;
- period 3 (1967-): introduction of data processing systems. Some shift work tasks transferred to daytime work.

4.3.1 Period 1: manual work

In this period, the heat-producing electrical charge in the ovens was regulated manually by raising or lowering the anode. An instrument for measuring electrical charges was installed on each oven. A worker could, by observing how the electrical charge changed over a period of time, evaluate when it was necessary to fill the oven with more oxide. This was done manually from a large cart filled with oxide. It required that the worker physically broke through the crust formed on the top of the oven with a hammer-like drill, drive by compressed air. When the oxidation process reached a critical point and "blussing" occurred, the worker was also required to break through the crust and stir the metal at the bottom of the anode. The pot was emptied according to an evaluation made by the operators.

4.3.2 Period 2: mechanisation

Small trucks for adding raw materials and trucks with crust-breaking equipment were added during this period. In addition, a long metal lance bringing compressed air directly to the anodes under "blussing" was introduced. Finally, an automatic "bluss" warning system was installed.

4.3.3 Period 3: automation

The key feature of this period was the implementation of a computer-based production control system. This system was based initially on an automatic control of the electrodes' height in order to maintain a constant level of electrical energy in the oven. In this way, a "bluss" could be achieved when desired (once each 24-hour period) by delaying the addition of oxide after partially draining the oven. At first, the system contained only one-way computer terminals for the production of operational data, but later two-way terminals were implemented so that workers could ask for the kind of information they desired. The automatic warning system was also developed further during this period to show which ovens needed what kind of treatment (such as crust breaking or additional oxide). The next steps in technological development will be an automatic "blussing" system and an automatic crust-breaking and raw-material transport system.

4.4 Changes in work organisation and working conditions

The following description is based on the developments in the factory from 1954 to 1980. These 26 years have been divided into three periods of technological change, as already described. We shall now describe the changes in work organisation that occurred during these periods. In general, we shall see that the first period was characterised by authoritarian management and many-sided work tasks which were allocated and completed in a co-operative way with relatively little mechanical equipment. In the second period, heavy mechanical equipment was introduced and the work with this equipment was allocated to special employees. In the third period, which is characterised by automation in the form of a computer-based process control system, there has been further specialisation between the workers on regular hours (daytime) and the shift workers. Let us now examine these changes in work organisation and working conditions more closely.

4.4.1 Period 1: manual work

Work in the production hall was organised in teams of four pot or oven monitors responsible for 44 ovens, who were supervised by an assistant foreman. The assistant foreman was responsible for the operation of the pot, made measurements and decided when additional raw materials should be added to the oven and how much of the metal should be drained off. The oven monitors were responsible for regulating the electrical charge of the ovens by dropping the anodes and for evaluating the state of the pot. If the pot was too hot, there was danger of a burning breakthrough. This could be coped with by packing the edges with cryolite. The oven monitor was also responsible for coping with a "bluss" and for preventing the crust on the edges from becoming too thick. These tasks, and especially the task of regulating the electrical charge, required a great deal of walking over long distances. In addition, one worker served the entire hall with oxide from a large cart.

The work of an oven monitor required both experience and technical competence. Within a work group, the work tasks of the oven monitors were rotated among the group's members so that everyone was able to manage the process. Workers were not tied to a particular location, and there was no problem in being away from short periods since those tasks which required constant attention, such as coping with a "bluss", rotated among the whole crew. In certain respects this "gang system" resembled a self-managed form of work organisation based on equality of task competence and a certain freedom of action. One limitation to this freedom of movement and self-management was the relatively authoritarian organisation with assistant foreman, foreman, etc., each of whom had responsibility for a set of ovens and supervised the respective work groups.

Basic training for the job of oven monitor consisted of a 14-day machine training course, complemented by extensive on-the-job training as part of the work group. Generally it required one to two years of experience as an apprentice in order to cope effectively with the diffferent tasks of the oven monitor.

4.4.2 Period 2: mechanisation

The implementation of mechanical equipment and trucks resulted in a number of tasks being removed from the job of oven monitors and becoming specialised. Work such as coping with the "bluss", which had previously been an integrated part of oven monitoring, now became routinised and controlled by the "bluss" warning system. This job now gave little opportunity for freedom and competence, and became a very low-status job which tended to be assigned to new employees. The regular breaking down of the crust on the edges of the pot and the sweeping of the floor were transferred from the shiftworkers to the workers on regular hours.

The "gang" basis of work organisation was abandoned when the trucks with the oxide crust-breaking equipment were brought into use. The assistant foremen were eliminated, and the gangs were split up so that two oven monitors now had the main responsibility for a series of 22 ovens in their own sections, so that each shift was responsible for a specific number of pots. For these pots, the oven monitors performed the necessary measurements, made decisions concerning the addition of raw materials and chemicals, and decided how much of the metal should be drained off at any time.

In that the assistant foremen had disappeared, the demands for technical competence and evaluative decision making increased for the oven monitors. Management was less authoritarian than previously, and communication, for example, from engineers and other specialists was often directly to oven monitors. On the other hand, the other monitors lost a number of work tasks as a result of increased specialisation. Furthermore, the new work organisation provided less opportunity for natural social contacts between workers than was characteristic earlier, but there were still so many people in the workplace that there were good opportunities for the workers to meet informally, for example at coffee and lunch breaks. Training was somewhat reduced in comparison with the first period, since the task of adding oxide to the ovens was removed from the job of the oven monitor. After a training period of approximately three weeks, a new employee began to work together with an experienced oven monitor in the particular section of ovens.

4.4.3 Period 3: automation

The company changed its shift plan from four to five shifts in 1975, as part of reducing the total working week to 36 hours. (In Norway a 36-hour week is a full working week for shiftworkers.) With the implementation of a five-shift plan and change to a computer-based process control system, considerable changes in crew size and work organisation occurred in the electrolysis hall. The new control system took over the task of regulating the electrical load, and operational data became available at output terminals which initially were placed in the foreman's office. Thus, in the beginning, it was only the foreman who received training and acquired understanding of the use of computer-produced operational data. After a while, oven operators also received the necessary training, so that they could go to the foreman's office to get necessary operational information.

In the event that the computer broke down, there was a back-up system of manual routines in reserve. This back-up approach was based on the assumption that oven operators had enough insight into the process and could evaluate the operation and the state of the ovens as a basis for regulating them according to their own professional judgement. At a later stage, a reserve computer was installed which cannot take over if the main computer breaks down.

In the transition to automatic process control, the job of the oven operator was divided so that control and evaluative tasks were transferred to non-shift regular day workers, while shiftworkers retained the routine control tasks. Regular day workers became responsible for the operation, control and maintenance of the ovens. Their tasks included measuring the metal, evaluating the amount of metal to be drained off, regular crust control, maintenance of the oven and the vacuum equipment for removing dust, and cleaning and maintenance around the oven. Daytime workers took over work tasks for oven monitors working shift, and in addition received a number of maintenance tasks. The principle of a work group or gang was again implemented during the day with the rotation of tasks within the gang. Work as a daytime oven monitor is now a job with variation that requires the exercise of technical judgement and provides good possibilities for social contact with co-workers.

Oven surveillance for shiftworkers, however, has been reduced to a kind of watchman's work. Work tasks, which require low levels of training, consist of keeping the ovens covered with oxide in order to prevent unplanned and uncontrolled gas and burn-out through the sides of the ovens, and routine inspection of oven operation. The only technical competence required is that workers can react in the proper manner should there be problems that are identified by the computer system. The number of jobs on shift were reduced so that one oven monitor became responsible for the same area that two had been responsible for before. This implies rather large distances between different oven monitors and few possibilities for informal social contacts as a part of the work. In addition, it is not possible to leave the workplace. Coffee and lunch breaks have to be taken on a rotation basis so that one oven monitor is temporarily responsible for twice as many ovens as usual when his counterpart goes to eat. This means additional stress for the monitor in the workplace and a lonely meal for the one who is taking his break. On regular daytime work there are three to four times as many workers assembled as there are oven monitors on a shift. In short, the possibilities for interesting work and social contact for shiftworkers on the evening and night shifts were reduced dramatically at the same time that the technology was improved.

The local trade union research committee summarised their analysis of changes in job descriptions during these three periods in a report.[37] The following data and discussion are taken from that report. We should note that the data presented here are objective job design changes based on a systematic analysis of company job descriptions over a 26-year period. This is not a study of worker attitudes or feelings about the consequences of technological change but a systematic analysis of objective data by the workers themselves.

4.5 Summary of the consequences of technological change

The data presented in tables 62 and 63 show that the consequences of technological change are quite different for various groups of workers at different stages. Generally, we can see that the implementation of new technical equipment has decresed the amount of physically heavy work, but this has often gone hand in hand with increased specialisation of work tasks (table 62). For example, work with the new equipment was separated from the work of oven monitor and allocated to one particular person. The specialisation and division of labour developed with the implementation of new technical equipment is amplified by the introduction of automation.

Nevertheless, it is quite clear in this plant that the consequences for the employees are dependent on how work is organised and how tasks are allocated. The workers on the regular day schedule (i.e. not on shift work) have much more varied and multi-faceted work tasks than before, while the shiftworkers have received more routine monitoring tasks which require much less technical proficiency than before. Both the social contact necessary as a part of the daily work and the possibilities for social interaction in the workplace are also modified by these technical and organisational changes. Again, the workers on regular day hours still have good possibilities for social contacts. This is because they work in gangs and allocate the work among themselves. Shiftworkers work individually and there are large physical distances between workers during the work process. Since coffee and lunch breaks must be rotated, there are few opportunities for social contact. Finally, there is no real need for oven monitors on the shift to contact each other as part of their normal duties. These changes to the relative advantage of day workers and the relative disadvantage of shiftworkers are significant when we bear in mind that shiftworkers are still the largest group of production workers.

In summarising the trade union research committee's findings and analysis of the company's formal job descriptions during these three periods, we find (table 63) that there had been both positive and negative changes of equal magnitude in the transition from period 1 to period 2 (eight positive changes and eight negative changes, with two criteria

Table 62: Aluminium plant: fragmentation of oven monitoring jobs

Type of work	Work task		
	Period 1 (-1958)	Period 2 (1959-68)	Period 3 (1969-78)
Attending smelting pots (shift work)	Covering pots Packing edges Anode regulation Sampling metal Cleaning soot Mixing aluminium oxide Controlling anode effect Adding oxide Crust breaking Chiselling edges Measuring metal level Judging pots' resistance	Covering pots Packing edges Anode regulation Raking Sampling metal Cleaning soot Controlling anode effect Mixing aluminium oxide Sweeping floors	Covering pots Packing edges Control Extra control if required by the data processing system
Specialised jobs caused by introduction of technological equipment		Crust breaking Adding oxide	Crust breaking Adding oxide Controlling anode effect
Attending smelting pots (day work)			Chiselling edges Sampling metal Measuring metal level Packing edges Cleaning soot Judging pots' resistance Sweeping floors Change/direction of floor Assembly/disassembly of resistance Preparation for framing Adjusting mantels Raking Regulating mantels Cleaning and tightening gas caps Chiselling overhang

Source: Edgar Gundersen et al.: Teknisk utvikling og arbeidsforhold i aluminiumselektrolyse [Technological change and working conditions in an aluminium smelting plant] (Trondheim, IFIM-NTH, 1980).

Table 63: Aluminium plant: changes in job criteria for oven monitors on shift work[1]

Job criterion	Period of change[2]		
	1-2	1-3	1-3 (daytime work)
Adequate training	-	-	0
Possibilities for learning and development	+	-	+
Knowledge, competence and experience	(+)	-	+
Skill	+	(-)	+
Age, demand for speed	0	(+)	0
Specialisation of jobs	-	-	-
Work areas	-	-	-
Isolation	-	-	0
Possibilities to leave the working area	-	(-)	+
Physical work environment	+	+	+
Repetition	-	-	+
Absence	0	0	0
Working/shift schedule	+	+	+
Opportunity to interfere with systems	-	(-)	(+)
Terminals, accessibility to workers	-	-	-
Control over individual performance	+	+	+
Possibilities of control	+	+	(+)
Total for each value	+=8	+=6	+=11
	0=2	0=1	0=4
	-=8	-=11	-=3

[1] All judgements are compared to the situation in period 1.

[2] Key: + = positive change; 0 = no change; - = negative change.

Source: Gundersen et al., op. cit.

showing no change). In the transition from period 2 to period 3 (i.e. from mechanisation to automation), the majority of changes have been negative, and most of these negative changes concern the socio-psychological aspects of QWL (knowledge, competence, experience, skills, social contact, etc.). As we can see, these changes adversely affect the shiftworkers because the day workers have a large proportion of positive changes (11 out of 18 changes). We note that none of these consequences were planned, since prior to this project the company did not consider the QWL consequences of automation. Nevertheless, few of these consequences are the direct and unavoidable result of technological change.

Our data analysis suggests at least three possible findings, supporting the following hypotheses:

(1) Technological development is associated with the fragmentation of work. There is a clear tendency in the historical material analysed here for technological developments to go hand in hand with increased specialisation.

(2) Technological development reduces skill levels in an uneven manner so that some workers (in this case a minority) experience an increase in skill levels and other QWL variables (especially contact with co-workers), while other workers (the majority in this case) experience a decrease in skill levels and possibilities for contact.

(3) QWL consequences are not simply a direct and automatic result of technological change, but also reflect to a large degree decisions on the management and organisation of the work itself.

In short, while we find evidence that technology does make a difference to QWL, we do not find strong evidence for technological determinism. Similarly, while we find that there is some possibility of organisational choice, we do not find that such choice can be exercised independently of the type of technological change being implemented.

The implementation of more mechanised equipment and automated control systems has clearly been an advantage for the company in that productivity has been substantially increased. There has been a fivefold increase in the amount of metal produced per employee in the period 1956-78 (from 55 tons to 250 tons). The number of employees in the electrolysis hall during this period was reduced from 450 in 1963 to 230 in 1977. In fact, there are now approximately the same number of workers in the plant as a whole as there were in 1963, because a new production hall was started in 1968.

4.6 Trade union influence on technological developments

We have already seen some preliminary results of an increased level of interest on the part of the local trade union as a result of their activities in this project. One such result is that the union has taken the initiative in promoting a discussion with other local trade unions within the same corporation about long-term technological developments. Another is that it has developed a common forum for discussion with local management, where all future automation projects will be discussed from the beginning. Finally, the local trade union has become engaged in two concrete technical projects. The first concerns the purchase and placement of computer terminals that would be used for monitoring production processes, while the other concerns the evaluation and possible purchase of Japanese production control technology. Partly as a result of the trade union's activity, plans for both these projects have already been modified by the company in order to satisfy to a larger degree the requirements of the Norwegian QWL law. Only the future will show to what extent the trade union has real influence in the company's technical and organisational development.

5. Conclusions: possibilities for technological choice?

An important goal of IFIM's research is to understand better the relationship between QWL and technology. We began by asking whether existing theories of technological determinism or free organisational choice were adequate explanatory frameworks. Within the context of the ILO research design, we chose three case studies of companies where both mechanisation and automation (including computer-based process control systems) were being applied to essentially batch production technologies in the Norwegian process industry. The result of this technological change was to create a type of hybrid technology that, while still based on batches, shared many characteristics of a continuous flow. Thus we call it a "batch-flow" technology. Our aim in this concluding section is to summarise our findings concerning job design and automation in the development of batch-flow technology.

Although closer examination of our case studies has revealed important task-technology differences, even within the same technology, we find a number of consequences common to all of them. Other consequences are more evident among the three cellulose production units than the aluminium smelting plant. The data from the latter are the only information we have on objective job design changes over a relatively long period (25 years), and these allow us to examine the polarisation hypotheses more closely. This is quite a diversified data base, which must necessarily be the case when comparing data from different projects which did not have the same research design. Although it is not an adequate basis for drawing firm and final conclusions, it is nevertheless useful in helping us think about work design consequences of automation.

Within these limits, our data show quite clearly that neither simple-minded technological determinism nor unbounded organisational choice are adequate to explain the broad patterns in the data. However, we find that each of these different explanatory frameworks can, given relevant data, contribute to a more complete explanation. For example, in the cellulose plants we do not have data over long time periods, and technological change is increasing the process (or flow) aspects of essentially a batch technology (requiring new computer-based process control systems which in turn lead naturally to physically small but functionally comprehensive control rooms). The job design consequences of automation in this kind of technology seem to follow what one would expect from the analyses of Blauner[38] and Taylor.[39] Essentially, they found increased automation to be associated with increased frequency of multi-skilled, self-managed work teams (e.g. C 2 in contrast to C 1 in case study 1), even where there was no organisation development effort and a modified foreman's role remained (as in case study 2). However, the organisational choice explanation is also relevant, as we see in the Hunsfos pre-case from the Norwegian Industrial Democracy Programme in the mid-sixties. The chemical pulp department at Hunsfos had older, well-used equipment comparable to C 1, but a work organisation like C 2.

Similarly, the analaysis of objective job description data over 26 years at the aluminium factory tends to support the polarisation hypothesis in that technological change supported many a job design improvements among regular daytime workers at the cost of de-skilling for shiftworkers. Nevertheless, this apparent de-skilling may have been related to intentions to enhance daytime work and reduce the total number of shiftworkers (since the Act on quality of working environment of 1977, shift work is only possible in Norway under limited permits issued by the State Work Inspection Agency for special circumstances).

In short, our analysis suggests some reservations concerning the general applicability of either technological determinism or free organisational choice. Different researchers with different theoretical interests seem to design research studies that are more relevant to advancing particular theoretical interests than to clarifying necessary conditions and limits in exercising organisational choice under different kinds of technology. Our analysis suggests that it might not be a question of whether, for instance, Blauner is more correct than Braverman, or whether both of these fail because of the polarisation hypothesis. We suspect rather that before these explanatory frameworks can be compared, there must first be an analysis at a more detailed level as to what kind of assumptions are made, what kind of data are relied upon and, most important, what kinds of technologies are examined under what kind of organisational, market, political and historical conditions. (Contrast the perspective in Mortensen[40] with, for example, that in Davis and Taylor.[41])

Despite these reservations concerning the adequacy of competing theories, we find several consequences common to the different case studies. In all cases there have been work design changes that occurred after and appear to be associated with technological change. By and large, these are the kinds of consequences one would expect with increased automation -

(a) improvement in the production process as evidenced by increased production and productivity, more stable, higher-level quality and more efficient use of raw materials;

(b) relatively stable manpower (or possibly slightly decreasing when seen over several decades) largely because, in a highly energy- or capital-intensive industry with good markets (e.g. aluminium), returns are increased more by marginal increases in process efficiency rather than by reducing an already relatively small workforce;

(c) relatively marginal QWL consequences directly attributable to technology because -

(i) much is already automated, so that new technology introduces nothing fundamentally different;[42] and

(ii) a number of different changes are made at the same time (e.g. a whole new factory built), thus confounding the effects of technological change alone;

(d) substantial improvements in physical QWL -

(i) in control rooms; and

(ii) also out on the production floor itself;

(e) little or no evidence of QWL criteria being involved in the choice or design of new technology, even where there have obviously been QWL consequences;

(f) either no wage earner influence in the choice and implementation of technology, or influence so limited (either on questions of manning levels and wages or on trivial decisions such as types of chairs or the colour of walls, etc.) as to be of only token significance for QWL. At best, in the design of new control rooms (e.g. in C 2 and in case study 2), some workers participated in solving limited and practical implementation problems (e.g. the layout of information on a VDU, equipment layout, placement of furniture, etc.), but none were involved in the planning and design of the basic technology itself.

This last point about the lack of worker influence in technological change may seem strange, given Norway's advanced laws and labour-management agreements guaranteeing such worker participation. We should note, however, that these rights have only recently come into being formally and are not yet widely practised. All of our case material either precedes the establishment of these rights or, in the case of the aluminium factory, involves early attempts to implement them. More recent data (e.g. from the aluminium factory) show that the local trade union has in fact delayed the introduction of a new process control system pending a fully fledged QWL consequence analysis. In both the first two case studies, workers are participating in new projects (e.g. the drying department redesign in case study 1), and this seems to reflect current company policy.

Another set of consequences are most evident in the kind of technology we have examined in the cellulose plants, simply because the technological change involved creating a physically small but functionally large and integrated control room. To build a conventional control room to monitor the same functions would require 20-30 meters of dials and gauges, forcing operators to move around considerably. By using the new VDUs, operators could sit in one place and call up on their consoles whatever functions they wished to monitor. The technology provided a better overview of the process and a better basis for learning, since operators could see the results of their interventions more quickly and in a larger context. Thus there was a basis for more interesting jobs.

In addition, the creation of relatively small control rooms led to more group-based social relations which in turn tended to reinforce learning and competence, especially where management's philosophy of organisation emphasised multi-skilling and self-managed teams. Even where this philosophy was not strong (e.g. case study 2, where management favoured increased flexibility through multi-skilling but maintained conventional forms of supervision and reward systems), there appeared to be a natural drift in the direction of more work group autonomy and of the foreman as more of a boundary regulator (as we would expect from the research of Blauner and Taylor).

We find it highly significant that these tendencies are greatly enhanced for those organisations that aimed expressly at creating a more group-based, self-managed form of work organisation. Interestingly enough, the Hunsfos experiment led in fact to the creation of a rudimentary form of control room which evolved entirely from the needs of the organisational change effort; there was no technological change. Conversely, in case study 2, where the same basic production technology as in the Hunsfos case became automated through a computer-based process control system, a modified form of hierarchical work organisation evolved to meet the needs of the new technology; there was no planned organisational change.

In other words, there appear to be at least two routes that lead to more group-based work system organisation. The one (as in Hunsfos) originates in new ideas about more self-managed forms of work organisation (e.g. autonomous work groups). The other (as in case study 2) originates in new computer-based forms of process control technology. Given the limits of our data we cannot say which of these is the most certain, best or most travelled route. Good arguments and well-known research finds can be marshalled for both. We can say, however, that when both routes are combined (as in C 2), then they appear to be mutually reinforcing. These routes and relations are presented graphically in figure 30.

In conclusion, we see that neither the technological determinists (both optimists such as Blauner and pessimists as Braverman) nor the organisational choice schools (such as Trisy et al.[43]) alone provide adequate explanations for our empirical data. If we simply concentrate on the case material around dimensions of old versus new technology and self-managed versus hierarchically managed work organisation for our cellulose plants, we find no coherent pattern based on technological change -

Hunsfos
(pre-experiment): old technology - hierarchically managed work organisation

Figure 30: Two routes to more autonomous forms of work organisation

ROUTE 1: technological change

ROUTE 2: organisation development

new process control technology

new concepts about more self-managed work organisation (e.g. autonomous work groups in socio-technical theory)

creation of compact, comprehensive control room

more diverse monitoring tasks and possibilities for stimulation

creation of more group-based social relations

organisation development aimed at increasing work group self-management

increased possibilities for learning among workers

more autonomous forms of work organisation

increased task competence among workers

foreman's role shifts in direction of boundary maintenance and developmental support

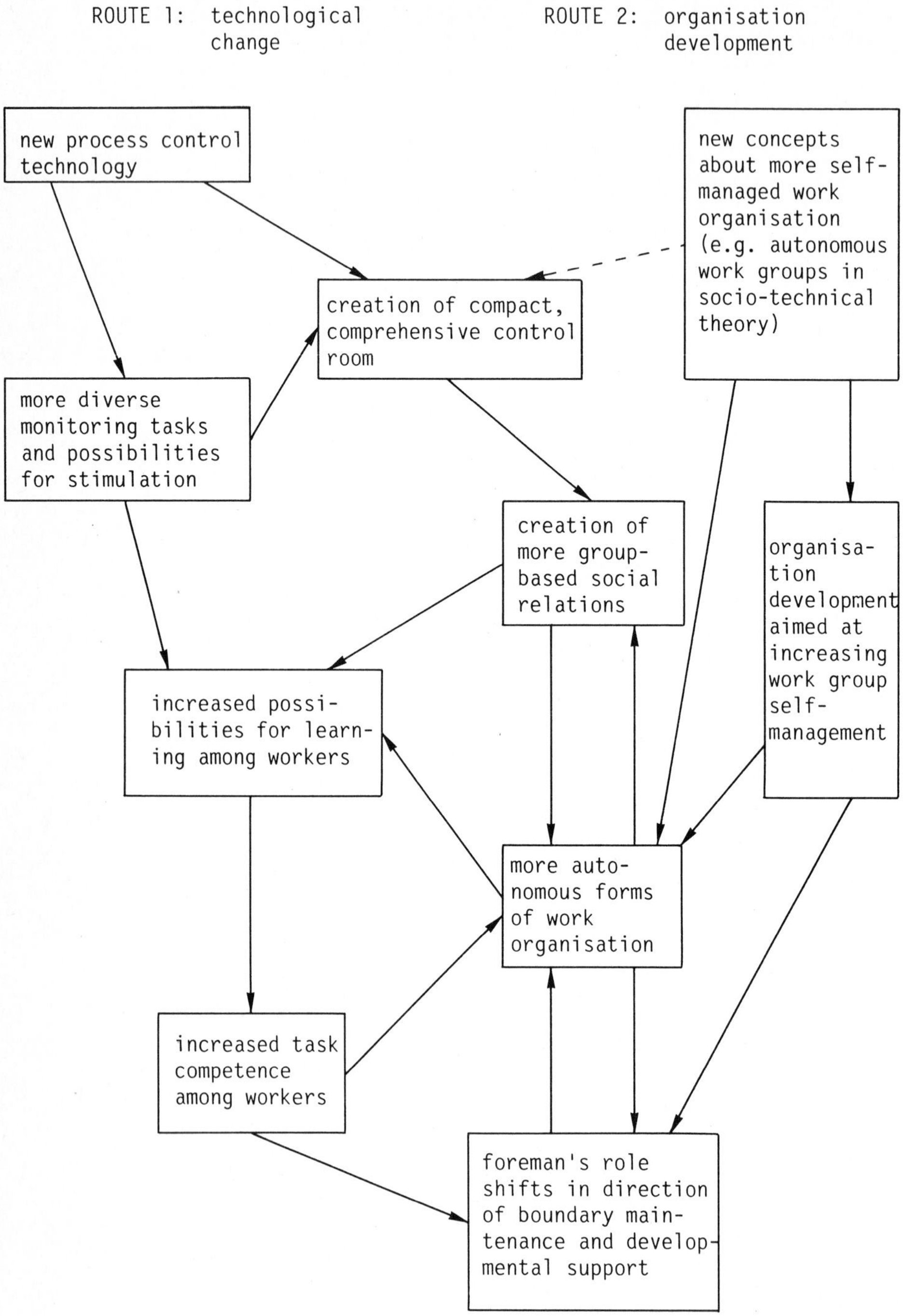

Hunsfos (post-experiment):	old technology - self-managed work organisation
Case study 1 (C 1):	old technology - hierarchically managed work organisation
Case study 2 (C 2):	new technology - self-managed work organisation
Case study 2:	new technology - modified hierarchy

The only consistent finding is that the clearest cases of more self-managed forms of work organisation are associated with planned organisational change (Hunsfos - post-experiment and C 2) rather than technological change alone.

This analysis verges on the over-simplified since these technologies are not exactly the same and, given capital-intensive process types of technologies, work organisation even before a technological change is not as hierarchical as for example in assembly-line technologies. Furthermore, we have not included other factors such as market conditions, company position in the market, labour market limitations, industrial relations and trade union traditions, degree of unit interdependence with the rest of the organisation, and other factors which we (and others)[44] suspect could be relevant. For example, we suspect that one additional reason why the work organisation in C 2 became so self managed is because the unit as a whole is much less integrated into the large company (raw materials are imported directly to the plant from abroad and the finished product is processed further into paper in an operation which is organisationally relatively idependent). In contrast, the cellulose production unit in case study 2 and in Hunsfos are highly integrated in the paper production process. Here we see that factors quite independent of technology or management philosophy can affect the impact of technology on work organisation and job design.

In summary, we find that for the type of modified process technology we have examined here, there seems to be a general drift in the direction of more autonomous forms of work organisation. Optimistic technological determinism (e.g. Blauner[45] and Taylor[46]) can explain this drift in some cases (e.g. case study 2), while organisational choice (under quite different degrees of automation within the same type of technology) seems to be the best explanation in other cases (e.g. Hunsfos and C 2). While this is our main finding, we cannot dismiss the polarisation hypothesis because, with the exception of the aluminium company case, we do not have adequate data.

To the extent that we have data over time from the aluminium case study, we find tendencies towards polarisation between day workers and shiftworkers, but this appears to result more from a desire to transfer workers from shift to day work than solely from technology. We suspect that data, for example, on skill levels for the entire organisation over a long time period (15-20 years) in a non-process technology would support the polarisation hypothesis. Unfortunately, our case studies were not designed expressely to test this hypothesis. Therefore, we are not able to evaluate the suggestive findings supporting polarisation under similar technological conditions.[47]

In production work itself there are new jobs which are not necessarily directly comparable to the older jobs, but this does not seem to be a result of subcontracting out the more interesting tasks to create the kind of de-skilled "patched together" jobs found by Cohen-Hadria.[48] The main job design consequence in our case material is the evolution of computer-based monitoring of larger parts of the process from a control room. These jobs require basically the same knowledge of the chemical process as before, but at a higher level of abstraction depending on the complexity and stability of the process. They tend towards typical control room jobs with good learning potential, but long passive periods (associated with boredom) while the process is stable. The problem is typical and well known in process technology and stands out more sharply than indications of polarisation. In our case studies we find what might be a transitional type of job where there is less boredom because the process is less stable and/or less automated.

However, our main finding supports our initial scepticism about uncritically relying on a single explanatory framework. There does not seem to be any all-ecompassing theory to explain critical factors that condition the effect of automation on work design. We have found evidence of increased workplace autonomy to be associated with technological change alone, with organisation redesign alone, and with a combination of both increased automation and explicit organisational choice. It seems reasonable to conclude that traditional paradigms explaining technology and organisation are shifting, if not breaking down. Further research is needed to develop new, more general theory from the fragments we have identified.

Notes

[1] The Norwegian contribution to the ILO international study is written by the "Technology Group" at the Institute for Social Research in Industry (IFIM). Although different authors are responsible for their separate parts, the product is the result of lengthy discussions within the group. Max Elden helped to initiate and co-ordinate the project. This report is based on research made possible by a grant from the Work Environment Committee, Royal Norwegian Council for Technical and Scientific Research (Arbeidsmiljøkomite, Norges Teknisk-Naturvitenskapelige Forskningsrad).

[2] Leonard: "Norway: Today, Europe, tomorrow the World", in The Economist, 15 Nov. 1975 (Survey 3).

[3] Described in Fred Emery and Einar Thorsrud: Democracy at work (The Hague, Nijhoff, 1976); later developments summarised in Max Elden: "Three generations of work-democracy experiments in Norway: Beyond classical socio-technical systems analysis", in Cary L. Cooper and Enid Mumford (eds.): The quality of working life in Western and Eastern Europe (London, Associated Business Press, 1979).

[4] Phillip Herbst: Socio-technical design (London, Tavistock, 1974); idem: Alternative to hierarchy (The Hague, Nijhoff, 1976): Fred Emery: The emergence of a new paradigm of work (Canberra, The Australian National University, Centre for Continuing Education, 1978).

[5] Kristen Nygaard and Olav Terje Bergo: "The trade unions - New users of research", in Personnel Review, spring 1975, pp. 5-10: Jan Henrik Bjørnstad and Torstein Bjaaland: "Oppbygging av kunnskaper: fagbevegelsen" [The development of knowledge in the trade union movement] in idem (eds.): Arbeidsmiljø og Demokratisering [Quality of working life and democratisation] (Oslo, Tiden, 1980), pp. 59-72.

[6] Bjørn Gustavsen: "A legislative approach to job reform in Norway", in International Labour Review, May-June 1977, pp. 263-276; Bjørn Gustavsen and Gerry Hunnius: New patterns of work reform - The case of Norway (Oslo, Universitetsforlag, 1981).

[7] Ministry of Planning: Langtidsprogrammet: 1982-85 [The National Plan: 1982-85], St. Meld No. 79 (Oslo, Centraltrykkeriet, 1981), p. 39.

[8] K. Nygaard and J. Fjalestad: "Group interests and participation", in OECD: Information System Development (Paris, 1979).

[9] ibid., p. 3.

[10] There is no up-to-date comprehensive and professional assessment of industrial relations in Norway available in English. Recent brief treatments are available in Norway, Komiteen for Internasjonale Sosialpolitiske Saker: Labour relations in Norway (Oslo, 1975); Joep Bolweg: Job design and industrial democracy: The case of Norway (Leiden, Nijhoff, 1976), pp. 1-18; Thoralf Qvale: "The Norwegian industrial relations system", in Industrial Democracy in Europe, International Research Group: European industrial relations (Oxford, Clarendon Press, 1981), pp. 11-33; and Thomas Sandberg: Norway: Autonomous groups and industrial democracy (Uppsala, University of Uppsala, Department of Business Administration, 1981; Working Paper 1981/5).

[11] Gustavsen: "A legislative approach ...", op. cit.; Gustavsen and Hunnius, op. cit.

[12] Gustavsen and Hunnius, op. cit.

[13] See, for example, Ministry of Local Government and Labour: Employment and working conditions in the 1980s - Perspectives on the significance of the technological and economic development for employment and working conditions, Norges Offentlige Utredninger (NOU) 1980: 33 (Oslo, Universitetsforlaget, 1980); Ministry of Finance: Økonomiske og sociale virkninger av ny datateknologi [Economic and social effects of new data technology], NOU 1981; 14 (Oslo, Universitetsforlaget, 1981); Ministry of Planning: Langtidsprogrammet: 1982-85, op. cit.

[14] Ministry of Finance: Økonomiske og sosiale virkinger ..., op. cit.

[15] There is still little available in English concerning the practical effects of these innovative agreements, which have also begun to appear in other industrialised countries. See Peter Docherty: "User participation in and influence on systems design in Norway and Sweden in light of union involvement, new legislation and joint agreements", in N. Bjørn-Andersen (ed.): The human side of information processing (Amsterdam, North-Holland

Publishing, 1980), comparing developments in Norway and Sweden; and Leslie Schneider: *A study of the implementation of data agreements in ten leading Norwegian concerns* (Trondheim, IFIM-NTH, 1982).

For further reading on the role of trade unions, see Morten Levin: "A trade union and the case of automation", in *Human Futures*, Vol. III, No. 3, 1980, pp. 209-216; and P.G. Martin: "Strategic opportunities and limitations: The Norwegian Labour Party and the trade unions", in *Industrial and Labour Relations Review*, Oct. 1974, pp. 75-88.

[16] R. Blauner: *Alienation and freedom: The factory worker and his industry* (Chicago, University of Chicago Press, 1964); Peter F. Drucker: "The promise of automation. America's next twenty years" (Part II), in *Harpers Magazine*, Apr. 1955, pp. 41-47.

[17] James R. Bright: "Does automation raise skill requirements?", in *Harvard Business Review*, July/Aug. 1958; Henry Braverman: *Labour and monopoly capital: The degradation of work in the twentieth century* (New York, Monthly Press Review, 1974).

[18] Anders Mathisen: *Produktion, kvalifikation, arbejdsmarkedspolitik* [Production, skills, labour market policy] (Copenhagen, Hunksgoard, 1978); Projektgruppe Automation und Qualifikation: *Theorien über Automationsarbeit* (Berlin, Argument-Verlag, 1978); Knut Veium: *Ny teknologi og kvalifikasjoner - hva vet vi?* [New technology and deskilling - What do we know?] (Trondheim, IFIM-NTH, 1980).

[19] Nils Mortensen: "Impact of technological and market changes in organisation functioning: The conceptual framework in bourgeois and Marxist theory and research", in *Acta Sociologica*, Vol. 22, No. 2, 1979, pp. 155-159.

[20] H. Kern and M. Schumann: *Industriearbeit und Arbeiterbewusstsein* (Frankfurt am Main, Suhrkamp/Kno., 1970).

[21] See, for example, Blauner, op. cit.

[22] Mortensen, op. cit., p. 138.

[23] See Emery and Thorsrud, op. cit.; Herbst: *Alternative to hierarchy*, op. cit.; Bolweg, op. cit.; Elden, op. cit.; and Sandberg, op. cit.

[24] Guvstavsen: "A legislative approach ...", op. cit; Thoralf Qvale: "A Norwegian strategy for democratisation of industry", in *Human Relations*, Vol. 29, No. 5, 1976, pp. 53-69; Gustavsen and Hunnius, op. cit.

[25] See, for example, Nygaard and Bergo, op. cit.; and Bjørnstad and Bjaaland, op. cit.

[26] See, for example, Nygaard and Fjalestad, op. cit.; and J. Fjalestad: *Some factors affecting participation in systems development* (draft), Paper presented to Arbeitstagung über Partizipation bei der Systementwicklung, Bonn, Feb. 1981.

[27] See, for example, Max Elden et al.: Fagbevergelsen og EDB i prosessindustrien - Slultrapport [The trade union movement on EDP in the process industry - Final report] (Trondheim, IFIM-NTH, 1980).

[28] ibid.

[29] Emery and Thorsrud, op. cit.; Per Engelstad: Teknologi og sosial forandring på arbeidsplassen - Et eksperiment i industrielt demokrati [Technology and social change in the workplace - An industrial democracy experiment] (Oslo, Tanum, 1970); idem: "Sociotechnical approach to problems of process control", in Louis E. Davis and James C. Taylor (eds.): Design of jobs (Santa Monica, Goodyear Publishing, 2nd ed., 1979), pp. 184-205.

[30] Emery and Thorsrud, op. cit.

[31] Einar Thorsrud: "Democracy at work: Norwegian experiences with non-bureaucratic forms of organisation", in Journal of Applied Behavioral Science, Vol. 13, No. 3, 1977, pp. 410-421.

[32] For similar findings also in process technologies, see Emery and Thorsrud, op. cit.

[33] Egil Skorstad and Knut Veium: Automatisering og kvalifikasjoner i prossessindustrien [Automation and skills in the process industry] (Trondheim, IFIM-NTH, 1982).

[34] See Elden et al.: Fagbevegelsen ..., op. cit.

[35] Edgar Gundersen et al.: Teknisk utvikling og arbeidsforhold i aluminiumselektrolyse [Technological change and working conditions in an aluminium smelting plant] (Trondheim, IFIM-NTH, 1980).

[36] See also Elden: "Three generations ...", op. cit.; and idem: "Varieties of workplace participatory research" (Trondheim, IFIM-NTH, 1980), published in Journal of Occupational Behaviour, Jan. 1983.

[37] Gundersen et al., op. cit.

[38] Blauner, op. cit.

[39] James Taylor: Technology and planned organisation change (Ann Arbor, University of Michigan, Institute for Social Research, 1971).

[40] Mortensen, op. cit.

[41] Louis E. Davis and James C. Taylor: "Technology effects on job, work and organisational structure: A contingency view", in Louis E. Davis and Albert Cherns (eds.): The quality of working life, Vol. I (New York, Free Press, 1975), pp. 220-241.

[42] See also J. Woodward: Industrial organisation: Theory and practice (London, Oxford University Press, 1965), pp. 198-9.

[43] Eric Trist et al.: Organisational choice (London, Tavistock, 1963).

[44] For example, Mortensen, op. cit.

[45] Blauner, op. cit.

[46] Taylor, op. cit.

[47] Yves Cohen-Hadria: Automation and work integration in the cement industry (Paris, Ecole Polytechnique, 1978; mimeographed).

[48] ibid.; see also research studies cited therein.

—

CHAPTER XIII

JOB DESIGN AND AUTOMATION IN THE POLISH MACHINE INDUSTRY: THE CASE OF NON-PROGRAMMED AUTOMATION IN A PLANNED ECONOMY

(by Marian J. Kostecki, Krzysztof Mreła and Wlodzimierz Pańków)

1. Industrialisation, technological development and society: the case of Poland

1.1 Introduction

For many years now, a number of Polish sociologists have been undertaking an investigation of industrialisation, its social elements and consequences. The subjects of their fieldwork and theoretical investigations include the following: the socio-cultural determinants of industrialisation;[1] the socio-cultural consequences of industrialisation;[2] the consequences for the class structure and social stratum[3] and mass culture;[4] changes in the countryside;[5] the socio-spatial consequences, including urbanisation;[6] local communities, the processes of stabilisation and migration;[7] the consequences of industrialisation for institutional structures and the mechanisms of functioning of institutions and human behaviour in the process of industrialisation;[8] and, finally, the development process and prospects for industrialisation.[9] The technical and economic processes of industrialisation and the socio-political and cultural processes connected with them provide the basic context for psycho-social and organisational aspects of automation.[10]

The social effects of technological progress and in particular automation have been the subject of study in Poland for many years, and especially after 1956. Numerous institutions are engaged in research into the social effects of industrial innovations, socio-technical design, safety at work, etc. Among the institutions dealing with different problems of mechanisation and automation are the Institute of Philosophy and Sociology (Polish Academy of Sciences) and the University of Łódź. Similar research work is done in the research centres attached to the Ministry of Machine Tools Industry, Chemical and Civil Construction Industries. These studies are particularly concerned with the problems of the adaptation of people to work, and with absenteeism and fluctuation of manpower. Many interesting results of this work are now available.

In Poland there are also a number of organisations which are engaged in research of a more concrete and practical nature, the results of which are intended for direct application in industry. Among them are the Central Statistical Board, the Institute of Labour, Remuneration and Social Problems (Ministry of Labour), the Central Institute of Labour Protection, etc. Much work has been done by the last two institutions in the sphere of the improvement of working conditions and job enrichment. Moreover for a considerable number of years there have been developments in the field of work organisation and management, linked with the process of computerisation and automation. Among the leading institutions engaged in this field are the Polytechnical Institute of Warsaw, the Institute of Organisation and Management (1974-78), the Praxicology Department of the Polish Academy of Sciences, the Management Institute of Warsaw University and many others.

Many journals are currently published in Poland on the problems of work organisation, working conditions and management.

1.2 Societal elements and stages of industrialisation

Drawing on a model for the study of industrialisation as a process,[11] we may assume that the beginning of industrialisation, its development and consequences are determined by elements which, although always present, take different forms according to the historical context. In this section we outline the main features of the situation in post-war Poland.

The first element is that of "entrepreneurs", i.e. individuals, groups or institutions which decide the character, pace, intensity and goals of industrialisation.[12] In this context, Kerr et al.[13] prefer to speak about industrialising elites and stress that, with the exception of Western Europe (where industrialisation was conducted by the middle class), entrepreneurs in the strict sense are non-existent at any other time or place.

In Poland, the hierarchic organisation of state institutions has become the "socialist entrepreneur". These institutions prepare plans, manage and control their implementation, and participate in economic and technical decisions, including those concerning the automation of production processes. The principal institutions concerned are the Political Bureau of the Central Committee of the Polish United Workers' Party, the Central Committee and its departments, the Council of Ministers, the Government Praesidium, the Planning Commission, ministries, Parliament and parliamentary committees, as well as numerous institutes and R & D centres attached to the Polish Academy of Sciences, to technical and economic universities, and to industrial enterprises.

The "socialist entrepreneur" is "a political entrepreneur which orients itself first of all by ideological guidelines ... The principles of this ideology define the strategic goals of industrialisation, as well as the direction, sequence and intensity of their implementation".[14] Kerr et al.[15] see these features, in particular the political character of the industrialising elite and the subordination of the technical and economic goals of industrialisation to political choices, to be the most essential specific features of socialist industrialisation. These features also have a decisive influence on the socio-economic profile of the model created in the course of industrialisation.

The next element is the group of technical cadres: i.e. managers, engineers and economists who carry out decisions taken by the industrialising elite. The significant features of the Polish post-war situation in this respect are summarised in Szczepański.[16]

Another element needed to describe the industrialisation process is the model of the enterprise which evolves in the course of this process. This model is determined by party ideology. The enterprise is state owned and is usually governed by a party-nominated (or at least endorsed by an appropriate party authority) economic administration official. Pointing to the fact that an enterprise of this type is "the constitutive institution of the new society", Szczepański writes that "every factory that is founded, every production unit and every institution must be smoothly accommodated in this general model whose functioning is to develop an appropriate macro- and micro-structure of the all-embracing socialist society.[17]

Two more elements in our description are the capital needed to pay the cost of industrialisation and the methods of acquiring it. In Poland, the most important sources of capital have been taxes imposed on private farmers, profits from nationalised enterprises already in existence, wage taxes and a wages policy ensuring a high level of accumulation.

The sixth element which deserves mention is the labour force, its characteristic traits, traditions and attitudes.

If by industrialisation we mean the whole of processes and phenomena arising in the course of industrial development, then automation is one of its vital components. The essence of industrialisation is determined primarily by technical processes, the introduction of new technologies in raw material processing, the mechanisation of working methods (automatic equipment included) and the launching of large-scale, serial and standard production.[18]

In terms of the chronology of industrialisation, automation concerns only the later stages of industrialisation. In Poland we may speak of at least three stages of industrialisation, namely -

- the initial stage (1944-49). This was the phase of industrial reconstruction after the devastation of the war, and of introducing basic social and political reforms conditioning future socialist industrialisation. It was also the stage of reconstruction of essential mechanisation;

- basic industrialisation (1950-70). This consisted in the transformation of Poland from an agricultural-industrial country into an industrial one. The national economy began to develop, with an emphasis on heavy fuel and power, and machine industries. In the second phase of this stage, in the sixties, the beginning of automation occurred, particularly in the fuel and power industry (e.g. the construction of an automated hard coal mine), in the chemical and in the machine industries. The introduction of automation in industry was preceded by the first post-war plenary meeting (1960) of the Central Committee of the Polish United Workers' Party, devoted entirely to technical advance;

- complementary industrialisation (1971- . The complementary character of this stage relates not only to filling the gaps created by one-sided pressure on the development of production but also to the modernisation of industry. By the mid-seventies, one out of every two machines in Polish industry was less than five years old. The essential directions of automation were, primarily, the electrical machinery, power and fuel, and food industries. At that period we note the beginnings of automation in the sphere of production and turnover, and attempts at the automation of services (e.g. posts and telecommunications), as well as the introduction of computerised record systems on a larger scale.

The data contained in table 64 point to the large and growing role since 1970 played by innovations based on new investment ventures and/or the purchase of new licences. As regards the outlays in question, the engineering industry was placed second among all industries. As regards the purchase of licences, it was far ahead of all other industries in the seventies.

In view of the fact that we shall proceed to examine the determinants and consequences of introducing automation to the machine-tool industry, we present some essential data on the production of NC machines against the background of the production of automatic control and adjustment equipment and computers and EDP equipment (tables 65-67).

Table 64: Poland: breakdown of financial outlays on technical development in state-owned industrial enterprises, 1970-79 (percentages)

Item	Year	Mechanisation and automation of production and advanced technologies	Other technical and organisational undertakings	Starting up production of new machines, equipment and commodities	Total
Industry	1970	47.1	9.8	43.1	100.0
	1975	53.8	3.5	42.7	100.0
	1979	42.5	3.8	53.7	100.0
of which:					
engineering industry	1975	49.5	5.1	45.4	100.0
	1979	25.2	4.6	70.2	100.0

Source: Rocznik Statystyczny Przemysłu [The Statistical Yearbook of Industry], table 13 (151), pp. 349-350 (Warsaw, 1980).

Table 65: Poland: production per head of machine tools, automatic control and adjustment equipment, computer systems and EDP equipment, 1950-79

Commodity	Unit	1950	1960	1970	1975	1979
Milling machines (per 10,000 inhabitants)	pieces	1.7	8.7	11.2	12.4	11.2
	tonnes	5.1	13.9	20.0	18.0	15.2
Automatic control and adjustment equipment	(Zl)	.	0.8	63.5	189	308
Computer systems and EDP equipment	(Zl)	.	.	20.2	199	333

. = Nil or negligible

Source: Rocznik Statystyczny Przemysłu, op. cit., p.506

Table 66: Poland: total output of machine tools, automatic control and adjustment equipment and EDP equipment, 1960-78

Commodity	Unit	1960	1965	1970	1975
Machine tools	'000	25.9	35.8	36.3	42.2
Automatic control and adjustment equipment	Zl millions	23.2	326	2 061	6 393
EDP equipment	Zl millions	.	72.3	656	6 867

. = nil or negligible

Source: Rocznik Statystyczny Przemysłu, op. cit.

Table 67: Poland: production of machines and automatic equipment, 1950-79

Commodity	Unit	1950	1960	1965	1970	1975	1976	1977	1978	1979
Milling machines	No. of items	4 173	25 855	35 803	36 299	42 229	34 629	32 209	39 558	39 318
	Tonnes	12 771	41 252	53 563	65 011	61 180	57 987	55 555	57 604	53 588
- Machining	No. of items	3 825	22 470	31 134	29 726	33 192	24 234	23 178	29 991	29 712
	Tonnes	11 385	32 381	43 326	52 827	48 821	45 421	43 278	45 259	42 482
of which:										
Numerically controlled	No. of items	.	.	.	13	180	231	307	310	327
	Tonnes	.	.	.	100	1 366	1 963	2 473	2 175	2 080
- Plastic working	No. of items	348	3 385	4 669	6 573	9 037	10 395	9 031	9 567	9 606
	Tonnes	1 386	8 870	10 237	12 184	12 359	12 566	12 227	12 345	11 106
Automatic control and adjustment equipment	Zl millions		23.2	326	2 067	6 442	7 897	8 900	9 927	10 844
of which (elements of block system):										
Pneumatic infinitely variable adjustment		.	-	-	309	961	995	924	841	909
Hydraulic infinitely variable adjustment		.	-	-	63.3	136	136	111	130	125
Electric infinitely variable adjustment		.	-	-	153	613	705	884	873	693
Computer systems and EDP equipment	Zl millions	.	.	72.3	656	6 760	8 565	10 267	10 748	11 756
of which:										
Main frame computer systems	No. of items	.	.	32	60	100	105	70	60	51

. = nil or negligible; - = data not available.

Source: Rocznik Statystyczny Przemystu, op. cit., table 25 (45), pp. 116 and 120.

Table 68: Poland: percentage share of manual, mechanised and automated jobs in selected industrial branches, 1965-76

Type of job (level of mechanisation)	Industry									
	Fuel		Non-ferrous metals		Chemicals		Glass-making		Textiles	
	1965	1976	1965	1976	1965	1976	1965	1976	1965	1976
Manual[1]	48.9	13.8	45.1	25.1	40.6	23.6	64.3	45.0	36.4	16.9
Mechanised/machine[2]	51.1	86.2	54.9	74.9	59.4	76.4	35.7	55.6	63.5	74.6
of which:										
Partly mechanised	32.6	18.1	37.5	31.8	31.1	33.9	25.5	32.9	55.8	68.4
Fully mechanised	16.6	61.9	15.3	35.7	21.2	33.5	6.8	14.0	7.6	10.5
Automated[3]	1.9	6.2	2.1	7.4	7.1	9.0	3.4	8.7	1.2	8.1

[1] Work may be (a) simple, where only simple tools or no tools at all are used; or (b) mechanised, where mechanised tools (electrical, fluid or pneumatic drive) are used, which are not a major investment, and simple machines and installations at the lowest technical level (i.e. without drive, but constituting a major investment).

[2] Work may be (a) partly mechanised (i.e. successive manual or partly mechanised work cycles, the worker performing manual operations, auxiliary, control and measurement tasks); or (b) fully mechanised (i.e. successive manual, partly or fully mechanised work cycles, the worker performing in part manual and auxiliary tasks, and in full control and measurement ones).

[3] Work may be (a) partly automated (i.e. automatic or manual succession of work cycles, mechanical or automatic control of work operations, either by conventional method or by NC program; the worker performs auxiliary, and partly control and measurement, jobs, and also monitors the process by way of control); or (b) fully automated (i.e. automatic succession of work cycles and complete automatic control, the job being reduced to watching the course of the work process by way of control or monitors).

Source: J. Auleytner: Zmiany w charakterze i trésci pracy w zwiazku z postepem naukowo-technicznym [Changes in the character of job content in relation to scientific/technological progress] (Warsaw, IPiSS, 1978).

"The great leap", as some called it, or the "speeding-up" of the seventies is illustrated by the data shown in tables 65-67. Automated production technologies, automated equipment and know-how resulted from imports in 70-100 per cent of cases. This was linked with an essential reorientation in the method of financing complementary industrialisation and modernisation, which was financed by credits granted by governments and banks of the Western countries. Table 68 shows the increase in the proportion of mechanised and automated jobs between 1965 and 1976.

Poland's "opening" to Western technologies and technical advance was in line with the official strategy of doubling the production capacity of Polish industry and of catching up with the industrialised countries. This "opening" was inevitable in order to meet the objectives formulated above.

As an integral element of industrialisation, the processes of automation of production are subject to the same organisational structure presented earlier in this chapter, i.e. initiated and imposed "from the top" by ministries and industrial associations. This, however, does not mean that manufacturing, servicing and other organisations have always been reluctant to accept the change. New investment ventures and/or purchases of new licences are usually the sole source of the enterprise's growth. As there exists a strongly hierarchical system of industrial management, the enterprise carries very little of the burden of investment costs. Moreover, an investment venture (which is founded on means allocated by the superior unit) enhances the enterprise's bargaining power in relation to the superior unit.

1.3 The structure of production and the administrative system of work regulation

The industrialisation process described above, whose main features Poland shares with other socialist countries, has certain specific traits. Some of them are specific to Poland (e.g. the relatively extensive private sector, particularly in agriculture, but also in trade and industry), while others are specific to all the countries of Eastern Europe (e.g. similar institutional structures, a similarly insignificant role of money and the market).

As regards the forms of ownership, the sphere of industry in Poland may be divided into what is known as the socialised sector (i.e. state-owned and co-operative industry) and the private sector. Data concerning these two sectors is given in table 69.

1.3.1 The socialised sector

The dominant socialised sector (and its relationship to the State) is a complex institutional structure consisting of the following elements (1980):

Table 69. Industry in Poland by type of ownership, 1980

Item	Socialised industry				
	State-owned	Co-operative	Sub-total	Private industry	Total
No. of plants	19 937	38 422	58 641	137 592	196 233
(%)	34[1]	65.5[1]	30	70	100
Labour force employed ('000 persons)	4 401.9	754.9	51 182.5	260.8	5 443.3
(%)	85	15.6	95.2	2	100
Gross output (Zl 1,000 millions)	1 916.5	246.5	2 173.2	44.2	2 217.4
(%)	86.4	11.1	98.0	2.0	100

[1] These figures do not add up to 100 per cent on account of 271 industrial plants belonging to social associations which are not included in this table.

Source: Rocznik Statystyczny Przemysłu op.cit. pp. 6, 48-49 and 70.

- 38 ministries (and equivalent units, including co-operative sector headquarters), most of them grouped by industrial branches; employing around 8,500 people;

- 163 headquarters of industrial associations, employing 20,177 people which are the intermediate level in the administrative hierarchy managing the economy. Industrial enterprises are subordinated to these headquarters. Industrial associations are further subdivided by industrial branch. State-owned industry employs some 40 per cent of the total labour force employed in the economy and this figure has been virtually constant since 1950. Of all investment outlay in the economy, 37 per cent is allocated to state industry. The value of fixed assets in socialised industry accounts for some 30 per cent of the total value of fixed assets in the economy, having risen tenfold in the years 1950-80 (as compared to only 3.3 times for the economy as a whole). The share of gross industrial output in national income is twice the figure for 1950, equalling 52.3 per cent. While the national income in 1979 was 6.8 times higher than in 1950, the value of industrial output was 12 times higher (in fixed prices). Those last figures show that industrialisation was not only designed to be, but actually was the main factor responsible for the growth of national income;

- over 4,500 enterprises comprising a total of some 31,000 industrial plants (and 4.7 million employees), about 4 per cent of which are economically self accountable;

- in addition, there are over 14,000 industrial plants (166,000 employees) subordinated to non-industrial enterprises and over 13,000 supplementary industrial plants (344,000 employees), whose output is used in the production of the enterprises to which they are subordinated;

- 271 industrial plants which are neither state owned nor co-operative, but belong to social associations.

1.3.2 Organisational structure of production units

The economy's institutional structure also entails organisational consequences at the level of individual production units. On the basis of comparative studies of organisational structures of 136 production units in industry[19] (all of them principal units with a legal identity) that the authors of the present paper conducted in the late seventies, certain conclusions may be drawn.

First, the relationships between organisational structural variables and organisational goals, technologies or size is very weak. Within each of these three categories, large differences in organisational structure were observed. It was found that size (measured by the number of workers employed, the volume of current assets and outlay on basic production) and technology (measured as the degree of diversification of product and of workflow technologies) were both strongly positively correlated, and the most significant contextual factors correlated with structure.

If we consider the fact of strong correlation between size and technology, the connections between variables describing the technology of the organisation and its structure seem to be apparent:

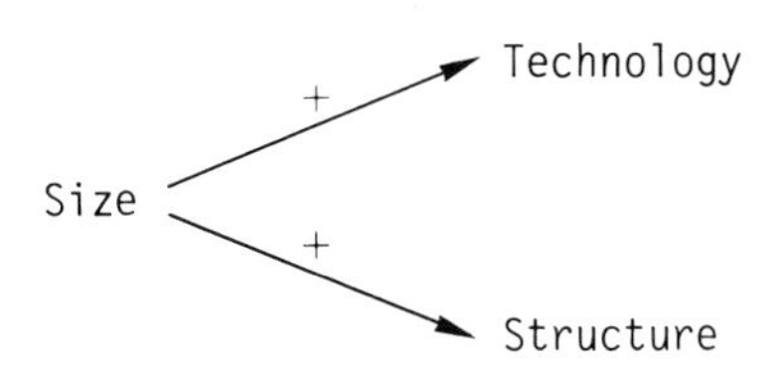

The degree of diversification of production and differentiation of technology (information technology included), as well as the degree of mechanisation and automation, are higher in larger enterprises in Polish industry. This means in effect that the choice concerning allocation of new technologies, new products and automated equipment, made at higher levels of the administrative hierarchy of industry, is primarily centred on large enterprises. From the viewpoint of the central administration, large enterprises are able to bear the high cost of technology. This is particularly noticeable in the case of computer-based technology: large firms can afford to buy hardware and to cover the current costs of operations. On the other hand, as mentioned earlier, larger enterprises have a greater "driving force" in bargaining for new equipment or advanced technologies.

The size of the organisation also furthers a bureaucratic structure (i.e. formalisation, standardisation and centralisation are higher, the number of levels in the management hierarchy is greater, etc.). Meanwhile, the growing diversification of production and differentiation of technologies, as well as mechanisation and automation, do not bring about a reduction in the degree of bureaucracy: formalisation and standardisation are not reduced, and neither is the number of hierarchial levels. Thus, changes in technologies do not, in principle, modify the organisational structures within which those changes were introduced.

Secondly, organisational structures and industry do not seem to be coherent and integrated, i.e. the individual organisational units do not communicate and co-operate with each other according to the formal organisational links which have been established. Instead, units form ad hoc arrangements based on, for example, recurrent problems which require co-operative solutions. The pressures which lead to these problems of industrial coherence include -

- the pressure towards formalisation coexisting with the pressure to apply less formal patterns of action, connected with the functioning conditions of the organisations. Adjustment to those conditions seems to call for an application of similarly informal patterns of action, reinforced by the pressures of entropy. The source of the latter lies in the style of management on a macro level, which consists in the application of individual, direct commands or instructions rather than reliance on fixed impersonal rules, which, in turn, is conducive to a development of the sphere of organisational "memory";

- the pressure to impose universal structural patterns (irrespective of the organisation's specific goals and technology) which filters down through vertical channels, and the simultaneous pressure towards specialisation which flows through the horizontal links among organisations;

- the pressure to standardise information flow, which coexists with the pressure to disorganise information flow: sporadic application of programming and controlling procedures, orientation according to the rationale of the moment ("instant rationality") and the increasing autonomy of information links.[20]

These pressures, which affect the structures of the organisations under study, reflect the contradictions between the necessary adjustment of each work organisation to the requirements of a uniform, hierarchical structure of economic administration, of which the enterprise is a link and the adjustment to specific functional requirements of the enterprise (specific technology, characteristics of the employees, type of operation, etc.).

Thirdly, there are no strong relations between structural features of organisation and measures of organisational effectiveness. The differences among structural features explain a mere 4-25 per cent of variations among organisations as regards net output, net profit, productivity per worker and other economic measure of effectiveness. In any case, measures of effectiveness are themselves weakly coherent, with the result that an organisation which is effective in one respect is consequently not equally effective in others. Links between organisational structure and effects are extremely tenuous, the only exception being the links between those structural variables which, like the number of sub-units, the span of control or the levels of hierarchy within organisations are correlates of organisation size.[21]

1.4 Level of employment and the structure of the labour market

One of the unchanging elements of the industrialisation process in Poland has been the policy of full employment. Not only is the right to work guaranteed constitutionally, but unemployment as a social phenomenon on the national scale is virtually non-existent. The proportion of people registered as looking for work was 0.6-0.8 per cent of persons employed in 1979 (the respective figures for 1948 and 1938 being 3 per cent and slightly under 16 per cent). At the same time, a number of posts remained vacant for each person seeking work. Naturally, this does not mean that there have been no local pockets of unemployment or job shortages affecting particular professions.

In 1978, over 51 per cent of Poland's population was economically active, of which some 30 per cent were in industry. This high proportion has been enforced by the doctrine of industrialisation rather than determined by technological needs. Table 70 shows the structure of employment in industry by branch between 1960 and 1978.

An important conclusion to be drawn from these data is the fundamental unalterability of the structure of employment in Polish processing industry over a period of nearly two decades, the only major change being recorded in the electrical-engineering industry. During the years 1970-79, this sector experienced the highest investment outlays, setting up new work places for 300,000 workers over 1970-75 alone (an increase of a quarter in comparison with 1970).

Table 70. Poland: the structure of employment in industry by branch, 1960-78[1]

(percentages)

Industrial branch	1960	1978
Production of electricity and power	1.7	1.7
Extractive industry	38.4	23.5
Processing industry of which:		
iron and steel	6.1	5.3
electrical-engineering	14.9	34.1
chemicals	6.5	6.8
textiles	20.6	17.7
food	11.8	10.9
Total	100	100

[1] According to the International Standard Industrial Classification.

Source: Rocznik Statystyczny Przemysłu, op. cit.

It is significant that in the years 1960-79 industry absorbed a huge increase of over 2 million workers (the labour force rising steeply from just under 3 million employed in industry in 1960 to slightly over 5 million in 1979). If we examine the increase in employment over successive five-year periods -

1960-65: 570 700

1966-70: 682 000

1971-75: 696 700

1976-79: 40 900

we may observe that until the mid-seventies the increase in employment was largely responsible for the rise in gross industrial output. If throughout the 1960-79 period employment in industry grew by 167 per cent while gross industrial output grew by 448 per cent, the share of employment increase in the rise in output may be estimated at some 37 per cent. During the years 1971-75, this share was lower (despite the quantitatively larger employment increase), equalling 26.7 per cent, whereas in the years 1978 and 1979 the whole increase in gross industrial output may be attributed to higher productivity. However, it should be noted that the average annual production growth rate in the years 1976-79 fell to about half the value in the years 1971-75 (from 10.5 per cent to 5.8 per cent). Starting from 1979, the production growth rate dropped so low that the rise in national income fell below zero.

The coexistence of these two trends shows that the development of Polish industry rests above all on extensive growth factors. A dramatic drop in the number of workers to be absorbed by industry brought it close to a standstill.

The basing of the growth of industrial output on the rise in the number of jobs was simple, owing to the relative cheapness of labour maintained through administrative decisions of the central authorities managing the economy. As the Polish currency is non-convertible, it is impossible to compare the costs of labour directly. However, the data at our disposal allow us to estimate that the costs of wages in socialised industry in 1979 amounted to 12.5 per cent of total material and non-material costs (the figure remained unchanged from 1975), while social insurance comprised 2.3 per cent of the total in the same year. By comparison, materials and fuels had the highest share in production costs (-64.2 per cent).

1.5 The educational system

The educational system in post-war Poland comprises the four following levels:

- elementary or primary (7-8 years' compulsory schooling);
- basic vocational education (2-3 years) aimed at training skilled workers;
- secondary (4-5 years), divided into general and vocational; and
- university (4-5 years).

Employment in socialised industry by educational level is shown in table 71.

Table 71. Poland: employment in socialised industry by educational level, 1970-79

Year	Total labour force in socialised industry	Workers with given educational level (%)				
		University	Secondary vocational	Secondary general	Basic vocational	Primary
1970	4 044 203	2.6	9.5	3.4	21.7	62.8
1979	4 681 582	3.6	16.0	4.1	30.4	45.9

Source: Rocznik Statystyczny Przemysłu, op.cit., pp. 200-201.

Two characteristics of the Polish educational system are relevant to the problems discussed here. The first is the fact that the enrolment in schools depends on forecasts of the demands of the economy. With the exception of compulsory primary education, the number of places at schools is subject to central planning. After projecting the future demand for skilled personnel in different jobs, decisions are made as to the number of students to be accepted by different schools. Every school is subordinated to a ministry, which sets its enrolment quota.

The second characteristic is the marked prevalence of technical vocational education. While Western European countries are experiencing decreasing proportions of students enrolling at technical universities, the six leading places are taken by the socialist countries (Poland came second in 1971, with 35 per cent of all students enrolled in technical faculties), with Yugoslavia placed eighth.[22] As all these countries are centrally planned economies, one is led to believe that this is a dominant factor. It is also interesting to note that there is no correlation between the prevalent type of education and the level of economic development.

As this situation has prevailed for a considerable period, the Polish economy is saturated with holders of technical university diplomas. Two-thirds of all the university-educated employees in Polish industry are graduates of technical faculties. Among graduates of secondary schools, the proportion is even higher, amounting to 72 per cent. With such an excess of university-trained engineers in the economy, the use made of their qualifications is far from satisfactory. This can be attributed to many factors, the most important of which are as follows:

- the pay and promotion system: the only way for an employee to improve his position and obtain higher pay is by seeking promotion on the ladder of administrative posts;

- the relationship between the number of university-trained engineers and technicians with secondary education: as a consequence of a shortage of technicians, engineers must be employed in posts which could be filled by skilled technicians. As a comparison, British industry employs an average of 3.4 technicians for one engineer (2.6 in Poland), and in the engineering industry an average of 6.4 technicians per engineer (2.6 in Poland);

- underdeveloped batch production in industry: in consequence it is necessary to conduct design and technological work on at least twice as many products as in countries whose industrial potential is comparable with that of Poland;

- inadequate facilities: a shortage of such essential equipment as calculators and photocopying machines forces people to use methods deriving from cottage industry for the preparation of technical blue-prints;

- delayed implementation of economic projects for which engineers have been prepared; this is the case especially in the electronics and chemical industries, which are ostensibly glutted with graduates of those faculties.

1.6 Innovation

The facts above described concerning the educational system are responsible for a situation in which two clear trends can be discerned. The first is a boom in the number of invention designs, utility designs and proposals for technological improvements submitted in socialised economic units. Their number reached 1,681,000 in the years 1973-78, i.e. 50 per cent more than in the years 1967-72.[23] The other trend is the drop in the number of patents obtained abroad for home inventions. A drop of over 20 per cent was recorded for the years 1971-75, and the figure for 1976 was nearly 10 per cent lower than that for the preceding year. This puts poland in one of the lowest places among European countries.[24]

Thus, Polish technical cadres are the source of what is in fact minor technological progress, its characteristic features being -

- a predominance of innovations in the product/design over technological innovations;[25]

- a preponderance of innovations of limited applicability: a mere 1.8 per cent of domestic patents are utilised at two or more plants;[26]

- a majority of minor innovations. Despite the above-mentioned 50 per cent rise in the number of patents, the share of inventions and improvements in national income dropped in the seventies from 16 per cent in 1970 to 10 per cent in 1979;[27]

- predominance of second-hand innovations. According to studies conducted by the Łódź University Institute of Political Economics in the years 1977-79, only a minor proportion of the inventions (under 10 per cent) are new, original contributions, some 30 per cent are innovations close to the leading world standards (albeit based on already known and used technical solutions) and most rely on solutions which are generally known and applied internationally.

This extensive use of qualificiations in the development of industrial technologies seems to fit in with the societal characteristics of industrialisation in Poland, as presented above. All the studies conducted in Poland on innovations in Polish industry have found employees to be very averse to all innovations. This aversion results from the fact that employees are rewarded first of all for the attainment of planned quota assigned by a superior unit, and not for the economic effect of the enterprise's activity. This attitude is equally widespread among manual workers,[28] engineers,[29] managers[30] and all other categories of employees. Naturally, the low innovative character of the management system, does not mean that there is no place at all for innovation. Above all, it means that the management and financing system lacks permanent mechanisms driving towards innovation.

1.7 From national background to automation within an enterprise

We may expect that the course of automation processes in specified enterprises will be determined, to a considerable degree, by more general causalities of technological development in the entire economy. We do not suggest that full technological determinism will occur, or even that technical factors will fully explain the phenomena taking place at the national level and at the level of particular industrial organisations. We expect, however, the national background to provide a set of basic restrictions within which the processes of technological development may progress. The most important of these are as follows:

(1) Many institutions participate in preparing and introducing a new technology (e.g. NC machines) with the institution directly concerned, which is the object of the introduction and not necessarily the initiator. The institutions participating in respective decisions include both economic administrative (headquarters of associations, business-line and functional ministries, foreign trade enterprises, etc.), scientific research (laboratories, design offices, R & D centres, institutes and technical associations), and strictly political bodies (departments of the Central Committee or of local Party committees). In general, the scope and character of participation of these institutions is not clearly specified.

(2) Production enterprises do not follow economic criteria in introducing new technologies. The introduction of a new technology is often determined by non-economic criteria such as prestige, outside pressure, fashion, temporary financial surplus, etc.

(3) Innovations, particularly the automation of production processes, are of domestic origin only to a small extent. They are, in most cases, foreign licences, which considerably limits the role of scientific-technical centres in the process of introducing new technologies. This situation is furthered by the general policy of catching up with the world front ranks, as industrial modernisation is seen in terms of the purchase of licences.

(4) A factor which furthers the introduction of automatic production processes is the relatively high level of workers' qualifications. However, the employment policy, which gives preference to full employment, and thus, with the limited financial possibilities of the State, low labour costs, is operating in the opposite direction.

(5) Although the rule of planning is being followed both in the national economy as a whole and within single enterprises, automation processes are often introduced without comprehensively prepared programmes.

(6) The introduction of automation in an enterprise is not as a rule associated with concomitant changes in existing organisational solutions.

We will show in the case study how some of the phenomena and limitations discussed above affect the introduction of automation at shop-floor level.

2. Case study - The introduction of numerically controlled milling machines

2.1 Design of the study

In choosing the subject of the case study from among several thousand instances of automation in Polish industry, we followed a certain basic criteria, as follows:

- an enterprise was chosen from the sector of means of production, since this is regarded as the main development vehicle in the industrialisation policy;
- a machine industry enterprise was selected because it is here that automation processes have been most widely introduced: moreover, many phenomena characteristic of the overall situation (such as the decision to base the automation on licence purchase) also occur there;
- a medium-sized enterprise was chosen;
- there occurred an unsatisfied demand for the production of this enterprise. Thus, we deal with the producer's market, as typical of the Polish economic situation;
- the enterprise experienced a permanent shortage of staff, particularly of manual workers, which is typical of the majority of enterprises in Poland.

From the point of view of the research scheme, it was also essential that the analysis should cover the process of production automation rather than automation of information processing. No less important is the fact that the enterprise is 100 years old, and that automation was introduced into those surroundings.

We do not claim that the choice of enterprise according to these criteria allows us to test general causalities described above, which refer to the entire industry. The results of the case study can only be of an illustrative and not of an evidential nature. The results may also (though not automatically) serve to modify statements on general causalities.

This illustrative character of the case study made us select a non-scientific methodology, based on informal interviews with people employed in this enterprise: administrators, engineers and manual workers.

Each of these interviews covered a different range of questions, the topic of conversation taking account of the knowledge of the respondent about events taking place in the enterprise in connection with the introduced automation. The starting point was the wish to obtain the fullest possible picture of the introduction of NC milling machines and to determine its effects. The person who indicated the most suitable informants was the chief of the data-processing section of the enterprise, while further interviewees were suggested by successive informants. A dozen or so persons were involved in the talks, which took from half an hour to four hours. The fieldwork took place between February and August 1981.

The choice of a non-experimental methodology means, on the one hand, that we do not present conclusive data that could authenticate more general findings. On the other hand, the data contained in the study do not create an unjustified impression that figures expressing social phenomena can be directly compared with data of other analyses collected in this volume.

Apart from these qualitative data, we also present data obtained from the documents available at the enterprise. The drawback of the latter is that, in most cases, they do not consider NC machine operators as a separate category. Wherever possible, this group was compared with operators of conventional machines.

2.2 The enterprise

Founded in 1881, the factory first made artillery shells and, from 1903, high-quality fire pumps. It suffered large destruction during the First World War. In 1919, after Poland regained independence, the factory was taken over by a new owner, and it was in that year that the production of machine tools was launched. The starting of production was difficult due to the economic crisis at that time. Up until 1930, 2,500 machine tools of 80 different types and variants were manufactured in the enterprise, and were known for their high quality.

In 1936 the factory launched the manufacture of a new family of milling machines based on technical specifications supplied by an American company. Simultaneously, the manufacture of arms (anti-tank guns on Bofors' licence) was started. However, the outbreak of the Second World War disrupted the operation of the factory. Some of the machines and equipment were transported to the east, and the technical records were hidden in a safe place. In November 1939 the Germans took over the management of the factory.

Immediately after the Liberation, the employees began to rebuild the factory (278 machine tools were repaired in 1946 and a year later the first 30 machine tools left the factory). In 1948 the enterprise was taken over by the State. Two years later, the factory's former technical office was transformed into a separate CBKO (central machine design office), whose task was to supply blueprints and technical specifications for the reconstructed machine tool industry. In the years that followed, the factory mastered the know-how and launched the manufacture of new types of milling machines.

The factory's newly acquired command of the NC technique made it possible to launch the manufacture of NC vertical and horizontal machining centres. The latter type of machines are manufactured on construction specifications elaborated by a Japanese company.

2.2.1 Organisation

The factory forms part of the association, which is part of this sector of industry subordinated to the Minister of the Engineering Industry. The design work is done partly in a separate institution accountable to PP, i.e. the PD (central machine design office), which is located in the vicinity of the factory. Foreign trade is dealt with by the Metalexport foreign trade enterprise accountable to the management of PP.

The division of the entire enterprise into sections is similar to that practised throughout Polish industry (figure 31). Immediately responsible to the managing director are two chiefs of the management and quality control departments, as well as several deputy directors of the main branch lines of the enterprise. The departments of interest to us, P 1 and P 2, are subject to the deputy director for production; within the branch line of the same director there are also other production departments and the section dealing with operations and the technical service. At the same time, the branch line of the chief technologist, the chief constructor and the programming section of NC machines are subject to the director for technical matters. Most of the sections connected with new investments come under the deputy director for maintenance of operations and investments, while all those handling employment, welfare and staff matters are covered by the deputy director for economic affairs. It may be assumed that the problems and difficulties involved in co-ordinating the activities of these dispersed sections did not facilitate the implementation of partial automation of the NC machine production.

2.2.2 Production

For many years the enterprise has been producing milling machines only. In the case of conventional machines, this production is primarily of a short-serial character, as opposed to single piece production in the case of NC machines.

At present the factory manufactures -

- three types of tool-room milling machines;
- three types of universal and horizontal milling machines;

Figure 31: Numerically controlled milling machines: a simplified organisation chart

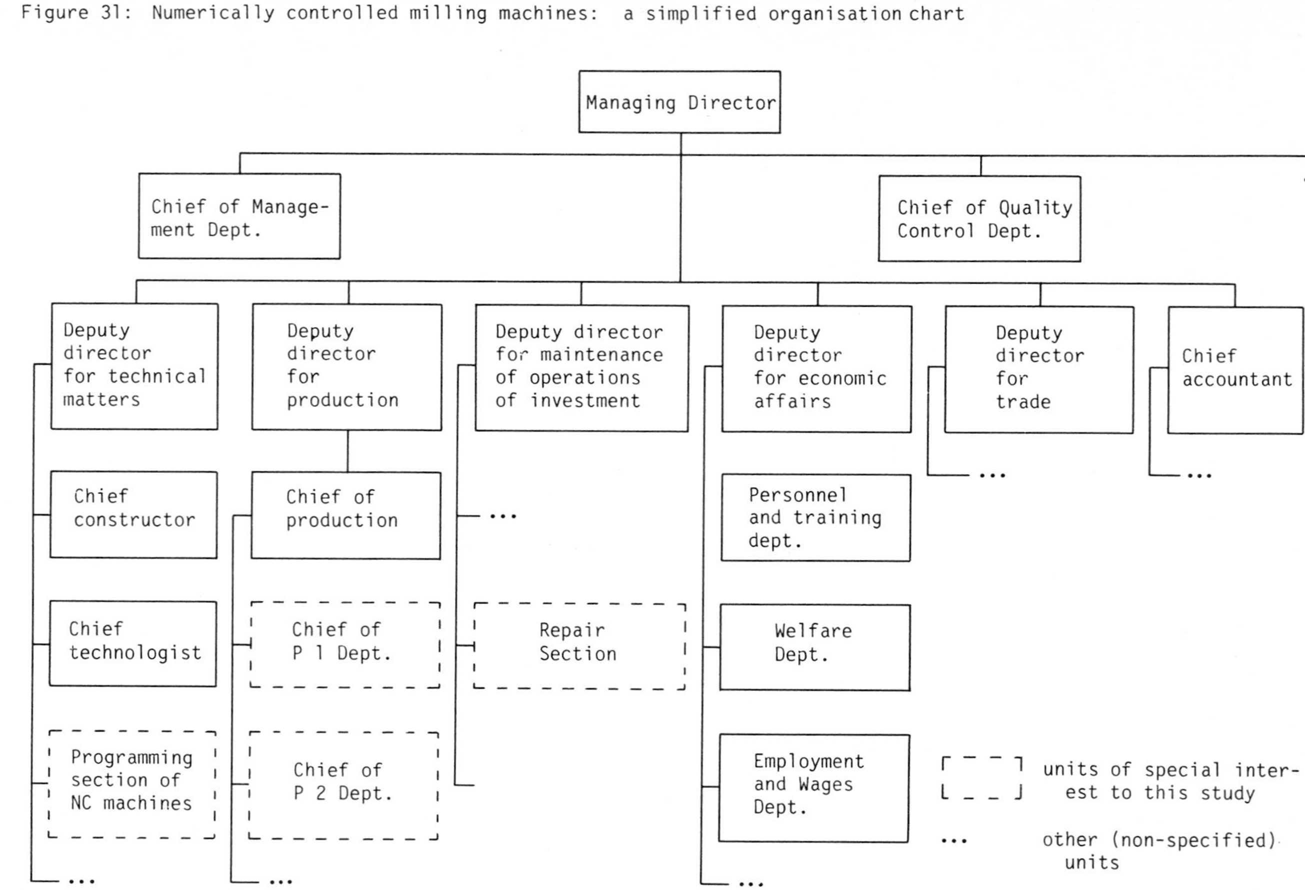

- five types of vertical milling machines (two of them NC);
- three types of NC machining centres;
- five types of transfer machine tools;
- components of machine tools and special equipment for milling machines.

The differences in the quantity of conventional machines produced (table 72) are primarily due to the degree to which the enterprise succeeded in coping with the shortage of raw materials and materials needed for production in a given year. As regards NC machine tools, the uniqueness of each machine produced, even if contained within a particular type, is a significant factor.

Table 72: Numerically controlled milling machines: volume of production at the enterprise, 1977-80

Year	Unit	Conventional machines	NC machines	Total
1977	No.	351	76	427
	Value[1]	343 051	337 228	680 279
1978	No.	389	71	460
	Value[1]	360 791	352 961	713 752
1979	No.	390	26	416
	Value[1]	358 340	284 735	643 075
1980	No.	323	45	368
	Value[1]	315 851	385 381	701 232

[1] Zl'000s, at current prices.

The differences in production value (table 72) are again caused mainly by changing prices for raw materials and semi-products. These changes are effected by the State, and the enterprise has practically no influence on them at all. According to the employees of the enterprise, the price for the milling machine is not connected with its market value, its technical level or its use parameters. However, it is impossible to check the correctness of this viewpoint since the enterprise does not make this type of calculation. Moreover, if the product is sold on the foreign market, the money obtained does not reach the enterprise. Instead, the enterprise receives only a part of the equivalent in convertible currency (i.e. in convertible zlotys), which is not sufficient to make any substantial purchases.

2.2.3 Employees

The workforce of the enterprise slightly exceeds 1,500, about 660 of whom are manual workers. The last few years have witnessed a steady, though not very fast decline in the employment level. The rate of decline is identical for manual workers and white-collar workers, though the reasons are somewhat different. In the former case it is caused by the shortage of manpower on the local labour market, while in the latter it is forced by pressure from higher economic administration.

The sources of manpower are various. The four following methods of recruitment, particularly in the manual category, may be singled out:

- free intake from the employment agency of the local municipal office. This is the most frequent method of recruitment, although it is limited by temporary blockades on engaging new employees. Its disadvantage is that it does not guarantee the employment of skilled workers whose qualifications suit the character of the production;

- employment of graduates of local vocational schools at the elementary and secondary levels. Until 1975 the enterprise ran its own school attached to the factory, which offered vocational training to young people, and after that date graduates of vocational and technical schools located in the same town were employed. Each school has one class under financial sponsorship and vocational guidance of the enterprise, where future workers and employees are trained. About 30 young workers come from this source each year;

- investment in a non-industrialised region. Since there is fairly strong competition in the local labour market, particularly from a large tractor-producing plant, a proportion of employees are required and transported from a small town about 60-70 km away, located in an agricultural area. In that town a production hall was built where many workers are trained each year, some of them then being transferred to the enterprise;

- retraining of white-collar workers with the necessary theoretical background. Recently, this operation was furthered by the fact that a proportion of the blue-collar workers from the enterprise were "exported" temporarily to a company in the Federal Republic of Germany. A retrained white-collar worker, after working for a specified period of time in the same enterprise, is given the chance of going to work abroad, and sometimes even priority. This produces some conflict among the workforce between the old and new workers but, simultaneously, it provides the firm with manpower.

The stability of the workforce at the enterprise is relatively high, and does not differ essentially from the average situation in Polish firms, the number of years of service averaging 16. Table 73 shows labour turnover for employees as a whole and for manual and non-manual workers. It appears that the number of employees resigning their jobs

Table 73. Numerically controlled milling machines: labour turnover in the enterprise

Year		Enterprise total	Non-manual workers	Manual workers
1977	Left	223	49	174
	Taken on	193	37	156
1978	Left	257	62	195
	Taken on	207	44	163
1979	Left	284	68	216
	Taken on	165	38	127
1980	Left	277	62	215
	Taken on	204	43	161

has been growing steadily over the last few years, particularly in the case of manual workers. On the other hand, the number of persons taking up employment is maintained at about the same level, except for 1979, when there occurred an obvious drop in the number of employees, especially manual workers. The effect of these tendencies is a systematic decrease in the employment level at the enterprise, as shown in table 74, which applies mainly to manual workers, production workers included. At the same time, we note that the number of supervisory workers slightly increased. At the time of writing the enterprise was short of 41 manual workers (i.e. over 6 per cent of employees in this category).

Table 74: Numerically controlled milling machines: labour force employed in the enterprise, 1977-80

Year	Total	of which:		
		Supervisory staff	Manual workers	Production workers
1977	1 702	40	713	673
1978	1 642	40	688	648
1979	1 617	42	684	642
1980	1 553	43	660	617

The number of days of absence of enterprise employees has remained for various reasons at a similar level for some years, with a slight downward trend in 1980 (table 75). In this year the number of days of non-certified absence from work, particularly among manual workers, declined significantly.

Table 75. Numerically controlled milling machines: number of days of absence in the enterprise, 1977-80

Year	Type of absence	Enterprise total	Manual workers
1977	Sick leave	31 535	23 625
	Holidays	36 254	25 761
	Non-certified incapacity	494	482
1978	Sick leave	30 250	22 417
	Holidays	36 334	25 783
	Non-certified incapacity	571	567
1979	Sick leave	31 436	24 837
	Holidays	36 283	25 937
	Non-certified incapacity	473	463
1980	Sick leave	27 750	22 247
	Holidays	29 085	25 545
	Non-certified incapacity	179	174

These personnel difficulties may, to some extent, account for the low level of use of the means of production available in the enterprise. On the one hand, one may assume that personnel difficulties provided the basic incentive for production automation. On the other hand, however, these same difficulties do not allow a full, or even adequate use of the expensive NC machines. It is possible that a partial automation of production would allow the achievement of the production level planned for the enterprise; but, as we have pointed out, the existing management system creates essential impediments (disincentives) to exceeding production plans. Moreover, the recent problems concerning energy and raw materials apparently call in question the rationale of production automation, even on a partial basis.

Finally, we briefly present data showing the educational level of the employees (table 76); it should be emphasised that these data have not changed essentially in the past few years. The educational structure of the labour force at the enterprise is rather favourable and slightly exceeds the average level of Polish industrial firms. The fairly high percentage of employees with secondary and higher education is striking; these two categories, together with employees with vocational elementary education, comprise over two-thirds of the total workforce.

Table 76. Numerically controlled milling machines: educational level of employees at the enterprise

Educational level	% of employees
Less than elementary	1.3
Elementary	29.4
Basic vocational	36.5
Secondary (general and vocational)	25.7
Higher and incomplete higher education	6.8

2.3 The process of introducing automation

The basis of the change was the introduction to the enterprise, previously equipped with conventional machine tools only, of NC machines and the launching of the manufacture of NC machines. This process had been delayed in Poland by a boom during the early seventies in conventional machines and by high sales both on the domestic and world market.

The installation in 1978 of the first NC machines was preceded by six-year preparations, which consisted of constructing and producing this type of machine in the enterprise. This was effected on conventional machines, and the first NC machines were produced in 1972, followed by further production in conjunction with a foreign company in 1974. Large batch production was achieved in 1977, and the first NC machines installed in the factory in 1978.

The effect of such a course of preparation is that different groups of workers acquired different experience connected with NC machines. The design engineers and program engineers worked longest on these machines, including both those employed permanently at the enterprise and those who were employed at the PD and who later on assumed work at the enterprise. As the firm ensures a guarantee service to buyers of the machines, groups of workers specialising in maintenance and repair were set up. Both engineers and the maintenance workers, had, in general, worked in the enterprise for many years. As a result of the sequence of events described above, the least acquainted with the new type of machines were the workers, who would have to handle them later on.

According to the employees of the enterprise, the years 1972-78 were not the time of actual preparation for introduction of NC machines in the enterprise, but became such a time after the decision on NC machines was made. This means that the whole process was performed in an unplanned, amorphous way, rather than being the outcome of the development strategy of the firm. No preparatory operations were carried out, such as familiarising the workers with the machines, breaking down psychological barriers, etc.

This sequence of introducing automation reveals the following characteristics:

(1) Programmes for automation were formulated in various institutions: research, administrative and manufacturing (including the enterprise in question). Most often they concerned one aspect or stage of automation. Thus, it is impossible to say whether the procedure for introducing automation had earlier been elaborated in detail at any particular place. The process was more a sequence of developments, with actors in the automation process responding to the situation that presented itself and in this way contributing to the shape of future situations.

(2) It is impossible to compile a complete list of actors in the automation process. The process described above involved the design of its own solutions by the enterprise's design office and PD, a contribution by other divisions of the enterprise (mainly the NC technique section unit), other enterprises manufacturing machine tools, the headquarters of the association, at least two foreign trade enterprises, at least three ministries (the Ministries of the Engineering Industry, Foreign Trade and the Maritime Economy, and Finance), research institutes implementing the government research programme for machine tools, the Planning Commission attached to the Council of Ministers and the Council of Ministers itself, to say nothing of different level party authorities and representatives of potential buyers, who commissioned modifications in the machines manufactured for them, etc.

(3) The effects of operation of each of these institutions was not restricted to the enterprise in question, as the automation process in the machine industry involved a number of different enterprises. No enterprise is self-sufficient as regards designing and production for the foreign, or even the domestic market, but operates under strict supervision by its superior unit.

(4) The cost-effectiveness analysis performed before, in the course of and after the conclusion of successive stages of automation does not reflect the cost borne nor the profit gained. Largely arbitrary, those calculations are rather a bid for more resources that the enterprise may obtain from the association headquarters, the association headquarters from the ministry, and the ministry from the Planning

Commission. If economic considerations have any influence on decisions to launch production, or to purchase machines or licences, they merely hinder (or open prospects for) the obtaining of resources which are administratively allocated. However, the allocation of funds alone means very little, as the superior unit must also grant the enterprise the right to employ extra labour and assign to it the so-called working limits (i.e. a detailed prescription on how to utilise these funds).

During the talks with the heads of departments and design engineers, a number of reasons were given for the introduction of NC machines. First, the increasing shortage of labour was emphasised most frequently. (Since one machining centre replaces six conventional machines and eight operators, the machines were installed.) Another reason was to achieve a higher precision of working which NC machines permit. This was of significance, it was said, in machines destined for export. The next aim to be reached was the modernisation of the enterprise, considered as an autotelic and not as a utilitarian objective.

Lastly, mention was made of the necessity of maintaining and/or winning new markets by, among other things, introduction of new products, the manufacture of which would not be possible using conventional machines. Some of our interlocutors argued, however, that the enterprise could have functioned very well for many years without changing the types of machines produced as, according to them, countries in a moderate stage of development and those of the Third World would need machine tools now produced by the enterprise for a long time to come. They also said that the installation of NC machines was rather the result of acquiring the knack of producing these machines than of any requirements of the market. They also perceived the main reasons for installing the NC machines as being the parallel actions on the part of some of the plant's engineering staff, the headquarters of the association, the ministry and the Planning Commission. These did not encounter any obstacles, since in the seventies an atmosphere furthering production modernisation has prevailed in Poland. It is beyond our scope to assess this argument; we merely present it as a point of view.

Beside these motives suggested by the engineering staff, there was another which recurred in almost all the interviews. This may be summed up by such formulations as: "it _had to be_ introduced", "_the whole world_ is introducing NC machines", or "since these machines are being produced, why shouldn't we use them in our country?". This motive, in spite of its ambiguity, was characteristic of the engineering staff. It was an expression of a mixture of professional ambitions ("we don't want to be worse than others") and of a sense of necessity of technological development, a necessity considered to be an inevitable historical process.

In our interviews with engineers and other workers, we found no trace of financial arguments (such as reducing production costs or an increase in value of products sold) or of an orientation on improvement of working conditions for manual workers, on raising wages, humanisation of work, etc. It was of course difficult to judge how far the statements of our interlocutors were a rationalisation _ex post facto_, and how far they were actual assumptions on the decisions made about the introduction of NC machines in the enterprise. It was also impossible to judge whether those assumptions had been taken into consideration by the policy-making bodies; some of our interlocutors were convinced that this was not necessarily so. According to one interviewed, the introduction of automation in the enterprise resulted from the fact that -

> ... tremendous resources were invested in the machine industry, regardless of whether there was a chance for the expected effect to appear or not ... The whole of the seventies were set on machine industry. The management of the enterprise was interested in strengthening its position through strengthening the firm, and the easiest way to achieve it was to accomplish our purchase of the licence. Therefore, although we had our own, fairly good designs, the licence was purchased. It was not incidental that it was purchased for us, since a well-qualified staff was available here. Owing to this, the purchase is not a disaster, after all.

The share of employees of the enterprise in making decisions on the installation of NC machines, particularly those resulting from the licence purchase, is evaluated by them as insignificant.

2.3.1 Technological and organisational changes

After their introduction in 1978, NC machines were allocated to two departments, light machining (P 1) and heavy machining (P 2). These two departments show many differences as well as similarities.

Each of the departments occupies one hall. However, the way in which the NC machines are arranged differs. In P 1, they were placed among conventional machines and located in the gaps left by conventional machines, now eliminated from production. In P 1, they were placed at the edge of the main production area and separated by a glass wall from the other part of the hall. The glass wall functions as a screen protecting the NC machines against dirt. In the case of P 2, this role is performed by light casing or artificial fibre, which protects only the operating parts of the NC machines.

Another differentiating factor is the assortment of manufactured products. In P 1 are produced smaller parts of machine tools and their equipment. These are generally small items with a weight ranging from a few grams to 50 kilos. In P 2 the manufacture of the main machine tool body is carried out, i.e. large elements sometimes weighing hundreds of kilos. Even considering the size of the products, the time of work is different. However, from the point of view of technology, degree of complexity, precision of performance, etc., the production of both departments is very similar. Both conventional machines and the NC milling centres simultaneously manufacture both types of parts.

Both departments also have a very similar system of internal organisation. In neither do the NC milling machines comprise a separate section guided or supervised by a specially appointed foreman. At the outset of the automation of production in P 1, a post of milling centres' foreman was established and the NC machines were placed in one group. After six months, however, the previous division into three main sections (light, medium and heavy frames) was restored. Most of the NC machines belong to the light frames section. The main reason for giving up this form of organisational separation of the NC machines was the awareness that these machines manufacture the same components as conventional machines do, and that the separation of operators of these machines disturbs the traditional division of production units.

The launching of NC machines did not effect a change in the way of organising the department. The possibility of setting up an NC machine group was initially considered but, having acquainted themselves with the organisation of other enterprises, the managers decided to preserve the existing division into five production units. These are distinguished by type of product (e.g. gear wheels, sleeves, levers, etc). In three units NC machines work alongside conventional machines.

In neither of the departments did the introduction of NC machines essentally change the way that the operators were serviced. The following people come within the range of job contacts of both NC and conventional machine operators:

- the production engineer, who supplies the software and chracteristics of the machine's tooling;

- the planner, who supplies the job card (completed at the end of every month according to the jobs done);

- tool room staff, from whom the operator receives the tools necessary for machining (a standard set of tools is attached to the work-stand);

- department transport, which supplies and removes the items;

- crane operator (in the case of machining centre operators), who lowers the frame on to the machine table, and removes it afterwards;

- repair gangs;

- supervisors, i.e. foreman and department manager.

Although no essential differences were noted between NC and conventional machine operators, some slight changes occurred. One of them is the need to repair NC machines with the help of electronic engineers as well as mechanics. Two gangs deal with the repair of machines in both departments: one from the Chief Mechanic's branch and the other from the Chief Electrician's branch. In the case of conventional machines, it is relatively easy to distinguish mechanical from electrical breakdowns, so that the operation of two separate gangs does not give rise to any major technical problems. However, in the case of NC machines, this situation leads to quarrels between mechanic and electronic engineers and between repair gang workers and machine operators, as the latter's wages depend largely on the machines' efficiency (and thus their ability to work). The factory has not so far managed to set up joint mechanical and electronic gangs.

As a consequence of the introduction of NC machines, a group of programming specialists had to be formed to develop programs for the control equipment of these machines. The group, consisting of specially trained production engineers already employed at the enterprise, also took over the programming of NC machines installed at the enterprise. These programmes were separated not only from the operators themselves but also from the organisational unit directly concerned with the NC machines.

2.4 Job characteristics and automation

As stated before, the introduction of the NC machines was not preceded by any preparation, involving programmes of humanisation of work, QWL or work safety. This absence of advance planning meant a lack both of programmes preceding the installation of the NC machines and of preparatory action. The NC machines were treated by engineers and managers as another exchange of machine stock. As may be seen from the analysis below, this viewpoint is correct with reference rather to P1 than P2.

2.4.1 Job content

The content of the operator's job comprises the following typical sequence of steps:

(a) fix the item for machining;

(b) set the machine, considering actual and projected measurements of the item ;

(c) "load the magazine" (i.e. fix all the tools necessary at successive stages of machining);

(d) start up the NC machine and watch (or watch and steer) the course of the machining;

(e) after the machining is over, perform measurements of deviation from the projected measurements;

(f) if the measurements are not correct, repeat operations (a), (b), (d) and (e);

(g) fix next item for machining.

The most significant difference between the job content of the NC machine operator and the operator of conventional milling machine is that the latter must set the machine manually, whereas the programming device does this work for the former. Otherwise, they key differentiating operation is step (d). In the case of operators of conventional machines, this is the time in which they change the setting of the steering devices of the machine and inspect partially finished goods. In the case of NC machine operators, this stage consists of merely monitoring the operation of the machine.

2.4.2 Monitoring: idle time

P 1 carries out the machining of small items. The time of automatic machining of a single item ranges from 2 to 55 minutes. In practice, it is necessary for the operator to be present all the time to watch the course of the machining, take out the finished pieces and fix the next items for machining. Constant presence and eye control over the whole machining process is largely indispensable due to low-quality tools; cracking or breaking of one of them would entail not only spoiling the item but, more important, a possible breakdown of the NC machine. Furthermore, an operator in P 1 works on two NC machines at the same time.

P 2 conducts the machining of large machine frames. The production time here for a single item may be anything from several to more than 100 hours. The operator watches closely the machining of only the first item in each batch and, when it is ready, measures the tolerance preserved and deviations from the projected dimensions, if any. If deviations occur, the operator introduces corrections to the program. The machining of the following pieces does not require the operators' presence but, despite this, they seldom leave the work-stand.

The basic difference here between NC and conventional machine operating is the time of monitoring (table 77). This idle time, provides the main criterion for evaluating the strenuousness of work on the NC machines and concerns only operators working in P 2, who experience up to 7 1/2 hours' monitoring during a shift. This can be extremely tedious. The centres are closed off from the production room (glass partitions), and operators have a table and chairs close to the work-stand, where they sit reading newspapers, some of them studying notes or textbooks (i.e. those who attend secondary vocational evening schools as, despite the formal educational requirements, not all are graduates of secondary schools). One of the workers, a machining centre operator who does community work for a district residents' organisation, was preparing material for the forthcoming meeting.

The advantages of the "free" time (the feeling of greater autonomy and the possibility of undertaking private tasks during work time) are not equalled by the degree of strenuousness experienced. Operators leave their workplace extremely rarely. The awareness of the high cost of the machine (and the high cost of breakdown seems to oppress them. Thus, if they happen to leave the workstand (in the course of automatic machining), they do it with a sense of guilt and fear of a breakdown, which might ruin not only the machine item but the machine itself.

Table 77. Numerically controlled milling machines: estimation of uninterrupted monitoring time during a shift for NC and conventional machine operators

Machines	Department	
	P 1	P 2
Conventional	2-20 min	0.5-1.0 hr
Numerically controlled	1-20 min	2.0-7.5 hr

2.4.3 Autonomy

In general, both operators and managers believe that the best operators of conventional machines have the highest level of autonomy, because they are the best-qualified specialists in their profession. Superiors do not interfere with the way in which they plan and perform their work. Products manufactured by the machines which they operate are only subjected to routine quality control performed by employees of the Quality Control Department.

Other workers, i.e. both operators of the NC milling machines and other operators of conventional machines, admit that their autonomy level is similar and also fairly high. The two groups differ rather in the content than in the level of autonomy. When operating conventional machine tools, NC machine operators had a sense of exclusive influence on the quality of the machined item, dependent upon their professional knowledge and craft skill. Although in the operation of NC machines this element of manual skill is no longer important, they still have a sense of influence on the conduct of the work through the setting of the sophisticated machines and the possibility of introducing corrections to the program. These operators are conscious that they have acquired "a higher level of initiation" and, although most of them were reluctant to leave conventional machine tools and start work on NC machines (fearing the complexity of the job or even being sceptical of the possibility of automatic machining), today none would like to go back to work on conventional machine tools.

The degree of isolation from the social environment can be regarded as one of the measures of work autonomy. Operators of the NC machining centres in P 2 stand out markedly in their feeling of isolation. The physical isolation (distance, glass partitions) is considered by them to be an essential disadvantage.

2.4.4 Job satisfaction

Here we consider three elements of job satisfaction, the first of which is satisfaction with activities performed. Operators of the NC machines have the feeling of their importance in view of the fact that they are few, and working with very costly equipment. This does not, however, give them a feeling of exclusiveness, and they are not perceived as an elite (in any sense of the word) by their colleagues.

This feeling of importance is directly linked with another measure of job satisfaction, namely, the level of wage satisfaction. The NC machine operators, like the operators of conventional machines, are not satisfied with their wages. This dissatisfaction seems to arise from exterior reasons (mainly the high inflation rate, reaching 15 per cent according to official data), though it is intensified by their comparing the wages they receive not with those received so far, but with the wages of the best operators of conventional machines.

In some cases a decrease of wages was experienced by operators changing to the NC machines. The operators concerned stated, however, that the fact of their current job being physically less strenuous is a compensation to them.

Our third measure of job satisfaction is the satisfaction with physical working conditions. In spite of the many changes described above, the NC machine operators were unaware of them. In fact, the only consideration emphasised was the lack of air-conditioning in the partition-separated space where the machining centres in P 2 are situated.

2.4.5 Stress

The only evident stress symptoms were found among the NC machine operators in P 2. The reason for the stress was the awareness of the high cost of the machine services and of the machine products. No stress was found among the NC machine operators in P 1. Potential factors eliminating stress, such as lower risk of work accidents, were not identified by the operators.

2.4.6 Social environment

As we have mentioned before, the NC machine operators do not constitutute a separate organisational unit or section, nor do they form a group in the social sense. They are not treated as a group, and they do not consider themselves to be one. The level of conflict among workers as regards performance of the job has not changed essentially, and remains very low.

The introduction of NC machines diminished the workload on conventional machines. This, in turn, reduced the wages of operators doing piece-work on conventional machines. Furthermore, shortages of materials have not been conducive to growth of production and full utilisation of the machines. As two more machining centres are being mounted, the conflict will surely be aggravated. It is additionally compounded by the fact that some of the good piece-work jobs are shifted to the machining centres, whose cost down-time is several times higher than that of conventional machines. Moreoever, the standard machining centre work rates in force are excessive, with the result that they must be in continuous operation in order to avoid losses. This source of conflict was indicated by management but not by the workers.

The only area indicated by the operators, in which their relations with the social environment have changed, was relations with repair gangs. So far, they had been dealing with mechanics only, whereas now they had to co-operate with electronic engineers as well. Between these two groups there is a permanent tendency to prove to each other that the breakdown occurred in their field. These conflicts prolong the repair time, which means a financial loss to the operator.

2.5 Correlates of production process automation

2.5.1 Personnel correlates

The two departments differ from each other both in employment size and in the number of NC machine operators working there. In both cases, however, the NC machine operators comprise a small percentage of all employees, about 4 per cent of manual workers and about 5 per cent of production workers. Table 78 gives employment figures for P 1 and P 2.

In P 1, 15 NC machines are operated by 12 workers (eight on the first shift and four on the second). In P 2, six machining centres are in operation, four of them Japanese and two Polish. These are operated by ten workers (six on the first shift and four on the second). Next to them are conventional machines in both departments.

Departments do not carry out recruitment of workers; the central administration of the enterprise performs this function. Meanwhile, the chief of each department is responsible for the best use of the workers: their allocation, work assignment, making use of their skills and qualifications, and action to encourage the worker to stay with the enterprise.

In view of the small scope and gradual method of introducing automation, the selection of operators for work on the NC machines is carried out individually. No qualification tests or other standard methods are used, but the redeployment of the worker on the NC machines is effected after talks held by the department manager and foreman on the one hand, and the operator concerned on the other.

An important factor in the selection process for work on NC machines is the wage system. Broadly speaking, the latter accounts for the fact that the best operators of conventional machines do not want to change over to work on NC machines for fear of financial loss. Thus, the task of the managerial staff is to prevent the poorest workers from operating the NC machines, and this is the purpose of the above-mentioned selection criteria.

The formal training requirements for NC machine operators are vocational secondary education (eight years of primary school and five years of secondary school) or basic vocational education (eight years of primary school and three years of basic vocational school), followed by a three-month specialist training for a given NC production job in each case. A candidate must have worked directly in production for at least three years. In practice, NC machines are operated by workers with vocational technical secondary education, while conventional machines are operated, as a rule, by workers with basic vocational or elementary education. Workers with vocational technical secondary education operating conventional machines are a definite minority (20-25 per cent).

In general, it is workers with relatively long years of service in the enterprise who become operators on NC machines. The average length of service in the firm of the operators of NC machines amounts to 19.4 years in P 1 and 17.3 years in P 2. It is worth noting that the number of years of service of workers operating conventional machines is also fairly high: 21 years in P 1 department and 24 years in P 2. Generally speaking, the workforce of both departments is extremely stable. On the other hand, it is in these two departments that the most acute shortage of workers occurs: P 1 was short of 21 workers at the time of the study, and P 2 of 15 workers.

NC milling centre operators received special training to prepare them for the operation of these machines. Two of the operators participated in a two-week training course at the enterprise of the Japanese manufacturer. The other five operators were trained by the technical service of the manufacturer during the installation of the milling centres. The remaining operators were trained by technical specialists at the enterprise or by those operators already trained.

2.5.2 Financial correlates

The time of work on every single item is prescribed by the production engineer. The worker is paid for the hours worked on items accepted by quality inspection (those rejected by quality inspection are not paid for). In this case, the department commission decides whether the item was spoiled through the worker's fault or through other causes (such as latent material flaws). If the spoilage is a consequence of the worker's negligence, he is financially liable for it.

The setting of a standard time of work on items produced is far more important for conventional machine operators, as these machines are used to make simpler items which require less time. Thus, the workers scope for discretionary timing is much smaller.

Table 78: Numerically controlled milling machines: employees in P 1 and P 2 departments, 1977-80

Year	P 1 (light machining)				P 2 (heavy machining)			
	Total no. of workers	Supervisory staff	Manual workers	NC machine operators	Total no. of workers	Supervisory staff	Manual workers	NC machine operators
1977	250	8	225	-	163	7	150	-
1978	239	8	216	10	166	7	153	2
1979	245	9	220	10	161	7	148	6
1980	241	9	217	12	141	7	130	10

Table 79: Numerically controlled milling machines: average monthly premiums paid to workers, 1977-80 (percentages)

Year	Enterprise as a whole	P 1 (light machining)		P 2 (heavy machining)	
		Conventional machines	NC machines	Conventional machines	NC machines
1977	20.5	18.0	-	26.0	-
1978	20.0	18.0	18.0	26.3	44.0
1979	20.0	19.8	20.0	27.3	44.0
1980	20.5	29.5	30.5	28.1	46.1

- = not applicable

The operators of conventional machine tools are paid only for the volume of work done, and wages are directly proportional to the amount accomplished. The only restrictions on quota performance are due to fears that if quotas are exceeded by a large proportion of employees, then they may be raised.

In the case of NC machine operators, there is a maximum and minimum limit on the volume of work done; every instance of transgression of those limits entails a reduction in premium, which amounts to 30 per cent of the basic wage in P 1 and 50 per cent in P 2. This method of paying NC machine operators is intended to provide for an optimum utilisation of the machines (lower limit) and to prevent a wasteful exploitation of the machines and a preference for quantity over quality of work (upper limit).

The pay of NC machine operators in P 1 has two components: the fixed component and a flexible component (the premium). The fixed component depends mainly on qualifications and work seniority. The flexible component depends on how much of the projected plan is attained. Workers who attain 100 per cent of the plan receive a premium amounting to 30 per cent of the fixed component. If they achieve less than 100 per cent, the premium is reduced by 1 per cent for every 1 per cent below the plan, and if they attain less than 80 per cent of the plan, they receive no premium at all. When workers exceed the plan, they receive 0.5 per cent premium for every extra 1 per cent, but if they exceed it by more than 10 per cent, no premium is paid. Special pay regulations apply in the case of a machine down-time which is not the fault of the worker.

Very good conventional machine operators take home the maximum of Zl 9-10,000 monthly, and NC machine operators around Zl 7,000. For this reason, the best operators of conventional machines do not choose to move to NC machines, which attract average workers.

In P 2, as in P 1, two pay systems are in force for machine operators: piece-work on conventional machines and day-work on NC machines. Apart from the fixed pay component, NC machine operators receive a maximum 50 per cent premium if they perform the assigned monthly quota. The final amount of the premium depends on the attainment of this plan. At the initial stage of operation of NC machines, when there were frequent breakdowns and down-times, the production chief and other members of enterprise's general management were anxious to introduce a piece-work pay system in order to get the workers interested in the machine's operation. However, these pressures were effectively curbed by department managers.

Data on the average monthly premium paid to employees in departments P 1 and P 2 against the background of that received by employees of the enterprise as a whole are presented in table 79.

2.5.3 The QWL effects of automation

The above-described features of the process of automation of production characterise its course and shape, primarily from the angle of the enterprise. These changes, however, are not insignificant for the

NC machine operators. According to former operators of conventional machines who become NC machine operators, the factors discussed had certain effects as summarised in table 80.

2.6 Conclusions[31]

(1) In order to arrive at the right conclusions concerning the case study, one should take into consideration two factors which are linked to each other. The first is that the change to NC machines was not included in the innovation plan of the industry and should therefore be considered as non-programmed, an exception rather than a rule in the Polish economy. The second is that the innovation was a marginal one, which did not make a significant impact on the overall production system of enterprises.

(2) As the change had not been foreseen by the plan of the enterprise, it had to be introduced in such a way that disturbance to the production system was minimised. This is why management and designers chose the strategy of minimum risk, a logical step as the main criterion for the evolution of their work is fulfilment of the production plan. In order to minimise the risk, efforts were made to reduce the amount and extent of change. The choice of tested and proven foreign-made technology also helped to reduce the risk.

(3) The accepted strategy was based on the method of trial and error. Care was taken to avoid any steps which would result in disturbances to the total production system, work organisation and management.

(4) As a result of the chosen strategy, the innovation process took the shape of a gradual introduction of changes and their "natural" slow adaptation over a relatively long period of time. The slow pace of change helped to secure gradual acceptance by workers and managers of new roles and procedures, and to reduce potential conflicts.

Table 80: Numerically controlled milling machines: changes in quality by working life experienced by NC operators.

Criterion of change	Department	
	P 1	P 2
Work shifts	No change	No change
Cleanliness of the work-stand	Slight improvement	Significant improvement
Isolation of the work-stand	Slightly more	Significantly more
Work-pace	No change	Significant decrease
Noise, vibration	No change	Significant improvement
Risk of accident	Improvement	Improvement
Physical effort	Slight decrease	Significant decrease

Notes

[1] See for example, A. Karpinski: Zagadnienia socjalistycznej industrializacji Polski [Issues in the socialist industrialisation of Poland] (Warsaw, PWN, 1958); W Markiewicz: Spoleczne procesy uprzemyslowienia [Social processes of industrialisation] (Poznań, Wydawnictwo Poznańskie, 1962); Jan Szczepański: Zmiany spoleczeństwa polskiego w procesie uprzemyslowienia [Changes in Polish society in the process of industrialisation] (Warsaw, IW CRZZ, 1973).

[2] See, for example, J. Turowski: Przemiany wsi pod wplywem zakladu przemyslowego: Studium rejonu milejów [Changes in the countryside affected by industrial organisation: The case of Milejów region] (Warsaw, PAN, 1965); Dyzma Galaj: "Zmiany spoleczne na wsi w Polse Ludowej" [Societal changes in the countryside in People's Poland], in A. Sarapata (ed.): Przemiany spoleczne w Polsce Ludowej [Social changes in People's Poland] (Warsaw, PWN, 1965); Socjologiczne problemy industrializacji w Polsce Ludowej [Sociological problems of industrialisation in People's Poland] (Warsaw, PWN, 1967); Sarapata, op. cit.

[3] S. Widerszpil: Skład polskiej klasy robotniczej: Tendencje zmian w okresie industrializacji [The structure of the working class: Changes in the course of industrialisation] (Warsaw, PWN, 1965).

[4] Antonina Kloskowska: Kultura masowa [Mass culture] (Warsaw, PWN, 1964); Kazimierz Zygulski: Drogi rozwoju kultury masowej [Ways of development of mass culture] (Warsaw, PWN, 1970).

[5] Galaj, op. cit.

[6] S. Ziółkowski: Urbanizacja, miasto, osiedle: Studia socjologiczne [Urbanisation, town, settlement: Sociological studies] (Warsaw, Wiedza Powszechna, 1967).

[7] W. Makarczyk: Czynniki stabilizacji w zawodzie rolnika i motywy migracji do miast [Factors affecting the stability of farming occupations and motives underlying rural-urban migration] (Warsaw, Ossolineum, 1964); Markiewicz, op. cit.; W. Mrozek: Rodzina górnicza [Coal-miners' families] (Katowice, Wydawnictwo Slask, 1965).

[8] Salomea Kowalewska: "Wzór osobowy i pozadane postawy pracowników w Polsce Ludowej" [Personal patterns and required attitudes in People's Poland], in J. Szczepański (ed.): Przemysł i spoteczeństwo w Polsce Ludowej [Industry and Society in People's Poland] (Warsaw, PWN, 1969); Adam Sarapata: Płynność i stabilność kadr [Fluctuation and stability of cadres] (Warsaw, PWE, 1967); Szczepański: "Wzór osobowy ... ", op. cit.; Aleksandra Jasinska and Renata Siemieńska: Wzory osobowe socjalizmu [Personal patterns of socialism] (Warsaw, Wiedza Powszechna, 1975);

Ł. Sajkiewicz: Zarys przemian w strukturze organizacyjnej i funkcjach organizacyjnych przemyslu PRL w latach 1944-48 [An outline of changes in the organisational structure and function of industry in the People's Republic of Poland, 1944-48] (Warsaw, 1960); Franciszek Krzykala: "Proces Kszłaltowania sie struktury i form organizacyjnych w toku butowy huty aluminium" [Process of structure crystallisation and form of organisation in the course of establishing an aluminium mill], in Zeszyty Badań Rejońw Uprzemyslawianych (Warsaw), No. 4, 1963.

[9] Szczepański: Zmiany społeczénstwa ...; op. cit.; Witold Morawski: "Społeczenstwo a narzucona industrializacja" [Society and imposed industrialisation], in Kultura, 7 Dec. 1980.

[10] Jolanta Kulpińska: "Socjologiczne: psychologiczne aspekty automatyzacji" [Sociological and psychological aspects of automation], in Spoleczne skutki postepu technicznego (Warsaw, IW CRZZ, 1969); Jan K. Solarz: "Spoleczne problemy automatyzacji w zakladzie przemyslowym" [Societal problems of automation in an enterprise], in B. Gałeski (ed.): Zmiany spoleczne i postep techniczny [Societal changes and technological progress] (Wrocław, Ossolineum, 1971); Andrzef Straszak: "Współczesne Kierunki automatyzacji i jej skutki dla teorii i praktyki zarzadzania" [Contemporary directions of automation and its effects on the theory and practice of management], in Spoleczne skutki postepu technicznego, op. cit.

[11] Szczepański: Zmiany społeczeństwa ... , op. cit., pp. 13-18.

[12] ibid., p. 13.

[13] Charles Kerr et al.: Industrialism and the industrial man: The problems of labour and management in economic growth (Cambridge, Mass., Harvard University Press, 1960).

[14] Szczepański: Zmiany społeczeństwa ... , op. cit., p. 14.

[15] Kerr et al., op. cit.

[16] Szczepański: Zmiany społeczeństwa ... , op. cit.,pp. 20-21.

[17] ibid., pp. 21-22.

[18] ibid., pp. 12-13.

[19] Krzysztof Mreła, et al.: Organisational structure and effectiveness in their context: Concepts and findings (Warsaw, IFIS PAN, 1980).

[20] Krzysztof Mreła and Jadwiga Staniszkis: "Degree of coherence and mechanisms of transformation of organisational structures", in Polish Sociological Bulletin, No. 2, 1977, pp. 49-62; Mreła et al., op. cit.

[21] Mrela et al.: Organisational structure ...; op. cit.

[22] See GUS: Mały Rocznik Statystyczny [Short Statistical Yearbook] (Warsaw 1975), p. 35.

[23] Przeglad Techniczny-Innowacje (PT-I), 1979, No. 32.

[24] ibid., 1980, No. 18.

[25] ibid., 1980, No. 15.

[26] ibid., 1980, No. 19.

[27] According to Lesłw Wasilewski, General Secretary of NOT Chief Technical Organisation: "For years, the type of construction innovation has become rooted in Poland. The community of Polish engineers has not acquired the features necessary for technological innovation ... the designing of an item of one's choice serving discretional, theoretically possible functions is a totally different thing from designing and constructing technological lines ... This was felt particularly in the 1970s, when we witnessed an import of technologies ..." (PT-I, 1980, No. 12).

[28] J. Tulski: Postep techniczny a wynalazczość pracownicza [Technological progress and labour invention] (Warsaw, KiW, 1973).

[29] Jan Hoser: Zawód i praca inzyniera: Kadra inzynierska w swietle ankietowych badań socjologicznych [Occupation and tasks of the engineer: Engineering cadres in the light of a sociological questionnaire (Wroclaw, Ossolineum, 1970).

[30] Marian J. Kostecki: Kadra kierownicza w przemyśle: Analiza socjologiczna [Managerial cadres in industry: A sociological analysis] (Warsaw, PWN, 1979).

[31] The following are references of general interest to this chapter: A. Achijezer: "Rewolucya nauczno- tieczniczeskaja i rukowoditelstwo razwitiem obszczestwa" Scientific-technological revolution and the management of social development , in Woprosy Filisofii, No. 8, 1968; Marian J. Kostecki and Krzysztof Mreła: "Comparing comparative studies on organisations: Patterns of incompatibilities", in Polish Sociological Bulletin No. 2, 1980, pp. 57-68; idem: "Incompatible studies on comparable organisations", in Barriers and perspectives: Studies in the Sociology of organisations (Warsaw, IFiS PAN, 1981), pp. 13-96; idem: "Revolt of the incapacitated: Organisational causes and consequences of the Polish Summer 1980", in D. Dunkreley and G. Salaman (eds.): The International Yearbook of Organisation Studies 1981 (London, Routledge and Kegan Paul, 1982), pp. 105-125; idem: "The incompatibility of Comparisons: Comparative studies on organisation and the cumulation of scientific knowledge", in Organization Studies, 1983, No. 4, pp. 73-88;

Włodzimierz Pańków: "Polskie lato 1980: Kryzys systemo władzy. Przyczny, przejawy, perspektywy" [The roots of the Polish Summer; a crisis of the system of power], in Sisyphus sociological studies, 1982/3, pp. 33-48; I.E. Rubcov: "Rol naucznogo i inzinierno-tiechniczeskogo truda w razvitii proizvodstwa," in Socyalno-ekonomiczeskije problemy techniczeskogo progressa w SSRR [Socio-economic problems of technological progress in the Soviet Union] (Moscow, Izdat Wysszaja Szkola, 1966); Andrzej Rychard: "Chleb i demokracja" [Bread and democracy], in PT-I, 1981, No. 5; Zmiany w warunkach i trésci pracy w zwiazku z postepem naukowo-technicznym [Changes in conditions and job content as affected by scientific-technological progress] (Warsaw, IPiSS, 1978); A.A. Zvorykin: Nauka, proizvodstwo, trud [Science, production, work], (Moscow, Izd. Moscow, 1965).

CHAPTER XIV

AUTOMATION IN USSR INDUSTRY

1. National background

1.1 Definitions and principles concerning automation

The introduction of comprehensive mechanisation and automation is considered in the USSR as a means directed at attaining two major goals: improving the efficiency of production and reducing the amount of manual, unskilled and arduous work.

Nowadays automated production is understood in terms of management: "information", "communication", "regulation", "control", etc. In other words, automation means machine control where the objects of management are also machines and processes. Automatic control systems exert influence on workers and intermediate mechanisms. The more advanced form of automation implies the principle of feedback, which considerably extends the area and kinds of its application for the needs of management of machine and complex worker-machine systems. Besides, the fact that information is not only the object of labour but also the product is becoming an ever more important component of the functioning of such systems. Apart from this, it results in a considerable decrease in the direct involvement of people in production process control.

Automation development leads to the formation of a whole complex of machines. In Soviet literature they are usually subdivided into the following three groups:

- automatic machines designed to influence the object of labour, technical control, handling of objects, etc.;

- cybernetic control equipment that models and optimises production and controls the machines in the assigned mode;

- servomechanisms and engineering that implement feedback, i.e. provide the automatic control system with information on process parameters and send the correcting impulses to the working units.

The following classification of automated production (by degree of automation) is current among Soviet experts:

- complete automation, i.e. automatic performance of the larger part of control of major parameters in technological processes and equipment, as well as of all operations;

- non-complete automation, i.e. automatic performance of technological process operations only;

- partial automation, i.e. some elements of the automated operations can be performed manually.

The degree of automation considerably influences the content of operators' labour.

The Research Institute on Labour under the USSR State Committee on Labour and Social Problems singles out the following six general directions of automation with respect to its technological improvement.[1]

- the growing technological complexity of automation, and the increased number of technological operations performed in the automatic cycle. These replace hard physical and machine-manual operations in inter-operational transportation and loading-unloading jobs, and reduce the need for unskilled and low-skilled labour;

- extended functions performed by automatic equipment without direct involvement of an operator. Creation of an active control system with automatic setting and adjustment of equipment considerably reduces the amount of labour required for equipment servicing. Automatic sorting and inspection machines eliminate monotonous manual and visual control and, thus, excessive visual strain;

- improvement of the interrelationship of the automatic production elements (i.e. more flexible communication between the machines and the appropriate parts of automatic lines) provides for more efficient organisation of adjustment and maintenance work;

- creation of easily adjustable production lines. In cases of batch and high-run production of similar products and parts, quick re-adjustment of equipment requires half as much effort by the adjuster;

- adaptation of machines to serve people. Alleviation of labour as a whole and reduction of nervous strain in particular are achieved by built-in automatic quality control and level of adjustment of the equipment, introduction of information boards indicating performance of the equipment and points of frequent failures on the line, the ergonomic placing and layout of control panels, etc.;

- computer application for automated production control, i.e. application of computers together with their peripheral devices to indicate the performance of the equipment and to supply information to adjusters, operators, mechanics and power engineers, as well as to shop and factory management.

1.2 The extent of automation in the USSR

In accordance with the 11th Five-Year Plan,[2] the productivity of social labour must be increased by 17-20 per cent. Labour productivity increases are to account for no less than 89-90 per cent of the growth in national income. The existing shortage of labour makes it extremely important to increase production at existing enterprises with the same or fewer personnel.

It is planned to speed up development and the introduction of new technology. In the current five-year period, the rate of machinery renewal is to be increased by about 50 per cent. Enterprises are not allowed to manufacture products of an obsolete design whose use yields no tangible gain in labour productivity. Scientific research is to be made more effective by substantially reducing the time needed to introduce innovations and by strengthening the links of basic and applied research with production.

The 26th Congress of the Communist Party of the Soviet Union stressed the need -

> ... to develop the production and ensure extensive use of automatic manipulators (industrial robots), built-in automatic control systems with the use of microprocessors and microcomputers to establish automated shops and plants; to accelerate the introduction of automated methods and means of quality control and of testing products as an integral part of production processes.[3]

This development of automation may be illustrated by the following figures: in 1979, 24,700 automatic production lines were introduced as compared with 10,900 in 1978 and 6,000 in 1965. Special attention is paid to the production of the most advanced types of automatic equipment with programmed control. In the middle of 1979, the output of such equipment amounted to 59,600 units. In 1960, 16 units of metal-cutting machine tools with programmed control were produced; in 1965, 49; in 1970, 1,588; in 1975, 5,545; and in 1979, 7,937 units.[4]

In the current five-year period, it is planned to expand considerably the production of control mini-computers as components of basic technological equipment, instruments, and various control and monitoring systems and facilities. Unit power of assemblies and installations is to be increased, their working life extended and their reliability and efficiency enhanced. The current trend, in accordance with which the rate of automation exceeds the rate of introduction of mechanised production lines, will be continued.

The traditional directions of mechanisation and automation do not always result in gradual elimination of manual labour. Though the larger part of the work is done by the machine, the worker often turns out to be a mere supplement carrying out monotonous auxiliary operations. Industrial robots are capable of taking over from a worker the bulk of this work. For the last decade an increasing output of robots has become one of the most important trends in USSR industry.

Since 1974, when the first Act on industrial robot production was adopted, more than 100 types of robots have been designed and full-scale production of more than 30 types of manipulators has been started. In 1980, the Central Committee of the Communist Party of the Soviet Union again considered measures for increased production and wide application of automatic manipulators.

By the end of the past five-year plan period, the park of industrial robots exceeded 6,000, and the number is still growing. This has led to important socio-economic consequences. It has been found that a robot is capable of replacing two or three operators and its introduction results in fourfold increase in labour productivity of two to four times. This is coupled with improvements in working conditions and production quality.

A new robotised assembly line at the Petrodvortsovsky watch-making factory with a yearly production capacity of 4 million watches had demonstrated the possibilities for considerable increase in output and for radical improvement of the job content. The change affected 500 women operators previously engaged in manual assembly operations on a conveyor line, with forced work pace. Similar results have been achieved by enterprises in Moscow, Kiev, Vladimir, Tula, Krasnoyarsk and other cities.

As in other parts of the world, the number of people with managerial duties has rapidly grown in the USSR during the last decade. This situation has aroused concern: some of the consequences even seemed alarming. Only a few years ago there were warning voices that "with the current performance of engineers, technicians and office personnel engaged in planning, management and accounting, it would have been necessary to involve in these spheres by 1980 the entire Soviet adult population".[5]

The first Soviet electronic computer was created in 1951. In 1953, the large electronic calculating machine (BASM) appeared, performing 10,000 operations per second. Then a number of non-transistor computers were produced that were utilised primarily for research. A new generation of transistor computers appeared in the second half of the fifties, with a capacity of 100,000 operations per second. The third generation of computers appeared in the late sixties. The seventies saw the appearance of micro processors, treated nowadays as the heralds of the future.

In the late sixties the concept of computer-based management systems (CBMS) was introduced and began to spread widely. This concept may be understood as a worker-machine management system designed to manage production which cannot be fully programmed or formalised due to uncertainties in key parameters. The system is therefore called "computer-based", not "computerised". The integrated CBMS links together various interacting subsystems, including control of organisational as well as of technological processes. Table 81 shows the rate of CBMS in the USSR.

1.3 Institutions concerned with introduction of automation

The USSR has three main types of scientific establishments: academic, industrial and educational. These are interconnected, while each still retains its specific angle and status; academic science is aimed at fundamental studies (i.e. definition of the most promising lines in the technological progress) and creation of a theoretical and methodological basis for industrial science which is, in its turn,

Table 81: USSR: introduction of computer-based management systems (CBMS), 1966-79

Item	1966-70		1971-75		1976-79	
	Number	Average per year	Number	Average per year	Number	Average per year
Total CBMS	414	83	2 309	462	1 677	429
CBMS of an enterprise	151	30	838	168	297	74
CBMS of technological processes	170	34	564	113	911	228
CBMS of regional entities	61	12	631	126	316	79
CBMS of ministries and agencies	19	4	168	33	61	15
Computer-based information system (CBIM)	13	3	108	22	92	23

Source: The USSR national economy in 1979, Statistical yearbook (Moscow, Gosstandizdat, 1980), pp. 107-114.

oriented towards applied research to satisfy the needs of developers and designers of new technology; higher learning institutions carry out both fundamental and applied research.

The academic system includes the USSR Academy of Sciences with its Siberian branch and its affiliations in the Far East, the Urals, etc., and 14 academies of the Soviet Republics. Research centres have been established in a number of regions. The scientific personnel of the USSR Academy of Sciences totals 46,000 among them 753 full academicians and corresponding members of the Academy. In fundamental science one can distinguish 2,000 directions of research. The allocations for fundamental studies in the USSR amounted to 13-15 per cent of total allocations for science. Among the academic institutions handling problems of automation are the Institute for Control Problems, Institute for Machine Studies, etc.

Applied research is conducted by numerous industrial and educational institutions. The USSR allocates 40-60 per cent of the total science budget to applied research; 80-90 per cent of the results can have direct practical application. Research institutions under the Ministry of Electrical Engineering, Ministry of Instrument Engineering, Automation Technique and Control Systems and others are the leaders in this field.

A wide network of scientific institutions handle social problems of automation and computerisation within the USSR Academy of Sciences, including the Institutes for International Labour Movement Studies, Sociological Studies and Economics Studies. About 800 scientific institutions study labour protection problems. Among them are seven institutions in the trade union system, 16 within the Academy of Medical Sciences, and Ministries of Health Care in all the Soviet Republics.

There is a large group of organisations engaged in applied research on automation of production. These include technological and design institutions and bureaux, as well as laboratories. Their output includes development of technological and organisational aspects of new products or improvement of existing ones, involving the preparation of drawings, projects, charts, standards and prototypes.

Two public organisations should be specially mentioned: the All-Union Council of Scientific and Technical Societies and the All-Union Society of Inventors, which co-ordinate the efforts of professional and non-professionals engaged in automation and computerisation. Both are sponsored by the trade unions.

A new hybrid form of organisation, the scientific and production amalgamation (SPA), has been widely established in the USSR. SPAs handle the scientific, design and production aspects of new product development. They are not so much aimed at full-scale production as at pilot production (prototypes), leaving development of mass production to plants.

The USSR pays much attention to the introduction of innovations, especially in automation. This is encouraged through a complex of organisational and economic measures and techniques with, first of all, centralised planning for technological re-equipment of production.

1.4 Automation and job content

As was mentioned before, in the USSR introduction of comprehensive mechanisation and automation is considered an important means of improving work content, upgrading working conditions and reducing the amount of manual and arduous work. This task is spelt out in Article 21 of the USSR Constitution -

> The State concerns itself with improving working conditions, safety and labour protection and the scientific organisation of work, and with reducing and ultimately eliminating all arduous physical labour through comprehensive mechanisation and automation of production processes in all branches of the economy.[6]

It is planned, for example, that in the course of the current five-year period, the number of people involved in materials handling will be reduced by 1.5 million to 2 million due to the introduction of mechanisation and automation.[7]

In accordance with the study of Academician Zvorikyn,[8] growth of the following occupations follows the speeding up of automation:

- workers servicing new technological processes. This group of workers is growing, but relatively slowly, because the technical level of technological processes, especially in large enterprises, is rather high and the need for labour is not great;

- workers servicing mobile machinery. They constitute a very numerous and growing group, consisting mostly of highly skilled workers. As a rule they perform diverse types of work which combine relatively little physical effort with mental work. For a long time to come they will comprise a considerable proportion of industrial workers,

- workers operating stationary machinery. They comprise in the first place a large group of operators of metal-machining tools and other machines; with the growth of automation the importance of this group will lessen;

- workers engaged in assembly, repair and adjustment jobs (i.e. adjusters, fitters of automated and machine tools, general fitters, electrical equipment fitters and electricians). They represent a large group of workers whose importance is bound to grow with the development of technology and whose work will be steadily perfected both through the use of more accurate instruments and higher standards of knowledge without which this skilled work cannot be performed.

- workers servicing continuous automated technological processes and automatic equipment - a relatively small but rapidly growing group fulfilling the work partly of workers and partly of engineers. These new occupations, brought into being by the scientific and technological revolution, are causing numerous problems;

- workers servicing electronic computers and computer stations. Their proportion is bound to rise in the course of further development of the scientific and technological revolution.

These conclusions show that the introduction of automation does not automatically result in the desirable change in character and content of work. New, sometimes even more difficult, problems emerge. In a number of cases, for example, a paradoxical phenomenon may be observed at industrial enterprises: the higher the comprehensive mechanisation and automation, the greater the role of physical work in the total labour structure. The point is that in the case of comprehensive mechanisation and automation, new technical means are being introduced. This sharply reduces the number of people employed in basic production processes, while the proportion of people employed in auxiliary processes (connected mainly with such subsidiary operations as supply of parts and raw materials or transportation of raw materials and finished products) increases relative to the total labour structure. By implementing a number of technical measures in the course of reconstruction of old enterprises, and especially the construction of new plants, this shortcoming is being eliminated.

Secondly, there is the important question of changes in the character and content of work resulting from the introduction of semi-automatic or automatic equipment and automated systems. This sharply reduces the amount of arduous physical work, thereby easing labour. However, another problem then arises, that of monotonous work.

It is found that many new occupations brought about by automation, as for example the monitoring of automatic systems at electrical power stations or in the chemical industry or operation of collating machines in offices, are characterised by monotonous work. As has been established by numerous investigations, this type of activity slows down work movements and causes early signs of reduced lability and excitability of the visual, motor and other functions. It also tends to reduce labour productivity and to increase work spoilage.

In many cases, as for example the operation of control panels of modern rolling mills, work is characterised by nervous tension due to the need for constant attention and swift and precise reactions. This places heavy mental loads on the operators, and such developments can result in an increase of cardiovascular diseases. Research conducted by the Kiev Institute of Labour Hygiene and Occupational Diseases shows that the incidence of myocardial infarction among workers carrying out mental tasks is equal to 1.23 ± 0.20 per 1,000 workers and among manual workers (occupations characterised by muscle tension) 0.41 ± 0.26 per 1,000 workers. It is characteristic that the highest indices of myocardial infarction are found in those occupations which contain elements of creative work requiring a high degree of responsibility and nervous and emotional tension, when the personality's psychic resources are mobilised but are either not systematically realised or realised only in emergency cases of breakdowns.

In accordance with Zvorikyn, occupational aspects of uniform monotonous mental work, as generated by scientific and technological progress, render the problem more acute than does monotonous but varied physical work. The negative effects of such mental work are more profound than those of monotonous physical work. This raises the problem of either enriching non-creative mental work with creative elements (which is difficult to realise in practice) or entrusting this type of work to electronic computers.

It has been found that considerable improvements can be achieved by alternating periods of work and rest, by rational division of the technological process, by alternating contiguous operations and also by broadcasting music and songs, by changing the speed of the conveyor line according to changes in work capacity during the working day, etc. It is becoming ever more obvious that it is not a matter of replacing one type of work by another - of physical labour by mental labour - but of finding new forms of combining physical and mental work, especially when physical work can be freed from arduous operations and becomes goal-oriented work closely combined with mental tasks.

It was proved in many instances that the negative effects of working at control panels can be reduced if operators not only take the readings of instruments and wait until they need intervene in the production process, but are also involved in various aspects of the production and social life of the enterprise collective through a number of channels. At the same time, their leisure activities should involve various tasks and aims. In these conditions, instrument watching will be supplemented by conditioned reflex reinforcements, taking the edge off the negative effects connected with the character of the work.

All these findings witness the fact that it is impossible to combat negative consequences of the new occupations connected with automation in isolation from the socio-economic conditions of production activity.

1.5 Work organisation practices in the USSR related to automation

1.5.1 Work brigades

The most significant trend in work organisation in USSR industry in the seventies was the large-scale introduction of semi-autonomous groups or, as they are called in the USSR, "work brigades". At the end of the decade, 48.6 per cent of the total industrial workforce were members of work brigades, and it is expected that by the end of 1982, the number will reach 53-55 per cent.[9] This trend is closely linked with the introduction of comprehensive mechanisation and automation. In many instances, introduction of the brigade method of work organisation is reinforced by the need to match the challenge put forward by new technology. At the same time, the switch to collective forms of work organisation provides a fruitful base for speeding up technological change.

The explanation of the fact that the percentage of work brigades is much higher in highly mechanised and automated production sections than in labour-intensive units is usually linked with the following advantages of brigade work organisation:

- orientation of the member of the brigade to the common final result of the collective work;

- higher flexibility in job allocation, self-adjustment, co-operation and collective responsibility for the whole production cycle;

- possibilities for easy adaptation of new workers;

- job enrichment and reduction of negative effects of monotonous work.

It has been found in USSR industry that design of work organisation in accordance with the particular requirements of highly automated technology usually results in 15-20 per cent productivity growth. According to some estimates, it can even be higher.[10] The brigade form of work organisation also provides possibilities for skill upgrading of young workers which are twice as great as in individual work organisation patterns. It helps to improve motivation and work satisfaction, to reduce labour turnover and to speed up the introduction of innovations.

It is well known that the growth of comprehensive mechanisation and automation results in considerable change in the occupational structure. The growth in the need for machine maintenance and repair aggravates the existing shortage of such personnel. The main reasons for the shortage is that it usually takes five to seven years for workers to become qualified repair specialists. Meanwhile, machine operators reach the same professional grade in much shorter time, and as a result receive a higher wage. The work of a repair worker is more demanding and more physically stressful.

It has been found that brigade work organisation helps to resolve this problem. Two different approaches are used. One is to incorporate the repair personnel into the existing work brigades. Another is to set up special repair brigades. Each solution has its own advantages and disadvantages. Inclusion of repair personnel into complex brigades fully responsible for the given production line or sector helps to eliminate the antagonism sometimes found between machine operator and machine repair personnel. As brigades possess control over the distribution of collective brigade earnings, the option is open to compensate with higher pay the inferior conditions of those engaged in repair and maintenance.

However, the limited scope of work in small brigades often does not permit the inclusion of repair workers into the brigade. The existing trend towards enlarging brigades by uniting work groups engaged in different segments of the production line helps to overcome this limitation. Bigger brigades are usually more efficient, and work in such brigades is more satisfying as they provide better possibilities for job rotation, job enlargement and job enrichment. The higher efficiency of enlarged brigades is partly attributable to social factors. As shown in a study conducted in Leningrad enterprises,[11] large brigades provide more possibilities for the realisation of the advantages of brigade work organisation which can be expressed in the factors reflecting the level of brigade development. The maximum size of brigades is limited by the design of the production cycle as well as by the growth of managerial overheads.

In many cases, it is found feasible, from the economic point of view, to set up specialised maintenance and repair brigades servicing the needs of the whole production process or its parts. Efforts are made to link the work of the brigade with the final results of the servicing unit of the plant. Often the amount of bonus paid to repair and maintenance personnel is directly connected with the idle time of machines due to unscheduled repair.

Collective forms of work organisation open new options for increasing worker participation. The actual pattern of participation can vary considerably, depending on the size of the brigade and the level of its development. The most important instrument of workers' participation is the workers' meeting. Such meetings are held regularly at least once a month and the agenda would usually include different aspects of production activities and the social life of the brigade. Work productivity, technological innovations and their impact on group work, problems of violation of work discipline, approval of new production schedules and many other issues are also discussed. One of the most important items of the meeting is the distribution of collective earnings.

Besides holding regular general meetings, brigades often elect a brigade council. The council acts as a link between the brigadier (team leader) and the work brigade. The most important decisions are usually taken by brigades after consultations with the council. One of the major tasks of the council is to evaluate the actual contribution of each brigade member in the fulfilment of the production schedule and to prepare suggestions for the distribution of collective earnings.

As was mentioned earlier, an increase in the scope of automation often results in growth of the size of work brigades. It is becoming quite common for small brigades of 10-12 workers to be enlarged into groups of 50-60, or even 100 members. Such growth is causing profound managerial changes in the brigade, as well as at enterprise level. Workers in the brigade will usually be split into a number of work groups, with group leaders reporting to the brigadier. The high autonomy of the work brigade "makes unnecessary many functions of external management, the actual scope of which is considerably reduced; it ceases to touch upon the internal problems of the life of the work brigades but is carried out in a few key directions".[12]

The considerable number of functions and responsibilities being carried out before by supervisors are now delegated to the brigadier, thus providing possibilities for increasing the scope of supervisory management. The organisational structure of enterprises has in many cases undergone considerable change, becoming less hierarchical and more efficient. The brigade form of organisation also makes it possible to reduce the number of production schedule indicators. Active worker participation in development, organisational and technological improvements takes different forms. One is the setting up of so-called "teams of creative co-operation" for resolving various problems. Such teams include workers, engineers and economists willing to participate in the work. Another important instrument of self-management is the council of brigadiers, which includes brigadiers of a given enterprise. Such councils work in close co-operation with the top management of the enterprise and possess considerable bargaining power.

It has been found that the proper planning of the introduction of work organisation influences to a large degree the final results. Many enterprises find it useful to set up special committees, or temporary advisory bodies, whose task is to monitor the process of change. At regular sessions, the committee considers the situation, evaluates positive and negative developments and draws up remedial measures; it also approves plans for the setting up of new brigades, etc. In some enterprises, as for example at the First State Bearing Factory the sessions of the committee are often held in the production shops and sections.

Such committees cease to exist as the brigade form of work organisation becomes dominant and they hand over their functions to the council of brigadiers.

In some cases, as for example in the Kaluga Turbine Works, the setting up of a permanent functional department on the brigade form of work organisation was found to be useful.

1.5.2 Social development planning

Promotion of QWL in USSR industry is closely linked with the practice of social development planning (SDP). The first plans of social development were introduced in a number of Leningrad factories in the sixties, and this practice now concerns the majority of industrial enterprises. SDP has become an integral part of the comprehensive system of short- and long-term planning. The social plans are developed jointly with plans for technological and economic development in order to secure a systematic approach to the social and economic growth of an enterprise.

The shape and structure of SDP can differ depending on the specific situation of the given enterprise. In general, however, the four following items provide the framework of a plan:

- development of the social structure of the enterprise;

- improvement of working conditions and protection of the health of workers;

- improvement of the remuneration system, development of welfare facilities and improvement of living conditions of workers;

- communist education of the individual and development of workers' social involvement.

The plan covers the whole scope of QWL problems, including living conditions and recreational facilities for workers and their families. Preparation and execution of the plan provides possibilities for further developing workers' participation.

In order to manage the process of plan formulation, a joint committee is usually set up of representatives of management, functional services, Communist Party and Konsomol organisations, and workers. At the first stage, numerous meetings, discussions and sociological studies are conducted at all sections of an enterprise in order to form a realistic picture of the existing social structure of organisation, psychological climate and workers' attitudes and desires. Next the planning department in co-operation with other functional departments and sociological (internal or external) groups, if such are available, designs the outline of the SDP and aligns it with economic and technological development plans. The plan outline, after being fully evaluated by functional and public bodies, is revised and - once approved by enterprise management - becomes obligatory for execution.

The practice of SDP, properly organised, is a solid means for efficient socio-technological design of new systems or redesign of existing ones. First of all, SDP helps to formulate social requests or the need to improve existing machinery or introduce new technology. Requests emerge at the initial stage of plan development, when steps are taken towards critical evaluation of the existing production process, work organisation and working conditions. In such cases, sociological study helps to determine workers' satisfaction and disssatisfaction with job content and working conditions. All these findings, as well as information on the dynamics of the workforce, social structure are used to design the request.

SDP helps to formulate criteria for technological choices, which are also influenced by a multiplicity of social and economic factors. At the design stage of new technological systems, a social development plan provides the basis for determination of possible social impacts, both positive and negative. This enables designers to improve design compatibility with the social system. Among the most important impacts of a new system on the personnel of the enterprise are -

- social stress: transfer of workers, change in work content, need for new skills, etc.;

- social costs: loss of status, disengagement of workers, growth of exterior control, etc.:

- social incentives: opportunity for promotion, improvement of working conditions, meaning of work, greater worker dignity, etc.

Special sociological procedures are used to determine the character and intensity of these effects of automation in certain cases. Among them are methods of estimating work satisfaction and the level of worker anxiety, sociometric tests, etc. Some of these impacts may be considered as socially unacceptable if there is no possibility to counterbalance their ill-effects or if they contradict fundamental social values (e.g. humanisation of work). In such cases, the project will be modified or abolished. On the positive side, SDP provides possibilities for conducting radical change in the whole social system of the enterprise in order to meet the requirements of a technological innovation.

1.5.3 Work organisation standards and models of work organisation

The need to take into account the human factor at the technological design stage was recognised in Soviet industry at an early stage of automation development. Available theoretical developments and empirical findings in the different branches of industry provided a basis for the preparation of a collection of "requirements and normative materials on scientific work organisation which should be taken into consideration when designing new and renovating existing production enterprises, and designing technological processes and equipment".[13] This document provided the basis for developing in 1978-80 similar guiding documents for specific branches of industry. They have obligatory status, and require that work content, working conditions and ergonomics are taken into account at an early stage of technological design. However, design institutions and enterprises do not always have a full understanding of the needs of giving consideration to human and social factors at the design stage of new technology. For example, in Moscow in 1980, out of 120 projects of new industrial enterprises, after critical examination 32 per cent were not approved by the state authorities because they did not give sufficient consideration to the above-mentioned factors.[14]

There is evidence that extensive use of ergonomic principles in technological design produces a positive effect not only on working conditions but on work efficiency as well. For example, the foundry section of the Baranovich factory, which was designed with special concern for QWL factors is superior in production per worker to the Centrolit machine factory by 30 per cent and to the Moscow Stankolit factory by 35 per cent.[15]

The search for effective methods of job design in the course of the introduction of new technologies and automation, or redesign of existing production systems, resulted in the development of so-called "models of work organisation". This innovation was first developed in the USSR in the early sixties, and has now become very popular there, as well as in a number of East European countries. More than 3,500 such models have been developed and introduced at 6 million work stations involving more than 30 per cent of the Soviet workforce. In some branches of industry with a high level of mechanisation and automation, such as machine tools and chemical processing, this innovation has affected the work organisation of 50 per cent of white- and blue-collar employees.[16]

Models of work organisation provide a systematic approach to design or redesign of individual or group organisation of work, conducted together with related changes in shop-floor management, systems of payment, training and so on. Besides considerable improvement in working conditions and the working environment, the introduction of work organisation models usually results in a labour productivity growth of 10-15 per cent.

The goal of this innovation is to derive social and economic benefits from diffusion throughout an industry of advanced patterns of work organisation which have been proven in practice. The first models were developed for individual working posts and certain work operations, but the need to introduce sophisticated technology and collective forms of work organisation resulted later in the development of models of work organisation in production sections and shops.

The development and introduction of work organisation models has been incorporated into the state economic plans of the USSR, and has become a powerful vehicle for progress based on radical change in work organisation. The percentage of workshops and work stations modernised in accordance with the models has become an important indicator of the actual level of work organisation and working conditions in enterprises or in an entire industry.

The models of work station organisation cover the following elements:

- definition of the scope of each scheme;
- equipment and physical layout of work stations;
- characteristics of the workpiece;
- specifications for supply and despatching arrangements;
- sequence of operations, financial and performance indices, quality standards;
- workers' qualifications and responsibilities;
- the human aspect of industrial design; and
- occupational health and safety, including fire protection.

The introduction of automation also generates a need for models covering the work arrangement of group operators united by a common task. Such models often include the following items:

- main technological-economic parameters of the production shop for which the model scheme is intended;

- expected benefits from model scheme introduction (such indicators as the degree of mechanisation of manual operations, working environment, safety, labour turnover, organisation of the workplaces in accordance with models, and so on);

- primary division and co-operation of labour (technological, functional and skill division of labour is given, together with descriptions of operations performed by certain groups of workers);

- work enrichment and work enlargement (description of the conditions needed for the changeover to the brigade method of work organisation - team work, composition of the brigades, choice of work organisation and payment patterns);

- changes in production unit management system (new shop organisational structure, division of functions between managerial staff);

- organisation and equipment of workplaces (general layout of the production shop, general description of the main items of technological equipment and functional furniture, workplace grouping in accordance with the level of mechanisation, number of workers per workplace, list of the appropriate models of work station organisation);

- organisation of servicing operations (type of operations, schedule, units and posts responsible for conducting service operations);

- working conditions (working time arrangements, organisation of shift work, rest and recreational facilities, working environment, ergonomic arrangements, protective clothing, safety at work);

- rate setting and remuneration system;

- training (establishment of the training needs to suit the new pattern of work organisation, training programmes for new workers, upgrading of workers' skills, special training schemes for multi-skill job holders);

- communications and signals (general layout and description of the shop communication network, organisation of the signal system, equipment required);

- costs and benefits (specification of the potential costs of implementation, methods of calculation of potential benefits);

- introduction of models (major steps, division of responsibilities, workers' participation).

Many hundreds of institutions in the USSR are engaged in the development, approval and implementation of models of work organisation. Among them are two main groups or organisations: research and design institutions attached to the different branches of industry, and centres of scientific work organisation. The Central Scientific Research Institute for Labour Studies and the All-Union Methodological Centre for Scientific Work Organisation and Management are the leading institutions in this field responsible for development of the methodology of model design and for monitoring this work all over the country. The total number of specialists in work organisation engaged in model development and application in the USSR exceeds 45,000.

Development of the models is budgeted from two sources - the state budget and the finances provided by enterprises. Model development is usually covered by the state budget funds and pilot implementation by the enterprise concerned.

At enterprise level, the model is subjected to study and evaluation by functional departments concerned in order to establish its appropriateness. Special attention is given to the changes which will be required in the technological system. Modification of the technological process lies outside the scope of the model, but at the same time it is often unavoidably the first step in the implementation of the model as a result of the strong dependence of work organisation arrangements on the technological process. If the model's content is found to be expedient, job redesign in accordance with the model is included in the enterprise plans. Particular attention is paid to the process of implementation.[17]

2. Case study - Social problems of automation in cement production

2.1 Introduction

The second part of this chapter comprises a case study of the integration of an automatic process control system at a typical cement plant in the USSR. The development of such systems has lately been gaining ground in this country, and increasingly attracting the brainpower of engineers, scientists and administrators. However, the problems that emerge at the interface of such systems and the social organisation within the enterprise have not been adequately studied so far. The solution to these social and technical problems therefore involves many complications which give rise to contradictions and to provisional, partial or erroneous conclusions that have to be faced. This paper covers one such complicated case, when the social research into these problems started only at the stage during which the innovation was to be integrated at the enterprise, rather than being conducted at the development stage.

The author's point of view reflects the basic principles of SDP (see section 1.5.2), a concept that has been formulated by Soviet sociologists over several years.

This case study follows the following pattern. First, we define the characteristics of the enterprise environment and technical processes under study, and study methods. We then examine in detail two particular cases (case studies 1 and 2) illustrating the emergence of socio-technical problems and their solutions. Finally, we attempt to draw certain general conclusions regarding such solutions,which are valid not only for the cases cited but also for similar innovations elsewhere.

2.2 General background and methodology

The study programme focused on the tasks, problems, and methods specific to a particular enterprise with its technological and social characteristics. In other words, the features common to similar studies were specifically interpreted in the context of the enterprise. This means that a repetition of the study at another enterprise differing in technology (even if it belongs to the same industrial branch), personnel and innovation introduced would require an appropriate adjustment in the study programme.

2.2.1 Specific features of the approach

There are now quite a number of studies dealing with the impact of automation on the nature and conditions of work. The majority of them, however, concentrate on the social after-effects of the introduction of new technology which, although undoubtedly essential, is not sufficiently constructive an approach. This study therefore emphasises the social conditions of automation that imply projections both of probable changes in work and in conditions of various groups of workers, and the outline of arrangements to prepare the above-mentioned groups for a transfer to work under new technological and organisational conditions. In keeping with the SDP concept, the projections make it possible to modify a number of parameters in the technological system according to purely social considerations, and better to co-ordinate the technical and human problems resulting from changes in particular production sections.

Some specific features in the technological development of the enterprise at the stage under review (see section 2.2.2 below) made it possible to compare various phases in the integration of the change. The approach adopted treated the introduction of an automatic control system as a type of innovation process (i.e. the emphasis was on the transition from one state to another, a process involving a conflict between alternatives, procedural changes and mutual adaption), as well as determining the structure of the case study for each group of workers. A comparison of the initial state, the preferred solution chosen and the result obtained facilitated a view of the dynamics of the entire process.

Our case study attempts to tie in the analysis of specific situations with the evolution of the enterprise as a whole. In examining the direct effects of process control automation on the nature and conditions of work, we also take account of organisational and socio-cultural parameters of the enterprise as part of its overall environment. This is particularly important because it provides framework for interpretation of workers' responses to specific decisions.

2.2.2 The enterprise environment

(1) Plant history. Slightly over 30 years old, the plant was designed as a new project absorbing major technical and technological developments of that period, and was constructed within a relatively short time. It is now a well-established enterprise, until recently subject to no major reconstruction or other modifications.

(2) Production. The basic materials - clay and chalk - are quarried at a distance of about 3 km from the plant and shipped by pipe and by rail to the raw materials department where the first process stage - grinding and mixing - takes place. The resultant raw mixture, called "slurry", is fed to rotary kilns where slurry combined with various admixtures undergoes the second, principal process stage, its product being clinker, which is identical to cement in its chemical composition. The third process stage - clinker milling with admixtures - produces various brands of cement differing in admixture composition and milling fineness. Each cement brand is stored separately in a huge circular tower. Finally, the marketing division and traffic department despatch products to consumers.

A specific feature of production at the plant is a second process line shortly to be put into operation. The second line takes account of the latest developments in the cement industry and, particularly, the experience of control automation implemented in the first line.

(3) The personnel. The plant has close to 1,600 workers employed in its production departments, plant management and auxiliary services. Men account for 60 per cent of the personnel, over 80 per cent having either secondary (complete or incomplete) or higher education. In view of the fact that the plant has been in operation for 30 years and that its production personnel took shape almost at once, with alternative places of work rather scarce and with labour turnover negligible, the ageing of the workers becomes obvious. In some years the proportion of workers of retirement age (or approaching it) has reached 30 per cent of the entire staff of wage and salary earners. The absolute majority of the workers originate in the locality. Approximately half of them have come from the nearby villages to become first-generation production and office workers. A considerable number of them fit into the category of so-called peasant workers (i.e. they work their personal subsidiary plots of land in the neighbouring countryside, or within the town boundaries if the plots adjoin their own houses). The plant has its own technical college from which most office workers and a large percentage of highly skilled production workers (in particular, kiln and mill operators) have graduated. The majority of engineers graduated from higher technological institutions, taking correspondence courses in the neighbouring regional centres. Not more than 15-20 per cent of all engineers were employed by the plant directly after their graduation from daytime courses at various higher educational establishments, and these are not local inhabitants.

(4) The settlement. The plant is located in a typical small town with a population of about 50,000. The majority of its inhabitants are first-generation town-dwellers retaining firm family and economic ties with the countryside. This is manifest in their way of life; the breakdown of their time shows a large amount of agricultural work done in their free time and on rest days on individual plots, either attached to workers' houses or allotted by the trade union in the countryside to grow fruit and vegetables; neighbourly and kinship ties, developing in a similar pattern, provide the basis for mutual assistance and active contacts; territorial migration is insignificant, since inter-urban changes of residence are extremely rare, while push-pull migration (daily arrivals from the nearby villages to work or to shop in town with return on the same day) is a common practice; divorces are rare, families have many children, and rural ritualism is maintained in wedding ceremonies, send-off parties for conscripts, etc.

The town is a mono-industrial one in which the entire industrial production is essentially confined to the cement plant, giving rise to characteristic socio-cultural effects. Of particular significance to the extension of intra-organisational relations that take shape in the sphere of basic production at the plant to social relationships outside work. In addition to neighbourly and kinship ties, production ties uniting workers employed in the same plant units (sections, brigades or shops) rank prominently, and are reflected in choice of living arrangements and, consequently, in leisure-time contacts, etc. On the other hand, the kinship and neighbourly ties affect labour relations both in a positive way (from the standpoint of production), as exemplified by mutual aid in work, and in a negative way, when allowances are made for each other's blunders.

(5) Traditionalism. The above characteristics of the personnel and the settlement create a specific socio-cultural environment of the enterprise. Generally speaking, it is characterised by a complex inter-twining of communal and administrative structures. Its specific manifestation can be found, for example, in the tradition of hereditary occupations when a son comes to work in his father's brigade, often placing himself under the latter's official supervision, or in a more extensive succession to trades and skills covering two or three generations. Such working teams are distinguished by a relatively low level of conflict, since labour disputes are often hushed up or settled through communal channels. At the same time, if a conflict does occur, the parties sometimes separate not so much on the basis of formal organisational criteria but rather in keeping with their "clannish" attachments.

Another important aspect of communal elements in the subculture of the plant may be seen in a high degree of integration of its social processes and in the inter-relationship of local events and general occurrences. If, for instance, one group of workers expressed a negative view on their experience of operating new machinery, the other units were promptly infected with the same attitude, and each working team in its own section responded as an integral unit. This type of subculture is characterised by a somewhat underdeveloped motivation towards attainment, resulting in a high degree of satisfaction with the existing nature and conditions of work. Thus, the workers were inclined to favour stability in the social and labour environment, which implies, to a certain degree, an inertness of the current system. Hence, they were suspicious of innovations and attempted to evade or even to transform them to fit their habitual patterns. This tendency is particularly important for the case study.

(6) Principal groups: In spite of the comparatively large number of workers (for a continuous process) and the complex personnel structure of the plant, some occupational groups play a particularly important part in the production process although numerically they are few. The key position within the groups affected by automation is assigned to teams engaged in raw materials kilning and clinker milling (see (2) above). Although the social problems resulting from the automation of their work vary significantly, they involved common difficulties, errors and solutions. We concentrated, therefore, on these two occupational groups.

2.2.3 Subject of the study

The subject of this study was the following technical innovation: the introduction of automatic control of engineering and labour processes in two specified production sections (see (6) above). In other words, a single general process is reviewed in terms of two particular case histories. Each is analysed first within the framework of interface engineering and various social solutions to the problems posed by the general process; secondly, it is studied in the socio-cultural context of the enterprise regarded as an integral unit. Still more specifically, the subject of study may be defined as the changes that take place in the transition from the old to the new production.

The main contradiction in the changes brought about in the course of the process under study may be found in the differences between the engineering and technological criteria of the solutions advanced by the designers, on the one hand, and the social value of various parameters of work and working conditions, on the other. Although contradictions between technological and social values may be considered as universal, in this particular case decisions made with respect to specific manifestations of the problem attempted to work towards appropriate mutual adaptation of the automatic control system and the work behaviour of workers affected by the system. The roots of the problem, thus interpreted, were found not only in objective differences between technical and human factors of production, but also in diverging value judgements of those involved in the innovation - development researchers, designers, sociologists and various groups of workers. Therefore, the mutual adaptation simultaneously implied a search for an interprofessional consensus.

The general task of the study therefore consisted in analysing various social provisions for the integration of the automatic control system. The provisions included the following forms of mutual adaptation:

- social assessment of technical decisions, both preliminary and based on experience;

- the search, in co-operation with development researchers and designers, for alternative technical solutions;

- training of workers for the change to new tasks and ways of operation;

- design of interpersonal relationships in the sphere of personnel organisation.

2.2.4 Study methods

(1) Specific methods of study. As is characteristic for this type of case study, we do not deal with mass processes. In the two case histories, the objects of study are small groups of workers numbering two to ten persons. The use of formalised methods devised to provide statistical data (questionnaire survey type) is inappropriate here. At the same time, the sociological study of such groups involves face-to-face dealings with each worker which may result in confidential contacts. These contacts facilitate the collection of diverse information about each individual (i.e. the worker's perception, feelings, opinions, desires and emotions) which constitute an exclusive advantage lacking in mass surveys. However, the information collected in this way, profound and reliable as it is, can only be formalised to a very limited degree.

(2) Data collection techniques. In view of the above, the principal study method was based on close observation. The method is characterised by a researcher's prolonged presence in the environment studied, by daily contacts with workers as well as with innovation developers and designers, and by discussions of various alternative decisions with them. The method has much in common with the incorporated observation technique, differing from it mainly in that close observation allows the sociologist to retain his research and advisory status, with no attempt to simulate an affilitation with a given group. In addition to the principal data collection method, the following techniques were employed:

- in-depth interview on a pattern common to both workers and automatic system designers;

- time and motion study;

- document analysis (design, administration, personnel and other documents).

(3) Parameters for analysis. In order to make a structural classification of the information obtained, two characteristic series were chosen to describe the technical and social components of the process under study. The first series included -

- the purpose of automation (determined through interviews with development researchers and designers); and

- automation hardware components machinery and equipment (planned and employed; determined in the same way as above).

The second series included -

- assessment of innovation by the workers involved (satisfaction/ dissatisfaction in general and in relation to individual components, proportion of down-time; determined through interviews, observation and document analysis);

- workers' technological dependence (amount of work time predetermined entirely by the production process and the time allowed for variability in workers' behaviour; determined through observation, time study and interviews with workers and system designers);

- complexity of work functions (stability/uncertainty characteristic of the production process, amount of work and routine work requiring initiative; determined expertly through observation and time study);

- social status (prestige and authority, identified through workers' self-appraisal and the researcher's evaluation of the objective prerequisites leading to status changes, as well as plant managers' evaluation; the use of mutual evaluations appeared to be a very complicated procedure because of inadequate information and the relative isolation of other workers);

- industrial health conditions (temperature, noise level and dust, determined through consultations with experts).

In some cases additional parameters (changes in relationships, in personnel, etc.) were taken into consideration.

2.3 Case study 1 - Raw materials kilning

We now move on to our case study of raw materials kilning, which was the first in the process chain to undergo process control automation.

2.3.1 The situation prior to the change

(1) Production process. Kilning of raw materials is of major importance in the process chain of cement production (see section 2.2.2 (1) above). Each of the two process lines consists of four rotary kilns that convert raw materials into an intermediate product (clinker). Each kiln is 150 m long and 7-8 m wide, with an incline. Raw materials (slurry) consist of clay, chalk with admixtures, and water are fed from one side. Slurry moves down the incline in the kiln undergoing three-hour clinkering. The process is continuous and the kiln is never turned off.

(2) Specific characteristics of the group. The principal personnel of the section consists of four rotary kiln operators and four assistant operators (with operators' sons in the latter subgroup). All pairs work in shifts. The age of the operators varies from 35 to 45, that of the assistants from 25 to 30. Among all categories of cement workers, this occupational group holds the highest status since it is the most complex and critical part of the entire production process, where errors may have repercussions in all production sections over a long period and at considerable cost. The section is therefore staffed with the most highly skilled workers, almost all of them having technical college, and one of them higher education. Their wages are not dependent on output, and reach the level of a senior production expert. The respect they command is also manifest in their easy admittance to the plant manager's and production manager's offices, and in their nomination for elected positions in local and regional soviets of people's deputies and in party and trade union executive bodies.

(2) Functions. Prior to the introduction of automation, operators and their assistants performed the following three basic functions in controlling slurry kilning:

- temperature control in varous kiln sections by mechanical adjustment of gas injections;

- air draught control for the rarefaction of raw compound in the kiln; and

- control of raw materials feed to the kiln.

The division of functions between operators and their assistants largely depended on their relationship (a feeling of confidence, mutual understanding, etc.).

At the beginning of their working day the operator and his assistant took over the shift and inspected the equipment. Their main concern in this inspection was gas leakage. Once an hour they put down all instrument readings in a special log. Three times a day the assistant operator, together with a laboratory technician, took raw material and clinker samples for a calcium-content chemical and physical analysis in the corresponding plant laboratory to determine the kiln underheating/overheating level. However, this analysis could not assess the process taking place in the kiln, a disadvantage that an operator had to compensate for by visual inspection of compound colour, density and spreading through a peephole in the kiln. An assistant operator also had the responsibility for wiping cement dust off the instruments and cleaning the section. The process control mechanism was carried out by means of two handles for gas feed and draught adjustment. An operator and his assistant attended to four kilns at the same time.

(3) Nature of work. An operator's task consisted in maintaining a stable kilning routine: raw materials underheating and overheating equally resulted in defective products. The maintenance of optimum

conditions depended on the assessment of kilning progress in different parts of the kiln and at different times. Moreover, the level of uncertainty in essential process parameters was very high. The chemical composition, slurry consistency, gas pressure and quality, the movement of compound in the kiln, etc., could not be strictly uniform, but varied considerably, thus requiring constant monitoring of the operation of the kiln. Strictly speaking, it was the uncertainty and instability of the process that essentially accounted for the great complexity of the work of rotary kiln operators. It was found that it took about 30 per cent of an operator's working time to remove the major divergencies of the process.

It should be noted that this uncertainty in the stated process parameters was actually beyond the operators' capacity to control. What were the effects of this?

First of all, an operator had to rely not so much on information about the process as on intuition acquired and accumulated over many years. The quality of kilning was assessed by colour tints and compound spreading. Laboratory analyses provided either prognostic data that outran the process, of ex post facto data that lagged behind. Besides, in visual monitoring a worker could see only a small part of the kiln interior through a dimmed peephole. Instruments indicating temperature in several outlying parts of the kiln, gas injection, air jet force and slurry feed could only provide an indirect description of kilning progress. When confronted with an information shortage, therefore, an operator generally counted upon his own experience, that is, upon his intuition.

This gave rise to a very significant sociological effect: the individualisation of the production process interpreted as a large-scale dependence of the latter's efficiency and quality on an operator's personal qualities, for each operator had unique experience, abilities and intuition. This "authorship" of the process provided broad opportunities for the worker's self-actualisation, for the realisation of his individual potential and for an awareness of his personal role in production.

On the other hand, the technological effect might imply a relatively broad margin of error, indicating a liberal spoilage tolerance. For instance, if there was a deviation in kilning temperature, an operator was disposed to increase gas injection and to maintain an extra heat reserve to stabilise the process for as long as he considered necessary. The inevitable burn (up to 15 per cent of clinker) and the overconsumption of gas were treated as a natural loss. The same thing took place when an operator needed to divert his attention from process control in a kiln to some other activity.

There is an objective clinker quality characteristic - calcium content - in cement production, the standard content being 0.21-1 per cent of calcium. The operator tried to stabilise the process, choosing an average condition in relation to the stated extreme values; this was achieved about 70 per cent of the time. The "stable" part of working time was occupied by the correction of insignificant deviations in the course of the process, visual inspection of kilning, functional check of devices, passive observation of instrument indications (generally performed by an assistant operator), logging, talks with the department process engineer, etc. However, chronological time did not coincide with production time, and the peak of an operator's primary activities fell during the period of major deviations.

During this period the high uncertainty of the process resulted in its poor controllability. An improvement in the latter could only be attained at the expense of the worker's increased technological involvement, which meant a greater dependence of the worker's tasks on a kiln operating mode. Both the process engineer and the kilning superintendent believed that an operator should spare more time and give more attention to visual inspection of the kilning process through the peephole, checking on temperature variations, steady rate of kiln charging, etc. All actually agreed that in this way one could prevent or eliminate more promptly up to 25 per cent of major deviations in the course of kilning. However, such continual monitoring would more than double an operator's technological involvement, require advanced skill on the part of the assistant and add to the fatigue of both.

(4) Working conditions

The physical working conditions were characterised by high levels of heat, noise and dust. Operators and their assistants were entitled to special money compensation for the unfavourable conditions of their work, retirement at an early age, free additional food, extra days of leave and other benefits. In the close proximity of a kiln, there was a decline in the physical parameters described above (section 2.2.4 (3)). Direct observation of the process bore witness to the detrimental effects for human beings.

Regarding the social conditions of work, the kiln operators had the highest status of all categories of workers, as has already been mentioned. The principal parameters of kiln operators' status, as acknowledged by the plant managers, were their singularity. Operators themselves emphasised the complexity of their functions and the comparatively high labour remuneration. This economic factor accounted for the appreciation of their working conditions by other categories of workers.

2.3.2 The changes introduced

The socio-technical problems posed by the changes, and the search for ways of resolving them constituted the most important aspect of the innovative process.

(1) Characteristics of automation. The kilning subsystem is part of the integrated process control system common to the whole part. The basic production task underlying its establishment in overcoming the uncertainty of the kilning process. This was to be attained by encompassing the maximum number of process parameters and check-points and by controlling them in a more accurate and selective manner. An electronic computer with an adequately high-speed response was installed and

equipped with about 20 gauges recording process conditions (temperature in the compound clinkering zone, draught intensity, gas flow rate, slurry feed, exhaust gas temperature, etc.). The same controls that kiln operators had used before were connected to the central control board. The processes in all the kilns were now controlled by eight central control board operators.

The social goal of automation in the kilning section was confined to an improvement in the physical working conditions of operators and assistant operators. This task was solved mainly by a sharp reduction in workers' contacts with the kiln. A system of instruments was intended to furnish the necessary information to the control board, and the operation of the kiln was thought to be possible mainly in a remote-control mode.

It should be noted, however, that the sociologist did not take part in the development and design of this control automation system, but joined the project only at the stage of its introduction. For this reason the social objectives of the innovation were rather vaguely defined at first, provisions for its interaction with the workforce were inadequate, and social problems were essentially figured out during the initial attempts at integrating the system. However, the solutions that were found could then be instantly applied in the construction of the second process line at the plant.

(2) Changes in functions. The basic set of kiln control functions remained unchanged (section 2.3.1 (2) above). Yet two problems arose: the change in the character of functional performance and the redistribution of functions.

A proportion of these functions came to be performed automatically. While formerly an operator first noted a deviation in the process and then decided to what extent he should add gas or increase the draught, in a number of cases a temperature change, for example, now immediately increased or reduced gas, air or raw material feed. In other words, an operator had shared his functions with automatic equipment.

The problem of another division of functions - between kiln operators and control board operators - turned out to be much more complicated. The introduction of the system was accompanied by the appearance of a new social group, that of central control board operators. The result was a problem of intergroup relationship. The new social structure appeared contradictory in its very essence, as described below.

The contradiction was underpinned by a major technological difference: the new system provided a steadier control of the process than a kiln operator did, and it reduced the period of major deviations in kilning by 20-30 per cent. However, an operator was still better able to manage the control in cases of major deviations or emergency situations. The question was to decide when the system should be switched on and when an operator should assume the control. Control board operators and kiln operators varied in their evaluation of the necessary extent of their participation in the process control: they were inclined to diminish each other's significance.

The technological difference was augmented by a socio-cultural difference. The point was that the group of control board operators was composed of non-local young graduates of higher engineering institutes, who had not yet managed to adapt to the new environment, local standards and traditions (see section 2.2.2, above), and who did not consider themselves bound by the traditional status distribution. Their subculture was characterised by a cult of technological rationalism, and they regarded their associates - kiln operators - as a vanishing occupational category. The kiln operators, on the contrary, were convinced of their own indispensability and of the prime role of their skill. Thus, there were also socio-psychological elements in their intergroup relationship that prevented them from reaching a consensus on the division of control time.

In addition, there was a purely administrative aspect in the relationship: the control board operators were obliged to increase the rate of system utilisation and thus exerted pressure on kiln operators on behalf of "technological progress". Thus a complex intergroup contradiction arose that demanded an appropriate settlement.

(3) Changes in the nature of work. The introduction of the automatic system provided a kiln operator with a control board with four display screens that enabled him to see the representation of the compound kilning zone in each of the four kilns. The instruments installed in the board indicated by pointer deflections or recorder movements the state of various process parameters. Besides, the board was equipped with a system operator communication device for control changeover to the central board.

The greater variety and volumes of information made it possible for a kiln operator to increase his predicting capability, particularly in relation to some adverse phenomena. For instance, in a rotating kiln a part of the compound is deposited on the walls and forms ever-growing "rings"; these "rings" block the passage of the rest of the compound inside the kiln and may result in clogging. Whereas formerly a kiln operator could see through a peephole only the later stage of "ring" formation, he could now detect the very beginning of it and prevent its appearance by increasing the draught. The same was true for other situations. Previously, a kiln operator had to wait until an adverse condition reached extreme proportions before taking action; now the system enabled such critical conditions to be avoided.

The introduction of the new system provided a kiln operator with a predicting capability of a much wider range. Recordings became more comprehensive and precise. For example, fuel consumption, temperature variations and other variables were now recorded hourly. Recorder charts gave an essentially complete retrospective picture. A kiln operator could now identify regular patterns and tendencies in the course of the production process and compare alternative control strategies.

Thus, kiln operators' direct technological involvement decreased considerably. In those cases when the switch of the system operator communication device depended solely on kiln operators, they could maintain the automatic control of the process for over half their working day, almost entirely withdrawing their attention from the instruments and operating only on recordings, i.e. performing an analysis. Moreover, during the half of their working time that they were "involved", they began looking for and comparing various combinations of means and actions, and were thus engaged in creative work during this period too.

The differences in the response of the kiln operators to these conditions were, of course, quite marked, and they made use of the new capacities in different ways. While the extent of direct technological involvement decreased with regard to all the kiln operators, the complexity of their tasks appeared to be a direct function of their motivation. Two of the four operators showed great interest in the analytical aspect of their work under the new conditions (when interviewed, they stressed the opportunity to do an analytical job as a factor of their enhanced satisfaction with the new nature of work). On the other hand, one of the kiln operators adopted a passive line of behaviour: making the utmost use of automatic devices, he spent the rest of the time as if it were free. Still another alternated his line of behaviour. Thus, a change in the complexity of tasks depended on the operator concerned. In the latter case, the complexity decreased as a result of a general reduction in the period of major deviations that necessitated an operator's involvement in the control process.

Still another specific feature in the nature of the kiln operators' work manifested itself. The period of major deviations in the production process, as mentioned earlier, was reduced, though not excluded altogether. The problem now lay in the fact that the remaining part of the period could not be formalised since the system developers failed to develop a control algorithm applicable to some specific deviations (for example, the entry of refractory components into slurry, breakdown, etc.). Not only did circumstances greatly reinforce kiln operators' arguments in favour of their indispensability, but it also gave rise to other far-reaching effects.

(4) Changes in working conditions. The physical working conditions were positively affected by the changes: the effects of noice, dust and temperature on kiln operators and their assistants fell sharply. The kiln operator's board was installed in a small booth with transparent walls to mitigate the action of these factors. This could not be done before, as an operator needed to maintain close contact with the object he controlled.

As to the social conditions of work, the kiln operators' status, in the opinion of plant managers, shrank in importance. When interviewed, the managers noted a possible transition of kilning to fully automatic control. Besides, the kiln operators' work now seemed much easier to their fellow workers. In the socio-cultural environment of the plant, work prestige was largely determined by its arduousness. The kiln operators' self-appraisals were divided, corresponding to differences in the assessment of work complexity; they went up for the "analytically" minded and down for the "passively" disposed.

The social conditions of kiln operators' work also included a new factor, an increased outside supervision of their work conduct. First, the central board registered the exact time during which kiln operators switched the process over to automatic operation by means of the system operator communication device. (The plant managers encouraged the large-scale application of automatic equipment.) Secondly, the recordings gave objective and graphic proof of the quality and efficiency of the way the process was controlled by kiln operators. Thirdly, in the context of a certain competition with central control board operators "to gain control over the kiln", the kiln operators found themselves the focus of their counterparts' close attention. This increased outside supervision developed into a factor of dissatisfaction manifest in all kiln operators and their assistants. Certain forms of outside supervision evasion appeared, such as concealment of records and distortion of accounts.

Finally, the individualisation of the production process (see section 2.3.1, above) declined; for reasons stated above it became relatively more impersonal, which also resulted in the kiln operators' lower self-appraisals.

The economic conditions of work also featured an innovation: bonuses (though temporary and small) were introduced for system utilisation rates. However, this change did not particularly affect workers' behaviour.

(5) System assessment. The character of system assessment was extremely important in working out specific alternative solutions for the emerging social and technical problems (see section 2.3.3, below). Therefore, the differences disclosed here deserve consideration (in accordance with section 2.2.1, above).

The assessment of any innovation is largely predetermined by the socio-cultural characteristics of the wider environment and by specific differences in its component subcultures, as well as being dependent on the changes it brings about in objective human conditions. Individual characteristics of participants - their place in the social structure of an enterprise, educational background, motivation, etc. - cannot be disregarded either. Assessment of the system introduced were therefore categorised in accordance with these differences. In addition, the assessment of different groups varied over time.

The system assessment on the part of plant managers was initially characterised by unrealistic expectations. Great hopes were placed on the system, still at the design stage at the time, primarily in relation to plans for the advanced management of the enterprise. Process control automation was believed to facilitate a centralised output quality and volume control. A number of difficulties and limitations encountered in the process of automation somehow diluted these hopes, which gave way to certain disappointment.

The greatest interest in the innovation was displayed by a numerically small group of young graduates of higher engineering establishments just assigned to the plant. Their active support was evident in their attendance of lectures delivered by the system developers at the enterprise, by their voluntary help in solving technical problems of automation both inside and outside their official duties, and by other manifestations of support. Their attitude towards the system was essentially uniform for the entire duration of its development and introduction.

It is natural that each of the above-mentioned groups differed in the influence it exerted in the process of innovation. However, the principal environment that was to absorb the system as a whole and its subsystems consisted of those categories of production workers that would be directly involved in automated processes.

The group of kiln operators and their assistants shared the traditionalism typical of the basic socio-cultural environment of the town and of the plant, with characteristic inertness and little inclination to innovations (see section 2.2.2(5), above). Therefore their first response to the introduction of the system was somewhat indifferent (a prevailing attitude at the time was expressed in the following manner: "As long as it doesn't get in the way, it's all right".). However, the integration characteristic of community consciousness in the given environment, noted earlier, showed itself in a manner that the system developers had not expected. The automatic metering of tank filling with raw components displaced two female workers who had measured the filling level with a plumb bob attached to a calibrated cord. The female workers were transferred to another job that they were less accustomed to. The repetition of similar cases changed the attitude of the workers to the innovation, for they now realised that the integration of the system resulted in a change of work stations. In the case of kiln operators, misgivings about any transfers could easily be overcome, because cement production had no other equivalent jobs.

The third stage in the evolution of the attitudes of the kiln operators was brought about by the appearance of a "rival" group of central control board operators. Gradually, the kiln operators assumed a negative attitude towards the innovation. The influence and the status of this category of workers were such that they soon infected with their mood other groups of plant personnel, including the managers.

This negative attitude also implied a purely psychological barrier stemming from an obvious underrating of the technological capabilities of the system and doubts about new technology. However, this barrier was comparatively easy to overcome through training and practical experience. As to the objective social and technical changes, these required a search for more conceptual solutions.

2.3.3 Solutions to social and technical problems

The resolution of the emerging contradictions proceeded from the interpretation of the principal problem as that of mutual adaptation of automatic systems and the social organisation of production (see section 2.2.3, above).

(1) Adaptation of the system. The evolution of the social and technical problems described and the formulation of their alternative solutions reflected a change in the relationship among the diverse value orientations of initiators of and participants in the innovation.

The initial objectives of development researchers and system designers can be defined as technical ones. In the kilning section they found their expression in maximising the role of the system and, in particular, in endeavouring to oust by means of technology the occupation of a rotary kiln operator, which remained as a structural vestige. The decision proved to be unrealistic for two reasons. First, the operators' exceptional technological and production status provided them with powerful means of opposing such a decision, since it infringed on their interests. The approach was bound to fail for objective social reasons, since no provisions had been made for transfer of these workers to equivalent jobs. Secondly, it became equally impossible for purely technological reasons. As kilning control automation was implemented, it became increasingly clear that in cases of major process deviations an operator's experience and intuition could not be replaced, since automatic devices could maintain only a stable performance process (albeit of prolonged duration) or a process with minor deviations. The conclusion was evident: the kilning process could do without a central control board operator, but not without a kiln operator.

The next variant solution provided for the retraining of central control board operators in the skills of kiln operators. In addition to the social contra-indications, as stated above, an important argument against such an approach was the great difficulty of mastering the art (non-formalised skill) of kilning control within a short period of time, particularly so without the kiln operators' willingness to co-operate; and it was precisely the art of kiln control that was considered the most valuable asset handed down in the course of time from kiln operators to their assistants.

Two ideas played a decisive role in the search for an acceptable solution: a remote control board; or an autopilot. We consider both of these in detail below.

It was thought to be socially expedient and technologically feasible to remodel the automatic process control system in such a manner that the kilning subsystem, by way of exception, was set up (both physically and operationally) right in the kilning section. The central control board of the system was no longer used to control kiln operations, but it retained the monitoring of other process stages and purely metering functions in relation to the kilning section. In this way the parallel control of kilning by the two groups of workers and their contradictory intergroup relationship (see section 2.3.2 (2) and (4), above) was discontinued. The operator now held the posts both of kiln operator and of central control board operator, with a corresponding change in title. As a result, the operators' status grew in importance, since their qualifications rose considerably through a course of retraining, they joined the category of so-called "worker intellectuals", and plant managers now held them in higher esteem. Nevertheless, the body of semi-skilled workers who had grown accustomed to have respect primarily for arduous work still regarded the apparent "easiness" in the new nature of the operators' work as a factor diminishing their former status.

The idea of an autopilot refers to applying the system algorithm to the occupational psychology and functional characteristics of an operator. Airline pilots are known to switch on an autopilot in flight when they encounter a specific set of standard, formalised situations. In high-risk situations (take-off, landing, etc.) they are inclined to assume manual control of the aeroplane. The kilning subsystem is similarly recommended to the operators as a new technological tool they can handle, and as an automated assistant. All types of outside supervision and bonuses for the application of automatic devices were discontinued, and the only performance criterion for the work remained the output (i.e. the quantity and quality of clinker produced in the kiln). Operators themselves come to realise that greater reliance on the system and more time for analysis resulted in better-quality indices of their work. In eliminating or preventing major deviations, they become completely involved in control.

The effect of such adaptation can be seen in the following example. Since the kilning process in a kiln is characterised by highly delayed action, each specific change requires the effect of increased force for the prolongation of its action. The operator pays no attention to the many incipient deviations, but instead concentrates on applying the effect at the time when the deviations grow in extent, thus requiring stronger action. The operating mode may be compared to a saw with deep but widely spaced teeth. Automatic equipment detects earlier deviations and, therefore, applies effects more frequently but in a lesser volume to maintain the stability of the process. Here the saw has shorter teeth, but the spaces in between are narrower. Examining the recordings and laboratory sampling data, the operators observe that there is less underburning and overburning in the automatic mode of operation, so that their confidence in the system grows.

(2) Personnel adaptation. As noted earlier, the rotary kiln operators' educational level was the highest among the plant workers, even prior to the introduction of the new system. Therefore, the task of retraining them to work under the new conditions was somewhat simplified. However, they still had to take courses of lectures and on-the-job training to operate the system, partly in working time but mainly in their own time. An understanding of the system gradually removed manifestations of the psychological barrier, such as doubts, misgivings, etc.

2.4. Case study 2 - Clinker milling[18]

The milling section has much in common with the raw materials kilning section in terms of technological and organisational characteristics. However, its social and technical problems and their solutions have a distinctive character.

2.4.1 The situation prior to the change

(1) Production process. Clinker milling is a process stage that follows raw materials kilning (see section 2.2.2 (2), above). There are seven cement mills installed in the section. These are drums up to 3 m in diameter and up to 14 m long, rotating at a speed of 17 rpm. There are two chambers inside each drum, equipped with milling bodies: the first contains heavy metal balls for grinding clinker compound, while the second has metal cylinders for a finer grinding. Clinker with gypsum and additives is charged into a drum from one side, and ready cement is drawn off by suction pumps from the other side and fed into storage towers. The type of cement is determined by the combination of components and the fineness of milling.

(2) Specific characteristics of the group. The mills are attended to by four operators, each having an assistant. Such pairs of workers supervise the concurrent processes in all the mills, changing shifts every eight hours. All the operators are men in the age range 36-50. Three operators have secondary education (ten-year schooling), and one graduated from the plant-affiliated technical college. Two of the assistants are men and the other two are women, all are approximately the same age as the operators. Their educational background is an eight- or ten-year schooling. All the workers in the group are locally born, their service records in the section ranging from six to 16 years.

(3) Functions. Operators were responsible for the normal course of clinker milling in all the seven mills, including the output of different types of cement in the specified volume and range. Assistants were responsible for the working order of equipment and instruments and for their adjustment, with repair being outside their functions. The operators and assistants sometimes substituted for each other.

(4) Nature of work. Operators spent as much as 60-70 per cent of their working time at the control board, where indicators and control handles were installed. They left the board to carry out visual inspection (together with their assistants) and to detect disturbances outside instrument control, such as grease or cement leakage, or in emergency situations caused generally by the ingress of foreign bodies, the choking of a clinker feed bin, the overheating of bearings in the mill shaft, etc. Once in a while an operator was introduced by his foreman or dispatcher to change the type of cement produced. Following the instructions, the operator manually altered, by means of control handles, the proportions of gypsum and additives in clinker compound and the fineness of its milling. Simultaneously, he switched over the cement suction pumps, diverting them to the particular storage towers. Every two hours operators received laboratory data on the chemical composition and

milling fineness of the latest batch of cement produced. The information lagged behind considerably and did not affect operators' current decisions. Still, it disclosed certain patterns that might be taken into account in choosing a process strategy.

The extent of mill operators' technological involvement could reach as much as 70 per cent of their working time. However, this involvement was quite passive because nearly 80 per cent of the time at the control board was spent merely in watching instrument readings (i.e. keeping an eye on the instruments in case the latter should indicate any disturbance in the production process, either by a red light or a pointer deflection). This was a kind of enforced idleness since, in fact, mill operators spent only 10 per cent of their time at most in making complex decisions. For example, if a laboratory analysis detected spoilage, it also indicated its cause, as a rule. Generally, spoilage derives from defective fineness, that is, the milling is either too coarse or too fine. Less frequently, the laboratory indicated a disproportion of essential components pointing, for instance, to an excess of gypsum. In such cases the operator faced several alternative decisions: a replacement of milling bodies; an increased air blow-out of the compound; or a complete readjustment of component feeders.

An element of intuition was also required in the operators' work, since some process parameters could not be adequately formalised. For example, the feeders were not calibrated, the process taking place inside the mill was invisible and clinker density was not uniform. Therefore the operator had to "feel the mill" (the expression used by the operators to define each other's skill). In other words, operators' work involved uncertainty resulting in the complexity of their labour (10 per cent of their time was spent in complex decisions in the context of considerable uncertainty).

(5) Working conditions: The physical working conditions were characterised by the unhealthy effects of high dust content and noise level. The effects were less in the control room, although they were still above the norm. Operators and assistants received extra wages and additional days of leave for the unhealthy conditions.

The social conditions of work were determined by the group's position defined as that of highly skilled workers. Their occupational status, from the standpoint of its objective importance in cement production and its prestige among the plant personnel, was second to that of kiln operators. The same applied to the relationship of their wages to those of other groups of workers. They were also paid stable wages.

2.4.2. The changes introduced

(1) Characteristics of automation. The milling subsystem is basically similar to the kilning subsystem. The feeders were fitted out with automatic batchers that changed a gate slope in response to a central control board instruction; the mills were also equipped with transducers of milling fineness, compound rarefaction in different spots inside the

mill, final control element positions, etc. Some of the transducers were of an analogue type, while others were digital. All transducer signals terminated at the data indication post where digital data on each mill were grouped on display units. The post was also equipped with paper-tape printers that recorded, at intervals specified by the operator, data on equipment and milling conditions, as well as deviations of various parameters from the specified standard, and all operator interventions. The control mode was computed by an algorithm which automatically transmitted the required effects to actuators. However, system operators had at their disposal remote controls by means of which they could monitor the milling process independently of the algorithm.

In a sense, milling control, like kilning control, was not completely automatic, but rather automated, since it had provided for a human element to be incorporated in the system. At the same time, a major difference between the two subsystems was also manifest here: in contrast to kilning, the milling process appeared to be much more easily formalised, and a central control board operator proved able, both theoretically and practically, to monitor mill performance.

(2) Changes in functions. The introduction of the kilning subsystem brought about a fundamental redistribution of functions in the process control. There appeared a group of central control board operators provided with the technological capabilities to take over the tasks of mill operators. However, mill operators were not relieved of their traditional duties. They even retained the original control board with handles and their right to use it. Finding no adequate social solution at the time, the plant managers proposed that the two groups should come to an agreement regarding the division of control time. Thus, a case of parallel control emerged. Still, none of these participating in the innovative process (plant managers, system developers, and workers) regarded the solution as final.

(3) Changes in the nature of work: In the context of parallel control, the structure of the mill operators' working time changed considerably: on average they were idle for about 70 per cent of the working day. Two reasons accounted for this. First, they had to share a part of control time, previously at their own disposal, with a corresponding system operator. Secondly, far fewer complex decisions required the mill operator's active technological involvement in the remaining time, since control board operators tried to exploit the system to the full in as many of these complex cases as possible. In this they were supported by plant management, and it also suited both algorithm developers and instrument designers. Gradually, elements of equipment maintenance crept into the mill operators' work. With the abundance of some instruments and with certain modifications in their design, the maintenance grew in complexity and demanded special attention in equipment adjustment. The instrumentation continually drew on the experience and aid of mill operators who found themselves naturally involved in their new technological environment. While the system was being integrated, the volume of their work grew continuously.

(4) Changes in working conditions: The physical working conditions changed to a certain extent. Although the overall dust content in the section decreased, mill operators had to spend more time near the mills where dust concentration and noise level were the highest. The introduction of system components designed to reduce significantly the leakage of ground cement and gypsum lagged behind, but the prospects here looked promising. Noise control produced no immediate results.

The social conditions of work were subject to considerable disarrangement. The main result was the loss of status on the part of the mill operator. The "rival" group of central control board operators, by the mere fact of their existence and activity, raised questions regarding the need for mill operators. An intergroup tension imparted a feeling of restlessness at work, because of disagreement on the division of control time. Mill operators' prestige fell in the eyes of plant managers, and their self-appraisal and satisfaction with work declined. The economic conditions, however, remained unaltered.

(5) System assessment. The initial attitude of the mill operators towards the innovation may be defined as restrained admonition expressed in the phrase: "Haven't we been equal to the job?" Fundamentally, their attitude did not change throughout the implementation of the system. In the order of events, the introduction of the milling subsystem followed the kilning subsystem. All the doubts on the part of the kiln operators spread wave-like through socio-cultural channels, reaching other personnel groups. Therefore the mill operators, aware of a possibly favourable solution to their problem, expressed appropriate expectations, and for this reason helped system developers and designers in working out a number of specifically technical decisions. Similar to all other cases, the attitude to transfers and retraining were correlated with workers' age: the younger they were, the more willingness they displayed.

2.4.3. Solutions to social and technical problems

Similar to the preceding case history, the study of the social problems, which emerged lagged behind. Social solutions had to be sought at the stage when contradictions had already been detected, and the search was conducted in conformity with the mutual adaptation principle.

(1) Adaptation of the system. Milling process control automation, as noted earlier, was distinguished by the fact that the system, as compared with mill operators proved to be more efficient technologically and could essentially replace them. However, since the discharge of mill operators was impossible for social and legal reasons, their coexistence at the initial stages was an accepted fact. At the same time it signified a delay in the all-round integration of the system.

(2) Personnel adaptation. New work stations for the mill operators and their assistants were not arranged immediately. It was advisable to find a way to include them in the system that would preserve as many

features of their former work nature as possible and require a minimum of occupational and psychological readjustment. Considering the current tendencies in changing the work of mill operators (see section 2.4.2 (3), above), the following formula was chosen: the central control board operator controls the process, while the mill operator controls the equipment. The new position of the mill operator was designated "equipment overseer". The mill operators underwent brief full-time retraining to master the instrumentation of the equipment. In this new capacity, they partly combined the functions of an adjuster and an electrician. Their wages remained on the same level. As to assistant operators, they were reassigned to various production units in accordance with their skills and previous wage levels.

2.5 Conclusions

The two case studies of work automation in cement production, reviewed above, provide a field for generalisations that may probably go beyond the framework of this study.

2.5.1 Knowledge obtained

First, it should be emphasised that a social study of work automation should not only deal with the after-effects of the process viewed from a human aspect, but should also facilitate the social conditions for its implementation. In our opinion, the most appropriate methodological paradigm of such a study is the treatment of the process as an integrated social and technological innovation. The implication is that the development of automated systems should go beyond mere invention to innovation planning; this would formulate technological and social objectives and develop organisational, technological and sociological procedures for its implementation.

Proceeding from this concept, we can identify the following characteristic social planning errors resulting from an underestimation of the stated methodology, especially at the early stages of innovation. First, the appearance of "rival" groups in each case was a result of inadequate research into the social conditions of the innovation; moreover, it was the emergence of intergroup tensions that presented the greatest obstacles. Second, there was negligence on the part of system developers and plant managers in paying attention to the assessment of the system both by the personnel involved and by the broader socio-cultural environment. The tendency of many engineers and managers to attribute a negative assessment by workers to a mere psychological barrier diverts the problem solution from the objective causes underlying definite group responses to the subjective sphere of consciousness, emotions, habits and the like.

2.5.2 Social costs

It must be admitted that even the most thorough social analysis of procedures for mutual adaptation of worker and machine cannot always guarantee against heavy social costs. The studies cited status losses, social tensions and other failings. Still, the most serious and widespread social cost of technological progress is the loss of skill, which in fact has a significant autonomous value and becomes a social property. For instance, with the introduction of program-controlled machine tools, lathe operators no longer need skills and experience. In our case there was also a danger that the new tasks of those involved in the system would make the work monotonous. Special efforts were made to counteract this negative tendency. In other words,labour automation can often give rise to contradictions between rational and human value orientations. Moreover, contradictions may also emerge even within values: for example, reductions in industrial hazards and measures to improve physical working conditions sometimes trigger a negative response from the workers involved because the measures deprive them of special supplements to their wages. Such problems cannot always be provided with an unambiguous solution.

2.5.3 Automation efficiency

The production and social efficiency of automation can be integrated by means of diversified mutual adaptation of the automatic system and the social organisation of the enterprise. In this sense the above-mentioned "autopilot" principle could be a promising conceptual trend in the development of forms of mutual adaptation.

It is also evident that a search for specific solutions to social and technical problems should be based on universally approved principles. Yet provisions should also be made for a unique solution in each particular case, taking account of the specific characteristics of the personnel, environment and work situations.

The efficiency of process control automation manifested itself in changes that affected the socio-cultural environment of the plant. This observations is interesting in view of the character of this case study (see section 2.2.1) and with regard to the role of socio-cultural factors in the implementation of the entire innovation (see section 2.2.2 and elsewhere in the case studies).

Emphasis should be laid on a change that took place in the balance between formal organisational and community elements of the plant's social organisation. This balance was changing in a maner that was characterised by the growing role of the former to a certain detriment of the latter. Let us consider its particular manifestations.

Kilning and milling processes made provisions, as noted earlier, for taking regular samples of the product for a quality analysis. The sampling intervals were not, however, strictly observed by laboratory technicians, with time lags amounting up to 40 minutes. The operators did not object, however, although such lapses in time impaired their process control capabilities, because neighbourly, kinship and other relationships between many plant workers overlapped their official relationships. With the introduction of the automated system, the exact time of sampling and its results came to be systematically recorded, thus requiring greater exactness from the laboratory technicians.

Formerly, workers often used oral, confidential channels for information exchange, even in those cases when the exchange of information demanded the use of official channels and documents. The negative results of this practice were inadequate execution of recordings, discrepancies in the content of information between the two channels, information leaks unacceptable to some managers, etc. The new system demanded a strict record of essential data, which enhanced the reliability and unambiguousness of information.

The distribution of official duties changed in the same manner. The overlapping of social and production relationships used to result in some cases in the shifting of official duty boundaries of operators and their assistants. Sometimes this could be conducive to extensive interchangeability of workers, but more often the practice resulted in lower quality of work and in vagueness of responsibilities. The introduction of the automated system brought about the revision of official duties of workers both directly involved in the innovative process and indirectly related to it (kilning and milling process engineers and shop superintendents). The criteria and parameters for evaluating their work were also revised.

Thus, the effects of the integration of the system in the plant proved to go beyond purely production objectives.

Notes

[1] Scientific Research Institute for Labour: "Mezhotraslevye metodicheskie rekomendatsii po organizatsii, normirovaniyu i oplate truda rabochikh v usloviyakh avtomatizirovannogo proizvostva [Interbranch recommendations on methods of work organisation, norm-setting and remuneration for workers engaged in automated production] (Moscow, 1979), pp. 8-10.

[2] N.A. Tikhonov: Guidelines for the economic and social development of the USSR for 1981-85 and for the period ending in 1990 (Moscow, Novosti Press Agency Publishing House, 1981), p. 19.

[3] New Times, No. 11, Mar. 1981, p. 26.

[4] Effektivnost narodnogo khozyaistva [The efficiency of the national economy] (Moscow, Nanka, 1981), p. 101.

[5] I.E. Rubtsov: Nauchno-tekhnichesky progress v usloviyakh razvitogo sotsialisticheskogo obshchestva [Scientific and technical progress in the developed socialist society] (Moscow, Misl, 1975), p. 141.

[6] Constitution (Fundamental law) of the Union of Soviet Socialist Republics (Moscow, Novosti Press Agency Publishing House, 1977), p. 12.

[7] Tikhonov, op. cit., p. 17.

[8] This section is based mainly on research of the Academy of Sciences, as described in A. Zvorikyn: "Automation and some socio-psychological problems of work", in Work and Technology, 1977.

[9] A. Osipov: "Uslovia effektivnosti raboty brigad" [Conditions for efficient brigade work], in Sotsialistichesky trud [Socialist Labour], 1981, No. 6, p. 60.

[10] "Tekhnika-organizatsia-proizvoditelnost truda" [Technology-organisation-labour productivity], ibid., p. 53.

[11] N. Lobanov et al.: "O chem svidetelstvyet opyt predpriyaty Leningrada" [Lessons from the experience of factories in Leningrad], ibid., 1981, No. 12.

[12] I. Kirillov: "Novosibirsky metod zhivet i rabotaet po povysheniyu effektivnosti" [The Novosibirsk method of improving efficiency works], ibid., 1980, No. 10, p. 53.

[13] Mezhotraslevye trebovania i normativnye materialy po NOT, kotorye dolzhny uchityvatsya pri proektirovany novykh i rekonstruktsy deistvuyushchikh predpriyaty, razrabotke tekhnologicheskikh protsessov i oborudovania [Scientific work organisation - interbranch requirements and standard-setting documents to be studied when designing new undertakings, reconstructing existing undertakings and planning technological processes and equipment] (Moscow, Goscumtrud, 1977).

[14] M. Begidzhanov: "Uchet trebovany NOT pri proektipovanii: nuzhna kvalifitsirovannaya ekspertiza" [The consideration of scientific work organisation requirements in planning and design: the need for expert advice], in Sotsialistichesky trud, 1981, No. 12, p. 56.

[15] ibid., p. 55.

[16] M. Glyantsev: "Vnedrenie NOT v 1978 g" [The introduction of scientific work organisation in 1978], ibid., 1978, No. 3, p. 69.

[17] For further information, see: A. Louzine: Work organisation models in the USSR (Geneva, ILO, 1980; doc. CONDI/T/1980/4; mimeographed).

[18] This case study will be presented in less detail than the previous one, since it involves mainly the application of principles developed in the case of raw materials kilning.

CHAPTER XV

JOB DESIGN AND AUTOMATION IN THE UNITED KINGDOM

1. National background

1.1 Introduction

During the seventies there was considerable interest in job design and QWL in the United Kingdom. The Department of Employment established the Work Research Unit, and numerous research projects, conferences and workshops focused attention upon the subject. However, the extent to which practice in industrial organisations changed as a result is probably marginal. Interest in the early eighties is focusing more on means of utilising new technology to improve performance. This includes consideration of both the processes appropriate for planning and implementing change, and the impact upon jobs and, more broadly, employment.

In the first part of this chapter we present information about the national economy, showing the shift from traditional to high technology industries and the considerable dependence upon relatively small enterprises. Government efforts to encourage the development of new technology and its adoption are described, as are changing employment patterns, along with rising unemployment and measures to alleviate its impact. Background information is provided on employment legislation and the development of trade unions, which in part explains the complex structure now existing. Finally, the development of ideas in job design is outlined, along with the more recent concept of technology agreements.

This background information is intended as a base from which the reader can interpret the case material presented in the second half of the chapter. The case studies presented are intended as indicative of developments rather than as examples of "successful" change adopting any one theoretical or ideological approach.

1.2 The national economy

Great Britain[1] exports a greater proportion of its production of goods and services than any other industrialised nation of comparable size - approximately one-third of GDP. About 28 per cent of total domestic output results from manufacturing, with services accounting for about 62 per cent.[2] Great Britain accounts for 5.5 per cent of total world trade. It is one of the world's largest importers of agricultural products, raw materials and semi-manufactures and is among the largest exporters of aerospace products, motor vehicles, electrical equipment, finished textiles and most types of machinery. It is also developing as an oil exporter.

The British economy is largely based on private enterprise, although some state-owned enterprises are amongst the largest employers. Some enterprises have both public and private participation. At present, direct state participation in industry is mainly effected through public corporations sponsored by government departments and responsible for particular sectors of industrial activity. Included in this category are most of Great Britain's energy and transport sectors, and several within manufacturing. Finance for capital expenditure not available from internal sources is mainly provided by interest-bearing loans from the Exchequer.

During the period following the Second World War, Great Britain enjoyed rising production and, until 1970, a low level of unemployment (2.5 per cent or less). However, economic growth, averaging 2-3 per cent up to 1971, was slower than in most other Western European countries. In addition, despite a considerable contribution from invisible earnings, a high balance-of-payments deficit on current account was seen as a major economic problem by successive governments.

In 1973, the economy grew by about 7 per cent. However, general inflation and sharp commodity price increases in 1972-73 led to recession in the United Kingdom as well as more generally. Recovery there appears also to have been weaker than elsewhere. For some time around 1979 there was a rise in SDP, partly as a result of the exploitation of North Sea oil and gas. However, GDP declined sharply in 1980 as a result of heavy destocking and reduction in fixed investment. Inflation has been high since 1973 (averaging 16 per cent per annum between 1973 and 1978) and unemployment has risen sharply since 1979.

Figures relating to the changing contribution to GNP from the various industrial sectors between 1969 and 1979 show a recent expansion in mining and quarrying (rising from 1.6 per cent to 4.8 per cent of GNP, due in the main to North Sea oil and gas) and in service industries, including insurance, banking and finance (from 6.8 per cent to 9.1 per cent) and health and education (from 5.3 per cent to 6.7 per cent). This was coupled with a sharp decrease in manufacturing over the period, from 33 per cent to 27.9 per cent.

Personal incomes before tax at current prices rose rapidly from 1969 onwards. The level of competitiveness, however, is estimated as being 40-50 per cent less favourable in 1980 than in 1978 on the basis of relative labour costs.[3] This results from British unit labour costs increasing faster than those of competitors, as well as from exchange rate appreciation. However, recent reports from some manufacturing companies would seem to indicate improvements in productivity, and hence a greater degree of price competitiveness.

1.3 Manufacturing industries

As indicated earlier, manufacturing is carried out mainly by private enterprise, exceptions being iron and steel, shipbuilding and aerospace industries, with some manufactured goods being produced also by British Railways and the Ministry of Defence. The Government also has a shareholding in a number of companies, while shareholdings in some other companies are held for the Government by the National Enterprise Board.

The Report on the Census of Production, 1977 reported that 69 per cent of the 108,028 establishments in manufacturing employed fewer than 20 employees and accounted for 7 per cent of total employment; 28 per cent had between 20 and 499 employees and accounted for 39 per cent of total employment; less than 2 per cent had between 500 and 1,499 employees, accounting for 21 per cent of employment; while 608 establishments employed more than 1,500 workers but accounted for 33 per cent of employment. The larger establishments tended to be in engineering, metal and vehicle manufacture.

Regarding the various sectors of manufacturing, expansion over the last decade has taken place in industries using advanced technologies (e.g. electronics, instrument engineering, chemicals and paper, printing and publishing). Decline has been most noticeable in the more traditional industries of coal, metal manufacture, mechanical engineering, shipbuilding and marine engineering, vehicles and textiles. The decline has accelerated in this latter group since 1979 (in metal manufacture, for example, output in the first ten months of 1980 was 28 per cent below the average level in 1979).[4]

Investment in manufacturing tends to follow the trade cycle, with a time lag between an upturn in trade and increased investment. In 1979 approximately 22 per cent of all manufacturing investment was apportioned to engineering, shipbuilding and metal goods, 18 per cent to chemicals, 13 per cent to vehicles, and 4 per cent to coal and petroleum products. Of the £6,583 million invested directly, 78 per cent went to plant and machinery. The total investment was enhanced by an increase in leasing rather than purchase to the value of approximately £750 million.

1.4 Scientific and technological innovation

In 1978, 2 per cent of GDP (£3,250 million) was expended on scientific research. Approximately half of the total was financed by the Government, about 50 per cent of this in its own establishments. Other government support is given to both universities and industry.

The Department of Education and Science is responsible, in the main, for scientific research, while the Department of Industry has responsibility for technology. Support for research into areas such as micro-electronics and computer applications (e.g. industrial robotics, information and communication systems) comes from the Science Research Council, one of five autonomous research councils funded mainly by the Department of Education and Science. Much of this research funding is used to support research in universities and similar institutions. The Government has also assisted companies with similar interests to form research associations, of which 38 exist.

A considerable amount of research is also conducted by private sector firms or on their behalf by research associations. In 1978, private industry expended £2,324 million on research.

The National Research Development Corporation (NRDC) was set up in 1949 as an independent public corporation to promote the development and application of new technology. It aims to exploit inventions from universities, research councils and government establishments and to provide finance for innovation by industrial companies. In addition to some 6,000 patents and 500 licensees, it has around 600 investments in development projects, of which about half are joint ventures with industrial companies. It has also been required to operate as a commercial enterprise. Projects include areas of development such as PCBs, computers and software, and CAD. The NRDC is now to be merged with the National Enterprise Board to form the British Technology Corporation. Priority is to be given to developing expertise in robot production and application.

1.5 Education

The most recent trend in education has been a substantial drop in the number of school children following a fall in the birth rate, a slowing down in the demand for higher education, and a reduction in public expenditure on education.

The proportion of pupils entering full-time education on leaving school is shown in table 82. The proportion of girls entering university increased between 1966 and 1978, but still fell short of that for boys. The proportion entering other types of higher education fell during the period, largely as a result of reduced provision for teacher training particularly at sub-degree level. However, there has been a marked rise in the level of qualifications, and in the number of young people who obtain some sort of qualification; in 1978-79, almost 60 per cent of the 25-29 age group held minimum qualifications (Certificate of Secondary Education, grades 2-5 or an apprenticeship), as compared with only 25 per cent of the 60-64 age group.

1.6 Employment

1.6.1 Structure of the labour force

The total working population of Great Britain in mid-1979 stood at 26.4 million out of a total population of 48 million, including 22.8 million employees (13.3 million men and 9.5 million women) and nearly 2 million employers or self-employed.

Table 83 gives an indication of the growth in employment during the last decade. A slight fall during the late sixties was due to increased proportions in full-time education. During the seventies there was a steady increase in the total working population. The proportion of women in the labour force continued to increase, particularly those working part time. At the same time, there has been a marked employment shift during the last decade from manufacturing to service industries, the proportion of total employees in the latter rising from 50.8 per cent in 1969 to 58.3 per cent in 1979. This reflects the decline in some sectors of manufacture referred to earlier, as well as increased productivity through automation and mechanisation.

Unemployment during the seventies increased at a fairly steady rate until 1979, since when the rate has accelerated considerably.

Table 82: England and Wales: destination of school-leavers by sex, 1966-78 (percentages)

Item	Boys				Girls			
	1966	1970	1975	1978	1966	1970	1975	1978
Pupils entering full-time further education as a percentage of all school-leavers (by type of course)								
Degree	8.9	9.0	8.8	8.7	4.3	5.3	5.5	6.0
Other higher education	2.2	2.0	0.9	0.6	5.3	5.5	2.5	1.2
OND/ONC/GCE Advanced and Ordinary level	3.3	3.9	4.2	3.7	2.6	2.6	4.1	4.3
Other full-time courses	3.6	4.7	5.0	4.1	8.9	10.8	12.4	14.4
Total	18.0	19.5	18.9	17.2	21.1	24.1	25.4	25.9
School-leavers seeking employment on leaving school	82.0	80.5	81.1	82.8	78.9	75.9	74.6	74.1

Source: Department of Education and Science: Statistics of Education, Vol. 2, School leavers, CSE and GCE (London, 1979).

Table 83: Great Britain: labour force trends, 1970-79 (thousands)[1]

Year	Employees in employment[2]	Employers and self-employed	Unemployed[3]	Armed Forces[4]	Total working population
1970	22 479	1 902	555	372	25 308
1971	22 122	1 909	724	368	25 123
1972	22 121	1 899	804	371	25 195
1973	22 664	1 947	575	361	25 547
1974	22 790	1 925	542	345	25 602
1975	22 710	1 886[5]	866	336	25 798
1976	22 543	1 886[5]	1 332	336	26 097
1977	22 619	1 886[5]	1 450	327	26 282
1978	22 666	1 886[5]	1 446	318	26 313
1979	22 825	1 886[5]	1 344	314	26 369

[1] Discrepancies between totals and the sums of their constituent parts are due to rounding.

[2] Part-time workers are counted as units.

[3] Excluding adult students.

[4] Including ex-Service personnel on leave after completing their service.

[5] Estimates.

Source: Department of Employment Gazette.

1.6.2 Government policy

The Government intervenes in the labour market by providing employment services, by supplementing the training undertaken by employers, by regional grant aid, by measures to alleviate unemployment and through legislation. Legislation, to date, has concentrated upon the regulation of terms and conditions of employment, of industrial relations and of the health and safety of the workplace. The Department of Employment deals with employment policy, industrial relations and pay policy, and unemployment benefits. The Manpower Services Commission (MSC), which is separate from the Government but answerable to the Secretaries of State for Employment and largely government funded, is responsible for advising the Government on manpower policy issues. The MSC's role is to assist in the development of resources and their placement.

A variety of schemes aimed at reducing the impact of unemployment have also been proposed or introduced by the Government. These include schemes to create jobs by encouraging employers to use short-time working as an alternative to redundancy, temporary employment programmes in special development and inner city areas, and a "job release" scheme to encourage early retirement. Training opportunities have also been created through a youth opportunities scheme and support for training places in industry. Measures taken to encourage the development of small businesses is also intended in part to alleviate unemployment.

1.6.3 Industrial training boards

When an individual is in employment, responsibility for industrial and commercial training rests with the employer. Since the Industrial Training Legislation of the early sixties, 23 industrial training boards and one industrial training committee have been in existence, and these now cover firms which include about 60 per cent of all employees. The boards receive government funding via the MSC to meet their advisory services and to enable them to make grants to employers to encourage training activities of national importance, where skill shortages have been identified. Some boards operate a levy scheme based on the employer's total wage bill and return monies where the standard and level of training meets the board's criteria. Others operate a levy-exemption scheme by conducting annual audits, which if satisfactory exempt the employer from paying the levy. The work of the training boards is currently being reviewed.

The MSC also promotes training for some 10 million employees in industries not covered by the industrial training boards.

1.6.4 Legislation

The Employment Protection (Consolidation) Act of 1978 consolidated earlier employment legislation and provided certain safeguards for employees in relation to their conditions of employment. Employers are required to give employees written information on their terms and conditions of employment, the disciplinary rules applicable to them and the procedure available where an employee has a grievance about his/her employee or is dissatisfied with any disciplinary decision, and minimum periods of notice to terminate the contract. Where employees have 104 weeks length of service, they are entitled to lump-sum redundancy payments if their jobs cease to exist and the employer has no suitable alternative work. Unfair dismissals and trade union rights are also covered.

Discrimination when recruiting workers or in their treatment in regards to terms and conditions of employment, promotion, transfer, training and access to other benefits or dismissal is forbidden under the Race Relations Act, 1976, and the Sex Discrimination Act, 1975. Under the 1970 Equal Pay Act, women are entitled to equal pay with men when doing work that is the same or broadly similar, or work which has been given an equal value under a job evaluation scheme.

Regarding occupational health and safety, employers have a duty at common and criminal law to take reasonable care of their employees and to provide safe working conditions, while employees have a duty of care towards each other, as well as for their own safety. A number of statutes lay down minimum standards of safety in certain kinds of workplaces or work. The Health and Safety at Work Act, 1974, reorganised the system under which safety and health at work was safeguarded, and extended it to cover everyone at work, as well as to protect the general public from industrial hazards. Under this Act, recognised trade unions may appoint safety representatives to represent the employees in a workplace. Additionally, two or more safety representatives may make a written request to the employer to establish a safety committee, which must then be set up within three months.

The functions of safety representatives include investigating dangerous occurrences and potential hazards at the workplace, and workers' complaints about health and safety matters; examining the causes of workplace accidents; making representations to the employer on health and safety matters; and carrying out routine and special inspections of the workplace. The employer must consult safety representatives in developing and checking the effectiveness of measures to ensure health and safety at work, as well as making available, with certain important exceptions, any information necessary for their tasks. Safety representatives must be allowed time for training and for their work without loss of pay. The industrial training boards include specific provisions for safety training in their recommendations.

1.7 The industrial relations system

Following the Second World War, white-collar trade unionism, particularly in the public sector, grew rapidly while membership of manual worker unions decreased. Trade union amalgamations took place. There was a general decline in the influence of employers' associations and a growth of plant bargaining with its increasing emphasis upon the role of the shop steward.

Concern in the sixties over the state of industrial relations led to the setting up of a Royal Commission on Trade Unions and Employers' Associations under Lord Donovan. This Commission identified two systems of industrial relations: a formal system based on official institutions and resulting in industry-wide collective agreements; and an informal system, normally at plant level and involved in local negotiations. The Commission recommended that the informal system should be more effectively regulated by formal plant agreements. An attempt by the Labour Government to introduce legislation in 1969, contrary to Donovan, proposed certain reserve powers for the Secretary of State, including enforcement of a conciliation pause, compulsory strike ballots and measures to deal with inter-union ballots. However, there was strong opposition to these proposals from the trade unions, who preferred voluntary regulation.

The Conservative Government in its 1971 Industrial Relations Act extended the influence of the Law. It established the National Industrial Relations Court, gave the recently formed Commission on Industrial Relations a statutory footing, and revised and extended the jurisdiction of certain other bodies. However, there was considerable trade union opposition which contributed to the fall of the Government in 1974. The new Labour Government formed an understanding with the trade unions, the "social contract". In return for co-operation from the trade unions in tackling the nation's economic problems, the Government committed itself to certain measures including the repeal of the 1971 Act, which was carried out under the 1974 Trade Union and Labour Relations Act, and followed by the Trade Union and Labour Relations Act, 1976 and the Employment Protection Act, 1975. The Conservative Government followed these with the Employment Act, in 1980. Between them these statutes make certain presumptions about collective bargaining; provide a legal basis for trade unions and employers' associations; establish procedures for determining the independence of trade unions; define for certain purposes union membership agreements; give rights to independent and recognised trade unions covering the disclosure of information by employers, consultation over proposed redundancies, time off for trade union officials, the appointment of safety representatives and the administration of pension funds; provide for conciliation and arbitration machinery; and give immunity for certain acts done in contemplation of furtherance of a trade dispute.

At the end of 1979 there were 477 registered trade unions, the long-term trend having been towards a reduced number. These unions can be classified as "general", "craft", "industrial" and "white-collar", although membership is no longer restricted purely to the one category of worker (e.g. general unions include craft workers). Total membership of trade unions stood at about 13.05 million at the end of 1978.

At the end of 1979 there were 191 listed employers' federations, but probably another 260 unlisted associations. These associations commonly negotiate with trade unions at industry level, provide machinery for the resolution of disputes, and give advice and assistance to their members on a range of employment and manpower matters. A few offer financial support to members affected by industrial action.

Collective bargaining is not universal throughout industry. The 1978 New Earnings Survey found that about 30 per cent of full-time employees in Britain were not affected directly or indirectly by any collective agreement. Nevertheless, collective bargaining is widespread, and can take place at national or industry, company or plant level. Normally in private industry, national agreements establish the minimum terms and conditions of employment, whereas in the public sector and nationalised industries, they set actual pay and conditions.

Bargaining at company and plant level is widespread. Where the company is not federated to an employers' association, these negotiations will be the sole determinant of pay and conditions. Local bargaining may supplement a national award. At workshop level, shop stewards and managers are often involved in bargaining over detailed conditions of employment such as the allocation of work, specific allowances, and overtime arrangements. Custom and practice has an important influence over such bargaining, where rules are rarely written.

Consultative machinery also exists in many organisations to provide for discussion about matters of common concern falling outside the scope of the negotiating machinery. Within the nationalised industries, such machinery is highly developed. In private industry this machinery is normally company based.

Regarding the number of industrial stoppages and workers involved, there was a decline in the number of stoppages beginning in each year over the period 1969-79. However, this was not the case generally in the number of workers involved in these stoppages. In addition, the number of working days lost through industrial stoppages has fluctuated during the period. Particularly noticeable are the sharp increases in metals, engineering, shipbuilding and vehicles and in the "all other industries and services" sector (the 1979 steel strike contributed significantly). The 1980 figures showed a sharp decline: during the second half of the year, the lowest number of working days was lost for any comparable period since 1966, and the number of stoppages reported was the lowest for over 40 years.

1.8 The quality of working life, job design and work organisation

In the previous sections we described the legal framework relating to employer-employee relationships, government measures to ensure adequate availability of skills and the industrial relations system. Many of the aspects covered by legislation relate specifically to QWL. The industrial relations system gives some scope for measures to be taken aimed at improvement.

Turning specifically to the nature of work, job design and work organisation, early initiatives in this field were evident during the First World War, in common with many other countries. After the War, the Industrial Health Research Board, on behalf of the Government and the independent National Institute for Industrial Psychology, continued studies relating in particular to occupational testing and selection of production workers.

Such work was accelerated during the period 1939-45, with particular focus on selection and training. In the late forties, the Government set up a Committee on Industrial Productivity, which in turn established a "human factors panel" to sponsor research in the area of work, and to a lesser extent work organisation. During this period the Tavistock Institute for Human Relations grew from the Tavistock Clinic, and from these beginnings developed the major stream of work focused around what is now known as the socio-technical systems approach. Similar work at the Glacier Metal Company focused upon organisational design, and later, work was to be started by the Department of Scientific and Industrial Research and the Medical Research Council (MRC).

In 1970 the Department of Employment commissioned a study of industrial and social changes affecting the behaviour and attitudes of people at work. This followed a period of fairly intense research activity, some of which was focused upon the attitudes of workers, problems of job satisfaction, labour turnover, absenteeism, etc. The study resulted in the publication of a paper in 1973 on QWL,[5] which recommended among other things the setting up of a body to co-ordinate and work out needs for a programme of development in industrial/commercial settings. From this developed the idea of the Department of Employment Work Research Unit which was set up in 1974 under a joint tripartite steering committee comprising the Trades Union Congress (TUC), the Confederation of British Industry (CBI) and Department of Employment representatives.

Around this time the Social Science Research Council (SSRC), through its Management and Industrial Relations Committee, was developing its interests and views in this area having, of course, previously funded research in organisations such as the Tavistock Institute and various university research groups. An Anglo-Swedish Conference was held and various other activities undertaken, leading indirectly to the establishment of the SSRC-MRC Work Organisation Working Party which made recommendations for the development of research in this area at the beginning of 1978. Thus, during the mid-seventies, the principal source of funding for research (mainly action research) in this area was the Department of Employment and, to a lesser extent, the SSRC and the MRC (influenced by the Department of Employment Work Research Unit).

In view of its prominence in the field during this period, it is perhaps appropriate to summarise the objectives of the Department of Employment Work Research Unit as follows:

- promotion of activities in the field of job satisfaction and restructuring;

- provision of consultancy and advice to companies;

- writing of reports, papers and notes to disseminate information on the subject;

- compilation of information on the activities of similar bodies in other countries;

- provision of training courses and appreciation courses for individuals and organisations; and

- development, sponsoring and control of research, particularly action research, to provide demonstration projects through university and other research groups.

This last item led to the development of ten such sponsored projects.

The SSRC Work Organisation Working Party in its 1978 report[6] identified several principal strengths in respect of recent experience in the United Kingdom, in particular research in work organisation. These were categorised as -

- work relating to the analysis of tasks and jobs using quantitative and qualitative measures and instruments;

- research using the socio-technical systems approach as developed by the Tavistock Institute;

- investigation of workers' attitudes and attitude changes, particularly with regard to job satisfaction, through the survey questionnaire method;

- work of a change agency or action-research type involving the participation of external change agents in situations within organisations;

- work associated with particular technological ventures, in particular relating to the nature of employment, jobs and workers' attitudes in the technological and changing technological situations (e.g. computers and aspects of engineering work, in particular mechanisation and automation in mass and flow production).

It was concluded, however, that current research was limited in certain respects. For example, there appeared to be a narrow range of situations in which research had been undertaken, the majority in manufacturing industry. Much of the work which had been carried out both in terms of action changes and more fundamental research was believed to reflect a managerial philosophy and had been undertaken with the support, but not necessarily the active collaboration, of trade unions. The focus of research reflected the particular disciplines of those external agents and researchers who were involved (mainly psychology and sociology), engineers and technologists having had limited involvement in work in this area, as was the case also with economists and anthropologists. It is particularly noticeable, in view of recent developments relating to total employment, the labour market, etc, that there had been little formal attempt to study the relationships of work, work design, work organisation and labour markets. Moreover, while pay restraint had been a major feature of the recent past in the economy of the United Kingdom, there had been little formal attempt to reconcile work organisation objectives with pay policy objectives.

As with work in other countries, there is an awareness of inadequacies both in terms of what has been sought and, more importantly, what has been achieved in the field of assessment and evaluation of work, in particular action research. Thus it might be argued that much of the work which had been done is of solitary significance and because of inadequacies in design and limitations in reporting, it is difficult to bring together past experience to form a broader and agreed view, for an assessment of the current state of art in this area. The SSRC recently earmarked funds for further research in the areas identified.

In a recent study of new forms of work organisation in the United Kingdom,[7] the authors reported that in their view the amount of significant change in work organisation aimed at improving QWL was very limited. From a survey of recent experiences, they were unable to identify any apparent relationship between effective procedures and approaches to work organisation change and organisational size, technology, ownership, etc. A variety of approaches had been employed and the organisations responsible in many cases claimed success for the particular strategy adopted. The authors of the study concluded that it seemed unlikely that the prescription of any particular approach, except in broad outline, would be appropriate or desirable. They also suggested that any change programme which required or necessitated the adoption of procedures, processes and attitudes which represented major breaks with traditional or existing practice or custom was more likely to encounter obstacles and problems and was less likely to be successful than a strategy derived from existing customs and practice, and employing existing frameworks and procedures. Further, it was put forward that in future work organisation as a specific field of activity was less likely to exist in isolation, but would rather constitute one of several complementary and necessary means towards the achievement of broadly stated ends. The emphasis would therefore be upon collective consideration of, and joint approaches to, more complex issues.

1.9 Technology agreements

It is widely agreed that the United Kingdom must adapt to new technology at least as fast as its competitors. A recent report by the Department of Employment[8] identified the constraints on change as shortages of key skills and industrial relations considerations, including management and trade union failure to adapt practices to the requirements of new technology.

The CBI and the TUC are in agreement on certain key issues relating to technical change. They state -

> A period of rapid and major technological change is increasing the need for a common understanding of such factors as world market requirements, investment and research and development, as well as the implications for employment. A joint approach is clearly the best way to develop such an understanding so that the fears and suspicions attendant on this process can be allayed, and the opportunities grasped.[9]

Further, they assert -

> Many of the key decisions which will determine the success or failure of adopting new technologies will be taken at the level of the

> enterprise and individual plants within it. It is at these levels that trade unions and employers have the responsibility to establish mechanisms and procedures which are sufficiently clear, comprehensive and accessible to allow the process of technological change to take place continuously and beneficially. Given the complexity and diversity of our industrial and representative structure, it would be impossible and indeed undesirable to attempt to draw up a single model for universal applicability.[10]

While emphasising the inappropriateness of a single model for decision-making and implementation, the CBI and the TUC advocate communication and consultation between management and trade unions over decisions, with the shared objective of increasing mutual understanding and common consent. With reference to changes to work organisation, effective manpower planning is seen as a means of identifying jobs which will be directly affected and "measures ... to change the direction of skills and avoid the spread of monotonous and unsatisfying work".

Over the past few years, many trade unions have spelt out their policy regarding new technology and many have also negotiated new technology agreements. In formulating these technology agreements, trade unions place emphasis on areas such as procedural aspects, job security, training, distribution of benefits and ongoing monitoring. To date most of the agreements reported have covered white-collar workers.[11] The Association of Professional, Executive, Clerical and Computer Staff (APEX) in its model agreement, one of the earliest published, states: "Full analysis, through joint union/management committees, will take place in respect of the effects of new systems on job content and job satisfaction. Jobs will be carefully designed to ensure that routine and monotony are minimised."[12]

As pointed out earlier, the key decisions determining the success or failure of adopting new technology will be taken at local level within the inidividual organisation. While the CBI and TUC may reach broad agreement, it remains to be seen to what extent individual employers or trade unions accept the broad framework, and also how the framework is translated into practice at local level.

1.10 Approaches to job design and automation

As stressed earlier in this chapter, the economy of the United Kingdom is a market economy largely based on private enterprise, and government intervention is relatively limited. Government policy is aimed at reducing inflation and creating an environment in which organisations will endeavour to improve on the recent poor industrial performance. Development is being encouraged primarily in industries using advanced technologies, although the Government has intervened in some declining industries to preserve employment.

While decisions about investment in new technology are influenced by government policies and action, they are generally made at enterprise level, as are decisions about implementation. The stances of the various interested parties are also affected by the views expressed by the Government, employers' federations, the TUC and individual trade unions. However, there is widespread recognition that no single model for decision-making and implementation has universal applicability. Although agreement prior to change on certain key issues is being advocated, it would be wrong to assume that this is necessarily seen as desirable at a local level, or even that compliance will be general where technology agreements are reached. Legislation relating to health and safety, however, provides employee representatives with the opportunity to review certain aspects relating to technical change. Traditional bargaining mechanisms at local, plant and organisation levels provide opportunities for employees to influence management decisions and have been extensively used historically for this purpose. Participative approaches to designing new systems have also been adopted, varying in degree of participation and stage in the decision-making process. However, these have most frequently been at a large stage in detailed working arrangements and commissioning.

The case material presented in the second part of this chapter was obtained from organisations where there has been interest expressed in issues relating to job design and QWL. Although a diversity of approaches is presented, in each case, participants would no doubt report the project as "successful" in relation to implementing technological change. In no case was a technology agreement in operation, this being too recent an innovation for results to be available.

2. Case study 1 - Automation in biscuit manufacturing

2.1 Company background

This case is based upon experiences in introducing automation in the processes of biscuit manufacturing. The company involved is a long-established food manufacturing company with a tradition of providing for the welfare of employees. The factory concerned is one of many, but manufactures the company's total biscuit output. The biscuits are mostly chocolate or half-chocolate. Other products from the factory include dried potato, cakes and packed tea.

The factory, situated in the Nort-West of England was opened in the early fifties as a "model" factory, at the same time that the local town developed considerably. Employing over 2,000, it is major employer in the area. The history of the factory, however, is somewhat chequered. It has experienced several threats of closure, resulting from fluctuations in the fortunes of products manufactured there.

While there are many trade unions, including the Transport and General Workers Union (TGWU) covering operatives and the Amalgamated Union of Engineering Workers (AUEW) representing craft engineers, the factory has a good industrial relations history. Certainly the problems experienced by many companies moving manufacturing into the area have not spread noticeably to this factory.

During the mid-seventies the company directors established a structure to facilitate increased worker participation. This consisted of the Company Conference, involving representatives from each group within the company in regular meetings with members of the Board of Directors, each group within the company having its own conference. At factory level there was an overall consultative committee, as well as departmental-level committees. In addition, initiatives were introduced aimed at increasing the involvement of all employees. These included briefing groups and several experiments in participative job redesign. In the particular factory such experimentation was undertaken with one specific aim, that of demonstrating the appropriateness of new forms of work organisation to the new facilities. Senior shop stewards from the factory had been involved in a company-wide study of job redesign. The study reported here followed from this experimentation and was undertaken by two of the people involved, an industrial engineer from the company and an outside researcher. The initial study of automation was undertaken in order to stimulate wider debate.

The new facilities were planned in the late seventies to replace earlier biscuit manufacturing production units. In general, they are not significantly different from those seen in other companies where new biscuit manufacturing equipment has recently been installed. Such equipment is purchased from one of several suppliers, with the configuration of equipment and small modifications to customer requirements.

2.2 Developments in the manufacturing process

The manufacture and packaging of chocolate-coated biscuits requires ten basic stages. The sequence of stages is ingredient weighing, mixing, shaping, baking, cooling, chocolate enrobing (covering), cooling, packing into bags, packing into cartons and finally packing into boxes. Other related activities include the preparation of chocolate and the assembly of boxes. At various stages in manufacture the product is inspected for weight, shape and appearance. These stages in manufacture are represented diagrammatically in figure 32.

The replacement facilities differ from former units in that -

- planned output per hour is now 166 per cent of former maximum capacity;
- enrober width is increased from 36 inches to 54 inches;
- baking time is 8 min compared to 12 min;
- cooling time after enrobing is 10 min compared to 30 min.

The level of mechanisation/automation at each stage in the process and the changes introduced are detailed in table 84. (The analysis is based on work described elsewhere.)[13] Significant innovations are apparent at four stages in the process, as follows:

Figure 32: The biscuit manufacturing process

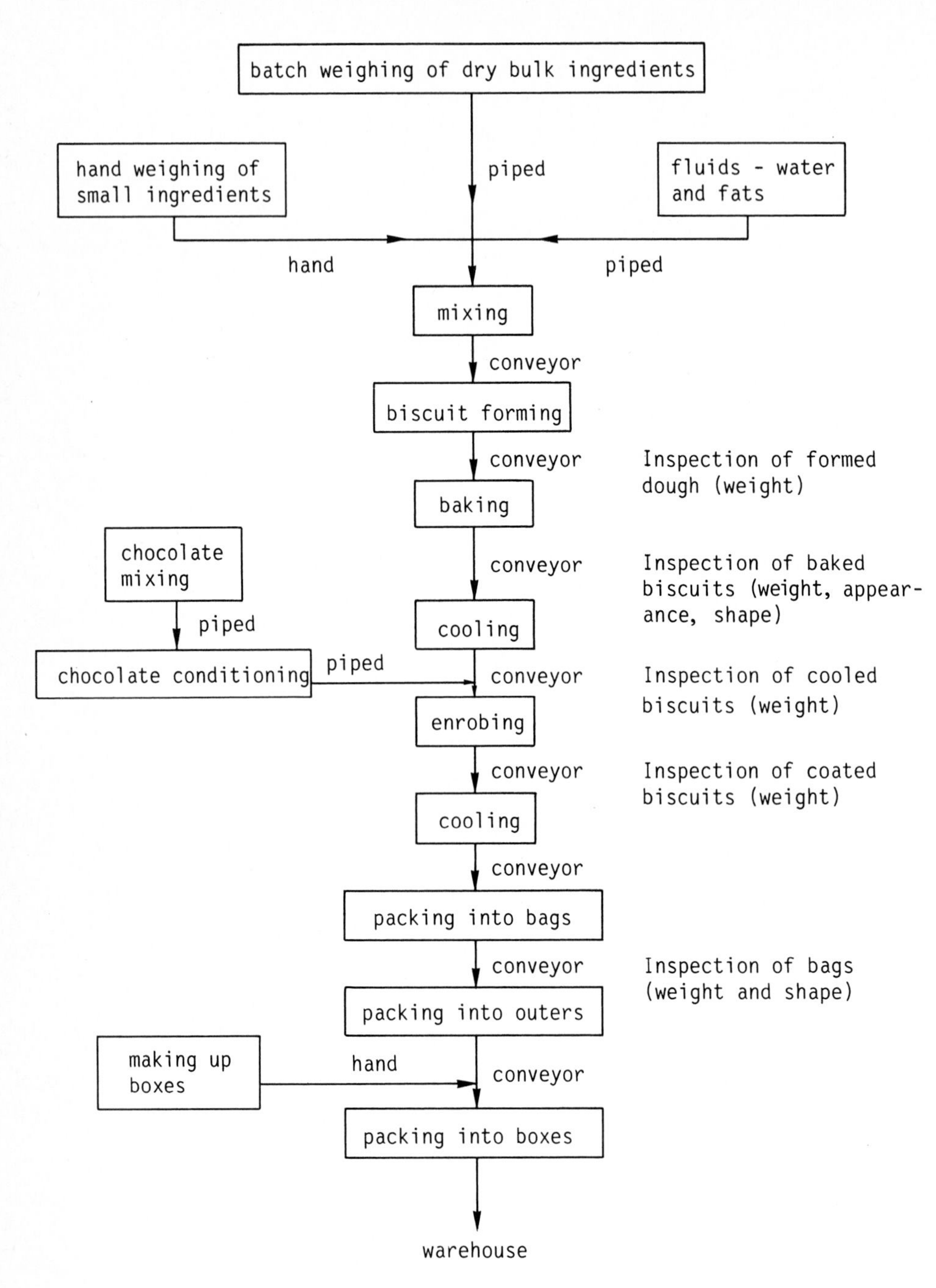

Table 84: Biscuit manufacturing: changes in the level of mechanisation/automation

Stage	Nature of material and material flow	Level of mechanisation/automation in former plants			Changes in level of mechanisation/ automation
		Process	Handling	Control	
Batch weighing	Powders; intermittent flow	Machine; automatic	Piped flow; automatic	Manual activation; automatic monitoring, regulation and rectification	Automatic monitoring of weight
Mixing	Solids and liquids; intermittent flow	Machine; automatic cycle	Conveyor; hand controlled	Manual activation, monitoring, regulation, rectification	Minimal
Forming	Semi-solid mass to discrete items; continuous flow	Machine; continuous cycle; automatic	Conveyor; continuous cycle; hand controlled	"	"
Baking	Semi-solid to solid (discrete items); continuous flow	Machine; continuous cycle; automatic	Conveyor; continuous cycle; automatic control	Manual activation; automatic monitoring and regulation, manual rectification	"
Cooling	Discrete items; continuous flow	Force-fed air cooling; continuous cycle	Conveyor; continuous cycle; hand controlled	Manual activation, monitoring, regulation and rectification	Air-cooling replaced by force-feed mechanical coolers; continuous cycle; automatic activation, monitoring, regulation and rectification
Chocolate mixing	Solid mass to liquid; continuous flow	Machine hand loaded; automatic cycle	Piped flow; continuous cycle; automatic control	"	Minimal
Chocolate conditioning	Liquid; continuous flow	Machine; automatic cycle	Piped flow; hand controlled	"	Automatic process; continuous cycle; automatic activation, monitoring, regulation and rectification
Enrobing	Liquid; discrete items; continuous flow	Machine; continuous cycle	Conveyor; automatic cycle; hand controlled	"	Minimal

Table 84 (contd.)

Stages	Nature of material and material flow	Level of mechanisation/automation in former plants			Changes in level of mechanisation/ automation
		Process	Handling	Control	
Cooling	Discrete items; continuous flow	Force-fed dual-flow air cooling; continuous cycle	Slat conveyor; automatic cycle; hand controlled	Manual activation, monitoring, regulation and rectification	Turbo-flow single-level cooling; contInous automatic activation, monitoring, regulation and rectification
Packing into bags	Discrete items; continuous flow	Manual and hand tool	Manual to and from conveyor	"	Process Machine-continuous cycle Handling Machine-continuous cycle; hand activated Control Automatic weighing machine; automatic activation, monitoring, regulation; manual rectification; sealing machine activated, manual monitoring, regulation and rectification
Sealing bags	Discrete items; continuous flow	Manual and hand tool	Manual to and from conveyor	"	
Carton assembly	Discrete items; continuous flow	Machine; continuous cycle	Machine; continuous cycle	"	None
Packing into cartons	Discrete items; continuous flow	Manual	Machine; automatic cycle; hand activated	"	None
Packing into boxes	Discrete items; continuous flow	Manual	Manual	"	None

- cooling between baking and enrobing. Baking and secondary processing in the earlier plants were carried out in separate areas with a connecting conveyor providing air cooling. The process is now continuous cycle with mechanical cooling which improves space utilisation;

- chocolate conditioning. Here fully automatic process control has been introduced, replacing manual control on former equipment;

- cooling between enrobing and packaging. A multi-level cooler fed from a central refrigeration plant has been replaced by a single-level cooler with integral refrigeration units. Process control is now fully automatic;

- the packaging operation. Hand operations including filling pre-made bags, weighing, weight adjustment and feeding bag-sealing machines have been replaced by automatic weighing, bag forming, packing and sealing.

Other technical changes have been introduced. The automatic monitoring of weighed dry ingredients will result in improved process control, since batches of incorrect weight will be rejected. The oven conveyor is now automatically cleaned during its cycle.

Many areas in which no significant changes in the level of mechanisation/automation have been introduced are highly mechanised, but still have high manual involvement in control operations. Automatic control of the process is being developed (e.g. the automatic monitoring of biscuits leaving the baking process and regulation of the forming machines), but the equipment is not yet readily available.

Modifications to a plant of this type would be possible at a later stage. At the final stages in packing, however, there is still scope for mechanisation as well as automation. The need for flexibility in packaging methods, as well as the cost of mechanical methods, appears to hinder change.

2.3 The design of new facilities

The initial feasibility study prior to the decision to invest involved both head office and local management and staff. Project engineering at the factory had responsibility for the technical aspects of the process design, and operations management for decisions concerning manning and productivitiy. Product development is factory based but marketing, at the time, was carried out in head office. The local director had the task of convincing other members of the Board of Directors that the investment should proceed.

The trade unions were kept informed of discussions and subsequent decisions through the consultative machinery. The local representatives were eager for the investment to proceed but recognised that representatives from other factories would like to see investment in their own factory. While these representatives were significantly involved in a later investigation into manufacturing facilities for tea production in the group, it does not appear that they had a major influence on the decision to invest in biscuit manufacturing facilities. The primary concern throughout was maintenance of the economic viability of the factory.

Following the decision to proceed with the investment, a project engineer was made responsible for the development of the project. Again, other interested parties were kept informed of progress and involved in decision making when felt appropriate. The trade unions did not become actively involved until negotiations at the commissioning stage concerning work allocation, manning levels and payment.

2.4 Impact of the changes upon jobs

The researchers had predicted that changes in the level of mechanisation and automation were likely to have an impact upon the numbers employed on the process, the supervisory ratio, levels of payment, career patterns and job security. The resulting changes were seen in part as a consequence of the decision to introduce more technically advanced equipment. However, these technical changes were also believed to offer the opportunity for both management and trade unions to re-examine aspects such as organisational design, job design, payment systems, structure and levels, manning levels and work practices. Change in both manning and payment levels is by tradition a subject for negotiation between management and trade union.

The technical changes made reduced the manning levels and affected the skills required for operation. Overall the process manning level was reduced in the ratio 3 to 1. The most significant changes occurred in the section concerned with the initial product wrapping. Here one machine with one machine operator at any one time replaced 20 operatives formerly engaged on predominantly manual work. This machine-operating job merited a higher grade in the job evaluation scheme than other wrapping-machine jobs in the immediate work area.

At the same time, supervisory ratios were increased. Formerly one supervisor was responsible for the mixing and baking operations and a second for the enrobing and packing activities. Both were responsible to a section manager. With the new system, supervisors were eliminated, a section manager being made responsible for the complete process from mixing to packing. Nevertheless, the ratio of plant supervision to employee was increased from a maximum of approximately 1:65 to 1:24. The increased technical complexity and the requirement for rapid corrective action when problems arose influenced management's decision to increase the level and quality of direct supervision.

The biscuit department produced a range of biscuit and wafer lines for both home and overseas markets. All products have fluctuating demand, influenced by both seasonal factors and the economic well-being of the community. Local management seeks optimum utilisation of available capacity. The replacement facilities provided high output from a comparatively small labour force. However, there was limited flexibility designed into the plant to allow rapid changes in wrapping methods employed, with the result that it was less versatile than older plants. Nevertheless, it was predicted that the new facilities would be employed fully throughout the working week.

While there was a reduction in manning requirements to operate the new plant, it was anticipated that overall there would be little change in the numbers employed, since the plant was planned to run on a three-shift basis. The employment pattern for women was to be on a part-time basis (a three-shift day) rather than full-time, as on the machinery being replaced. For men, three full-time shifts were to operate. Since emphasis was expected to be placed upon full utilisation of the new equipment rather than on other facilities, such jobs were believed likely to offer a higher degree of security to job-holders (i.e. there was less likelihood of transfer to other production units or lay-off).

The increased complexity of technology placed new demands upon personnel, especially those carrying out maintenance and supervisory functions. Changes in the nature of the operating tasks have been measured using a scheme described elsewhere,[14] and these focus attention upon the extent to which jobs possess attributes believed to contribute to job satisfaction.

Jobs on the old and new equipment were assessed according to ten characteristics, as follows:[15]

- level of variety;
- (non-)repetitiveness;
- attention with accompanying mental absorption;
- discretion and responsibility;
- employee control over aspects of the job;
- goals and achievement feedback;
- perceived contribution to a socially useful product;
- social opportunities;
- dependence upon others for task achievement;
- skill utilisation.

Tables 85 and 86 present the scores obtained for old and new jobs. A nine-point scale was used for measurement, low scores represent low levels of a particular attribute, which may therefore be seen as potentially lower in its contribution to job satisfaction. Those jobs in the former system which were identified as particularly low scoring were mostly associated with two stages, i.e. wrapping in bags and packing into cartons and outers.

In general, those jobs scoring at the mid-point or above on many of the job attributes were further enhanced as a result of the technical changes.

Table 85: Biscuit manufacturing: job and work attribute scores obtained on old plant on a nine-point scale[1]

Job title	Job/work attribute									
	Variety	(Non-) repetitiveness	Attention	Discretion/ responsibility	Control	Goals/ feedback	Perceived contribution	Social opportunities	Interdependence	Skill
Moulder operator	3.00	1.00	5.00	1.67	3.17	4.33	7.00	7.33	1.00	2.00
Oven worker	5.67	4.00	5.00	4.67	5.33	4.67	5.00	6.00	1.00	3.00
Chocolate mixer operator	6.33	5.00	2.00	4.33	3.17	6.67	9.00	6.33	1.00	5.00
Enrober operator	2.00	1.00	1.00	2.33	1.33	6.50	5.50	4.67	2.00	5.00
Cooler inspector	1.00	1.00	1.00	3.33	1.17	3.00	1.00	3.67	1.00	1.00
Bag filler	1.67	1.00	1.00	1.00	1.33	2.00	6.00	7.33	1.60	1.00
Service worker (packing)	3.67	3.00	7.00	1.00	3.17	3.00	6.00	7.67	3.67	1.00
Weigher/adjuster	1.00	1.00	1.00	1.67	1.33	2.33	5.00	7.67	1.00	3.00
Sealing bag machine operator	1.00	1.00	1.00	1.67	1.33	4.00	6.00	8.33	1.00	1.00
Aligner/seal inspector	1.00	1.00	1.00	1.33	2.17	4.00	6.00	6.67	1.67	1.00
Packing machine operator	4.00	1.00	5.00	3.33	2.17	3.67	9.00	6.33	1.00	3.00
Feeding bags to cartons	1.00	1.00	1.00	1.67	1.00	3.67	6.00	5.00	1.00	1.00
Filling outers	3.67	2.00	3.00	1.33	1.33	3.00	6.00	7.00	1.00	1.00
Labelling	2.33	1.00	4.00	1.00	2.83	2.33	5.50	5.00	2.00	1.00
Service worker (pallets)	5.67	5.00	7.00	1.00	3.33	3.67	2.00	7.67	3.00	1.00

[1] High scores indicate high motivational potential.

Table 86: Biscuit manufacturing: job and work attribute scores obtained on new plant on a nine-point scale[1]

Job title	Job/work attribute									
	Variety	(Non-) repetitiveness	Attention	Discretion/ responsibility	Control	Goals/ feedback	Perceived contribution	Social opportunities	Interdependence	Skill
Machine operator[2]	7.67	4.00	5.00	4.67	4.33	6.33	7.00	3.33	3.00	4.00
Chocolate mixer operator	6.33	4.00	6.00	5.33	3.50	5.00	7.00	2.67	3.00	1.00
Wrapping machine operator[3]	7.67	5.00	8.00	5.00	5.67	6.33	9.00	5.00	1.00	5.00

[1] High scores indicate high motivational potential.

[2] The machine operator displaced the moulder operator and oven worker.

[3] The wrapping machine operator displaced the cooler inspector, bag fillers, service workers (packing), weighers/adjusters, sealing bag machine operators, and aligner/seal inspector.

Note: The enrober operator was no longer needed as a result of improved transfer between conveyors and improved equipment. Other jobs did not change.

The wrapping machine operator on the new plant had a job scoring significantly higher on most attributes than the jobs eliminated. This job required the application of new skills including an understanding of information relating to the control of the process, and a command of procedures which enabled the worker to respond in varying conditions. The operator had a greater freedom of movement than those formerly performing related tasks by manual methods. This facilitated social interaction, while compensating for the reduced opportunities available during machine operating (noise and relative isolation being restricting factors).

Following management/trade union negotiations, the jobs of moulder operator and oven worker were integrated into one job. This change became feasible as a result of technical advances, particularly in oven temperature control. To some degree this technology was in use, however, on the former plant, and the change resulted from a re-examination of the organisation initiated by management. The new job involved the operator in controlling a larger section of the process. Additional skills and work variety resulted, but the degree to which the operator controlled his own work activity was reduced. Difficulties were experienced between management and trade unions in negotiating this change.

Chocolate mixer operators had a more repetitive task, requiring attention but being less mentally absorbing, with less direct feedback concerning achievement of goals, a reduced perceived contribution and generally lower in skill demands. Automatic control of part of their former duties accounts for the diminution of these attributes. Overall responsibility had increased, however, as had operators' ability to control their own work activity. They became responsible for a larger section of the process, were freer to move around, but had access to fewer workmates with whom to socialise.

The jobs remaining largely unchanged involved packing into cartons and outers, where there was no technical change. Additional sources of work variety for those in the low-scoring jobs on older plant included limited rotation between work stations, change resulting from redeployment when production was interrupted, or changes in product resulting in different work methods. The opportunities on the replacement plant for increases in variety through the latter two means were more restricted.

While many of the jobs on the former plant were low in a number of attributes, opportunities for social interaction were good. The equipment was relatively quiet and work stations were generally close together. Employees on the wrapping section were situated on both sides of a conveyor facing each other. At the packing end opportunities were slighly lower. The new layout afforded similar opportunities for social interaction for this packing group.

In summary, the technological changes have had a considerable impact upon the jobs performed: reduction in manning levels but increases for some workers in skill demands and responsibilities. In order to obtain a satisfactory return on investment, the plant was required to operate throughout the working week. This should have led to little overall change in employment levels but increases in shift working. Payment levels for those engaged on the new plant were higher in many cases. However, some jobs on the new plant, which were not changed as a result of technological development remain low in variety and responsibility/discretion, repetitive, highly controlled, low in goals/feedback, low in interdependence and in skill requirements.

2.5 Outcome of the project

As indicated earlier, the fortunes of the product of this factory are influenced by changes in the general economic situation. The particular product manufacture initially was newly developed. In addition, the presentation of the product in bags rather than more conventional wrappers was a departure from former practice. However, this change facilitated the high level of automation introduced in wrapping.

After a time, product sales were not maintained at the predicted level and three-shift working was discontinued. The manufacturing facilities remained underutilised and hours of work were reduced for existing staff for a period, prior to a decision to modify the facilities so as to increase flexibility in wrapping. More conventional wrapping machines were introduced alongside the high-speed wrapping machine, thus enabling other products to be manufactured for part of the working day.

One of the more significant changes in work organisation was the elimination of the traditional supervisor. This change was stimulated by a desire on the part of management to reduce factory overheads and by the need for a greater degree of technical competence from first-line management. The former responsibilities of the supervisor were retained by the section manager.

Even though changes had to be introduced to accommodate the new manufacturing facilities, these seem to have been limited to those essential for achieving operation in the shortest time following completion of the new plant. No radical changes in organisation, payment systems or working practices were attempted by management nor sought by trade unions.

Automation of processes in this form of manufacture may be considered as piecemeal. Much of the early effort focused on the first stages of the process (i.e. weighing ingredients and mixing). The stage most affected in this phase of automation was the labour-intensive stage of wrapping, while the final stage of packing remained as previously. Automation at the early stages in the process did not limit the flexibility of product manufactured, whereas the wrapping machinery introduced imposed constraints.

Even though the company had introduced a sophisticated structure for employee participation, there was a low level of employee involvement in decisions concerning the new technology. Industrial engineers and line management determined the management's preferred job and work organisation structure and negotiated with the trade unions where this was contrary to custom and practice. This discussion and negotiation took place once the facilities were in place, during the plant's commissioning. This approach was adopted despite the interest expressed at various stages by both management and trade unions in the initiatives in job redesign and participative problem-solving.

From the employees' viewpoint, a major concern during the period following the introduction of the new plant was job security, particularly since increasing unemployment was a feature of this area of England. In terms of jobs performed, the gains and losses in the nature of work tended to be evenly balanced for most workers. Where there was significant changes which probably added to the interest of the work, the operatives received more money for the increased responsibility but also probably lost the socialisation and group spirit associated with hand wrapping.

3. Case study 2 - A new ore terminal[16]

3.1 Background

This case describes the development of the work structure at a new facility designed to handle iron ore. The project was undertaken on the west coast of Scotland, commencing in July 1976, some time after many of the basic design decisions had been taken. It involved consideration of all the work associated with the process, and predicted a total of over 200 employees.

The new ore terminal formed part of a major restructuring exercise carried out by the British Steel Corporation (BSC). Overcapacity in production facilities and the changing economics of steel production led to a major reappraisal within BSC and, within this context, any work restructuring exercise was expected to produce improvement in overall operating effectiveness. It was assumed that improved job satisfaction would result where work was clearly well organised and operational effectiveness high.

The new ore terminal included more sophisticated mechanical handling and control systems than earlier plants. No detailed description is presented of either the technological change nor the impact upon work; the focus here is upon the process adopted for designing the new system and the form of work organisation which resulted. The principles of work structuring which formed the basis of the design were developed by Schumacker and are described in detail elsewhere.[17] In addition, a more detailed report describing other personnel-related matters is available.[18]

3.2 Objectives of the work structuring exercise

The following objectives were agreed between the Scottish Divisional Management and the Work Structuring Team from BSC:

- to structure the work of all a employees in the ore terminal so as to ensure maximum motivation and job interest consistent with high standards of efficiency.

This was to involve advising on -

- determination of optimum manning levels for the ore terminal;
- definition of the work to be performed by each individual and/or work group;
- incorporation of human and social considerations into ore terminal control systems;
- reporting relationships and organisational structure;
- information flow and communications procedures;
- work structuring implications for selecting procedures and training arrangements;
- payment systems and locally negotiated conditions of service;
- amenities and welfare provisions.

3.3 The decision-making structure

A Work Structuring Team was set up for the project. This consisted of a specialist in work structuring from head office, a senior industrial engineer from the Scottish Division, an ergonomist and a behavioural scientist. The team reported in turn to a Steering Committee chaired by the Director of Personnel and Social Policy from the Scottish Division, and comprising senior management from the Division as well as divisional organisers of the two main trade unions involved in the development (Iron and Steel Trades Confederation - ISTC - and TUC Steel Committee).

3.4 The ore terminal

The terminal became operational in November 1979. It supplies ore and coal to the Ravenscraig Works, as well as pelleted or lump ore to the direct reduction plant at Hunterston. There is also the facility to re-ship ore or coal to other sites. The terminal has been designed to meet initial requirements of 7-8 million tonnes per annum (ore equivalent), with a potential future capacity of 12 million tonnes per annum or more with minimal further capital development. The jetty is able to unload ships up to 350,000 tonnes capacity, although most vessels will be in the 150,000 d.w.t. category, as well as loading ships with up to 150,000 tonnes of ore.

On the jetty there are two rail-mounted Grab-type unloaders, and one rail-mounted skip-type loader with a rated capacity of 4,000 tonnes/hr ore at the same density. Materials are conveyed from the jetty by one conveyor with a capacity of 6,000 tonnes/hr from jetty to stockyard and one conveyor with a rated capacity of 4,000 tonnes/hr from stockyard to jetty. Transfer of material between the conveyor and the stockyard is by three stacker/reclaimers, each mounted on independent rail tracks.

The terminal will handle up to ten ore and three coal trains per day for six days a week for Ravenscraig, and further capacity is available. These trains run to a British Rail schedule and any delay in filling a train results in that "pass" being missed, with consequent financial loss.

Selection of the conveyor routes and stopping and starting of the conveyors is made from the control tower (belts can be stoppped in an emergency by a "communication cord" running alongside all of them).

Three distinct areas exist on the terminal site -

- jetty to wagon-loading station (approximately 3 1/2 km);
- control tower to wagon-loading station (approx. 2 km);
- stockyard to wagon-loading station (approx. 1 1/2 km).

3.5 Principles of work structuring

3.5.1 Main principles

The approach to work organisation adopted by the Work Structuring Team was based on the following seven principles related to practice in any given situation, following an analysis of the production system:

- the basic organisational unit should be the "primary work group" (i.e. 4-20 people);
- all work group members should be placed on the same payment system and conditions of employment;
- flexible working arrangements between members of the same work group should be maximised;
- each work group should be led by a designated supervisor;
- each work group should, as far as possible, be responsible through its supervisor for planning its own work;
- each work group should have the opportunity to evaluate the results of its performance and to compare these results with standards;
- each work group should perform a relatively independent and significant set of activities which together form a "whole" task.

3.5.2 Maintenance sub-principles

In parallel with these main principles, which relate primarily to production, the following series of sub-principles relating to maintenance was developed to facilitate collaboration between these two activities:

- maintenance activities can form a whole task in their own right;
- the organisation and location of the departmental maintenance function should mirror that of the whole production task it serves;
- production operatives should carry out their own first-line maintenance and maintenance personnel should monitor first-line production work;
- the separation of individual crafts and trades should be kept to a minimum and group working should be encouraged;
- the system of payment and conditions for maintenance workers should be compatible with those of production personnel in their area.

3.5.2 Documentation sub-principles

A further series of sub-principles was developed when the team, in co-operation with management and the Management Services Department, considered the documentation requirements for the ore terminal. These are outlined later but the basic theme is that, compatible with the work structuring approach, each work group should be able to plan, carry out and evaluate a whole task and that documentation should reflect the distinction between planning and evaluating.

3.6 Application of principles

A central task of the Work Structuring Team was to apply the principles listed above to the organisation of work at the ore terminal, and on this basis produce recommendations on its manning and operation.

To do this a systematic procedure was followed, of which these are the main steps -

(1) On the basis of knowledge of plant layout, workload at other ore terminals and fixed job stations, a rough estimate was made of overall manning requirements for the ore terminal.

(2) The production system was analysed and clusters of activities which formed complete tasks identified.

(3) Production work groups were allocated to discrete parts of the production system so as to secure the best possible "fit" between people and the technical inter-relationships in the process.

(4) An optimum manning level for each work group was decided, based on estimates of opportunities for job flexibility.

(5) On the basis of the group structure, a management organisation was built up.

(6) Through a role analysis, the job requirements of supervisory grades were determined.

(7) The information and communication systems were adapted to ensure that management, supervisors and work group members all had the necessary information to enable them properly to compare performance with standards.

(8) The proposed documentation for the ore terminal was analysed to ensure maximum delegation of authority to first-line supervisors, as well as overall operational control of planning and scheduling by management.

(9) A payment system was devised for work group members designed to maximise flexibility of working, encourage effort and unify the workforce.

(10) Compatible promotion lines, job descriptions and conditions of employment were recommended.

(11) The implications of the recommended work organisation for selection and training of operatives, craftsmen and supervisors were outlined.

(12) Recommendations were made covering amenities and working conditions, including the design and layout of the central control room.

3.7 The organisation of production

The Work Structuring Team carried out a transformational analysis of the production system in order to determine an optimum "fit" between the technology and geography of the site and the organisation of the people working there. This method of analysis classifies each step in the production process according to the degree of change taking place in the product at that point. By so doing, it is possible to identify clusters of production activities which are _technically_ highly inter-related and which should therefore be carried out by the same work group, which is _socially_ highly inter-related. Essentially, the analysis is of the several stages through which a raw material passes on its journey to becoming a finished product.

At the ore terminal, the activities which are technically most inter-related are those which are based on each of the conveyor systems, namely -

- unloading, transfer by conveyor and stacking, on the one hand; and
- reclaiming, transfer by conveyor and dumping into hoppers at the wagon station, on the other.

All activities within each of these systems are interlinked, since ore taken out of a ship's hatch must then be conveyed and stacked. Once it is stacked, however, the causal link with the next step is weakened because storage provides a buffer between production systems. Again, after the ore has been reclaimed from the stockyard, it must either proceed to the wagon-loading station hoppers, return to the jetty for reloading or, in the case of pellets, be conveyed to the direct reduction plant for conversion into sponge iron in the direct reduction furnace.

This analysis of the process identified three ideal groups of production operatives at the ore terminal, as follows:

(a) those operating the first transportation system from the ship's hold to the stacking of ore in the stockyard;

(b) those operating the transportation system from the reclaiming of ore from the stockyard to the wagon-loading station; and

(c) those operating the transportation system from the reclaiming of pellets from the stockyard to the direct reduction plant, including the major conversion in the direct reduction process.

Had the scale and layout of plant and the likely numbers of people involved at each process stage allowed this configuration, each work group would have been both relatively autonomous externally in respect of other groups and fairly cohesive internally in respect of the production system it was controlling.

Unfortunately, however, neither the geography nor the labour intensity of the plant was compatible with this technological solution for the following reasons:

(1) The engineering concept of a machine which both puts down and picks up ore (the stacker/reclaimer) would mean that the driver would have alternatively to be a member of one group - (a), above - while stacking, and of another group - (b) or (c) - while reclaiming. This could lead to a conflict of priorities between unloading or reloading ships and loading wagons or supplying the direct reduction plant. It would also create the undesirable situation where one worker belonged to two groups and had two supervisors, hence weakening the allegiance to either.

(2) The geography of the site weakened the case for a work grouping along process lines. Not only are the stockyard and the jetty, and the stockyard and the wagon-loading station a considerable distance apart from each other, but the stockyard itself is not laid out in a way which is compatible with its being operated jointly by two groups. An alternative to a process-based work group configuration had therefore to be sought, and a number of options were considered.

3.7.1 Option 1

One obvious possibility was to base work groups on the geography of the site. Option 1 also envisaged three groups of production operatives -

(a) a jetty team (responsible for loading and unloading ore from the ship and for operating the conveyor system up to the end of the jetty);

(b) a stockyard team (responsible for stacking and reclaiming and for all conveying of ore within the stockyard area up to the direct reduction plant bunkers); and

(c) a wagon loading team (responsible for conveyors to the wagon-loading station and for all activities within the wagon-loading station itself).

Under this option, each work group would have been situated in a relatively compact geographical area which would have eased supervision. Also, groups would have been of a manageable size, which facilitates good social relations and opportunities for flexible working.

On the other hand, geographical groupings also had disadvantages. First, neither the stockyard nor the wagon-loading station would have been technologically autonomous. The quantity, timing and sequencing of stockyard work is determined by events taking place on the jetty, since the conveyor system, which is single and indivisible throughout its length, crosses the group boundary. Similarly, the wagon-loading team would have been situated at the other end of a conveyor system which started in the stockyard. Neither group could have scheduled its own work, so that decision making would have had to be centralised away from the group, and delays and poor motivation would have been more likely to occur.

In addition, preliminary assessment of the manning required in the stockyard (4-5 workers) suggested that these numbers were too small to warrant the setting up of a separate work group under its own supervisor.

3.7.2 Option 2

Some compromise therefore had to be found between a process-based and geographically based structure. One possibility was -

(a) a jetty stockyard team (responsible for all jetty and stockyard operations, with the group boundary at the junction house where the conveyor system turns into the wagon-loading station); and

(b) a wagon-loading station team. This option recognised the geographical separateness of the wagon-loading station. It also embraced the jetty/stockyard process link within a single work group.

On the other hand, its disadvantages were -

- the two work groups were of very unequal size;
- the wagon-loading group was not technically autonomous since reclaiming forms part of the jetty/stockyard team's duties;
- there is a large geographical distance between the jetty- and stockyard-based members of the jetty/stockyard work group, which would make supervision, social cohesion and flexibility of working difficult;
- problems of priority conflict in the stockyard remained unsolved.

3.7.3 Option 3

Another option was to retain a two-group system, but to merge the stockyard with the wagon-loading station instead of the jetty, as follows:

(a) a jetty team (responsible for all jetty operations and the conveyor system up to the junction house prior to the stockyard); and

(b) a stockyard/wagon-loading station team (responsible for all stockyard and wagon-loading operations).

This option would have created work groups of roughly equal size and retained the advantages of a socially integrated jetty team on the one hand, and those of a technically integrated stockyard/wagon-loading team on the other. On the other hand, its disadvantage was that the stockyard/wagon-loading team was physically dispersed over a wide geographical area.

The problem faced by the Work Structuring Team when assessing these options was that, given the existing ore terminal layout and the fact that stacker/reclaimers were to be used, <u>there was no way of obtaining an ideal match between group size, geography, chosen technology and the process.</u> Every option had its advantages and disadvantages. No perfect solution was possible and the organisational choice became essentially an empirical one taken in the light of local circumstances.

On the basis of a reasoning process of this kind, the Work Structuring Team finally recommended Option 3 to the Steering Committee. This was accepted by them as the most acceptable solution in the circumstances for the following reasons:

(1) There is a perceptible difference in "culture" between those working in close proximity to maritime personnel (on the jetty) and those working on land.

(2) There was the need for closer co-ordination between the stockyard and the wagon-loading station than between the stockyard and the jetty, since dispatch of ore and coal by rail wagons was more frequent, varied and more continuous than the arrival of ships.

(3) This organisation of work groups more accurately reflected the difference between "supply" and "dispatch" in terms of the priorities of the site as a whole.

In the event, although the Steering Committee accepted Option 3, subsequent objections by the Transport and General Workers Union resulted in ship work being made the responsibility of the Clyde Port Authority. The implications of this arrangement are discussed later.

3.8 <u>The organisation of maintenance</u>

From the discussion of the production process in the previous section, it is clear that the major activities in the ore terminal are of a transportation and storage nature. The process is essentially machine determined. On completion of the production plan by the manager, and once the required process route has been selected and activated by the control tower operator, actual control of the process by managers, supervisors and operators is limited. The priority thereafter becomes one of ensuring that, from an engineering point of view, the plant operates efficiently.

This is not to underestimate the value of good operating practice (particularly in unloading), which results in less spillage, good throughput and less wear and tear on machinery. However, the unavailability of one unloader can lead to serious delays in ship unloading and hence involve the BSC in expensive demurrage charges. Furthermore, the breakdown of the belt from the jetty to stockyard stops unloading operations altogether.

Maintenance responsibilities are therefore equally as important as those of production, and the Work Structuring Team therefore recommended the appointment of supervisors with responsibilities in both area.

With respect to the maintenance organisation in the terminal, the work structuring principles (see section 3.5) gave rise to the notion of a Central Engineering Workshop for major repairs covering the whole terminal. It was also recommended that the shift maintenance team cover the whole terminal and not be split between individual work groups. This team was to carry out both planned maintenance and emergency repair work.

Finally, to encourage close working relationships between both production and maintenance functions, the Team recommended that production operatives carry out their own first-line maintenance, and craftsmen be encouraged to monitor the production process, draw management's attention to bad operating practice and make suggestions on operating practice where this would prolong plant life and improve capacity. The Work Structuring Team felt that it was essential that production and maintenance be fully integrated not only at managerial and supervisory levels, but also on the shop floor.

3.9 The organisational structure

3.9.1 The shift supervisor

Having decided to organise the ore terminal into two production groups and a single shift-based maintenance team, it followed that three shift supervisors were required at the ore terminal. Whether these supervisors themselves needed to be supervised by a shift manager depended on the number and nature of decisions which had to be taken by second-line management during shift.

To ensure that the role of the first-line supervisor was compatible with the work structuring principles, the team suggested the following specific responsibilities which in turn had implications for recruitment and selection, and indicated particular training needs:

- ensure that the work group plans, performs and evaluates its work effectively;

- agree targets with superior and work group, and be accountable for meeting these targets, against which performance would be regularly monitored. The targets set should include technical and human resources (e.g. production volume, material utilisation, maintenance performance, together with accident statistics, grievances, level of flexibility, etc.);

- delegate as many responsibilities as appropriate to the group to optimise participation and individual development;

- select and induct new members of the group, allocate them to their first job and ensure their attainment of agreed performance standards through necessary training;

- ensure the continuance of on-the-job and external training of group members to permit increasing flexibility within the section and possible progression outside it, recommending promotions and transfers where appropriate;

- help and advise members of the group, in particular in respect of their training and development, and personal problems;

- seek to resolve grievances, using a problem-solving approach within agreed procedures, before referring them to the superior;

- ensure that members of the group are aware of the financial implications of new work, work in progress, and work completed;

- hold regular briefing meetings with the group to permit adequate two-way communication about all work-related activities;

- recommend changes in manning standards, plant layout or equipment, etc., which would increase the effectiveness of the group;

- relate the group's activities to those of other sections/departments to increase overall organisational effectiveness;

- ensure adherence to safe working practices within the section, liaising with other specialists inside and outside the organisation as appropriate;

- the first-line supervisors should be encouraged to develop themselves through further education and special projects.

Within this general framework, the task-related role (the first responsibility listed above) was examined in detail for each of the three first-line supervisor's jobs identified at the ore terminal using a role analysis technique. This is a systematic way of analysing the supervisory task on the basis of -

- the main objectives of the group being supervised;

- the number and nature of decisions which have to be taken to achieve these objectives;

- the actions required to implement these decisions; and

- the incoming and outgoing information required to evaluate successful implementation and to communicate with the wider management systems.

When this technique was applied to the supervisors' tasks at the ore terminal, it became evident that an essential objective in all areas was top-quality preventive maintenance. This view was reinforced by managers at other ore terminals which were visited, who emphasised the inherent problems experienced in carrying out planned maintenance and repair work in circumstances characterised by a heavy and sometimes upredictable shipping programme coupled with the severe financial penalties incurred if unloading is delayed or wagons lost.

It was therefore decided to recommend to the Steering Committee the appointment of production shift supervisors who also had specific responsibilities for maintenance in their areas. The benefits of this "double-function" approach were that it could -

- help eliminate unnecessary delays in diagnosis of plant faults and breakdowns;
- promote more rational decisions on shutdown of plant; and
- encourage the adoption of an overall operational strategy which takes full account of the importance of maintenance to the successful running of the terminal as a whole.

Shift supervisors would be expected to relate to maintenance in that they should -

- have an engineering background and training;
- normally be expected to diagnose faults in their own area and be able to specify the appropriate craftsman needed to carry out repairs;
- be responsible for inspecting completed repair work and issuing certificates of "fitness for use";
- be expected to participate in the preparation of planned maintenance schedules and give agreement before plans were implemented; and
- be empowered to authorise their own production personnel to carry out simple, first-line maintenance work when necessary.

The Work Structuring Team acknowledged that the integration of these maintenance responsibilities into the production shift supervisors' role would inevitably alter the relationship with the shift maintenance team and its supervisor. Consideration was given to the possibility of omitting a shift supervisor from the maintenance team, but this would have resulted in craftsmen reporting to different bosses depending on which part of the site they were working in. This was considered unsatisfactory, and it was decided to appoint a shift maintenance team leader, whose job was to plan, supervise and evaluate engineering activities, in order to minimise disruption of operations and to meet scheduled plant availability and standards.

The Steering Committee accepted these recommendations as a basis for drawing up first-line supervisors' job descriptions at the ore terminal.

3.9.2 The control room operator

When the Work Structuring Team began its studies, the job of an ore terminal shift controller was generally regarded as perhaps the key job on the site. In early discussions with Scottish divisional management, the assumption was made that the job would be a relatively senior one, and in some quarters it was even suggested that this job should be coupled with that of shift manager for the terminal as a whole.

However, a role analysis showed that, both in terms of the nature and frequency of decisions to be taken and actions to be performed, the job is in fact a relatively low-level one. Most decisions concerning conveyor stop/start times, product routing, loading and unloading requirements, scheduling of ships or wagons, etc., cannot be taken by this individual alone, who is more often than not in the position of having to respond to someone else's instructions. The supervisory content is also minimal, and so the Work Structuring Team recommended that the job should be excluded from the supervisory system. This was accepted by the Steering Committee, and it was agreed that the title for the job should be "control room operator".

Nevertheless, the control room operator does have an important role in communicating and compiling information, with informing and recording tasks featuring large in the job. As such, the job might well be considered a suitable preparation for subsequent promotion to supervisor.

3.9.3 Shift management

The work structuring approach adopted endeavours to build organisational structures from the bottom upwards. Thus, several duties traditionally falling to the shift manager are justifiably allocated to his subordinates. A role analysis of the shift management and supervisory requirements endorsed this.

Having allocated first-line supervisory functions in the jetty, stockyard and engineering areas, the question remained as to what second-line controlling and co-ordinating functions still needed to be performed on shift.

The following emerged:

- monitoring and controlling ore terminal operations through the shift supervisors;
- resolving priority conflicts between supervisors;
- advising/resolving shift industrial relations problems which cannot be resolved by supervisors or wait until next day;
- meeting production variances by, for example, authorising overtime, transferring labour between groups, requisitioning transport on vetting expenditure as required on shift;

- resolving problems arising on shift with external agencies (e.g. Clyde Port Authority, British Rail) which cannot be handled by shift supervisors or shift controller.

After examination of the anticipated workload arising from these duties and following discussions with the Steering Committee, it was decided to include these responsibilities in the job of the stockyard foreman. This position thus becomes the senior post on the site in the absence of day management.

3.9.4 Day management

With regard to day management, the following supervisory jobs were identified in addition to the post of Ore Terminal Manager:

- Section Manager, Ore Terminal, who was responsible for -
  - co-ordinating the ore terminal production and engineering function;
  - providing specialist engineering advice when necessary; and
  - deputising for the Ore Terminal Manager;
- Workshop supervisor, responsible for supervising the central engineering workshops; and
- Accounts and Administrative Officer, responsible for supervising local accounting, production planning, primary recording and other administration arising in the ore terminal office.

3.10 Work group manning

3.10.1 Job descriptions

The traditional concept of one worker-one job is not always compatible with the work structuring approach with its emphasis on team working. Job descriptions should not be so rigidly defined and circumscribed as to restrict flexible working and multi-skilling. On the other hand, a forced system of total job interchange (as with some systems of job rotation), which ignores skill differences and individual preferences is seen as possibly equally counter-productive.

Work structuring advocates a combination of the two, the key concept being "relative specialisation". Job descriptions should be such as to encourage as much flexibility and job interchange within each work group as is consistent with co-ordination, skill and safety limitations.

The following steps were used to establish job descriptions:

- determine the group/task using transformational analysis;

- on the basis of the given technology, establish which job stations require regular specific manning, and which aspects of the work can be performed by small groups working together;

- using task analysis, calculate manning required for each group;

- draw up job or team descriptions.

Flexible working and multi-skilling were considered equally important in the maintenance area. There had already been some moves towards introducing flexible working arrangements in maintenance in the Scottish Division, and the topic was discussed with the Chief Maintenance Engineer and local management. This resulted in the adoption of a strategy which entailed the employment of three basic categories of craftsmen -

- mechanical craftsmen;

- electrical craftsmen;

- fabricators.

Each of these craftsmen, in addition to the duties associated with the particular trade, is required to cover a range of common duties.

After visits to other terminals, it was decided that there should be a control technician on shift, whose area of expertise was electronics. Other terminals had experienced problems, particularly with unloaders, which conventional electrical foremen had been unable to solve. They had adopted the practice of employing an electronics expert on shift, whose chief role was that of fault diagnosis. The extensive use of electronic equipment throughout the terminal amply justified the inclusion of this post in the shift team. The consequent key nature of this post is compatible with the supervisory role of a chargehand reporting to the maintenance supervisor.

In the workshop (on days), it was decided that there should be a mixture of craftsmen with similar qualifications to those of shift teams. It was also agreed that there should be provision for training single craft personnel to dual craftsmen, and then to multi-skilled craftsmen on a similar basis to the group working practice arrangement introduced elsewhere in BSC. The workshop would be under the control of the Workshop Supervisor reporting to the Section Manager, Ore Terminal.

It was agreed that the jobs of plumbers, diesel fitters and riggers were not required as single trades on the site.

3.10.2 Ore terminal manning schedule

On the basis of an analysis of the workload in each of these areas, and in anticipation of flexible working resulting from the application of work structuring principles, a manning schedule was compiled.

It had been agreed with the Steering Committee that this work, together with the compilation of job descriptions of all personnel, would be carried out with assistance from two persons nominated by the trade unions (one by ISTC, one by the TUC Steel Committee). In the event the work was done by local management.

This manning schedule compared favourably with the figures produced by the Scottish Division's Finance Department in December 1975 and by the Division's Personnel Department in April 1976, as shown in table 87.

3.11 Information flow and communications

When the Work Structuring Team came to consider the communication system for the terminal, it was apparent that the project engineer's underlying philosophy in designing the plant was one of a highly centralised control system. The assumption was that the central controller would direct the operation of the terminal from his tower, which was to be the hub of the system.

This was reflected in the proposals submitted by contractors for the "communication" system for the terminal, which was characterised by being highly centralised with communication between stacker/reclaimers, unloaders and loader, wagon-loading station and junction houses having to be routed through the controller, who acted as an exchange for the whole system. The controller was to have been a management position. No consideration had at that stage been given to the implications of this for the operational and managerial control of the plant.

It is worth making the general point here that the existing procedure in BSC is to carry out communications systems design concurrently with that of other plant and hardware requirements, but before any but the most cursory consideration of the manning and managerial requirements of the plant. Decisions have been firmed up and orders placed well before any discussions concerning manning take place. This creates problems for anyone concerned with the design of jobs in new plant. The failure of design engineers to recognise that the nature and location of the communications hardware they install has considerable implications for the subsequent control and organisational structure of the plant results in job designers having very little room for manoeuvre.

3.11.1 Communications philosophy

In considering the communication needs of the ore terminal, the Work Structuring Team's objective was to ensure that audio-communications media (i.e. two-way radio, telephone) together formed an integrated, comprehensive system compatible with the work structuring approach. The criteria used were that -

- the position of supervisors in the work group should not be undermined by the adoption of a system that would bypass them;
- each work group should have sufficient information to plan, perform and evaluate its own work and enable its members to participate fully in the work of the group;

Table 87: Ore terminal: comparative manning estimates

Personnel category	Work structuring proposals	Personnel Dept. estimate	Finance Dept. estimate
Management and day staff	10	34	22
Shift staff	13	20	39
Production operatives (excluding existing stevedores)	93	98	113
Maintenance and others	41	47	133
Total	157	199	307

- the communications system should mirror and support the overall work group system; and

- communication between work groups should therefore, wherever possible, be via their respective supervisors.

The Team consequently recommended that the "all channels open" and "star" approaches should be adopted so as to fulfil these criteria. The system is a highly centralised one, in which all communications between individual outposts have to be via a communications centre. However, every member of the group can communicate with all other group members. The operational advantages of the network are that -

- within each work group maximum information is available to members to enable them to work flexibly;

- foremen have maximum information to enable them to plan the work of their respective groups; and

- foremen are the focal point for information coming in from outside the work group, so that they can then handle the problems which occur on the boundaries between work groups. This supports the position of the foreman as leader of the group. Dispersion of incoming information among group members would undermine his position and hinder effective planning.

3.12 Documentation

In considering the documentation requirements for the ore terminal, the Work Structuring Team did not face the problems of decisions having been pre-empted, as was experienced with audio-communications. A set of documentation subprinciples was developed compatible with the work structuring approach. These are outlined below.

3.12.1 General

(1) Planning and scheduling data should be clearly distinguished from evaluation data on all documentation.

(2) Documents can usefully be ranked in terms of their time span, which in turn relates to the organisational level of their originators and of the part of the organisation for which they are responsible.

3.12.2 Planning documents

(1) Each planning document should cover the whole area over which the originator has managerial responsibility and should be subdivided into the areas supervised by immediate subordinates. Planning of work must coincide with the work group structure. Supervisors should only be responsible for planning the work of individuals within their control.

(2) Each planning document should only relate to the external objectives of the area covered. The internal objectives of the group, which involves the allocation of tasks, etc., should be left to the respective work group supervisor.

(3) Planning documentation should only be used when there is a time lag between the formulation of the plan and its execution. Where a plan can be immediately executed (e.g. a plan for moving ore over a shift), then the supervisor can transmit this to the work group members.

(4) Each level of personnel should participate in the formulation of the next higher level plan. This will help ensure that realistic and achievable targets are set and will enable employees to have a clearer view of wider organisational objectives.

3.12.3 Evaluation documents

(1) Each planning document should have a corresponding evaluation document which covers the same time period.

(2) All variable information recorded on planning documents should appear on the evaluation documents covering the same time period.

(3) All evaluation documents should contain sets of information relating to -

(a) the area covered by the equivalent planning document;

(b) actual performance data;

(c) standards or targets against which to compare the performance data;

(d) significant variances;

(e) reasons for variances.

(4) Evaluation documents should cover those operational variables which are critical to the area's performance (e.g. rate of work, maintenance, yields, etc.).

(5) All operational variables should be evaluated in financial as well as in operational terms. This avoids unnecessary duplication of paper by eliminating the parallel existence of production and financial documentation which essentially arises from the same operational source. Performance evaluation information about variables over which the personnel concerned have no control should be avoided.

3.12.4 Implementation

The Work Structuring Team in conjunction with the Plant Manager formulated some preliminary ideas on documentation for the ore terminal. These were relayed to the O & M Department at Ravenscraig Works which, with reference to the principles outlined above, produced a draft set of documents. These were discussed with members of the Team, representatives from O & M and Scottish Divisional Finance, and the Plant Manager. This helped ensure that the documentation met, as far as possible, operational requirements and any broader divisional financial requirements, and were commensurate with the work structuring philosophy before final versions were produced.

3.13 Personnel considerations

Consideration was also given by the Work Structuring Team to establishing employment policies for the ore terminal acceptable to BSC and the trade unions, but also forming an integrated and compatible part of a work structuring approach to work organisation.

The Team also developed a series of work structuring subprinciples for pay, promotion and conditions of service.

3.14 Conclusions

In this concluding section, we attempt briefly to answer three questions -

- to what extent was it possible to introduce an "ideal" work structuring solution to the Hunterston site?
- to what extent was the scheme which was introduced successful?
- what lessons can be learnt for future applications of work structuring in new plants?

3.14.1 Constraints

As the Work Structuring Team quickly discovered, the invitation to introduce work structuring principles to a "green field" site such as Hunterston did not give carte blanche to the approach. Many constraints existed, the major ones being the following:

(1) Technological constraints

By the time the Team had been commissioned, the Hunterston site was already under construction. Decisions had been taken as to layout, scale, type of machinery to be used, the control system and office accommodation. These decisions precluded the possibility of obtaining an "ideal" match between the technology, the geographical area and the work group, as advocated by the work structuring approach. The major inhibiting technological constraints may be summarised as failure to -

(a) match work groups to tasks;

(b) match scale of development to optimum manning;

(c) match site geography to technology; and

(d) match organisational structure to communications and control structure.

(2) Political constraints

Early management decisions made by the BSC also precluded certain organisational options on the Hunterston site which would undoubtedly have led to a more efficient operation of the terminal and more complete implementation of the work structuring principles. Main constraints included the following:

(a) failure to secure a "single union" agreement for the site as a whole. Before the Work Structuring Team was appointed, BSC had signed two agreements (with the Clyde Port Authority and with ISTC) which were not wholly compatible with one another. As a result, the homogeneous work groups advocated by the Team had to be broken up, leading to increased industrial relations and operational problems, higher than necessary manning levels and reduced flexibility of working. There was also a major inter-union dispute which delayed commencement of operation of the terminal by some six months;

(b) failure to co-ordinate the development of the ore terminal with that of the direct reduction plant. This stemmed from the time when the two parts of the site were under the control of two separate divisions within BSC. Had a unified control structure existed, it is possible that the layout of the site might have been much more compact (e.g. wagon-loading station, railway facilities, nearer stockyard) and the total capital and operational costs lower.

(c) failure to integrate project engineering with operational management during the site development. Many decisions (e.g. use and number of stacker/reclaimers) were based on the need to minimise initial capital costs and not on the concept of total life costs, which would have been more compatible with a work structuring approach. In addition, a lack of "line" representation on project committees and delays in recruitment of a site manager and supporting staff led to a neglect of operational, industrial relations and social issues during the early design stages (e.g. there was inadequate regard to the location and scope of amenity buildings, group supervisors' offices, the communications structure, phasing of recruitment and training programmes).

3.14.2 Degree of success

At the time of writing it is too early to evaluate the success or otherwise of the exercise. However, initial impressions are very favourable, both in terms of the speed and smoothness of the initial start-up period and in the spirit of co-operation prevailing amongst the workforce on the site. It is clear that local management valued the exercise for its help in developing the work organisation and procedures for the terminal.

An attempt, however, was made to determine the advantages of the work structuring approach in terms of its effect on manning levels. These are now the best in Europe. The management structure is also exemplary and compares very favourably with that of the other recently constructed BSC ore terminal at Redcar.

A crucial determinant of success will be the extent of the financial penalties incurred on account of excessive demurrage charges. As yet the terminal has not been in operation for long enough to make a comparison with other sites. However, preliminary indications suggest that losses due to demurrage will be substantially lower at Hunterston than elsewhere.

Finally, the advantages of the joint management/union approach adopted during the work structuring exercise should be mentioned. The entire scheme described here was introduced with the full participation and support of the major unions involved. There were no major disagreements or points of friction between the parties during any phase of the exercise.

3.14.3 Lessons to be learned

Although the approach adopted and the committee structure set up to supervise and execute the work structuring exercise worked well, the following useful lessons were learnt which might benefit such projects elsewhere:

(1) Timing of involvement. It cannot be emphasised too strongly that the earlier in a new capital development that work structuring principles are followed, the greater will be the benefits. Ideally, there should be work structuring involvement from the outset of the planning stage when strategic decisions such as the product route, scale, layout and

configuration of the plant are made. Once these options are closed, many advantages of the work structuring approach are mitigated.

(2) Involvement of design engineers. It appears to be common practice in BSC that when major new investments are made, the control of the design and construction is left to design engineers. While acknowledging the major role they have to play, there would appear to be many advantages in constituting a new plant development steering group with a wider representation, including a chairman committed to a work structuring approach. The work structuring Steering Committee for Hunterston did not contain anyone from the design engineering site.

(3) The Work Structuring Team. Our experience at Hunterston suggested that while part-time members of the Team are valuable, there should be adequate full-time representation on site to ensure continuity and proper supervision of the detailed application of work structuring principles. With better resources, we felt, a better job could have been done.

(4) Training. The amount of training needed both to convince local line management that a work structuring approach was desirable and to help them understand the principles in depth was underestimated. More training should have been available to ensure that the site supervisors and workforce could cope adequately with their enhanced work content.

(5) Pay and conditions. It would have greatly assisted the implementation of the exercise had there been an initial clear commitment by senior Head Office management in the Corporation (notably in the Personnel Division) to the measures which the work structuring principles were likely to entail. The lack of commitment to "single status" was a case in point.

4. Case study 3 - Technological changes in two breweries

4.1 General background

4.1.1 Structure of the industry

The United Kingdom is the third largest producer of beer in the world following the United States and the Federal Republic of Germany. Beer production in 1980 reached 39.6 million bulk barrels (roughly 11.4 billion pints), and in relation to the manufacturing sector as a whole, the industry accounts for slightly less than 2 per cent of net output, 1 per cent of employment and almost 5 per cent of capital investment.

Prior to the development of a national beer market in the post-war period, local breweries dominated the industry. However, their numbers shrank from 247 in 1960 to 80 in 1979. During this period British brewing was taken over by six large brewing groups, namely Allied, Whitebread, Grand Metropolitan, Scottish and Newcastle and Courage, which supply three-quarters of Britain's beer. This restructuring of the industry was achieved almost entirely by merger. Emphasis on the latter as a means of corporate growth is due to the vertical organisation of the industry. Brewers own the majority of retail outlets through which their beer is sold, and since the total of these is restricted by licensing policies, the acquisition of competing firms with their tied retail estate has been the only feasible growth path.

4.1.2 Economic performance

The industry is one which has been able to maintain its profitability in line with that of manufacturing as a whole, and while retailing has shown low returns on capital brewing and wholesaling have been very profitable. However, as beer sales have been falling over the last few years, it has been shown that many regional and independent brewers have been less affected. A Price Commission Report of 1977 found that rates of return on capital employed in production and wholesaling stood at 46 per cent for regional breweries and 53 per cent for local breweries, significantly higher than for the larger ones (32 per cent). These findings led the National Economic Development Office's Brewing Sector Working Group to identify several differences in operation between the "big six" and the regional and local groups.

It was found that large brewers tend to spend more on research and advertising, have in general spent more on investment, and on the retail side have modernised their public houses more quickly. Large brewers produced a far wider range of products, and in the important larger area have developed production to a far greater degree than smaller breweries. In the last few years, the lager brewers have come under pressure from the Campaign for Real Ale, which aimed at boosting local loyalties. The recent downturn in the United Kingdom beer market, and the phenomenal increases in energy costs have obviously affected the major breweries, with their large plants, more strongly. They also suffer most from the industry's problem of over-capacity. In addition, fuel price rises have meant heavy distribution costs which can account for 2-3p of the cost of a pint of beer.

The extra costs and their effects on rates of return have been accentuated by the increasing competition in the "take home" and free trade areas. Take-away trade boomed in the seventies and accounts for more than 8.5 per cent of the total market. Aggressive pricing rather than brand loyalty counts with the consumer, and the low margins of the "big six", which produce a large proportion of these canned beers, has been reduced even further by the enormous buying power of the supermarkets. In the free trade area, which now accounts for 36 per cent of all drink sales in "on licences" and 10 per cent for "off licences", the large brewers have offered loans to clubs, often at very low rates of interest to ensure that their beer is stocked.

4.1.3 Technical change and microtechnology

Extensive reorganisation in the industry from the mid-fifties provided the basis for technical innovation. These technological developments have had five principal aims, as follows:

- a reduction in production time by speeding up the process;

- an increase in capital and labour productivity by more intensive use of equipment;

- the substitution of labour by capital through increased mechanisation and automation;

- reductions in variations in product quality as the markets for particular brands of beer expanded;

- the production of both ales and lager in a single brewing system.

Over the last decade, microtechnology has played a major role in technical developments in the industry. It would seem that the brewing industry is particularly suited to the introduction of micro-electronics, for despite seasonal peaks, the industry enjoys high and stable levels of demand, which permit continuous year-round production. In addition, the size of many breweries in the United Kingdom will enable them to utilise widely the new technology. Micro-electronic controls afford flexibility, large data handling capabilities, reliability and relative cheapness in relation to the tasks they carry out, and the consequent improvements in performance that can be expected.[19]

These many functions of the "microchip" would therefore suggest that there are a number of areas in the brewing process which are suitable for micro-electronic applications. A survey was recently conducted of the extent of present and future applications in 12 operational areas of the industry. An analysis was made of 74 cases, which consisted of 36 small, 27 medium and 11 large breweries (tables 88 and 89).

Not surprisingly, a significant difference was found between size and the level of micro-electronic applications in most areas both in terms of present and future applications, except in the cases of present applications in the areas of stock control and accounts. To date it would seem that the smaller companies have made as much of an investment in these areas as the larger ones.

In terms of current applications, over 50 per cent of large breweries surveyed have implemented microtechnology in four operational areas of the brewing process. Moreover, in estimating future applications, over 70 per cent believed they would have made microtechnological changes in eight of 12 possible areas. Differences in present and future applications were also found to be quite substantial amongst medium-sized breweries. Whereas in terms of current applications there was only one area where over 50 per cent of such companies had adopted microtechnology, in the future it

Table 88: Brewing industry: present applications of micro-electronics

Unit/department	Size of brewery[1]					
	Small		Medium		Large	
	No. of breweries	% of total	No. of breweries	% of total	No. of breweries	% of total
Maltings	1	2.8	1	3.7	3	27.3
Brewing	2	5.6	11	40.7	9	81.8
Large containers	4	11.1	9	33.3	6	54.5
Small containers	3	8.3	7	25.9	7	77.8
Distribution	4	11.1	11	40.7	3	27.3
Stock control	11	30.6	10	37.0	1	9.1
Accounts	18	50.0	18	66.7	6	54.5
Maintenance	0	00.0	10	37.0	2	18.2
Warehousing	0	00.0	5	18.5	1	9.1
Retail control procedures	4	11.1	6	22.2	2	18.2
Administration	7	19.4	10	37.0	5	45.4
Energy audit	0	00.0	3	11.1	1	9.1

[1] Large: in excess of 1 million barrels p.a.; medium: between 100,000 and 1 million barrels p.a.; small: less than 100,000 barrels p.a.

Table 89: Brewing industry: future applications of micro-electronics

Unit/department	Size of brewery[1]					
	Small		Medium		Large	
	No. of breweries	% of total	No. of breweries	% of total	No. of breweries	% of total
Maltings	2	5.6	2	7.4	5	45.4
Brewing	7	19.4	19	70.4	11	100.0
Large containers	6	16.7	15	55.5	11	100.0
Small containers	5	13.9	10	37.0	8	72.7
Distribution	6	16.7	19	70.4	6	54.5
Stock control	17	47.2	18	66.7	7	77.8
Accounts	22	61.1	24	88.9	18	72.7
Maintenance	3	8.3	16	59.3	7	77.8
Warehousing	3	8.3	14	51.9	6	54.5
Retail control procedures	5	13.9	13	48.1	4	36.4
Administration	7	19.4	18	66.7	7	77.8
Energy audit	1	2.8	13	48.1	7	77.8

[1] See note to table 88.

was estimated that more than 50 per cent would have microtechnology in eight out of 12 possible areas. In the smaller companies the level of application is lower, but one can still observe estimates of substantial changes in the future.

Figure 33 presents the data in a more general form, illustrating the level of present and future micro-applications for the combined sample. The graph illustrates clearly how the rapid fall in the percentage of companies in the present, as the extent of micro-application increases, becomes much less dramatic in the future. There is indeed a substantial rise in the total percentage of companies with over 75 per cent of their brewing operations implementing new technology.

4.1.4 Reasons for implementation and employment implications

A survey conducted by the Brewing Sector Working Group[20] showed that company objectives did not differ fundamentally according to the size of the brewery. Improved quality control, reliability of product and decision-making assistance were the main reasons for considering micro-electronic applications. Five of the large companies had sought cost savings, but only two large and one medium-sized brewery had hoped to save labour. Four of the large companies aimed to receive better management information and one expressed the hope that greater safety might result. In addition, the overwhelming majority of companies were clearly making efforts to ensure that their staff had an opportunity to learn about micro-electronics through attendance at conferences or some formal training on the subject. One of the most disturbing findings of the survey was that, in spite of the current vogue for discussing micro-electronics and the recent government-sponsored training courses, publicity and information schemes, companies appear to rely only on limited sources for their information. Only one respondent claimed to have received information from the Department of Industry, while 11 depended on manufacturers and nine on trade journals.

As for the implications for employment in the industry, there will be a need for companies to rethink the way work is performed and organised. The number and type of employees needed will obviously be affected. A study group from the Department of Employment investigating the manpower implications of micro-electronic technology[21] visited a brewery in Japan which had installed a new automated palletising and warehousing system based on mini-computers, which employed five workers as compared with 25 on a conventional manual system. The study group also noted that the companies that had used redeployment and retraining in order to ease the process of change, and had accepted natural wastage as a means of reducing the number employed when this was necessary, had the fewest problems both in terms of industrial relations and in recruiting skills which were in short supply. The report thus gave general agreement to the suggestions raised by the TUC, in particular emphasising those relating to consultation and access to information.

There follow two case studies of how different brewing companies dealt with the challenge of technological change.

Figure 33: Brewing industry: present and future applications of micro-electronics

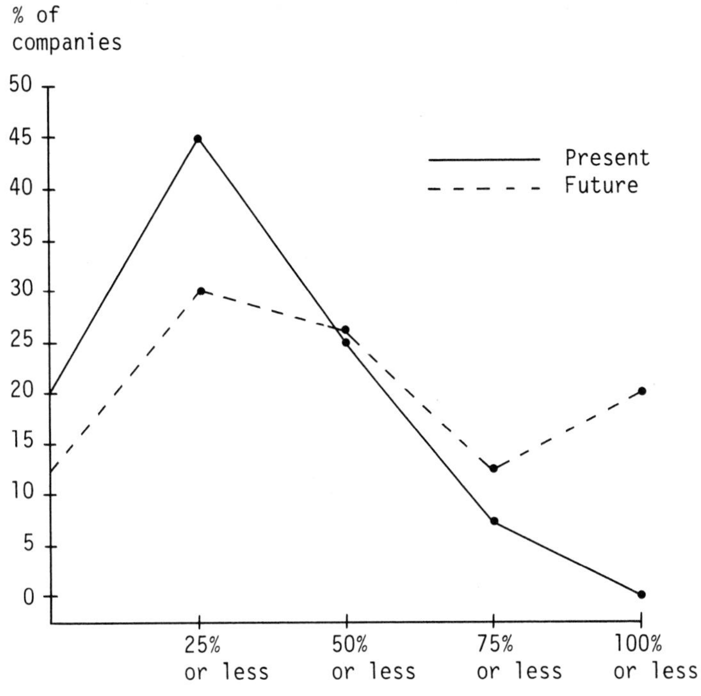

4.2 Plant 1

The new brewery of this company combined a move to a "green field" site with the introduction of a high-technology production process. The new development comprises the production centre for one of the four regional areas of this large brewing group. It replaces an old brewery and canning plant which were very labour intensive, with the brewing operation similar to that used about 150 years ago. The new plant was thought necessary to match the predicted growth in beer sales forecast in the mid-seventies and the rapid rise in demand for lager beers. It was estimated that the new brewery would have a capacity of 1.5 million barrels a year, which is approximately three times the capacity of the old brewery.

4.2.1 Manpower requirements

The technological change led to about a 30 per cent reduction in manpower. All employees were invited to transfer, but those who did not wish to could opt for redundancy. Under an agreement negotiated with the unions, compensation was based on 150 per cent of the statutory state payments. Early retirement was also available to staff over 50 years old, and pensions were not offset against redundancy payments. As is traditional in this industry, many of the employees had long service with the company and decided to take the options which were being offered. The result was that eventually 40 per cent of the people employed at the new brewery had to be recruited. A survey carried out by the company of employees' reasons for accepting the voluntary package highlighted the fears of insecurity, both in terms of changes in the type of work with the new technology and regarding the new three-shift system which was to be operated.

Part of the company's philosophy in carrying out the change was that machines should be used to do the "heavy labouring work" and that people should retain their decision-making capacity. Fuller automation could have been achieved, but management recognised that some manual aspects should be maintained to prevent boredom and improve job satisfaction. Nevertheless, the changes in job content and skills required were quite substantial. The company aimed for a "flexible worker" who could be moved from one job to another as need arose. There was complete abolition of many supervisory grades, resulting in only about three levels from the lowest category of worker to management. No evaluative study has yet been carried out to determine whether jobs are more satisfying as a result. In order to achieve objectives such as flexibility of working and an elimination of traditional demarcations, moves were made to harmonise staff and manual employees' terms and conditions of employment.

4.2.2 A new pay and grading system

It was believed by both management and unions that the old payment system failed for three main reasons -

- there were too many grades;

- the differentials between the grades had become distorted by overtime working;

- the nine-grade structure did not fit the new technology, as grades were too rigidly defined and both sides anticipated enormous change over the next few years.

Management and unions therefore decided on a new system of job evaluation which would provide the framework for the new structure. The system chosen was Paterson's decision-banding technique, under which grades are based not just on job content but also on the individual's ability and merit in that job. The difference between one grade and another is based on the level of decision-making required in jobs. In a particular job, this will depend on the type of work done; the nature of the decisions made (i.e. whether the job demands the employee to make the same or different decisions on a recurring or occasional basis); the level of discretion allowed; the amount of training or experience required; and the consequences of the various decisions. It was decided that, for each of the factors, there should be four "degrees" or levels of decision-making and at each level there should be two skill bands. The higher skill differential is payable when an employee becomes more proficient in a particular job and more versatile by acquiring job-related skills.

It is believed that the major advantage of this system is that all jobs can theoretically be evaluated in terms of just one factor, resulting in less complex wage bargaining with fewer areas open to conflict and dispute. In fitting in with this scheme, the company also felt that everyone working shifts should receive the same payment for the inconvenience, regardless of grade or type of work done.

4.2.3 The process of job redesign

The company set up what was termed a "humanities committee" to consider a whole range of human aspects related to the new site. Twelve major areas were identified for consideration, ranging from security, catering, manpower, and terms and conditions of employment to the initial occupation stage and early brewing trials. Through this committee the personnel department of the company was fully involved in the commissioning of the new brewery.

The committee was set up approximately four years after the initial decision to invest, and at about the same time a series of discussions was begun with the divisional officers and convenors of the three trade unions involved. As a result of these discussions, a consultative document prepared by management was presented to the unions for review before formal negotiations started. These negotiations concentrated on the following six areas:

- a procedural agreement;
- a new code of conduct to replace the former book of rules;
- a revised package of terms and conditions of service;
- the introduction of a new sick pay benefit scheme;
- the development of a payment structure;
- a suitable redundancy agreement.

Apart from these formal negotiations, a communications structure was agreed with the three major unions to ensure that all employees affected by the change were consulted. This was achieved through departmental consultative groups, co-ordinated by a central consultative committee. It was the responsibility of these groups to pass on views and recommendations to the central consultative committee. However, no formal joint union/management decision-making committees were set up. Nevertheless, the company did organise joint visits to other breweries which had recently carried out technological change in order to gauge the necessary changes in job design for the new brewery.

It was believed by management that there had been few problems with the commissioning of the new brewery, and it was thought that one of the main reasons for this was the good relationship between management and unions, with the latter voicing very few objections to management proposals. However, it cannot be overlooked that the change was also made easier due to the fact that the company had moved the whole operation to a new "green field" site rather than altering existing technology on the same site.

4.3 Plant 2

The new brewery of this company is a highly capital-intensive plant utilising very sophisticated technology. Most operations are process controlled, with brewing, fermenting, conditioning, dispatching the beer, and cleaning the plant and its pipework handled largely by operators at centralised control panels. The new plant produces up to 2 million barrels a year, which is achieved with only half the labour force that was employed at the old medium-sized plant. There has also been closure by the company of a number of small breweries.

4.3.1 Manpower requirements

There has been a major effect on manpower levels, particularly in the brewhouse, fermentation and process departments. In 1974, the old brewery employed a total of 83 operators on these functions, while in the new plant only 36 are needed. In addition to this reduction in jobs available, the nature of the work has also been changed. The technology placed new pressures on employees, such as the ability to work without direct supervision, and to take quick decisions on a continuing basis for eight hours at a time. These requirements have led to a decline in the number of semi-skilled and unskilled brewing and processing grades, and an increase in the number of skilled operators.

In order to achieve these changes in manning levels, the company adopted a policy of natural wastage, voluntary redundancy and some redeployment. When the new plant was under construction in 1973, management planned to freeze all recruitment at the old brewery and to recruit only temporary employees. Initially employees were recruited on this basis for one-and-a-half years, but the length of possible employment shortened as the date for commissioning the new plant and the changeover approached. As the temporary workers became familiar with the semi-skilled tasks involved, the permanent operatives selected for future employment were in turn released for intensive training and familiarisation with the new plant.

While all permanent employees were invited to apply for positions in the new plant, the selection procedures were very rigorous. The company stated that appointment would depend largely on ability rather than on length of service. In addition to more usual selection processes, all applicants for the senior operator posts were expected to take a 16 PF test, which yields scores on 16 personality factors such as shyness, submissiveness, boldness, etc. The answers were analysed by a consultant psychologist, and then the decision was made as to suitability for a given job. All other operators were also subjected to rigorous selection and recruitment procedures, and the company believes that they were an important feature in explaining the successful operation of the new plant. The company estimates that more than 70 per cent of its appointments to the new plant were in fact drawn from its existing employees, with the remaining proportion drawn from outside applicants including some of the temporary workers who were offered full-time employment.

At the new plant, supervision has become much more technically qualified. The company now recruits technical graduates for these posts rather than the old-style supervisor who worked themselves up from the lowest ranks. The operating groups on the line have become much smaller due to decreased manning levels, and this has resulted in a more participative style of management. The highly qualified line managers have taken on much of what is regarded as the traditional personnel role in terms of industrial relations, formulation of job descriptions, recruitment and training.

4.3.2 The pay and grading system

In the new brewery, the chargehand grade was abolished and the chargehand's rate was paid to the operators concerned. Two agreements relating to maintenance of earnings were also signed. These agreements identified four groups of individuals affected by the change -

- those earning more on transfer to the new site;
- those temporarily moving from shift to day work;
- individuals working fewer overtime hours with a corresponding drop in earnings;
- those required to work more hours but receiving lower gross earnings on transfer to the new site.

There was no problem with the first group. The second, it was agreed, would continue to be paid shift premium for up to ten months and would adopt a flexible approach to working hours during the commissioning of the new plant. The third group was compensated with an immediate cash payment of 50 per cent of the shortfall occurring within the coming year, and the fourth group was to be compensated in full for any shortfall.

4.3.3 The process of job redesign

The driving force in the whole change process was a specialist management team set up by the site manager at the outset. It was this team which had total responsibility for making decisions regarding the manpower implications of the technological change. The specialist team comprised mainly technical and production management, and the only representative from the personnel department was the training manager. The involvement of the personnel function in the whole process was minimal, which owes much to the company philosophy that line managers should be responsible for all personnel issues.

One of the main tasks of the specialist team was to devise a new training programme. The team members were able to familiarise themselves with the new technology by testing similar equipment already being used by other companies in the group. Visits were also arranged to observe similar processes at work in breweries outside the group, and to major suppliers of equipment and brewing materials. Members of the management team then started producing training manuals for operators, and during this process certain features of the plant design were altered to incorporate practical suggestions aimed at reducing some of the operating problems discovered. Formal training courses and sessions at the company's residential training centre were supplemented by working visits to other plants and suppliers to see the technology operating in practice. Manual tasks were intentionally built in during this job-design process in order to give the operators a specific degree of responsibility for the smooth operation of the plant, and to enhance job satisfaction.

No formal joint union/management committee was at any stage established to deal with matters concerning the introduction of the new technology. Formal negotiations were held with the unions concerning various selection, training and payment issues, and on matters outside the collective bargaining sphere the workforce was kept informed by management by means of posted notices. No consultative machinery was set up to deal with problems, and whatever consultation occurred was on an ad hoc basis as needs arose and as the workforce requested more information.

5. Summary and conclusions

Managers in the United Kingdom manufacturing industry during recession have turned their attention to the problems of short-term survival and long-term productivity improvement. Increased mechanisation and automation are seen by the Government, employers and trade unions as essential if Britain is to remain competitive in world markets. At the same time, recession has brought about increased levels of unemployment, a problem also of great concern to all parties.

In general, decisions about investment in new technology are taken at enterprise level. Similarly, once financial sanction is granted, decisions about the process for planning and implementing change are generally made within the enterprise. Consequently, views expressed by the Government, CBI and TUC about the management of technological change may have only limited impact on practice within organisations. In any case, it is widely agreed that no one formula exists and that it is appropriate for those most directly involved to establish their own procedures and working practices. Hence there is considerable diversity in approach.

There has been a long tradition in the United Kingdom of management/trade union bargaining over technological change. The emphasis in such bargaining has tended to move from payment, redeployment and manning levels to encompass a wide range of issues of concern to employees. The model new technology agreements formulated by some trade unions illustrate these issues, and would also be formulated prior to decisions to introduce any automated equipment. This would provide trade unions with an opportunity to be involved to a higher degree at early stages in the decision-making process. Consultative machinery has long been in existence in many organisations but has tended to have little impact, often being criticised for dealing largely with issues such as "tea and toilets". During the seventies, many organisations attempted to make the consultative process more meaningful, partly in response to interest in industrial democracy and a government enquiry on the subject. Technology agreements can specify procedures for consultation over decisions relating to new technology and may increase the effectiveness of consultative processes.

The case studies described in this chapter illustrate different aspects of the impact of automation on jobs. Case study 3 offers two examples from one industry, brewing, an industry which experienced considerable growth during the seventies. Growth in manufacturing output was achieved by technological change and rationalisation. Case study 1 was taken from the food industry, which has shown little growth in recent years and a relatively low return on investment. The company concerned also had a low market share in the product manufactured on the new facilities. Case study 2 was taken from a large publicly owned enterprise forced to rationalise manufacturing in the light of severe world-wide competition and general over-capacity.

The food industry case illustrates an incremental approach to the application of new technology, only certain parts of the total process being radically different from facilities elsewhere in the factory. The factory management and labour force had considerable experience of small-scale technological change, but limited experience at the time of the installation and commissioning of major new plant. Decisions concerning the new plant had to be taken within the wider context of the factory and its operation. Here much of management's emphasis was upon achieving deadlines for the plant being operational, and it would appear that there was avoidance of areas of possible conflict with trade unions and an emphasis upon minimisation of other changes. Nevertheless, several former jobs were merged and the role of supervision changed.

Opportunities for a radical rethink were available on "green field" sites (e.g. the steel industry and one brewery). Emphasis here was placed upon elimination of some of the former job demarcations and the development of a more flexible labour force.

In general, there seemed to be a predominance of design and/or project engineers in the decision-making process, following the initial decision to invest. The initial decisions were undoubtedly based on return on investment and economic viability of the total manufacturing facility. At the design stage, emphasis seems to have been placed on meeting specifications at the lowest capital cost. With low levels of involvement of operational management in this decision-making process, operational aspects seem not to have been a major concern to designers.

Decisions about job design and work organisation generally seem to have been taken at a late stage in the process of change. In the first case, for example, detailed consideration commenced during the installation of the plant, and agreement was reached with the trade unions concerning manning and pay during plant commissioning. In general, work organisation was considered at a stage in the change process when only minor changes in the design were possible. Despite this, many of the jobs resulting required increased operator skills and offered improved job attributes. However, the first case illustrated a deterioration in those jobs not directly involving work on the technically more advanced plant, where working methods changed little but the rate of working increased.

In case study 1, management introduced the new plant into an existing factory and appeared to seek to minimise disruption of the wider operation. In case study 2 we saw how, even in a "green field" site, earlier decisions of a political as well as of a technical nature could impose constraints on the job designer. However, even though the Work Structuring Team in this case was operating with constraints, it was successful in identifying alternative structures.

With high capital investment in new facilities, management sought means for ensuring continuous operation or at least high utilisation. The supervisory role appears to have changed to include a higher technical content. Operators in several cases were required to make decisions formerly the responsibility of supervisors in order that faults could be rectified more speedily. Attention, particularly in case study 2, was paid to the role and organisation of maintenance.

In most cases management agreed to increased earnings for work on new plants. This included a redesign of the job evaluation scheme, and in one case a completely new payment scheme. Agreement on flexible working and a reduction in job demarcations was accompanied by harmonisation of conditions of employment. Some aspects of decision-making were also retained in the jobs in one case in order to make them more satisfying, but this was perhaps only a temporary measure. In another case, operators on the new plant were thought to have higher job security; in the event this was not the case. These features of the new jobs, however, may still not make them attractive to the labour force. In the first brewing example, a large number of employees preferred to accept the terms offered for job severance rather than face the fears of insecurity in terms of changes in the type of work and shift patterns.

Different levels of employee involvement in the decision-making process are apparent in the case studies, but conclusions cannot be drawn about the relative success resulting. The level of involvement appears to reflect management's philosophy as much as trade union pressure. The use of "experts" also varies between cases. Only in one case was a work structuring specialist involved, but even here the need for the commitment of operational management to principles of work structuring was stressed.

Case study 2 illustrates clearly the formulation of a strategy for the application of agreed work structuring principles. In addition, a review of the experience gained was undertaken with a view to improving the procedures for future projects.

Job design is one of many considerations in the complex decision-making process of designing and installing technological change; it is one of a number of personnel-related issues to be resolved through management/trade union negotiations. Decisions have to take into consideration factory and company-wide implications. One of the prime objectives of management when installing new equipment is to achieve operations on programme and to budget. This seems to result in attempts to develop an acceptable "package" of employee inducements at minimum cost so as to give stability of operation, particularly during commissioning. Job design aspects often seem to be little considered until late stages in the total process. However, in the cases presented here, problems in start-up and subsequent operation did not generally appear to be the result of late consideration of jobs and work organisation.

Notes

[1] Great Britain consists of England, Wales and Scotland.

[2] Central Office of Information: Britain 1981 - An official handbook (London, 1981).

[3] Department of Employment: "Trends in labour statistics", in Employment Gazette, Jan. 1981.

[4] ibid.

[5] N.A.B. Wilson: On the quality of working life (London, HMSO, 1973).

[6] Social Science Research Council (SSRC): Report of the Management and Industrial Relations Committee's Work Organisation Working Party (London, 1978).

[7] D.W. Birchall et al.: A study of experience with new models for work organisation in the UK (Dublin, European Foundation for the Improvement of Living and Working Conditions, 1978).

[8] Department of Employment: Manpower implications of micro-electronic technology (London, HMSO, 1979).

[9] See "TUC and CBI inch towards new technology agreement", in Industrial Relations Review and Report, No. 232, Sep. 1980, p. 2.

[10] ibid.

[11] See "New technology: An IR-RR review of agreements", ibid., No. 227, 1980, pp. 2-9.

[12] Association of Professional, Executive, Clerical and Computer Staff: Automation and the office worker (London, 1980).

[13] See D.W. Birchall: Technological change in biscuit manufacture and its impact upon jobs (Henley, The Management College, 1979).

[14] C.A. Carnall et al.: "The design of jobs - An outline strategy for diagnosis and change", in Management Services, June 1976, pp. 48-51.

[15] D.W. Birchall: Job design: A planning and implementation guide (London, Gower Press, 1975).

[16] The information is summarised from P.C. Schumacker et al.; Work structuring at Hunterston (London, British Steel Corporation (BSC), 1980; internal document).

[17] P.C. Schumacker: Principles of work structuring (London, BSC, 1976; internal document).

[18] ibid.

[19] National Economic Development Office: Microelectronics in the brewing industry, Brewing Sector Working Group Report (London, 1980).

[20] ibid.

[21] Department of Employment: The manpower implications of micro-electronic technology (London, 1979).

CHAPTER XVI

AUTOMATION AND WORK DESIGN IN THE UNITED STATES

1. National background

1.1 Introduction

This chapter discusses issues associated with the nature of automated production systems and their impact on QWL. Case study material from a variety of manufacturing settings in the United States is presented with a view to conveying the "state of the art" concerning the introduction of new technology on terms which jointly enhance human and economic well-being.

In order to place these activities in a national context, the first part of the chapter presents an overview of the following: the industrial relations system; automation and collective bargaining; historical developments associated with automation and QWL; current developments in automation; and automation and job design. This is followed by introductory and case material derived from discrete and continuous process manufacturing industries, with a brief conclusion. The second part of the chapter is devoted to a case study of socio-technical design of printed circuit fabrication.

1.2 The labour force

In 1981, the civilian workforce totalled 107 million out of an estimated population of 228 million.[1] Women accounted for 43 per cent of the workforce, reflecting their continued increase in labour force participation (projected to rise to between 44 and 47 per cent by 1985).[2] The occupational composition of the labour force in 1981 was as follows: white collar: 51.9 million; blue collar: 29.9 million; service: 13.4 million; and agricultural: 2.8 million. In November 1981, approximately 9 million members of the civilian workforce (8.4 per cent) were unemployed. Undocumented foreign workers (more popularly known as "illegal aliens") residing in the United States have been estimated to range from 2 to 12 million,[3] with 30-45 per cent of annual labour force growth possibly consisting of newly arrived aliens (legal and illegal).[4]

Past and projected changes in the occupational composition of employment are presented in table 90.

It is apparent from table 90 that occupational growth tends to be concentrated in the white-collar (with the exception of managers and administrators) and service areas, with blue-collar occupations in manufacturing industries continuing their downward trend as a proportion of total employment. This is likely to be reinforced by the increased use of more automated production processes, although how trade-offs between improved productivity growth and employment are resolved during the eighties will be a central policy concern.

Table 90: United States: occupational breakdown of the labour force, 1968, 1978 and 1990

Occupational category	Proportion of employment (%)		
	1968	1978	1990[1]
White-collar:	46.9	49.8	50.9
Professional and technical	13.6	15.9	16.8
Managers and administrators	10.3	9.0	8.8
Sales workers	6.1	6.6	6.7
Clerical workers	16.9	18.3	18.6
Blue-collar:	36.2	32.6	31.5
Craft and similar	13.2	12.0	12.0
Operatives	18.4	14.6	13.7
Non-farm labourers	4.6	6.0	5.8
Service workers	10.1	14.8	15.8
Farm workers	4.6	2.8	1.6

[1] Projected.

Source: United States Department of Labor, Bureau of Labor Statistics: Employment projections for the 1980s, Bulletin No. 2030 (Washington, DC, 1979), p. 36; and Max Carey: "Occupational employment growth through 1990", in Monthly Labor Review, Aug. 1981, p. 45 (low-trend projections for 1990 used here).

1.3 The industrial relations system

The mainstay of what is traditionally described as an American "national labour policy" is the National Labor Relations Act (NLRA, 1935) and its subsequent amendments, the Taft-Hartley (1947) and Landrum-Griffin (1958) Acts. These laws regulate the procedural aspects of trade union organisation, collective bargaining, the administration of agreements and internal union affairs. Their basic provisions are the principles of free choice by workers of their collective bargaining representatives through fair elections, and the opportunity to engage in collective bargaining. The specialised National Labor Relations Board (NLRB) is the administrative apparatus for the Act, which has reinforced those institutional aspects of labour-management relationships regarded as unique in the United States: decentralised bargaining in industry-wide and individual workplace contexts, and both minimal and extreme involvement of the legal system in different relationships.[5]

With respect to the public sector, a majority of states have collective bargaining legislation applicable to employees of local and state governments, and regulation of collective bargaining for federal employees has existed under a combination of Executive Order and Congressional Statute since 1962. Despite important differences, most of the state and federal procedures have been modelled upon the principles of the NLRA.

Although the NLRA remains a basic foundation of national policy governing labour-management relations, since the sixties it has coexisted with major social policies aimed at direct regulation of the substantive terms and conditions of employment.[6] This has been interpreted as a major departure from federal emphasis on the promotion of free collective bargaining in conjunction with a laissez faire stance towards the definition of outcomes. The Civil Rights Act of 1964 prohibited discrimination in employment based on race, colour, religion, national origin or sex. Title VII of the Act sets forth rights and protection concerning the allocation of job opportunities, which directly affect seniority, promotion, transfer and other terms of employment for all workers. The Occupational Safety and Health Act (1970, 1976) and the Employee Retirement Income Security Act (1974, 1976) represent similar thrusts towards regulating substantive aspects of the employment relationship for union and non-union members alike.

In 1980, trade unions represented 22.4 million workers, or 20.9 per cent of the 106 million members of the labour force.[7] The 102 unions of the American Federation of Labor-Congress of Industrial Organisations (AFL-CIO) represent 15 million American workers. Organised labour reached its peak in 1953, when unions represented 17.9 million workers (25.5 per cent of the total civilian labour force). The major area of union growth in the seventies was in the public sector, a trend which has now diminished.

Trade unions continue to have difficulty in establishing new bargaining units. In 1979, they won only 45 per cent of 8,043 elections for collective bargaining representation conducted by the NLRB - the smallest proportion since 1935. Included in these figures are losses in 75 per cent of the 777 decertification elections held that year. In 1980, unions won 49 per cent of 8,198 representation and 27 per cent of 902 decertification elections.[8]

Shifts in the mix of the labour force appear likely to reduce the proportion of workers belonging to trade unions or covered by national agreements. Changes in industry, sex and regional composition suggest that groups with historically low unionisation rates will account for a larger portion of the labour force in the eighties.[9] These trends, coupled with the extensive application of sophisticated union-avoidance techniques by employers, pose a serious challenge to collective bargaining, which was punctuated in 1978 by labour's national defeat in its attempt to make changes in labour laws which would facilitate union organising efforts (the unsuccessful Labor Law Reform Bill).

A recent study of exclusively or predominantly non-union large companies revealed the following profile of preferred characteristics:[10]

- a corporate philosophy affirming that unions are unnecessary to advancing and protecting employee welfare;

- small plant size, ideally ranging from 200-1,200 employees, with large concentrations of women and professionals;

- plant location in non-urban areas and not in close proximity to large, unionised companies;

- profitable, growing companies, many with a high technology base and dominant market positions;

- highly centralised personnel departments which tightly control local operations; and

- close association (if not owner managed) between owners and managers in company decision-making.

In addition, contemporary "positive personnel" strategies for reducing worker incentives to unionise include the following:

- wage and fringe benefits comparable or superior to those paid to unionised workers in the industry or local labour market;

- high investment per worker in training and career development programmes;

- advanced methods of communication and information sharing;

- favourable support for worker participation in job-related decision-making;

- a formal grievance procedure (usually without binding arbitration);

- rational mechanisms for performance appraisal, merit-based promotion, wage and salary administration, with accompanying recognition of seniority principles where relevant;

- cultivation of a psychological climate which engenders and rewards employee loyalty and commitment; and

- in some cases, careful screening of job applicants to eliminate those who might be pro-union.

As one researcher has noted -

> ... "free collective bargaining" is now being subjected to a more severe test in both the private and public sectors. Collective bargaining is no longer viewed as the most preferred mechanism for setting conditions of employment in all contexts, on all issues or for all workers and employers. Instead, the costs and benefits of the bargaining process and its results are being weighed against the costs and benefits of alternatives.[11]

1.4 Automation and collective bargaining

Generally speaking, and contrary to popular conceptions, American trade unions have not resisted the introduction of advanced, computer-based technology. In countless instances they have supported and even stimulated it, if accompanied by adequate economic and physical safeguards for workers.[12] However, the ultimate right of management to introduce new technology has pervasively remained intact. This is confirmed by a variety of recent surveys of technological change clauses in collective agreements.[13] Contract provisions in the late seventies focused on smoothing worker adjustment to changes initiated and controlled by management. In no instance did a union explicitly have the right to influence the nature and content of the innovation itself.

Collective bargaining agreements, in varying intensity and degree, contained provisions in the following areas:[14]

- the introduction of technological change (management rights, advance notice and consultation);

- regulating job changes (classification, wage rates, work rules, union jurisdiction);

- smooth transition for workers (retraining, income maintenance);

- controlling reduction in the workforce (seniority, transfer, relocation, reduction in hours, layoffs, severence pay, early retirement, preferential hiring).

One of the major surveys,[15] covering 400 contracts representing a cross-section of industries and trade unions, found technological change provisions in 17 per cent of the selected contracts. Half of these required advance notice and consultation with the union. The other provisions described above varied widely by industry and were significantly less frequent than advance notice provisions, being evident in only 11 per cent of the contracts examined. The highest concentrations were in publishing, textiles and the retail trade. A third survey, which examined a broad range of technological change and permanent lay-off provisions in 100 major contracts,[16] found advance notice provisions in 31 per cent of cases. This was the most commonly existing provision, followed by provisions concerning severance pay (18 per cent), retraining and retention of seniority rights for fringe benefits (both 14 per cent).

Despite the paucity of contract provisions which explicitly address technological change (and while drawing no conclusions regarding their actual effectiveness), a wider blanket of potential protection extends through contractual language governing changes in "working conditions". Management customarily must discuss such changes and bargain over their impacts, but here, as before, the actual form and content of innovations which alter working conditions ultimately remain their prerogative. It is important to keep in mind that because only 21 per cent of the United States workforce (public and private sectors) belong to trade unions, collective agreements providing contractual protection of any kind are restricted to that group.

In the manufacturing sector, a significant step towards jointly managing the introduction and impact of technological change was taken by the United Automobile Workers (UAW) and Ford Motor Company in their 1979 contract.[17] Key elements included the following:

- formation of a National Committee on Technological Progress to meet regularly to discuss the "development of new technology at the corporate level and its possible impact upon the scope of the bargaining unit", and "other matters concerning new or advanced technology that may be referred by local unions or by the introduction of new technology";

- provision by the company of specialised training programmes to enable workers to perform new or changed work arising from advanced technology;

- obligation of the company to provide advance notice and a detailed description, as well as to hold discussions with the union at local level "as far in advance of a technological change as is practicable";

- appointment at local level of an existing union representative (and management counterpart) to be responsible for discussing anticipated introductions of new technology and evaluating their impact;

- assurance by the company that new technology would not be used as a means of eroding the bargaining unit, and an obligation to abide by the decision of an umpire in cases of disagreement.

The last point was recognised as an important gain for American trade unions in preserving bargaining rights over incursions from the growing use of computer-based hardware and software in the motor vehicle industry. The union was unsuccessful, however, in its attempt to secure the right to strike over new technology during its three-year agreement with Ford.

This has not prevented the UAW and other unions from persevering in their attempts at a more comprehensive regulation of the use of new technology. In September 1980, the International Union of Electrical Workers (IUE) passed a resolution on the use of robots and automated factory management systems which sanctioned future negotiation to secure contractual protection in the following areas:[18]

- limiting the use of robots to hazardous or dangerous jobs and guaranteeing worker protection from malfunctioning robots;

- prohibiting the surveillance of individual workers by television cameras and remote or computer-based monitoring systems (the UAW had attempted unsuccessfully to gain a similar provision covering time study, monitoring and discipline in the 1979 Ford negotiations);

- establishing "specific language on technological change providing for pre-notification with full disclosure of system capabilities, training or retraining, wage protection, attrition with no lay-offs and the prevention of the erosion of bargaining unit skills";

- continued pressure for reduced working time in order to create more job opportunities.

In a similar fashion, the International Association of Machinists (IAM) drafted a "new technology bill of rights" in 1981.[19] Its preamble notes that "the choice should not be new technology or no technology but the development of technology with social responsibility". The bill of rights states that new technology must be " ... used in a way that creates or maintains jobs; ... used to improve the conditions of work; and ... used to develop the industrial base and improve the environment".

It concludes by proclaiming that -

> The introduction of new technology is no longer the exclusive prerogative of management or an automatic process. Moreover, users of technology that violate the rights of workers and the society will be opposed. Instead of only responding to management actions, unions will seek full participation in the decisions that govern the design, deployment and use of new technology. The goal will be machines that fit the needs of people rather than the other way around.[20]

The IAM's bill of rights represents an extension of their previous policies concerning technological change. It is an ambitious statement of intent which is likely to find at least partial expression in future negotiations with major users of computer-based manufacturing systems.

Other forums have recently been established through collective bargaining wherein the development and introduction of new technology may be eventually tempered by human and social considerations. Among them is the creation of a Joint Working Conditions and Service Quality Improvement Committee by the American Telephone and Telegraph Company (the Bell System) and the Communication Workers of America in their 1980 negotiations. This national committee will oversee the initiation and progress of labour-management efforts at local level to jointly fulfil operational and human objectives through worker participation.[21] It is too early to discern how this programme will take shape, and whether or not it can serve as a viable mechanism for disclosing plans to introduce new or modified systems in this already highly automated field, jointly anticipating the consequences, and modifying designs and applications to enhance QWL.

1.5 Automation and quality of working life: historical developments

The period from the early fifties to 1966 witnessed an enormous production of literature and public debate about automation and its consequences.[22] Central to this outpouring was a consideration of the prospective roles of collective bargaining and government in cushioning automation's adverse effects and exploiting its potential benefits. Critical opinion became deeply polarised. Alarmists could draw much of their resolve from Norbert Weiner, who initially popularised automation through his books, Cybernetics and The human use of human beings, and who implored in 1950 -

> Let us remember that the automatic machine ... is the precise economic equivalent of slave labour. Any labour which competes with slave labour must accept the economic conditions of slave labour. It is perfectly clear that this will produce an unemployment situation, in comparison with which ... the depression of the thirties will seem a pleasant joke.[23]

By 1965, sceptics could argue, not without justification, that -

> Fifteen years after the concepts of "feedback" and "closed-loop control" became widespread, and ten years after computers started coming into common use, no fully automated process exists for any major product in any industry in the United States. Nor is any in prospect in the immediate future ... In their eagerness to demonstrate that the apocalypse is at hand, the new technocratic Jeremiahs seem to feel that any example will do; they show a remarkable lack of interest in getting the details straight, and so have constructed elaborate theories on surprisingly shaky foundations.[24]

In the intervening period, technological change was accommodated in imaginative ways through collective bargaining in a variety of industries, notably motor vehicles, long-shore work, meat-packing, rubber, steel and, to some extent, the railways. In 1955, for example, the UAW led the way by securing supplementary unemployment compensation benefits from the motor vehicle industry, in recognition of automation's displacement effects. In 1960, the International Longshoremen's and Warehousemen's Union and West Coast employers successfully renegotiated the continuation of their 1960 Mechanisation and Modernisation Agreement. The original agreement cleared the way for the introduction of labour-saving devices on the docks in exchange for compensation for surrendered job property rights, assurance against lay-offs, early retirement benefits and a minimum hours' guarantee. Also in 1966, the findings of the National Commission on Technology, Automation and Economic Progress were published. The Commission, established in 1964 by Congress, was generally acknowledged to have provided the definitive work on automation. It concluded that automation did not pose a serious threat to employment at the macro level, but emphasised its selective displacement effects at the level of the firm.

The Commission recommended an array of social programmes, including one which sanctioned the Government to be "an employer of last resort" for those incapable of finding work in the conventional labour market. By the time of the Commission's report, the "automation hysteria" had subsided, in part due to rapid growth in the economy as a result of expansionary fiscal policies and escalation of the Viet Nam War. As the Commission's staff director later noted, "the product of a year's rather intensive effort was some marginal increment in the education of the Congress and the country to the realities of its manpower problems".[25]

During the sixties, federal "manpower" policies, though generally successful, were only marginal in their impact. The Area Redevelopment Act of 1961, the Manpower Development and Training Act (MDTA) of 1962, the Vocational Education Act of 1963 (and 1968 amendments), the Civil Rights Act of 1964 and the Economic Opportunity Act of 1965 were the main legislative mechanisms for promoting training, access to the labour market and, to a very limited degree, temporary job creation for the period. Emphasis was almost totally on the problems of minorities and other groups unable to compete in urban labour markets. Basic elements of legislation to fight the "war on poverty" in the sixties were extended to new legislation for the seventies in the form of the Comprehensive Employment and Training Act (CETA) of 1973. Manpower programmes of the sixties were governed by the philosophy of "human capital investment" to combat chronic unemployment, but this was a preferred option only after the aggressive pursuit of employment-generating monetary and fiscal policies and anti-discrimination enforcement. Although Congress had limited control over general economic policy, it could directly enact structural and counter-cyclical job-creation programmes through subsidies to local governments and public agencies. CETA provided this vehicle, and the 1978 reauthorisation legislation added provisions focused on private-sector employment and training, with the more publicised public service employment legislative titles emphasising job creation in the public and non-profit sectors.

Meanwhile, as automation receded from public consciousness in the early seventies, a concern about worker alienation and QWL gained attention. Initially, debate on these issues was confined primarily to a relatively small group of intellectuals from academic institutions and various public policy bodies, as well as a diverse sprinkling of "pop sociologists" and the media. Much media attention was prompted by the strike of young motor workers in 1970 at the highly automated General Motors assembly plant at Lordstown, Ohio, as well as by growing reaction from blue-collar workers, who felt their status threatened by excessive governmental concentration on the economically disadvantaged. The publication of Work in America in 1972, commissioned by the Department of Health, Education and Welfare, fuelled the debate and was attacked as being partisan.[26] A somewhat similar reception from labour and management greeted the publication that same year of Where have all the robots gone?,[27] a study of blue-collar worker alienation and dissatisfaction. Those who actively or otherwise became identified with the nascent "quality of worklife movement" were labelled, depending upon the vantage point, "utopians", "do-gooders", "touchy-feelies", "liberal elitists", "ivory tower eggheads", "company stooges", "servants of power", "union busters", "time study men in sheep's clothing", "communists", or some combination of the foregoing.

Nevertheless, continued interest, controversy and, perhaps more importantly, experimentation at plant level with both unilateral and joint labour-management attempts to confront the detrimental aspects of impoverished jobs and work environment led to institutional developments. Through funding from the Economic Development Administration of the Department of Commerce and the Ford Foundation, three non-profit centres were established in 1974-75: the Center for Quality of Working Life at the University of California at Los Angeles, the National Quality of Work Center (in conjunction with the University of Michigan) in Washington, DC,[28] and the Work in America Institute in New York. These organisations performed a variety of distinct and overlapping functions which included information dissemination, technical assistance, research, education and training. Since 1975, over three dozen centres of varying magnitude and point of view towards QWL and productivity issues have been established.

As a focus of sustained institutional activity and support, the Federal Government has had a weak role with respect to QWL. A National Commission on Productivity was created by Congress in 1971 through an amendment to the Economic Stabilisation Act, and undertook research and information dissemination. When the Economic Stabilisation Act expired in 1973, the Commission took shelter as the Office of Productivity under the Cost of Living Council. The following year, it continued its metamorphosis as the National Commission on Productivity and Work Quality, with the additional charge of assisting in "the organisation and work of labour-management committees which may also include public members, on a plant, community, regional and industry basis".[29] In 1975, Congress replaced the Commission with the formation of the National Center for Productivity and Quality of Working Life, mandating four major responsibilities, as follows:

- to focus attention on the importance of productivity and QWL to the nation's economic strength;
- to stimulate the active co-operation of labour, industry and government agencies in voluntary productivity growth programmes;
- to establish a national policy for continued productivity growth; and
- to review the impact of government on productivity.[30]

The prevailing emphasis remained on productivity, with QWL as an incidental consideration. The Center lasted for three years on an average annual budget of 2.5 million US dollars ($) and was dissolved in 1978 due to lack of political and financial support, as well as disagreements between the Departments of Labor and Commerce. These departments subsequently absorbed a number of the Center's activities, with certain labour-management co-operation functions assumed by the Federal Mediation and Conciliation Service.

In late 1978, the National Productivity Council was established under Executive Order with the primary function of co-ordinating and guiding federal efforts to improve public and private sector productivity, as well as serving as the Government's principal contact with the private sector on productivity issues. As an inter-agency body, the Council did not receive budget appropriations. In a 1980 review of the Council's activities after two years, the General Accounting Office found that the Council had "largely ignored the functions assigned to it ... seldom met ... not provided guidance to federal productivity programmes, and not become recognised as the federal productivity focal point".[31] Because it was not supported by the Carter Administration, the Commission was allowed to disappear, with chances of any future revival extremely remote in the immediate future.

By 1981, however, QWL had captured the serious attention of managers and a growing number of trade union leaders. Although advocates in the early seventies had proclaimed that "quality of working life is an idea whose time has come", its drift into mainstream management thinking was to be confirmed in 1981 by the business magazine, Fortune, in the following synopsis:

> The cultural upheaval of the late 1960s and early 1970s brought a flurry of interest in work reform ... Some of the results were good ... More often, though, the results were dismal. In scores of companies and plants, hopes were aroused only to be dashed. Some programmes set up as controlled experiments did nothing more than turn the control and experimental groups against each other. Others were hit-and-run exercises by consultants pandering to the imperishable managerial fantasy of a patented quick-fix solution to complex problems. Even the best social scientists were divided over principles and tactics ... The current wave of interest in the quality-of-worklife concept is something else again. It has come on so suddenly that it has caught many long-time partisans off guard; some wonder dubiously if it is not a passing fad. But the climate is strikingly different from the past. Today's corporate leader is as often as not reviewing with dismay his stagnant productivity and substandard quality while glancing nervously across the Pacific for salvos from the competition; he is apt to be relatively open minded about changing his organisation.[32]

Those who had advanced the notion at the beginning of the seventies that long-term commitment to improving QWL meant more than the elusive pursuit of "job satisfaction" have found a certain irony in the belated pragmatism described by Fortune.

Much of the QWL-related activity during the seventies was directed towards restructuring work with respect to existing technological systems. Although usually well intentioned, occasionally successful and sometimes misguided, the predominant emphasis of these change efforts was on fitting people to technology in ways which improved intrinsic job rewards and economic efficiency. In isolated instances, most notably in new ("green field") plants, a design process occurred which enabled an atypically broad spectrum of organisational interests to plan together, with QWL factors among the explicit design criteria. These interests included engineers, architects, operations and personnel managers, process technicians and, in rare instances, workers and union officials. In an assessment of efforts up to 1979, one researcher observed -

> There are many signs that a growing number of United States managements are no longer going to let the working environment just happen. Some are going to try to manage that sector of their business just as they attempt to manage markets and finance, costs and assets ... What will be the effect of changing technology on working environment in the 1980s? Managements will slowly begin to react to existing technologies and to whatever technologies come along with better operating policies and practices; they will gradually produce better working environments. Many will not be content merely to react, however, and will instead begin to select and innovate new

technologies that offer advantages for the working environment. They will be stimulated partly by changes in ideology and intention, and, finally, perhaps most of all by the recognition that achievement of profitable, competitive productivity is steadily demanding a more competitive, dedicated and committed workforce, free of the resentments which fester and restrain in a poor working environment.[33]

1.6 Current developments in automation

Increased attention is now being given to intensifying the introduction of automated technology in the mass and batch production of discrete parts. Using continuous process industries as a model, major American manufacturing companies have begun to invest heavily in CAD-CAM equipment as a first step towards evolving integrated production systems, largely free (other than maintenance) of direct human intervention. This activity is being mainly undertaken by aerospace, motor vehicle and heavy equipment companies, which already have extensive experience with unintegrated ("stand alone") forms of computer-aided technology.

A key element in the shift towards more comprehensive spans of factory automation is the industrial robot. The definition of a robot varies widely between foreign and United States manufacturers, the main difference being the extent to which the equipment is programmable and versatile. The most frequently cited definition of a robot in the United States is "a re-programmable, multi-functional manipulator designed to move material, parts, tools or specialised devices through variable motions for the performance of a variety of tasks".[34] Thus, for instance, United States and Japanese statistics are not completely comparable, as Japanese definitions embrace a wider variety of machines which are not re-programmable.

Approximately 3,000 industrial robots have been installed in the United States. Sales of robots by American manufacturers in 1980 approached $100 million, up sharply from estimated sales of $65 million in 1979. However, this figure still represents less than 2 per cent of the $4.7 billion machine tool industry (of which robotics manufacture is often mistaken as a component). Robots are currently being produced at the rate of about 2,000 per year, and various forecasts estimate between 20,000 and 70,000 robots in use on American factory floors by 1990. Based on this range of estimated growth, it has been suggested that robots will not have a significant impact on the overall productivity of the nation's economy before the turn of the century. However, the impact could be felt sooner if large computer firms enter the market. Under this condition, as many as 200,000 robots could be in service by 1990.[35]

The great majority of robots presently operating in the United States represent the first generation of relatively primitive technology directed towards operations not requiring sensory or "intelligent" machine capabilities. Nevertheless, the current generation of robots perform a variety of industrial tasks, such as spot welding, tending die-cast machines, loading and unloading machine tools and presses, spray painting and materials handling. Most industrial tasks remain beyond the capacities of current robot technology, and a significantly heavier investment in R & D than present levels will be necessary before robots can assume the more complex and unstructured tasks which characterise the broad spectrum of American manufacturing. The most promising technical developments in the eighties are anticipated in arc welding and assembly operations. Other forms of CAM technologies are described later in this chapter.

The development of automated manufacturing technology has been the province of equipment suppliers, user R & D laboratories, non-profit-making technical centres and universities. While it is not uncommon for foreign companies in the forefront of technology also to be part of the organisation which makes most use of the equipment, this has rarely been the case in the United States. With increased automation, however, some American companies have noted the value of this link between equipment design and implementation. In the area of robots, for example, General Electric, Texas Instruments and International Business Machines now produce robots for internal use, employing the research to develop in this field as a new business opportunity. Companies and universities frequently work jointly to develop new technology under co-operative agreements. In other instances, companies donate funds to universities, such as the recent formation of a Robotics Institute at Carnegie Mellon University through a grant from Westinghouse. Government funding has also played an essential role in stimulating R & D, as well as in implementation at factory level, through such agencies as the National Science Foundation, the National Aeronautics and Space Administration, the National Bureau of Standards and the military.

The Air Force has played a significant role in stimulating the development and diffusion of CAM. Because of its reliance on the aerospace industry for military weapons, it has a strong interest in creating compatible and flexible manufacturing techniques. In 1977, the Air Force, together with a coalition of aerospace and electronics firms, technical centres and universities, initiated the ICAM (Integrated Computer-Aided Manufacturing) programme, a seven-year, $100 million effort aimed at unifying disparate production methods.[36]

> The project was spawned, according to the people who spawned it, by manufacturing's tireless neglect of efficient manufacturing methodologies. For years, it seems, development in the use of the computer as an aid to manufacturing has been disjointed ... Hardware and software systems have been designed, irrespective of one another, to address given problems on given days. The systems are myopic, rather limited-scope creations that suffice very well in so far as addressing a particular manufacturing function, but which never address their inter-relationships ... System integration has been tried, but never with premeditation ... "Afterthought" integration simply hasn't worked ... It has resulted in a proliferation of software/hardware and software/software matings that have in many ways tended to magnify problems.[37]

The ICAM programme supports the development of systematically related software modules for manufacturing management and operations. Some ICAM technology has been individually implemented in industry, with short-term gain, but primary benefits of the modular structure are dependent on a fully integrated system.

1.7 Automation and job design

Prevailing approaches to the design of work organisation for automation do not appear to differ substantively from those of other technical or procedural forms. In other words, the fundamental building block of the organisation continues to be the person/task ("job") rather than, for example, a system of roles filled at the discretion of a work group ("self-maintaining organisational unit").[38] Thus, with the person/task as the focal unit, the integrity of classical organisation design - job specialisation under hierarchical control - tends to be preserved. One writer has sardonically characterised the dynamics of this approach to system design as follows:

> What we need is an inventory of the ways in which human behaviour can be controlled, and a description of some instruments that help us achieve control. If this provides us sufficient "handles" on human materials so that we can think of them as one thinks of metal parts, electric power or chemical reactions, then we have succeeded in placing human materials on the same footing as any other materials and can proceed with our problems of systems design. Once we have equated all possible materials, one simply checks the catalogue for the price, operating characteristics and reliability of this material, and plugs it in where indicated. For the engineer or industrial engineer, these are precisely the terms upon which human beings must be considered ... There are, however, many disadvantages in the use of human operating units. They are somewhat fragile; they are subject to fatigue, obsolescence, disease and death; they are frequently stupid, unreliable and limited in memory capacity. But beyond all this, they sometimes seek to design their own system circuitry. This, in a material, is unforgivable. Any system utilising them must devise appropriate safeguards.[39]

The "appropriate safeguards" are manifested in an organisation's control system, which attempts to ensure reliable worker performance. Most important, the implications of automation for organisational structure and the work environment are not usually explicit in the terms of reference of the technical design process. Rather, they remain, tacitly or otherwise, illegitimate areas of concern. Engineers and systems designers themselves must function within a framework of control, custom and practice and overarching values which do not support the simultaneous consideration of technology, organisation and QWL factors. As a consequence, a familiar technological determinism guides the choice process in the evolution of factory automation. In this connection, the following remarks by a trade union official are germane:

> I feel that the manufacturing engineer plays an important role in determining the industrial relations climate in a particular plant ... Specifically, when the manufacturing engineer is engaged in such activities as setting batch sizes, determining plant layout, establishing process specifications, ordering changes in the priority of work to be done, or when making any decision about matters involving shop operations, he influences the behaviour patterns on the shop floor ... Typically, however, these production systems give too little

> consideration to the human beings who run the machines. All too often, it appears that these designers are guided by the feeling that technological advances in manufacturing engineering relegate the worker to the status of a machine, designed to perform small tasks, precisely specified on the basis of time and motion studies. He is assumed to be motivated primarily by economic needs and classified by a known degree of strength, dexterity and perseverance. The worker is often considered by management as incapable of dealing with variables in the production flow; any unplanned occurrences are to be handled by supervisory personnel.[40]

There have been exceptions, however. Some of the work restructuring efforts within a variety of American organisations during the seventies provided management and labour with insight into ways to improve poor working conditions generated by technologies chosen without regard to the work environment.[41] In these instances, technology itself remained unaltered, and the human system was adjusted to create a better fit. In addition, as will be described later, further inroads into aligning technical and human considerations have been made in the design of innovative work structures in new organisations. Although increased attention has been given to these developments, they are comparatively small in number and largely isolated within their respective parent organisations. It is too early to discern the extent to which these innovations will diffuse and penetrate conventional orthodoxies of job design.

At industry level, even quite modest opportunities to encourage the joint consideration of human and technical factors prior to the design of automated systems have not been seriously pursued. The first involved a segment of the Air Force's ICAM programme, devoted to studying "human factors affecting ICAM implementation". The completed study (by social scientists of a defence industry contractor) was based in part on a case study of the introduction of a fully automated manufacturing cell in an aerospace company. It addressed such issues as organisation design, maintenance, labour relations, employee acceptance/resistance, motivation, selection, training and the work environment, and was completed in 1981.[42]

Another arena for exploring these concerns relative to the ICAM programme was under the aegis of its advisory body, the Committee on Computer-Aided Manufacturing, based at the National Research Council in Washington. The Committee was comprised principally of industry and university experts from a range of disciplines and, in its final year (it existed between 1977 and 1981), a representative from organised labour. Part of the Committee's mandate was the identification and study of critical issues affecting the introduction of advanced factory automation.[43] It identified a number of organisational and labour-related issues, and was in the process of commissioning independent case studies when its budget was not renewed by the Air Force. Further exploration of these issues on a systematic basis by the Air Force and its aerospace subcontractors was abandoned.

The last example involves a project jointly funded by the National Science Foundation (NSF) and Westinghouse in connection with the development of an automated assembly system from 1972 to 1980. The technology, known as APAS (automated programmable assembly system), is designed for the complete assembly of small electric motors through the use of robots. A committee to define worker-related research, consisting of university engineers and social scientists, Westinghouse staff and a representative from the IUE, was formed in 1980 by NSF and Westinghouse. They developed a research agenda for investigating a broad range of human issues during and after the initial testing of the APAS system in Westinghouse's Research and Development Center. This was an opportunity to influence the initial design of this technology and its eventual implementation at a designated plant. The top four (of 11) research areas defined by the committee to accompany the first APAS tests were, in order of importance: task allocation; manning variations; input variables; and worker participation. A more comprehensive agenda was defined for research beyond the APAS laboratory tests, of which the top three items were: effects on individuals and groups; manpower requirements; and, tied in third place, coping with change and human factors.[44] Although both NSF and Westinghouse acknowledged the importance of this kind of research, no studies or further review activities have been undertaken nor are planned.

The above examples suggest that, even in instances where some recognition is given to the importance of individual and organisational issues in the development of new technology, it appears axiomatic that they remain secondary and have negligible impact.

At the level of public policy, there is a conspicuous absence of formal debate on the relationship between evolving forms of factory automation and their QWL implications. Similarly, issues such as the impact of the microprocessor revolution on employment and possibilities for retraining the blue- and white-collar workforce have not evoked widespread and focused public interest. No sustained concern appears evident in the foreseeable future, and the prospect of forming anything resembling a national manpower policy is more remote at the present time than in the seventies.

1.8 Automation in manufacturing: small batch and transfer line technology

For the purposes of this chapter, it is convenient to divide manufacturing into three broad categories: continuous process; high volume or mass production of discrete parts; and batch production of discrete parts in low volume. This section will concentrate on the latter two types as a prelude to specific case studies in those areas. Continuous process manufacturing will be taken up later in connection with case material representing that mode (section 1.9).

In instances where standardised, discrete parts or products are manufactured in high volume (customarily runs of 500-20,000) for a single production order, mass production techniques are economical. High volume over an extended time period justifies the use of integrated, relatively inflexible forms of mechanisation such as the automated transfer line. Product substitution or design changes are highly expensive and require lengthy time periods for configuration.

In contrast, batch production is required when several discrete products are manufactured at low levels of volume and standardisation. Approximately 75 per cent of all metal-working parts are produced in this manner. Diversity in products and manufacturing sequences requires flexibility, and batch production plants typically require a wide variety of machines, with higher unit costs as a consequence. Much of the extra expense is accounted for by time-consuming procedures of parts routing, setting up, loading and unloading machines, and manual control. Authoritative estimates suggest that machine tools in batch processing are actually cutting metal about 15 per cent of the time.[45]

Present-day CAM has its central origin in the development of NC machine tools in the late fifties. This was achieved at the Massachusetts Institute of Technology under contract to the United States Air Force.[46] Essentially, NC machines are controlled by coded instructions. An electronic control unit, originally (and still most commonly) with paper tape as the input medium, directs the machine's motions and operations. This system replaces certain functions of the skilled craftsman/machinist, most notably the determination of machine feeds and speeds, and relating operating conditions, as well as the control of machine motion.

Although tape is the most common input device used, later software developments enabled numerical data to be fed into NC machines via an accompanying mini- or micro-computer which directly controls a variety of machines (direct numerical control - DNC), or distributed numerical control, wherein a complete program is communicated from a central computer to an NC system. The advent of microprocessors has permitted NC systems to operate a single machine, as well as to be linked to larger computers in a hierarchical system.

Early NC machines controlled the operation of only a single type of cutter, such as a drill. As they evolved, these machines were equipped with automatic tool changers which contained a variety of cutting tools, thereby increasing the versatility of a single machine. These are called "machining centres". Systems are currently in use which integrate multiple machining centres and automatically control the movement of workpieces between NC centres through the use of computer-controlled material handling and loading systems. Thus, a single part can go through a sequence of cutting operations between machines completely automatically, under the overall control of a single human operator. Unlike mass production transfer lines, these "flexible manufacture systems" (FMS) randomly handle different types of parts and select and load them on to the machines most suited to perform the required operations.[47] Systems are also now in use which, under computer control ("computer-managed parts systems") automatically select workpieces from storage and move them to machining centres; transfer semi-finished parts to intermediate storage and return for final processing; and move finished workpieces to storage pending shipping and distribution.

These types of integrated systems (collectively labelled "computer integrated manufacturing" (CIM))[48] are intended to yield sufficient flexibility and precision to enhance a range of economic outcomes, including faster response capability to changing market demands (less costly retooling, set-up and down-time for customised changes), higher machine utilisation, lower work-in-process inventories, and more efficient conservation of materials and energy. A corresponding expected benefit of CIM, advanced by advocates, are its capabilities to reduce exposure to physical injury, noise and contaminated environments. It has also been one of the expressed objectives of the United States Defense Department, in stimulating the use of NC and its integrated variants, to reduce operator skill requirements to ensure that in the event of a military mobilisation, training times could be minimised. With NC as a base, the ingredients of CIM, coupled with CAD ("electronic drawing boards" linked to the machine tool) form the building blocks of what has come to be described as the unmanned or "automatic factory".

Users of NC remain a rather exclusive group. In 1978, 2 per cent of all machine tools (50,000 out of 3.3 million) in the United States were numerically controlled.[49] A 1976 survey noted that the high perceived acquisition and operating costs of NC machines would prevent wider diffusion in the ensuing decade.[50] As for FMS, it is estimated that approximately 40 are currently operating in the United States.[51]

Contrary to the aspirations of equipment suppliers and their customers, where CAM and some semblance of CIM actually exist, they are typically accompanied by a familiar web of labour-intensive activities at shop-floor level. The hoped-for wholesale transfer of planning and control away from the shop floor has remained an elusive management ambition. In this connection, one analyst of the history of NC has trenchantly observed -

> Perhaps the single most important, and difficult, task is to try to disentangle dreams from realities, a hoped-for future from an actual present. The two realms are probably nowhere more confused than in the work of technologists. Thus, criticism of existing, or past, realities are typically countered with allusions to a less problematic future; the present is always the "debugging phase", the transition, at the beginning of the "learning curve" - merely a prelude to the future ... There is no better reason to believe the engineering and trade journals today, much less the self-serving forecasts of manufacturing engineers ... judging from past experience, there is little reason simply to assume that the new experimental or demonstration systems will actually function on the shop floor as intended, much less perform economically. The author has visited four plants in the United States with FMS systems and found their economic justification suspect, their down-time excessive, and their reliability heavily dependent upon a highly skilled force of computer operators, system attendants and maintenance men; there was also little sign of further development ... DNC is simply another name for the automatic factory, the supreme fantasy of the industrial technocrats, now heralded by self-serving computer jocks, supported by beleaguered corporate managers (whose far-sightedness is more rhetorical than real) and, as usual, funded by the military (in this case, the Air Force ICAM programme).[52]

This view is partially supported by other empirical research on the application of computer-integrated technology in several companies. Using a socio-technical perspective, the authors conclude -

> We believe that there is growing imbalance in sophistication between the technical and social aspects of production systems ... New manufacturing technology is being designed with little regard for the skills, attitudes, and systems and procedures necessary to support it ... An overwhelming majority of companies do not have infrastructure that can properly evaluate whether to purchase a CIM, or that can control a CIM's operations once it is implemented ... It is likely that the problems we have identified will grow as automation becomes more comprehensive. As machinery becomes more complex, it also becomes less reliable. Hence, the ideal of the automated factory may be illusory. It is often depicted as a smoothly operating system free of human variability. Instead, it may well be a nightmare of uncontrollable problems requiring unanticipated upheavals in the workflow structure. While the number of employees will undoubtedly be reduced, those remaining will enjoy greater power over the production process. It is time to give serious consideration to less complex manufacturing concepts which stress interaction between human and technical components.[53]

On the subject of skill requirements, no uniform pattern has thus far been identified. Some NC systems (both standing alone and integrated) have continued to require skilled operators because of machine and program reliability problems. In addition, trade unions have persistently fought for higher job grade assignments to those working in computer-based technology, sometimes resulting in the payment of bonuses to the most skilled people operating the machines, but who nevertheless remain in a lower classification. In other instances, unskilled and semi-skilled workers operate NC equipment, relying on specialised "set-up" workers to intervene during difficult operational sequences. Separate studies of FMS in two companies revealed that in one, operators felt underpaid in receiving the same rate as conventional skilled machinists. In the other case, management assigned a higher classification to those working on FMS than the highest skilled machinist on traditional equipment. Because FMS positions offered no prospects for future job progression and were filled by younger, unskilled workers, this was viewed by many as a "bribe to act responsibly",[54] a theme which echoes Blauner's characterisation of industrial evolution from the "able worker" to the "reliable employee" in continuous process operations.[55] Typically, skill requirements in CAM systems are heavily influenced by management assumptions and operating philosophies. In a comparative study of skill requirements in NC and conventional machine shops, it was found that -

> Some companies regard NC machines as highly specialised, to be treated in ways related to their particular characteristics, while others regard NC as normal machine tools and subject them to the same work programming, planning and loading systems as conventional machines. Policy decisions about the number of hours per day the machines are to be worked, supervisory training, number of men per machine, whether to use setter-operators or operators, and so on, depend upon the depth of understanding by managers of the machine's fundamental characteristics, and how managers decide to use NC to deal with their own products and circumstances. All these decisions affect the job skills demanded of operators.[56]

The case studies which follow illustrate different types of accommodation to computer-aided small batch and transfer line technology. They represent variations on the central idea that the design of technical systems is, of necessity, governed by social choices which either restrict or nourish opportunities for improving QWL on the shop floor.

1.8.1 Example: numerical control

In mid-1968, plant managers at a major manufacturer of aerospace products were exhibiting great alarm over low levels of efficiency in the production of aircraft engine parts. More specifically, a crippling bottle-neck existed in a machine shop where NC lathes were in use. Tension and ill-feeling between workers and supervisors had been chronic in this unit during preceding years, and were reflected in grievance figures - the highest in the plant. Other symptoms of disorder included high employee turnover, abnormally low production speeds, unreliable quality, high levels of scrap and reworking, and frequent materials review inquiries. The section was especially strategic to the plant's highly integrated production system. If engine parts were not produced at sufficient rate or if quality slipped, schedules were delayed, other workers were idle and contracts were placed in jeopardy.

Resentment among workers was still nurtured from past struggles with management to upgrade the payment classification for the NC lathes. Strikes over this issue had been very damaging to the company, which remained steadfast in its position that hourly wage rates were commensurate with the skills required. The company took the position that it was simply providing button-pushing jobs. In the words of one company spokesman: "Any monkey off the street can be trained to operate the computerised lathes".[57]

However, in reality, button pushing was not enough. Because of the extremely close tolerances involved, the engine parts produced had to be absolutely flawless. In some cases, the tapes were not pre-programmed correctly, resulting in scrapped parts and breakdown of the machines. Frequently, the tools would undercut as a result of wear which was not compensated for in the tapes. Resets had to be made for second cuttings. Minor disturbances in the complex circuitry of the NC equipment was another regular occurrence. These disturbances were aggravated by the attitudes of the operators -

> Look, I could slow it down, and it would look completely above board. Say the machine stops running. I go to the foreman and say "I pushed the button and nothing happened". He calls in the maintenance man who brushes some lint out of the machine. All that takes time. But if I had wanted to, I could have brushed the lint out myself.[58]

The deteriorating situation resulted in a management initiative to begin a pilot programme of job enrichment. Management's rationale was revealed, at least in part, by the following statement at the programme's inception:

> The principal reason for a good many of our difficulties is that our hourly employees are lacking in motivation. They perceive themselves as being treated as immature, irresponsible, incompetent people who are relegated to a button-pushing status. A detailed analysis of their duties and responsibilities indicates considerable justification for their feelings. Because of the way their jobs have been structured, these men are not challenged: they have no sense of involvement in the total manufacturing scheme and they appear to derive little or no personal satisfaction from their employment ... [59]

The trade union was initially apprehensive but, with management's offer of a 10 per cent pay bonus to participants in the project, it accepted the proposal and played an active role in overseeing the programme from its inception in October 1968.

The programme was launched in a section of the plant where seven NC lathes were clustered. Recruitment to the project was carried out on a voluntary basis, subject to union seniority rules. Each operator was assigned to a single machine. Over the three shifts, this amounted to 21 workers, including a single working leader per shift. The foreman and unit manager were transferred to other units. The pilot unit reported to a single manager responsible for NC equipment, who was assisted by a manager who co-ordinated the special programme, measured its outcomes and provided feedback. In the initial phase, the pilot programme had the following additional features:

- flexible starting times;

- elimination of time-clock punching for informal lunch breaks; and

- extra responsibilities for operators, including preventive care of equipment, minor machine adjustments and repairs, and debugging of new tapes, tools and fixtures, as well as trouble-shooting existing ones.

The company sent a number of participants to programming school in the hope that they would be able to devise their own tapes. This proved to be too ambitious and was scrapped, although training was given in the areas of record keeping and related management paperwork. The company's wish that the operators carry out their own repairs was dampened by union job demarcation restrictions. During the first three months of start-up, productivity and machine utilisation dropped to low levels and accompanying attitudes were poor. There was confusion over roles and responsibilities. Absenteeism had not improved.

However, things changed for the better after January 1969. The next three months evidenced significant increases in machine utilisation and group productivity. Attitudes and levels of individual involvement also improved. The group became more cohesive with the development of more stable roles, greater understanding was established with management, and activity within the team became more orderly as efforts were made to harmonise work assignments. Group goals concerning machine utilisation, quality, scrap, reworking and costs also began to develop. In general, the pilot group had improved its internal communications and level of mutual trust. It also received stronger support from outside departments on such concerns as incoming quality, voucher control, machine maintenance and planning.

During the next four months, as activities became more routine and doubts faded about the continuity of the programme, general operator attitudes appeared less clear, with corresponding declines in machine utilisation and productivity. Reasons contributing to this were issues within the group about its informal leadership and its methods of disciplining members, diminished management attention to the project, the appointment of a new programme manager (there were eight during the programme) and holidays. The unit's performance was also affected by a strike of maintenance workers in the plant.

From September 1968 to the beginning of a 100-day industry-wide strike in November, performance and attitudes in the pilot programme deteriorated rapidly. The strike marked the end of the first phase of the experiment. Until this point it had yielded some encouraging but unsettled results. Machine utilisation and productivity had increased only slightly, if at all. Scrap and reworking had decreased significantly, as had the frequency of materials review disputes. Lateness and absenteeism had also fallen markedly. The overall effect on unit cost remained more or less unchanged.

With the return to work in early 1970, the pilot programme resumed with a different orientation. Whereas efforts were made in the first phase to enrich individual jobs within the context of a relaxed supervisory climate, the second phase focused on the collective group task. Rather than pursue the improvement of jobs in isolation, attention was devoted to making the group a viable, self-maintaining unit, capable of managing its boundary relative to other groups and hierarchical levels. A division of labour in the group was established whereby workers assumed responsibility for various co-ordination and support functions. Thus, group members conducted liaison duties with production control, planning, quality control and materials managers, as well as with the maintenance engineers and payroll officers. They also performed "housekeeping" and safety functions, training for new operators and record keeping.

Production control activities included scheduling jobs within the unit by the workers themselves, who also determined work assignments within and between shifts. When a bottle-neck was encountered at the main tool crib, the workers set up their own and stocked it to suit their requirements. When breakdowns or equipment difficulties were experienced, maintenance staff were usually required. However, operators were often essential to a specific diagnosis, as in the case of a program fault, because maintenance staff were spread over a ten-mile area and infrequently encountered the same machine. Some operators even became proficient at making their own programs.

Greater interest was also kindled in tool fixturing and design. One instance of this was when operators developed methods of cutting two dimensions simultaneously, thereby saving time and set-up costs. In general, despite limitations in technical knowledge, workers in the pilot unit became highly motivated to meet the challenges raised by multiple role requirements within the group. An important feature of the pilot group was that individuals were not coerced into accepting responsibilities they did not want, aside from basic time-keeping and house-tending duties. Peer discipline was applied to individuals who violated basic group norms.

By all reported accounts, the second phase of the pilot programme was a success. By September 1972, the project had expanded to include 63 operators (it was to reach its peak at 72). That month a national newspaper featured a story describing the experiment. The company did not disclose details of productivity or cost savings to the newspaper, though they did acknowledge that scrap and reworking had declined, accompanied by gains in product quality.

During this period, the union had assumed almost complete responsibility for monitoring performance and sustaining the programme, following local management's expression of concern about the implications of the project and their wish to discontinue it. This was prompted at least in part by management's fear that the union would attempt to extend the pilot programme and the accompanying 10 per cent pay bonus to other areas of the plant. This unsettled state of affairs persisted until the union actually expressed its wish to negotiate on extending the pilot programme, which also coincided with corporate-level management's apprehensions about its implications for other plants. Corporate management's concern also included the following:

- it had not been adequately informed by local management regarding the project's expansion;
- the 10 per cent bonus scheme was disturbing the corporate-wide pay classification system;
- the notion of enriched jobs raised the issue of whether or not they could be fairly paid relative to other company jobs;
- more fluid work group roles might not conform to National Labor Relations Board definitions of exempted jobs carrying management responsibilities;
- they did not approve of flexible starting times and unclocked lunch breaks; and
- longer-term reliability of the pilot project as a means for securing better productivity had not been demonstrated.

At this point company management, having already lost control over the programme's direction at local level, unilaterally terminated it. The union took no official position on the project's termination and, by early 1973, foremen had been reinstated in the pilot area. The company had to continue payment of the 10 per cent bonus until it was negotiated out of the contract three years later.

1.8.2 Example: flexible manufacturing system

The next study was undertaken in the tractor-producing division of a large, diversified manufacturing firm. The plant is trade unionised, employs 1,900 people, and in the seventies introduced an FMS. During a five-month period in 1980, a study was made involving direct observation, intensive interviews and survey questionnaires from a sample of foremen, machine repair workers, operators, loaders and tool setters.[60]

The FMS consists essentially of a network of computer-controlled transfer lines which transport individual raw castings in wheeled pallets, and present them to work stations in the requisite order for processing. The workpieces are loaded on to the metal-cutting machines by shuttle carriages, whereupon a second computer initiates the direct NC transmission which causes the machine to process the part. After the processing sequence is completed, the pallet is extracted from the machine and routed to its next work station or returned to a loading station for removal of the finished part. This procedure is undertaken simultaneously for several dozen parts of various "families" that may be in process at one time, and routed in random order among the work stations. The system consists of ten machines and three loading stations spread over approximately 100,000 square feet of floor area and joined by 12 tow chains which propel each four-wheel cart.

System utilisation rates (hours actually worked relative to theoretical capacity) had been projected by the equipment suppliers to be in the order of 80-90 per cent, but actual rates ranged between 50 and 60 per cent. One source of underutilisation stemmed from the system's complexity and interdependence. Because all elements were indispensable, pressures to minimise down-time during the system's frequent failures prompted hastily made "jury-rigged" repairs. As a consequence, each machine developed its own idosyncrasies which eventually required further rectification by the operators or maintenance workers. Workers also directly influenced system utilisation by controlling the pace at which they loaded and unloaded casting pallets. Finally, worker intervention was essential for monitoring the machines for blunt or broken tools, responding to program stops and trouble-shooting minor faults.

Analysis of the survey data provides insight into one of the underlying sources of sub-optimisation in the advanced FMS. Of the 20 employees working in the FMS, 18 participated in the study. The sample included responses from two foremen, two machine repair workers, six operators, six loaders and two tool setters. The questionnaire included items concerning the motivating potential of jobs and personal motivation of workers, as well as questions on job satisfaction and work stress. Results from the motivation questions were compared to data from a normative sample of 16 machine trade employees, while results concerning satisfaction and stress were compared to those of a national sample of over 1,500 adult employees.

In general, the sample perceived work on the FMS as having low motivational attraction. For the combined sample of 18, the composite score on motivational potential was below that of the normative sample. This was the case for four of the five job classifications in the sample, especially the loaders and operators. With respect to particular elements of motivation, all five classifications rated opportunities to exercise discretion (autonomy) below the normative sample's score. Respondents in four of the five grades scored task identity (the degree to which the job requires completion of an identifiable piece of work) below that of the normative sample. In contrast, perceived needs for personal growth and development were higher for the combined sample, and for four of its five job classifications, relative to the normative sample's score.

Job satisfaction was measured in terms of relations with co-workers, job challenge, resource adequacy, job comfort, promotion opportunities and financial rewards. The combined sample of FMS workers ranked below the national sample on every aspect with the exception of financial rewards, where there was parity. Of the five classifications, tool setters, operators and loaders were the least satisfied; foremen and repair workers felt reasonably satisfied.

Respondents from the FMS indicated that the work was stressful. Measures of stress were based on five factors: the extent to which -

- conflict existed over how the job should be done;
- employees had skills they were not using;
- workers did not know what was expected of them;
- resources to perform jobs were inadequate; and
- employees felt uncertain about their job security.

The combined sample scored below the national sample on all factors. With the exception of repair workers, every job classification had scores below the national sample on at least four of the five factors.

Finally, the absence of meaningful pay incentives, along with inequities in perceived earnings relative to workers in the same pay grade operating conventional equipment, were cited as further sources of dissatisfaction by a majority of the sample.

The above and other data from the open-ended questions indicated that the basis was lacking for generating worker commitment to the effective utilisation of the FMS. A familiar irony was manifest here: while the FMS was explicitly designed to remove the human element as much as possible from the manufacturing process, the precise opposite (human intervention) was required to bind the technical system together. By superimposing on to the FMS a traditional structure of specialised job classifications, an essential source of flexibility appears to have been lost. This example typified the prevailing tendency to concentrate on maximising the technical properties of the work system at the expense of equally important social and organisational factors. The study concludes -

Despite growing statements that we are entering the era of the automatic factory, available evidence indicates that such facilities are a long way from being a practical reality. The human element is still very much a factor to be considered, and appears to be making a significant difference in the utilisation of the few advanced systems which are not on line. The major effect humans have on system performance appears to stem from the manner in which they respond to the system's requirements for raw materials, maintenance and the clearing of faults brought about by operating anomalies. Until equipment can be brought on line which is totally automatic and completely reliable at acceptable costs, which can procure its own raw materials and dispose of its finished products, the behaviour of human beings will continue to be important.[61]

1.8.3 Example: automated transfer line

In 1974-75, a comparative study was undertaken of work attitudes in automated and non-automated settings in the motor vehicle industry.[62] Interviews were conducted with 200 employees from the highly automated engine block transfer lines and conventional machining areas of three factories (belonging to two different companies). The study included all major types of work performed in both types of settings - operators, stock handlers, inspectors, "set-up" workers, maintenance workers and supervisors; all shifts were sampled. A central objective of the research was to determine whether automated work significantly increased job satisfaction and self-respect.

The study's findings indicated that the attitudes and behaviour of workers under automated and manual conditions were quite similar. The same was true for maintenance workers and supervisors. The only exception was the "set-up" position in the automated engine transfer lines, which required installing and changing tools, and monitoring machine performance. Workers in this category reported higher degrees of satisfaction than their immediate workmates and manual counterparts in non-automated settings.

Similarly, attitudes towards automation did not differ between the two principal groups. Supervisors and maintenance personnel registered the strongest support (96 per cent) in contrast to the rest, who were less enthusiastic, though positive (70 per cent). Only 17 per cent of the total sample indicated that they would actively or passively disapprove of the introduction of "work-saving new machines and equipment".

Of all the variables measured, perceived job security was the most closely linked to support for automation: 80 per cent of its strong supporters were confident that, if displaced by new machinery, the company would provide them with "acceptable work". Another 17 per cent thought that the company might possibly do so. In addition, the findings suggested that if workers felt that management was concerned about their welfare, communicated openly with them, and ran the operations efficiently, they would be more favourably disposed towards technological change. The most important aspect of "organisational climate" was open, two-way communication between management and workers. Of those workers who felt adequately informed in advance of technological changes, 51 per cent endorsed automation. Alternatively, of those who felt that they had not been adequately informed, only 31 per cent were in agreement.

The findings from this study supported those of others which noted strong links between job interest, challenge, autonomy and career possibilities with job satisfaction. However, the research team was quite pessimistic about existing opportunities for automated processes to fulfil these demands -

> More automation may add a few more set-up positions and pose new and interesting problems for the already highly challenged maintenance personnel. But routine remains routine for many of the tasks that still require the presence of a human, even if technology has succeeded in substituting electronic control panels for manual manipulation.[63]

The research team had also hypothesised a strong relationship between opportunities for participation in decision-making, along with its practice, and job satisfaction, productive behaviour (attendance, punctuality and intent to remain in the job) and acceptance of new technology. No such relationships emerged from their data. The following interpretation was offered:

> Why was this so? One possibility is that the actual participation was not large or important enough - or apparently not connected closely enough to [workers'] immediate jobs and situations. Those few workers who had the most significant participation - whether working with the company's policies or against them - turned out to be the most concerned, and generally the most satisfied. Another possibility is that, because of the whole collective bargaining process of working through the union, the workers don't see a close connection between individual efforts and the results.[64]

These findings on worker participation brought a critical reaction from a senior UAW official, who observed that it is difficult for workers to evaluate the meaning and significance of worker participation programmes unless they actively experience them, and that workers in the study had not had the opportunity to be involved in a carefully designed programme. He further noted that -

> In this regard, it is important to distinguish between programmes of so-called job enlargement or job enrichment (which normally relate to redesigning of individual jobs) and programmes of improved QWL (which embrace all aspects of the relationship of the worker to his job and to the management). Involvement of the worker in the decision-making process is not simply a matter of enlarging the number of operations he performs or establishing work groups, etc. It is not a matter of imposing new engineering techniques upon the worker. Rather, it means finding better ways to afford the worker the broadest opportunities for input and innovation as to job design, methods and processes of operation, quality of product, plant layout and a host of other decisions regarding the worker's relationship to the total workplace.[65]

1.8.4 Example: automated transfer line and transitional tasks

The case of Factory 81 illustrates an attempt to establish the type of comprehensive approach alluded to by the UAW official quoted above. This has been under way in one of the car divisions of General Motors (GM) since 1975, and is currently concentrated in the early operational phases of a new facility.[66] The name "Factory 81" is derived from the fact that since 1972, GM has initiated operations in 81 new plants - either at "green field" sites or through total reconstruction of existing facilities. Many of these undertakings involved deliberate attempts to experiment with socially innovative management techniques under the aegis of the corporation's "quality of work life" programme. Several of these initiatives, as well as a large number within existing factories of the corporation and its subsidiaries, were carried out in collaboration with the UAW. They originated from contract language in the 1973 GM-UAW agreement which enabled labour and management jointly to pursue programmes at the local level specifically aimed at improving QWL. This was the first contract language of its kind in the United States.

The plant under consideration is part of a complex employing 22,000 (16,000 hourly workers). In 1975, with the company at the nadir of a prolonged drop in sales and market share, which included a gradual 30 per cent reduction in labour force size over a 15-year period and a conflict-ridden history with the trade union, a truce was established between labour and management. The company began to share planning and cost data on the next generations of small cars with the local union, and a company-wide labour-management committee was formed to improve two-way communication at all levels.

The next two years witnessed the formation of "quality circles" and joint problem-solving teams in departments throughout the division, and a dramatic reduction in wild-cat strikes and union-management conflict. Since 1977, the division has successfully regained corporate confidence through aggressive marketing and product development efforts, in parallel with its QWL activities. Sales increased by 77 per cent over the 1975 low point, and the division successfully competed with other groups in GM for exclusive manufacture of rear axles, engine cradles, torque-converters and the V-6 engine. This generated 5,500 additional jobs, some at the expense of workers elsewhere in GM. Labour-management committees currently exist in each of the division's adjacent plants, along with over 200 quality circles. A heavy investment was made in training (2,500 workers have received QWL-related training), with the division taking a longer-term view of its payoff in terms of workforce stability, quality, commitment and performance.

The genesis of Factory 81 occurred in 1979, when the decision was made to close the division's foundry, long responsible for producing a variety of heavy metal components, including engine blocks. This arose from shifts towards lighter fabrication materials for all future cars. During the next year and a half, the foundry building was gutted and $200 million in new equipment began to be installed for the production of torque-converters (with a planned maximum output of 18,000 per day by late 1982).

By mid-1981, the new facility employed 550 hourly workers, with eventual workforce size targeted for 1,200.

With respect to overall design, division management was encouraged to conceive of Factory 81 as a "green field" site. Management and the local unions have collaborated extensively on layout and work structure decisions. Actual equipment design has remained the preserve of engineering and development staff, in association with outside vendors. However, a division policy requires that final delivery of new equipment cannot occur until formal approval has been given by the workforce. This has frequently involved visits by workers to equipment suppliers to test and recommend modifications.

It was not possible to determine the substantive impact of these changes on the alteration of new technology and on worker morale. In general, it is apparent that labour and management of Factory 81 have collaborated significantly more than before on the assimilation of new technology. Work groups range in size from 8 to 14, and all have been trained in problem-solving, communication and conflict resolution. Workers do not punch time clocks; they rotate jobs within groups and operate under less supervisory control. The underlying rationale of Factory 81 is to develop and encourage the capacities of groups to undertake work-related decisions at the lowest possible levels.

In mid-1981, absenteeism in the plant was 2 per cent, compared with the GM average of 5 per cent. Since start-up in 1980, there have been three union grievances in Factory 81, compared with a total of 3,600 for the entire division. One sign of union membership approval of these developments was the recent re-election of the local president to a fourth term.[67] Summing up, one senior spokesman from GM observed the following connection with efforts like Factory 81:

> Admittedly, the nature of the technology does indeed limit the extent to which jobs can be redesigned or enriched. However, it does not limit the degree to which the roles of workers can be enriched. Despite the technology, means are available by which people can participate more actively and creatively in problem-solving, and in other activities, wherein they can make more significant "contributions" and thus derive greater satisfaction from their "work".[68]

1.8.5 Example: islands of partial automation and assembly

Packard Electric Division of GM manufactures a variety of motor vehicle systems and employs 15,000 people in the United States and Mexico.[69]

In 1972, the company was among the first in the corporation to build a new facility in a rural area of the Southern States. The new plant employed approximately 500 people and provided an opportunity for experimentation with more participative management styles in a non-union setting. In retrospect, company managers candidly refer to this early effort as having been a "benevolent dictatorship". A second plant, built a few miles away, went further than its predecessor in developing an organisational philosophy which supported the creation of semi-autonomous work groups, a payment system based on demonstrated skills and knowledge, and related social mechanisms aimed at increasing worker job satisfaction, involvement and commitment to production objectives.

In 1977, a confluence of circumstances at the company's main complex in the North, employing 8,000 people in three adjacent locations, led to the following developments. First, sustained labour-management conflict in the presence of a declining workforce, no new capital investment and the conspicuous emergence of the corporation's "Southern strategy"[70] led to substantial pressure from local members of the IUE to seek alternatives to a protracted union holding action. This found expression in the election of a new team of local officers who advanced a platform of exploring possible areas for increased collaboration with management. Consequently, both company and union initiated a number of co-operative ventures during the following year, culminating in the formation of a top-level "Jobs Committee" to search for more productive ways to operate and generate new jobs.

The second coinciding element was the need, resulting from developments in product and process technology, for the creation of new manufacturing facilities. Based on a local decision to scale down the size of new plants, the company was faced with the requirement of spreading operations among four new facilities. Confronted with the choice of starting up four new branch plants near the existing complex, using employees from the same labour force, or starting afresh in a Southern State or Mexico, the company chose to remain in the area, given the labour-management co-operation that had been attained. This was a significant achievement for the union and its members; 2,400 jobs had been retained.

The stage was thus set in 1978 for a co-operative planning process between local company engineers, line and personnel managers, union officials and shop-floor workers. It consisted of a labour-management committee which guided the work of a "design team" that applied socio-technical concepts to planning the new work systems. The principal design objectives were to reduce job fragmentation and its attendant boredom, improve communication among workers and problem-solving capabilities between work groups, develop greater workforce flexibility, and control key production variances as close to their sources as possible in order to meet strict standards of volume, quality and cost.

The four branch plants commenced operations during 1980, each employing between 600 and 800 workers. Three plants manufacture wire harnesses for engine emission control systems. Each plant is a self-contained, totally integrated operation with a one-day inventory system. The fourth facility is also totally integrated and produces other wire harnesses. The latter had actually begun operations in 1976, but its work structures were redesigned to be consonant with the "branch concept", as it was called, at the other three plants.

Manufacturing in each plant is divided between two basic operations: the preparation of wire circuit leads ("lead prep") and final assembly, with a buffer area between the two. Each finished wire harness may contain between 60-100 wires, depending upon the unit. Before lead prep operations were significantly automated, wire circuits were prepared by six people in a series of short-cycle operations on different machines. A few batch processing operations of this type remain, but the predominant mode is for a team of two operators per machine to load raw stock and take off prepared circuit wires. There are 113 of the lead prep, wire-indexing machines spread over the four plants. The lead prep operators rotate at will, conduct their own inspections, and are responsible for trouble-shooting their equipment and performing minor maintenance tasks. In contrast to the previous division of labour, all lead prep operators are in the same wage classification. Although these "islands" of automated activity in lead prep represent the state of the art in manufacturing technology for this product, the overall technical system remains low on the automation scale. Labour accounts for 50 per cent of total costs.

Final assembly operations are performed along 18-20 stations, each with an average job cycle of 60 seconds. In some areas, stationary fixtures are available off-line to permit the assembly of whole units. All workers in final assembly are multi-skilled, and rotate on a regular basis. They are paid the same hourly rate (there were formerly five job classifications and rates). Final assemblers are also responsible for inspection, repair work and servicing. Each assembly group is accountable as a whole for their unit's finished output.

In each plant, the production layout was designed to place workers in lead prep and final assembly in close proximity (12-15 feet) in order to expedite inter-group communication and error rectification. The possibility remains open in the future for multi-skilling to occur between the two production areas; its desirability and feasibility are presently unclear. Supervision has been retained in all areas of the plants, with control spans of 1:27.

From the beginning of this enterprise, the company invested heavily in training. All 2,700 workers (including foremen) received eight hours of orientation towards the new approach being pursued by labour and management, plus 40 hours of technical and social skills training.

It was also emphasised that each plant would be permitted to evolve its own form of accommodation to the general "branch plant" concept. Much effort thus far has been devoted to socially integrated work teams and encouraging greater worker participation. Groups have developed different norms regarding job rotation and decision-making, and latitude has been provided for individuals to opt out of group-oriented production methods. More training and support will be devoted to developing greater technical co-ordination within and between work groups. Supervisors are continuing to receive training in participative management and group problem-solving. An inter-plant task force of managers, technical staff, supervisors and union officials has been formed to identify and develop cost reduction projects.

Detailed information on the outcomes in human and financial terms is not available. The company has disclosed, however, that during the first year of operation, performance of the plants was sufficient to pay comfortably for all of the substantial training costs. The four branch plants also lead their company in cost reduction and, for a six-month period in 1980-81, were among the very highest in quality within the entire corporation.

Accounts given by the trade union have been positive in nature. Apart from preserving jobs, the improved working conditions have been acknowledged by the membership, in part, by the re-election of all local officers. Job cycles are still relatively short, but have been ameliorated to a certain extent by better environmental conditions and the participative climate that is being established. Individual differences regarding job rotation and involvement in teams will continue to require thoughtful accommodation. Each plant has thus far evolved unique customs and practices to address this issue, as well as others.

Difficulties have also arisen from some of the groups of skilled trades, which felt that their job territory had been eroded by the operators. However, the elimination of routine minor maintenance duties was conceded by many as an opportunity to engage in more challenging and important work. The job territory issue may remain thorny in the absence of more engaging (and perhaps overtime-consuming) work for the skilled tradesmen. Analyses of data from worker attitude surveys conducted before the branch plants started up in 1980, and again in 1981, have not yet been completed.

1.9 Automation in manufacturing: continuous process

In contrast to the manufacture of discrete parts, where, essentially, the geometries of raw materials are altered for assembly into finished products, process industries manipulate the composition of materials by chemical reaction, blending or purification. Typical process industries include petrochemicals, pulp and paper, and electric power. The recent use of advanced computer control technology in these industries has been stimulated by process and product innovations requiring closer integration between operating units, and needs for environmental protection, health and safety, and conservation of energy and resources. One technologist has summarised current policies on the appropriate role of the operator in these systems as follows:

> There are two extreme approaches: (a) to automate the process as much as possible and use a relatively skilled operator who could do little more than shut the plant down in case of an emergency; and (b) to use a better trained, perhaps college-educated operator who would also serve as manager. The ultimate direction industry will take is not clear at this time. On balance, the social and political forces for environmental protection, energy conservation, safety and health and better working conditions favour greater automation.[71]

The case examples which follow (sections 1.9.1 and 1.12.1-3) discuss traditional and alternative forms of organisational design and operation in continuous process industries. The conventional paradigm is epitomised by the design and functioning of the now-infamous nuclear power plant at Three Mile Island near Harrisburg, Pennsylvania. Apart from its intrinsic interest and relevance, it has been selected for discussion here because of the extensive documentation and research made available to the public after the accident. The remaining case material presents selected examples and composite assessments of different types of automated work systems based on socio-technical concepts. As in the previous manufacturing cases related to small batch and transfer line technology, they illustrate the importance of social choice in shaping the technical possibilities which affect the quality of human experience in the workplace, and economic performance.

1.9.1 Example: nuclear power plant

Early in the morning of 28 March 1979, several water pumps shut off in the Number Two Unit of the Three Mile Island (TMI) nuclear power plant. This prompted a series of equipment failures and human errors during the next few hours which resulting in the reactor core becoming exposed from lack of cooling water. Severe overheating and consequent damage to the reactor core led to the escape of radioactive gases and liquids into the surrounding containment building. Because coolant water also circulated to an auxiliary building, most of the subsequent radioactivity released into the air over the next few days was later attributed by experts to have occurred through that channel. Hydrogen gas was produced in the core after it became partly uncovered and, three hours into the accident, radiation levels had increased throughout the plant. With the exception of the outlet described above, the containment building was automatically isolated by the fourth hour.

Nearly ten hours into the accident, there was a hydrogen gas explosion in the containment building which was not detected until late the next day. Its realisation, coupled with the discovery of another hydrogen bubble in the reactor system, led to the profound fears that another and perhaps more serious explosion was imminent. However, these fears proved to be unfounded after the Nuclear Regulatory Commission (the federal agency which oversees the industry - NRC) announced on 2 April that it had miscalculated the extent of the problem. At this time, the major emergency was declared over.

From the beginning of the accident, the plant operators did not understand what was happening in the Number Two Unit and consequently were unable to control its recovery. Delays occurred in informing state and local authorities about the situation, and the public received vague and conflicting reports about the hydrogen bubble. The minor malfunction of a set of water pumps had precipitated a chain of events which had nearly ended in catastrophe.[72]

A number of major inquiries were initiated in the aftermatch of TMI with varying terms of reference. Of relevance here are the findings which shed light on the relationship between the design and operation of TMI-2's technical and human systems.

A distinguished commission appointed by the President of the United States determined that "the accident at Three Mile Island occurred as a result of a series of human, institutional and mechanical failures".[73]

In its first investigation of the accident, the NRC cited operator error as the principal cause of the accident. The term "mindset" was used frequently by the NRC in this report and in its testimony to the President's Commission. The NRC suggested that operator "mindset" in connection with a critical diagnostic procedure shortly after the accident began prevented them from responding correctly. Another "mindset" related to values, usually implied but sometimes expressed, which governed the actions of equipment suppliers, Metropolitan Edison (the public utility which operated TMI) and the NRC. This was articulated by an NRC member to the President's Commission as follows: "I think [the] mindset [was] that the operator was a force for good, that if you discounted him, it was a measure of conservatism."[74]

In other words, reliability and safety could only be achieved through concentration on equipment. The presence of human operators would not be problematic and, indeed, would be a bonus. In response, the President's Commission noted that -

> The most serious mindset is the preoccupation of everyone with the safety equipment, resulting in the down-playing of the importance of the human element in nuclear power generation. We are tempted to say that while enormous effort was expended to assure that safety-related equipment functioned as well as possible, and that there was back-up equipment in depth, what the NRC and the industry have failed to recognise sufficiently is that the human beings who manage and operate the plants constitute an important safety system ... In conclusion, while the major factor that turned this incident into a serious accident was inappropriate operator action, many factors contributed to the action of operators, such as deficiencies in their training, lack of clarity in their operating procedures, failure of organisations to learn the proper lessons from previous incidents, and deficiencies in the design of the control room. These shortcomings are attributable to the utility, to suppliers of equipment and to the federal commission that regulates nuclear power. Therefore, whether or not operator error "explains" this particular case, given all the above deficiencies, we are convinced that an accident like Three Mile Island was eventually inevitable.[75]

At the heart of this conclusion was the structural issue of divided responsibility and decison-making which adversely influenced TMI's technical design, operating procedures, control room design and operator training. Here, albeit in a rather extreme form, is revealed the prevailing paradigm within which traditional organisation design occurs. Although the NRC and the General Public Utilities Corporation (GPU - of which Metropolitan Edison was a subsidiary) played an especially prominent regulatory role, it is not atypical of a variety of technical and social performance requirements set for private sector organisations by regulatory bodies (i.e. environmental protection, occupational safety and health, equal employment opportunity).

Similarly, the segmented nature of the planning process undertaken by the public utility and its equipment suppliers and architect-engineers is also characteristic, to varying degrees, of the organisation design process elsewhere. In this instance, the utility relied heavily on the technical expertise of a major equipment supplier and the NRC. In this connection, it is noteworthy that engineers and designers from TMI's major equipment supplier (Babcock and Wilcox - B & W) had no first-hand knowledge and experience of nuclear reactor operations -

> Neither the head of the Engineering Department, a B & W engineer for over 20 years, nor the head of licensing, a B & W engineer for 25 years, had ever observed a B & W-designed nuclear steam supply system in operation at power. Nearly all engineers asked made similar responses. The head of the Design Section stated that he did not know if engineers from design had received training from [B & W] Training Services. The head of the Engineering Department did not know if any engineer had ever observed courses given to the customers. He further estimated that less than 20 per cent of B & W engineers had received simulator training.[76]

As a result, a technical system was purchased which had embedded in it particular assumptions about human performance and organisational requirements. For example, the control room was designed to be operated by a single person under steady state conditions. It was evident that the instrumentation in the room was not well situated for accident conditions, which raises the important point of whether an alternative or complementary design specification should not have been the rapid detection of and recovery from abnormal conditions. The President's Commission noted that "the design of the control room seems to have been a compromise among the utility, its parent company, the architect designer and the nuclear steam system supplier (with very little attention from the NRC)".[77] This issue goes beyond the need for better human factors engineering, although this was lacking in the overall reactor design, control room design and instrumentation.[78] It highlights the absence of a planning process in which the technical, organisational and human requirements are considered simultaneously in an integrated fashion.

A final illustration of how the segregated approach to organisation design and management was typified at TMI is in the area of operator training. Although legal responsibility for training operators and supervisors rested with the utility, it was necessary for Metropolitan Edison to contract out a significant portion of training to B & W. The latter only had responsibility for its components and no responsibility for the quality of the total programme. Co-ordination between the training procedures of the two companies was loose, an example being that B & W instructors were unaware of the precise operating procedures in effect at TMI. B & W's training simulator, a mock control console, differed significantly from the actual console at TMI. The President's Commission observed -

> We found that at both companies, those most knowledgeable about the workings of the nuclear power plant have little communication with those responsible for operator training, and therefore, the content of the instructional programme does not lead to sufficient understanding

> of reactor systems ... Since during the [NRC] licensing process, applicants for licenses concentrate on the consequences of single failures, there is no attempt in the training programme to prepare operators for accidents in which two systems fail independently of each other.[79]

The last point is revealing because it indicates the heavy emphasis given in training to recovery from a particular class of system failure - "large break" accidents - where a key technical element is involved and its potential impact has been previously anticipated and included in training. In the main, such accidents require extremely fast reaction and therefore must be automatically controlled by the equipment. This is in contrast to "small break" accidents which may occur in isolation, simultaneously, or in sequence much more slowly. Their potential consequences have not been analysed (through some form of probablistic risk assessment) and not prepared for through training.

The important point is that TMI's training programmes, as in other conventionally designed automated process plants, reinforced a limited appreciation of system interdependence. This was also compatible with the conventional operator and supervisory role definitions which focused exclusively on individuals as the principal learning unit rather than as part of a group (both work team and total workforce) as a mechanism for developing, sharing and applying practical knowledge on a continuous basis. Training at TMI embodied the inherent characteristics of a fragmented and bureaucratic process of organisation design and subsequent management. Underpinning every stage was, in the words of the President's Commission, "the persistent assumption that plants can be made sufficiently safe to be 'people proof'".[80]

1.10 Alternative approaches to continuous process technology

In contrast to the methods and characteristics of traditional organisation design exemplified by Three Mile Island, a number of new organisations in North America have been created during the past decade which are departures from prevailing conventions.[81] Many have embodied highly automated continuous process technology and are noteworthy because of the explicit consideration given during the design phase of QWL criteria. To a varying extent, these organisations have been devised and managed on the basis of socio-technical concepts. While there is a scarcity of published materials which document the evolutionary stages and performance of these new organisations, available information (published and unpublished) suggests a fair degree of eclecticism in the application of socio-technical ideas. Common to most have been the driving forces for better utilisation of materials, energy and overheads.

Although labour represents a fixed and relatively insignificant proportion of total cost, managements have recognised the economic implications of increased dependence on their workforces in highly automated systems, where high multiples of value added per worker are at stake. A high premium is placed on the capacities of people in these systems to perform reliably under both normal and abnormal operating conditions. Under the latter, the rapid reduction of uncertainty and the capacity for recovery require the adaptive qualities of a knowledgeable and committed workforce. In order to induce such qualities, management in these organisations has attempted to design jobs which offer attractive amounts of variety, challenge, interest, responsibility, social support, opportunities for career progression, security and, in some instances, sharing in the gains of increased productivity beyond normal wage increases. Job design was a product of a deliberate planning process where the technical and social requirements of the enterprise were jointly considered and accommodated to yield the blueprint for an internally consistent work culture. In the case of some new organisations, this has been a recondite process relative to their large corporate bodies, as well as to outside observers. In others, the elements of the design process have been documented and can be outlined here.[82]

1.11 The design process

Once the primary decisions are taken about plant size and location, the question is posed by senior management as to the desirability and efficacy of designing the new organisation along non-traditional lines. Typically, this is stimulated by the promptings of internal management, external consultants, or a combination of the two. A feasibility study is often commissioned, which includes seminars and workshops, discussion with managers in other companies which have attempted workplace innovation, and visits to plants designed on the basis of socio-technical concepts. Participants in this study may include a cross-section of the technical and managerial interests which will be responsible for developing the ultimate plans for the new organisation. If, after a review of the feasibility study, the decision is made to pursue the integration of social and technical aspects of the prospective organisation from the outset, an ad hoc design group is designated, consisting of a steering committee, a project design team and related specialists.

The steering committee is comprised of senior managers who review proposals from the design team, give advice, link the project to policies of the parent organisation, protect the project from hostile internal groups, provide support for implementation, and act as a bridge to the larger organisation for the diffusion of project-related knowledge. Under the steering committee, a design team is responsible for developing and recommending alternative courses of action. This group usually changes in composition over the course of the project to accommodate the specialities needed. At the start, it consists primarily of functional experts and, towards the end, tends to be comprised mainly of managers responsible for the plant's operation.

The design team is a forum for reaching agreement on the social and organisational values to be embodied in the new plant, and is a socialising mechanism for managers and other leadership associated with it. As the group addresses the various phases of the design task, outlined below, it customarily breaks into sub-teams whose members work as experts for a given sub-task and develop recommendations. However, only the design team as a whole has the authority to decide on how these proposals are integrated into an overall scheme, in social, technical and physical terms -

> No engineer, architect, personnel manager or plant manager is permitted to decide alone on the basis of his or her functional expertise. This means that each technical proposal is examined for its direct and unintended effects on the social system, and vice versa, before the team makes a design choice.[83]

This is in marked contrast to traditional design teams, whose scope is subdivided at the outset and whose terms of reference remain exclusively technical throughout the physical completion of the plant.

The design team begins by broadly assessing planning and implementation issues, and develops strategy. This is undertaken in conjunction with an overview of constraints posed by the parent organisation and their potential impact on the internal environment of the new plant. Next, the external environment of the prospective plant is assessed in terms of its geographic location, surrounding community, cultural traditions and local labour market in order to define any unique requirements which the socio-technical design must take into account. This phase is completed by the development of a statement of guiding objectives and principles, sometimes known as an "organisational philosophy", concerning the human, social and economic values which will be manifest in the operation of the new plant. While these statements frequently appear platitudinous and idealistic in tone, they are inevitably the result of protracted and heated argument and negotiation among their framers. Aside from serving the important purpose of clarifying values and reaching normative consensus among the design team and its steering committee, the philosophy statement henceforth serves as the benchmark against which design proposals, implementation plans and suggested performance measures are strictly to be employed.

Next, the design team analyses the technology to be employed in the new plant. Where highly automated continuous process technology is involved, close scrutiny is given to the stability and predictability of the system's inputs, outputs and intermediate transformation processes in order to ascertain its dominant characteristics. This invariably has important implications for the eventual specification of work roles, which will have as a cornerstone an emphasis on human learning. The design team frequently then analyses the technical system in depth in order to identify key potential workflow disturbances which, if unregulated, will critically affect desired outcomes. The analysis focuses on the nature and source of these disturbances (also known as "variances") and their potential means of control. This information provides insight into social system requirements, as well as those of secondary technologies such as instrumentation -

> Take something as simple, yet as complicated as deciding whether to place a meter on a particular machine. To make this decision, the designer must answer questions concerning the information or feedback that the meter would give the machine operator. Does the operator really require the information? How important is ease of physical availability? What about the timeliness of the information? If the information is not to be obtained from the machine directly, then from

> what other source in the social system organisation? If the meter is not available to give the operator direct feedback, what are the consequences for the kinds of decisions the operator can make or is prevented from making? Who else might have to apply them if the operator doesn't? In what ways would the operator then become dependent upon supervisors or others who, because they had the information, could make appropriate decisions at the workplace? What are the social consequences of this dependence? How would these social consequences affect economic consequences for the organisation? As these questions illustrate ... the choice goes deeply into the structure of the organisation and a consideration of the ways which people will be required to work in it, how they will be supervised, and whether, in fact, much supervision will be necessary if they themselves accomplish the required activities.[84]

Following analysis of the technology, the design team begins to generate preliminary alternatives for the design of the technical system. Depending upon the complexity of the conversion process and the variances identified, specialists may augment the team to aid its expert sub-groups in developing alternatives that support the organisational philosophy. These alternatives are reviewed by the overall design team in terms of their implications for social system functioning.

Successive steps involve delineating organisational units (sometimes called "boundary location"). The technical throughput processes are broken into manageable domains where key disturbances can be controlled closest to their points of origin. It is here that possibilities are explored for the creation of autonomous work groups, and the contours of job content, skill levels and career ladders are outlined. Also at this point, co-ordinating mechanisms and support groups are specified, as well as hierarchical levels. The number of supervisors, foremen and superintendents is contingent on the devised structure of the work system at the bottom of the organisation.

After a preliminary overall design for the organisation has been completed, more detailed job and role requirements can be specified, taking into account existing characteristics and conditions of the local labour market. Typically, this process is continuous, involving adjustments to the other elements of the design. Depending upon lead times and the availability of qualified manpower, the design may need to devise a start-up or "transitional" organisation capable of launching the new plant into a stable operating posture. When this step is completed, the design team develops systems of social support for the organisation: schemes involving pay and other financial rewards, job promotion and transfer, discipline, constitutionality and due process, training and recruitment. As before, all proposals are assessed against the organisational philosophy statement and must be compatible with other elements of the adopted socio-technical design.

Trade union involvement in the design of new organisations using socio-technical concepts has thus far been minimal. The only previously published account of union participation will be discussed below in connection with the design of a Canadian chemical plant. Even in this instance, the union was invited to participate at a late stage, although its commitment to the concepts did become firmly established and it played a key role in the plant's subsequent success. In this case, the new plant was built adjacent to an existing refinery with an establishing bargaining unit of the same union (in a densely unionised area). Consequently, there was little question of the new plant also becoming organised. A separate and novel contract was negotiated for the new plant and refinery workers were given the opportunity to transfer.

In the United States, the great majority of new organisations during the past decade were established in geographic areas where the workforces would not qualify as extensions of an established bargaining unit or local union branch. Without this condition, United States labour law prohibits the designation of an appropriate bargaining agent without an official election by the workforce in question. Since recruitment is undertaken at the end of the design process, union involvement is therefore legally preempted. In addition, most companies which have established new plants along socio-technical lines (as well as traditional or any other) have manifestly intended to remain non-union.

1.12 Characteristics of innovative new plants

Practically without exception, jobs in the new organisations have been designed on the basis of semi-autonomous teams responsible for the production of a well-defined service or product. The groups are flexible and highly trained, and undertake such activities as production scheduling and goal setting, task allocation between individuals, quality control, purchasing, record keeping, discipline and training of workers. Depending upon the process technology, single teams of up to 30 people may be responsible for the entire work flow on a shift, from raw materials to final product. In others, a shift may consist of 2-5 semi-autonomous teams responsible for discrete outputs along the transformation process. Similarly, maintenance functions and operations are sometimes combined, with each team capable of performing both, while in others a separate maintenance unit exists on a plant or area basis, with operators responsible for troubleshooting and minor maintenance tasks.

Frequently, the foreman's role has been eliminated, with several work groups reporting to a shift supervisor. This is often accompanied by the work teams designating a group leader or co-ordinator, whose role is to communicate with the rest of the organisation. In some cases, the coordinator performs supervisory functions, but remains a team member.

Pay in the new plants is linked to demonstrated skills and knowledge connected with the production process. Specific job classifications usually do not exist. In their place is a voluntary progression system tied to incremental clusters of skill and knowledge which provide equal opportunities of advancement for all workers. In conjunction with job rotation, this means that while a worker is performing a relatively low-level function in the team, his pay remains based on his earned position in the qualification hierarchy. Since a significant component of skills is acquired on the job, care must be taken by management to ensure that those who wish to progress are not blocked by counterparts wishing to remain at a given level. Because work teams are self-managing and can function best with highly flexible members, potential barriers to sharing job knowledge and training between the more and less experienced are reduced.

In some of the new plants, work teams play a central role in deciding whether or not a member has qualified for a higher pay level. In some organisations, workers are paid on a salaried basis, with primary responsibility for attendance remaining a work team function. Finally, plant-wide profit-sharing or cost-saving plans exist in a few of the new organisations. While some companies have reserved judgement on this until stable base periods for productivity gains can be established, others have rejected this approach as an undesirable precedent.

A strong emphasis in the new plants is placed on creating an egalitarian atmosphere. Conspicuous status differentials, such as separate parking areas, cafeterias, rest-rooms and the location of administrative facilities, have been reduced. Typically, time clocks have been eliminated. Meeting rooms for work teams are located adjacent to production areas and are utilised at their discretion.

In most of the new plants, management and work teams are jointly responsible for employee selection. Prospective employees are given detailed information about the organisation and its non-traditional features with a view towards sharpening self-selection judgements. Screened applicants are often interviewed jointly by management and their prospective workmates, who then make hiring decisions together. A similar emphasis is given to training, where the initial employee group receives extensive induction in business principles and social skills (group problem-solving, conflict resolution, communication and related human relations), as well as technical training. This is subsequently supported by in-house and paid off-the-job training programmes. A substantial portion of learning is a function of on-the-job training, and a combination of objective tests and peer review is used to assess a worker's abilities to progress to higher pay levels.

In general, the prevailing management style in the new organisations has been the delegation of decision-making to the lowest possible levels and the preservation of a plant-wide orientation to overall performance, information sharing and social integration. Maintenance of such an enterprise requires regular plant-level meetings to complement those among work teams. New plants have a variety of internal review bodies and permanent task forces to oversee the integrity of their participative systems, usually consisting of management and worker representatives. In some cases, members from the original plant design team periodically meet with plant-level review groups to address emerging problems and modify structural and procedural arrangements.

The following subsections consist of selected examples of how socio-technical considerations found expression in plant operations utilising automated continuous process technology, and concluding with a discussion.

1.12.1 Example: chemical plant

A $200 million Shell Canada chemical plant located in Sarnia, Ontario, commenced operations in 1979 after over four years of planning.[85] The plant employs approximately 150 people and is divided into six shift teams with one co-ordinator each, a small journeyman maintenance group and technical support staff, two operations managers and a plant superintendent. Each team rotates through a shift sequence devised to maximise work on days (53 per cent), and is capable of running all aspects of plant operations, including its administrative, laboratory, shipping, warehousing and maintenance activities.

The plant's design task force made important changes in the systems originally recommended by company technical specialists, notably the following:

- the original two control room design was changed and a single, combined control room was adopted;

- the final bagging operation was automated;

- use of separate quality control laboratory located in an adjacent refinery was abandoned in favour of placing one next to the plant's control room for use by operators;

- separate worker and management parking lots were combined to reduce status differentials; and, finally

- use of the process computer was significantly altered to support operator learning and control.

With regard to the latter change, studies of recently built plants making the same product revealed low capital utilisation rates (approximately 50 per cent), which made the capacity for quick recovery from downtime a highly desired objective. Furthermore, although the system was highly automated with relatively stable inputs, a large number of variables in the transformation process remained uncontrolled because of a limited knowledge of cause-effect relationships. The result was unpredictable yields relative to a variety of performance specifications in the product mix. The process was also dangerous. Aside from high pressures and temperatures, exposure of semi-finished materials to the atmosphere could have explosive results. Consequently, the design team changed the original technical recommendations for on-line computer control and, instead, developed software packages for utilising the computer off line -

> The necessary computer programs were designed so that operators could use the computer in the mode of evolutionary operations. In terms of what was known about relevant variables, the closed loops programmed in the computer actually are the maximum from an optimising point of view, as contrasted with a controlling point of view. The computer answers queries put to it by the operating personnel regarding the short-term effects of variables at various control levels, but decisions are made by the operators. Operating personnel are provided

with technical calculations and economic data, conventionally only available to technical staff, that support learning and self-regulation. In this manner, operator learning is enhanced. By utilising the experience of operators thus obtained, computer programs can be updated to further enhance learning and so on interactively.[86]

In the late stages of the design process, the Oil, Chemical and Atomic Workers Union (now the Energy and Chemical Workers Union of Canada) was invited to collaborate in various aspects of organisational planning, including job content, reward systems, training and implementation. This led to the negotiation of a unique, one-year collective bargaining agreement which strongly supported the need for group norms and work practices in the plant to evolve with minimum a priori specification in contract language. As a facilitating framework, the agreement sanctioned the following principles and values in its preamble:[87]

1. Employees are responsible and trustworthy, capable of working together effectively and making proper decisions related to their spheres of responsibility and work arrangements - if given the necessary authorities, information and training.

2. Employees should be permitted to contribute and grow to their fullest capability and potential without constraints of artificial barriers, with compensation based on their demonstrated knowledge and skills rather than on tasks being performed at any specific time.

3. To achieve the most effective overall results, it is deemed necessary that a climate exist which will encourage initiative, experimentation and generation of new ideas, supported by an open and meaningful two-way communication system.

After one year of the plant's operation, the contract was renegotiated and remained fundamentally intact, with the exception of salary changes.

1.12.2 Example: paper mill

In the seventies, a United States forest products company opened a new paper mill founded on socio-technical concepts.[88] The mill was totally integrated (including its own chemical pulping process), employed 240 workers on rotating shifts and was non-unionised. As in the previous example, many variables in the conversion processes were difficult to control because their cause-effect relationships were incompletely understood. Consequently, trial and error, based on the operators' experience and intuition, was heavily relied upon. This, together with the inherent unreliability of the equipment, placed a premium on operator access to relevant information, technical support and control to ensure product quality, and quick recovery from randomly occurring breakdowns. The process consisted of four separate technologies serially related, each yielding a semi-finished material or product. All were simultaneously linked to pollution control technology, which meant that disturbances in one system could rapidly affect the others. In contrast, other aspects of the technology had the opposite property: any adjustment to a conversion process itself would be delayed in its impact down-line. Finally, many of the workers were spread over a wide geographic area because of the size of the equipment.

The workforce was divided into teams, one for each functional area; each team had four crews responsible for their respective shifts. Teams were composed of operators, maintenance and laboratory workers, with jobs structured to emphasise cross-training and multi-skilling. The reward system reinforced this by paying workers on the basis of demonstrated knowledge. Each team was responsible for managing its own subsystem, and reported directly to a functional manager. The workforce consisted of experienced operators and maintenance craftsmen, along with workers not previously employed in the paper industry. There was no separate maintenance department, although a small engineering group provided technical support to the crews.

In order to support dual operational and learning objectives, traditional information systems were abandoned. Because the teams were widely dispersed and highly independent, each control room and instrument centre was equipped with appropriate indicators to monitor all key variables (irrespective of where they were controlled) which had a major influence on activity within its functional boundary. Thus, cross-instrumentation would permit operators to anticipate characteristics of the product or material being sent to them, as well as the effects of their product on the next unit. Equipment and instruments were grouped to ensure a clear separation between conversion steps, so that team authority and accountability were clear. Manual override capabilities were provided in areas where critical elements of the process had been previously designed to function under automatic control. Because immediate feedback concerning the process under a team's control was essential, each group had its own chemical laboratory for performing functional tests, rather than relying on a centralised testing department. A central laboratory was used, however, for sampling raw materials and final products before shipment.

1.12.3 Example: diversified products company

This company has had considerable experience since the sixties in experimenting with organisational innovations. Many have been undertaken in settings where advanced forms of continuous process technology were introduced. The company has pursued a pragmatic combination of behavioural science and business concepts which, because of their range of application, have acquired legitimacy throughout the organisaton as a viable employment strategy. Its philosophy is summed up as follows:

> If we give employees the same ability as management to focus on the business outcomes and let them see how their contributions pay off for them personally, we have achieved common objectives. And what's really great about all of this is that one of the pay-offs for the individual is the feeling of being a part of the business; by that

> we mean a feeling of ownership, which is exactly what we are striving for. We give every employee a chance to participate in an "adult business deal" ...
>
> The newer plants write guidelines rather than rules. They teach principles. They encourage flexibility in dealing with individual and organisational problems. They stress informal problem-solving rather than formal grievance systems. Discipline is administered on the basis of what is appropriate, rather than some predetermined code of crime and punishment. Employees in the new systems are taught about value - about producing a superior product at a competitive price. They are taught about our commitment to excellence and our expectations of ethical conduct. All this knowledge is brought to bear on everyday operating decisions and helps ensure that our customers get what they should and that the company remains competitive.[89]

A number of new organisations in the company were designed and apparently continue to function according to the characteristics of innovative new plants described above. Concerning self-managing work teams, the following has been gleaned from experience:

> We have learned that a word of caution is advisable here because occasionally there is an effort to make teams out of all groups of employees, expecting that to be a cure for many organisational problems. That is not always sensible and profitable. Teams must be built around interdependent tasks associated with achieving a specific common end point. Not all tasks are interdependent and not all are aimed at achieving the same end point. Our new plants, by and large, understand this situation and form teams only where teams make sense and have a pay-off.[90]

Multi-skilling in the new plants has produced the following result:

> The employee in the traditional system can perform all the operating tasks in one process. The employee has been taught no communication skills, no problem-solving skills, no maintenance or co-ordinating skills. By contrast, the employee in the new plant can perform operating tasks in several areas, and as such, is able to take over in emergencies, move to permit someone else to develop, fill in to prevent overtime costs, and contribute to problem-solving in the areas other than where he or she is currently working. This employee can perform some maintenance tasks, perhaps reducing maintenance emergency calls and certainly improving efficiencies. The employee can perform some administrative tasks and can co-ordinate the activities of the entire team, perhaps reducing the need for direct floor supervision, in turn allowing the manager time to do other things, such as train other employees, work on the method projects, conduct experimental orders and much more. It is perfectly reasonable to expect this level of development from a significant portion of the workforce within the five-year time frame.[91]

Finally, in connection with the ability of the innovative new plants to remain productive and adaptive, management has recognised the importance of continuity -

By continuity, we mean keeping employees' attention on the business purposes of the organisation. The newer organisations greatly increase the organisation's odds of achieving continuity by eliminating or reducing work stoppages, work slowdowns, general protests, legal action by employees and employee-instigated governmental involvement. Employees who have made the "adult business deal" are less likely than others to disrupt the organisation's attention towards the common objectives - as long as the organisation is living up to its part of the deal.[92]

1.12.4 Discussion

Whereas the above remarks and examples reveal the broad contours of the design and intended consequences of certain innovative work systems, it is more difficult to specify in detail the results obtained. Based on the few published reports extant, together with informal accounts, the following comments can be offered.

Because much responsibility for day-to-day management is delegated to work teams, corresponding pressures on their ability to function effectively can stem from a variety of sources. In some cases, inadequate advance training and/or unforeseen technical difficulties during early start-up phases of new plants have disillusioned their workforces. Regardless of the cause, when initial high expectations of management and workers have not been matched by reality, this has sometimes led to the imposition of management controls, such as more direct supervision, and retrenchment in multi-skilling programmes towards job specialisation.

Alternatively, early periods of crisis have often been successfully navigated through management's (and, where applicable, the union's) patient adherence to the original design philosophy and, where necessary, the renegotiation of performance expectations and standards, both internally and with the parent organisation. A central dimension of effective team functioning is the ability to evolve internally and externally acceptable work norms. Because teams are responsible for work assignment, training and member discipline, among other functions, issues of equity, fairness and respect for individual differences must be continuously applied. Predictably, these have been sources of strain. Groups have experienced difficulty in judging others in connection with pay rises, and in disciplining members. Tension arises when a member chooses not to broaden his or her skills and blocks another's training opportunity. Often, issues which cannot be resolved within a work group are referred to plant-level worker review bodies, and, if necessary, to management. In some cases, management has permanently absorbed these functions, either by design or default, leaving some form of multi-skilling as the core feature of the work system. In one company, multi-skilling, a supporting payment system and mechanisms for encouraging open, two-way communication have been the main "package" of techniques it has chosen to support in the design of other new plants.

In plants with first-level supervision, difficulties of role ambiguity, confusion over power and responsibility, and discomfort with the group dynamics of self-managing work teams have undermined the effective development of work norms which support the design philosophy. Management has contributed to this awkwardness by not adequately training supervisors for their new roles, and in those instances where first-level supervisors were assigned to a new plant on a mutually acknowledged, short-term basis, the untenability of their position has been further increased. Some plants with formally designated "co-ordinators" and "team leaders" as quasi first-level supervisors have faced similar difficulties of role ambiguity and uncertainty about job mobility prospects.

Some companies have started up new plants with traditional management controls as a transition until production is running smoothly, phasing in self-managing groups thereafter. Others have commenced operations with the rudiments of self-managing work units, some coping successfully, others requiring direct management guidance. Once operating on a stable basis, there are strong tendencies for work teams to become isolated within and between shifts, despite formal mechanisms of co-ordination. Correspondingly, management may neglect to develop or sustain plant-wide mechanisms for problem-solving, performance review and goal setting. At the level of work groups, production demands create pressures on their ability to attend to their own mechanisms of self-regulation.

Little objective evidence is available concerning the actual performance of the open job progression or multi-skilling systems. Depending upon the complexity of the technical process, optimal periods of time for mobility between entry and top-level jobs may range from approximately three to eight years. Although theoretically all workers are eligible to progress to top pay rates, in practice skill profiles in the internal labour market are typically expected to resemble a normal distribution based on abilities and interest. Derived mutual benefits are viewed as preferable to those under strict job seniority systems. The decision to acquire more skill and knowledge is voluntary (although recruitment interviews usually stress that multi-skilling will be expected of everyone), but there can be strong pressures on workers to cross-train; this has led to labour turnover. In those plants where operators of the process technology are encouraged, and in some cases required to develop maintenance skills, experience to date has suggested that this has taken considerable more time than anticipated. Nevertheless, the job progression systems appear significantly to have opened the range of opportunities for workers to enhance both financial and intrinsic job rewards.

In some organisations, the above and related problems have been accepted as teething troubles in an evolutionary process. In others, they have led to the deterioration and eventual abandonment of innovative work structures. Their continued existence and proliferation in a variety of North American companies suggests that, at a minimum, the derived economic benefits have been satisfactory. It has been reported that positive experiences to date have played a significant role in resisting trade unionisation, and as partial models for the revision of older work structures. Because there is a scarcity of published information of a detailed nature on the operation of the new plants, little is actually known on a systematic basis about how these work systems have matured and adapted to changing circumstances, such as economic downturn, lay-offs, management and worker turnover, the levelling off or peaking of opportunities to improve effort-reward ratios, shifts in political support from the parent organisation, and, where applicable, the trade union.

1.13 Conclusion

In the light of some of the case material discussed in this chapter, it is important to note that what is sometimes labelled a search to "jointly optimise the technical and social systems through a participative design process" may disguise a predetermined management commitment to install work teams or autonomous work groups as the "one best way". This has become known among sceptics as the "cookbook" approach to organisation design.

"Cookbook" applications of socio-technical ideas without regard to underlying principles are not without precedent. Ironically, Frederick W. Taylor earlier in the century was busy defending his scientific management from those for whom efficiency had become a craze. Taylor disassociated scientific management from "installing"[93] (to use the word of one of his concerned followers) piecemeal, faddish programmes, insisting that scientific management was -

> something that varied as it was adapted to particular cases, but always involved a mental revolution of employer and employee toward their work and toward each other ... [it] fundamentally consists of a certain philosophy which can be applied in many ways ... recognising as essential the substitution of exact scientific investigation and knowledge for the old individual judgement or opinion in all matters relating to the work done in the establishment.[94]

Despite the evils of scientific management, Taylor made a valid distinction between formulating a thought process sensitive to the unique aspects of a system undergoing or requiring change, and the mindless application of recipes and instant solutions.

Among the current candidate-substitutes for the "exact scientific" legacy of Taylorism are an array of behavioural science techniques. On the most superficial level, their application may be merely semantic and may mask an orthodox work structure or proposed design with the vocabulary of social science. Alternatively, they may be selectively (and mechanically) applied, as with some so-called "socio-technical" programmes and others involving pseudo-worker participation. These can result in strengthening managerial control over workers, albeit a few notches looser than under classical scientific management and cloaked in the language of pseudo-mutuality. In those work systems where authority and responsibility have actually been delegated to autonomous or self-managing work groups as a result of a participative design process, management could be presented with a serious conflict between its perceived need for total control over the production process and its increased dependence, as automation increases, on a highly knowledgeable, skilled workforce, capable of independent thought and action. Assimilation of these work systems into their parent company structures and cultures remains to be demonstrated.

Although some managements are less parochial with respect to the control issue, appreciating the benefits of a more technically and managerially sophisticated workforce, other aspects may seem threatening. A dramatic example is in the nuclear power industry, where greater operator knowledge and inquisitiveness could lead to embarrassing discovery and revelation of basic procedural and design inadequacies. In manufacturing settings, deeper worker understanding of the business aspects, as well as expanded work authority with respect to the production process, may challenge the quality (and perhaps legitimacy) of management decisions concerning such matters as resource allocation, distribution of economic gains and losses, employee safety and health, product content and design, corrupt practices and plant relocation. These areas may be presently beyond the scope of the "adult business deal" described earlier.

In the United States, many forces are at work to support managerial definitions of what is "problematic" and worthy of shared scrutiny and collaboration. Within this perimeter, we can expect continued experimentation with alternative forms of work organisation as computer-based automation extends deeper into the functioning of factories and offices. It is difficult to predict how extensive this will be. Thus far, organisational innovations have been relatively incidental to the dominant and largely unquestioned acceptance of bureaucracy and scientific management as the "state of the art" relative to the design, introduction and management of new technology in the workplace.

The proposition advanced by Blauner in 1964 that there was a necessary relationship between automation and social integration within the enterprise has been convincingly laid to rest.[95] Questions of social integration and its attendant consequences for QWL have less to do with automation and more to do with the formation and regulation of substantive and procedural norms in the enterprise. Only some of these norms are readily susceptible to what has become known as "organisational choice". We conclude with the observations of a British researcher -

> It could be argued that while automation may well produce significant changes in the objective work situation, it is none the less likely to be indeterminate in its effects on social integration. The degree of social integration in the highly automated sector will more probably depend on the cultural values prevalent in the wider collectivity to which the workers belong, and on the nature of the institutional structures characteristic of the society in which the automated sector emerges.[96]

2. Case study - Socio-technical design and new forms of work organisation: integrated circuit fabrication

2.1 Introduction

The introduction of the microprocessur has dramatically lowered the cost and expanded such applications of industrial automation as cybernetics, robotics, NC machines and CAD-CAM. However, apart from the automatic control of production processes, automation has eliminated the need for certain controls, because of a radical change in product technology.

The continued development of the microprocessor since 1970 has relied in turn on developments in photolithography which have permitted a remarkable miniaturisation (called large-scale integration - LSI) of circuits; and of achievements in physics and chemistry, especially the metal-oxide-silicon (MOS) process of fabricating transistor-based semiconductor circuits. Such technical breakthroughs as these suddenly moved computer electronics from a labour-intensive to a capital-intensive industry. A decade ago, workers with soldering irons meticulously hand assembled and wired separate electronics components such as transistors, condensers and resistors on to metal frames; today, a few workers monitor sophisticated equipment to fabricate computer chips in batches of up to 4,000 LSI circuits at a time. No soldering of separate components is done because with MOS and similar technologies none is required; such new semiconductor technologies have permanently altered the fabrication of computers and other similar integrated circuit devices. It is as if the new fabrication facilities were automated in the conventional sense (for example, many workers have been displaced or made redundant) but it is the product technology itself that has changed the production process, rather than merely the computer control of the process. Indeed, the MOS process involves the use of computers for automatic control, but the basic worker activities of handling the silicon wafers often requires human intervention and override rather than mere monitoring of machine activities.

Automation and worker displacement are connected, but the major automation question in the United States during the seventies has been not whether, but how, the jobs that remain after automation can be changed for technical effectiveness and improved in terms of QWL. The American dilemma has been how to reconcile the industrial era assumptions of technological determinism with the post-industrial requirements for coping with unpredictable, stochastic events.[97] Technological determinism leads to the conclusion that technology is the most important element and that jobs, work and organisation must be subordinated to support it; it assumes that there is "one best way", and that way is called for by the technology. Although the dilemma is not yet resolved, North American experience is accumulating to show that new forms of work organisation, designed to address social and psychological requirements as well as technical ones, are better able to provide for the flexibility of organisational response which currently characterises a successful system. American managers are beginning to realise that work teams or groups are more appropriate units of organisational leaning and system interdependence than are individual workers and supervisors. Attempts to trace the developments of the American industrial revolution, fathered by F.W. Taylor, backwards from 1910 to the factory system of the British industrial revolution, and forward to the United States human relations and job enrichment movements of the fifties and sixties, lead to the conclusion that the focus on individual jobs and workers is responsible for the success of the American industrial system from 1910 to 1960, and also responsible for the decline in productivity since then.[98]

The United States electronics industry has grown up literally since that period of shift in American fortunes. It is an industry which was developed both in large laboratories at IBM and AT & T, and in backyard workshops in Cambridge, Massachusetts and Palo Alto, California. Because so much of the industry started on a small scale and grew so quickly, the amount of concern for organisation design at all has been minor in relation to desire for technological innovation, and the pursuit of private fortunes. Thus the organisations we see in current United States electronics firms are inadvertent copies of the Tayloristic model which contributed to industrial success between 1910 and 1960, and to decline between 1960 and 1980. Although the capital investment in micro-electronics technology (e.g. LSI and MOS) has provided a marked improvement over the earlier products in cost and availability, the American microprocessor industry has not been able to compete favourably with the Japanese when both are using similar production processes for similar products. American managers in micro-electronics have become attentive to Japanese production and managerial methods, but have done little except emulate the employee suggestion plan embodied in quality control circles.

In North America, socio-technical systems as an organisational design approach provided a marked improvement in organisational effectiveness and QWL during the seventies.[99] Socio-technical systems design is a conscious effort to examine and evaluate the structural aspects of organisations (including division of labour, job design, work group definition and performance measures), and to guide and justify subsequent changes in structure. Socio-technical designs often involve teams of workers, and the nature of these groups (whether face to face, group size, time boundaries, supervisory roles, multiple skills and specialities) are determined by the objectives and philosophy of the system, employee capabilities, prior labour agreements and technical requirements.[100] Over periods of up to 20 years, socio-technical systems designs have proved durable in their continuance and ability to adapt.[101] The socio-technical systems concept has its roots in Europe rather than Japan, and it represents in the United States a way to begin to substitute a strong social organisation for the individual worker/supervisor model of the earlier American industrial designs.

The present case study illustrates the use and effects of socio-technical design methods in microprocessor fabrication. It represents an excursion into organisational design, a new area for the industry, and it shows how effective both in speed of application and in permanence of effects socio-technical designs can be. The semi-conductor industry is not widely trade unionised; no experience thus exists which deals with union-management co-operation in the industry. The identifiable audience for immediate change is therefore management, but it is hoped, in addition, that the material and its conclusions will also encourage actions taken directly by workers.

2.2 Lessons from Japanese competition

The Japanese excellence in micro-electronics manufacturing is a powerful stimulus for the American electronics industry to consider new ways of working. This is demonstrated in the new Nampa, Idaho, semiconductor plant of Zilog, Inc., a subsidiary of Exxon Enterprises and the producer of the Z-80, the industry standard for 8-bit microprocessors.

If asked whether Japanese methods of production had been the model for organisation, Zilog management would reply that their design techniques are largely based on those of the United States and Europe. It is maintained here, as elsewhere, that United States organisations cannot simply uproot the Japanese management style and transplant it to American soil. However, there are several ways in which the traditional American management approach can benefit from the Japanese.

In American micro-electronics, as in other industries, workers are regarded as liabilities. The result has been an increasing disconnection between American employees and their products. Workers are assigned to pieces of equipment where they perform a repetitive, narrowly focused job such as pushing a button or looking through a microscope. They are taught to make a decision whose impact they do not understand, based on information they do not understand. The button-pushing becomes an endless series of meaningless tasks. In Japan, occasional mistakes, particularly those made by lower-level employees, are considered part of the learning process. If employees are not successful in a particular job, they are transferred to a more suitable position in the company. In the United States, an employee mistake more often results in punishment; an employee who makes the same mistake twice is often fired. The rationale is that if that kind of behaviour is tolerated, everybody will be doing it. Therefore an example is made of the worker concerned.

Another characteristic of American industry is that organisational structure is shaped like a pyramid, with all the decision-making power at the top. This has remained true especially in the semiconductor industry, with its rapidly changing technology, highly complex products, and relatively few people with broad knowledge of those products. At the lowest levels of the organisation - the levels where the products are actually built - there is almost no information available beyond a specific directive to push a button on a machine. The product itself has no meaningful context, and so the job has no meaning. Average turnover in the American semiconductor industry has ranged from 50 per cent to 100 per cent annually between 1978 and 1981, during a period when the national average has averaged half that rate.

2.3 The project

The company was preparing to open a new microprocessor chip manufacturing plant in Nampa, Idaho, scheduled for 1979. All of those who were involved in the project were managers, and had taken part in previous plant openings. They had all experienced the high level of problems involved in start-ups, with the lengthy period needed to bring operators and process engineers up to "speed", and a considerable amount of early personnel turnover. They therefore wanted to improve on this performance. In addition, they set themselves some "unreasonable" goals. Table 91 presents these start-up expectations, compared with industry standards and actual achievements. To meet production needs, management had to get the new plant running in about two-thirds of the normal 18-month period needed to start up such an operation. At the same time, they had to achieve higher-than-normal yields of product quality. Since they realised that these goals were very ambitious and that they might not be able to meet them fully, they began looking for a new approach to organisational design.

Table 91: Integrated circuit fabrication: start-up expectations and actual achievements

Indicator	Industry standard	New plant goals	Actual achievement
Start-up (months):			
First silicon wafers	18 from start of construction	12	12
Good "chips" or "dies"	21	15	12
Standard yield	25	18	13
"Fab" process yield (%)	75	85	90-95 since start-up
Annual turnover of direct labour (%)	50-55	24	6

2.3.1 Background

In late 1978, the company hired an experienced plant manager, who had already established an assembly plant in Indonesia, for the express purpose of starting and running the new plant. In approaching the design of a new plant organisation, the new plant manager observed that the role of management in the semiconductor industry is traditionally based on the idea that manufacturing workers are a liability: "You tell them what to do and how to do it, but there is no prerequisite for explaining why". In the more traditional plants in the industry, people are placed at equipment bases. They are taught how to push the buttons, but they are not taught what they are doing from a product output standpoint.

In the spring of 1979, after working for several months at Zilog's factory in Cupertino, California, the plant manager began searching for a successful model in start-up ventures. Not only was he looking for a method of structuring the plant to produce a good-quality product with control of costs, but there was pressure to have the plant running in less than the normal one-and-a-half-year period.

In 1979, semiconductor firms in "Silicon Valley" (as the stretch of California between Palo Alto and San José is now called) were beginning to expand outside the area. Most had an established, wide base of operation in the "Valley" to support new ventures in other states. At that time, the company's semiconductor manufacturing group in Cupertino was small, and it was management's decision to leave it small, expanding manufacturing outside California. For a small company with an adequate but by no means ample technical and managerial staff, it was a risky venture. The new plant represented an investment of several millions of dollars 700 miles away. The site selection had been made in 1977. The area offered a stable labour base, and water resources to support a reasonably sized semiconductor plant, while the specific location was in close proximity to a major airport, had good educational facilities and was a growing centre for new companies.

The prevailing assumptions of manufacturing management in the Cupertino factory were orthodox, and typical for 1979. Jobs were reduced to their minimum requirements; and specialisation, job behaviour and compliance were valued more than success or results. A conventional "wafer fab" environment is organised around two distinct technologies: diffusion of elements or dopants, which deals with growing or constructing the silicon wafer's surface; and photolithography, which deals with the circuit geometries of that surface. Figure 34 shows the flow of a silicon wafer from raw material to finished circuits ("chips" or "dies") in a conventional plant, and graphically illustrates the 40 or more times a wafer changes hands (and ownership) during the process.

Fragmentation of work at Cupertino was manifest not only for the basic hourly workers (called "fab operators"), but for all others involved in manufacturing as well, including process engineers, engineering assistants, maintenance workers, inspectors, supervisors and managers. Jobs in the conventional "wafer fab" (manufacturing) facility vary little throughout the industry. In evidence of this, the job descriptions for the Cupertino plant correspond closely with those of the salary survey service bureau, Radford Benchmark, which are familiar to most manufacturers in Silicon Valley.

The basic fab operator's job description reads as follows: "under general supervision the operator performs, during peak periods, one or two specific fab skills at a certain level". Specific skills might include alignment of photolithographic masks on wafers, or operating a diffusion furnace. Other activities expected of an operator include "... maintaining work area and equipment consistent with good housekeeping standards, notifying work leader or supervisor of any repair or maintenance requirements, and completing production records as required". Higher-graded operators are expected to be competent in several skill areas, to be able

Figure 34: Integrated circuit fabrication: conventional "wafer fab" production structure

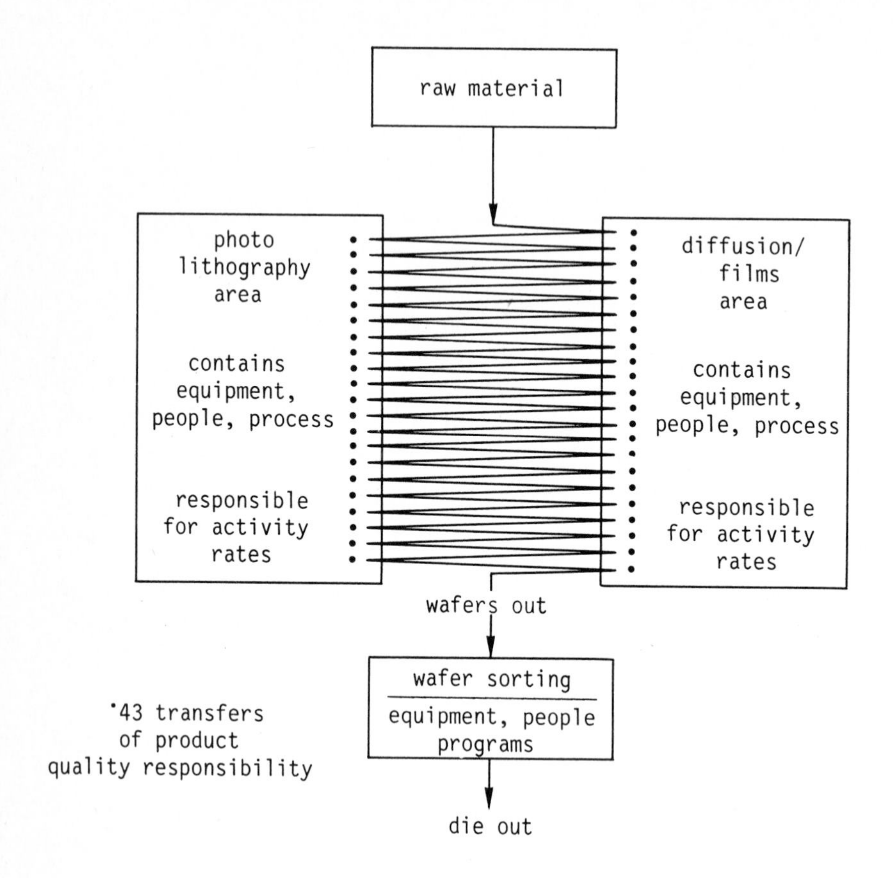

to fill in for absent co-workers as required, although they are generally limited to performing a single repetitive task throughout a working day. The degree of fab operator task variety, control over work, involvement in decisions, opportunity to learn about the product and about other skill areas (e.g. maintenance, engineering or training) is virtually nil, and the degree of communication between operators and these specialist groups is very low.

Operators report to supervisors whose Benchmark job description includes: "working on problems of limited scope, following standard practices in analysing situations from which answers can be readily obtained ... assigning and checking subordinates' work ... and performing the same work as group members". In a typical fab plant at least three additional levels of management preside above the supervisor. Engineering assistants, maintenance technicians and training technicians all have their own pyramids for hierarchical reporting and promotion within the manufacturing plant organisation.

This fragmentation and separation of jobs in Cupertino in 1979 was clearly in evidence even to the casual observer. The clear distinctions between the manufacturing specialisations listed above were evidenced not only by where people worked, but also by the language they employed. At every turn, casual outsiders (if they were allowed into the "clean room" environment of semiconductor manufacturing) would hear a babel of specialised languages. The engineers spoke to one another of electrical evaluation ("eval") "thresholds", and to their assistants of capacitance-voltage ("CV") "plots", but neither could talk to the fab operators or even the fab supervisors about such matters. The fab operators doing photolithography talked about "mask alignment", and those in diffusion spoke of "pushing boats", and "loading elephants". With their conversation so specialised and their jobs so machine-specific, the fab operators had no ability to talk about work-related matters with one another if they worked on different equipment. The fab supervisors spoke a language of "moving wafers" from station to station which was comprehensible neither to operators or engineers. Production management spoke of "balancing the line" and cumulative ("cume") fab "yields", and the maintenance workers spoke the electronics and hardware language of the factory's sophisticated equipment. These different "languages" created a barrier which extended even into the cafeteria, where fab operators who knew only their own limited machine-based dialect did not sit at the same tables with other operators who spoke in the tongue of their own activities.

What is most significant about this proliferation of esoteric "languages" is not that these occupational groups in the Cupertino plant could not (and thus would not) speak to one another, but that nobody spoke of the circuit "dies", or "chips", or in any way referred to the product or output of the factory. Everybody was encouraged to do their best at their specialisation, but no one was encouraged to consider the microcomputer circuits which were their product. Such a limited vision, of course, created and perpetuated a provincialism not only within the Cupertino plant, but between it and the new plant starting in Nampa. Because some of the middle managers for Nampa were drawn from this location, some distrust and misunderstanding was carried out of the Cupertino plant, both because of suspicion of the "new ways" at Nampa by some of those with more traditional values, and because of allegations of "soft-headedness" towards those who transferred by those who remained behind.

2.3.2 The design model

The plant manager's search within the semiconductor industry turned up few models for organisational design that were satisfactory for setting up a new plant in a short span of time. Thus in early 1979 he sought help from the organisation development consultant of the parent company. The result was the use of an "open systems model" of socio-technical analysis which several companies outside the electronics industry had been using for a number of years. The plant manager also became part of a network of plant managers of so-called "innovative plants" who met together several times a year to share work innovation experiences.

Shortly afterwards, the plant manager's subordinate managers (who would later make up the management team in the new factory) attended, with other company managers, including the President, a week-long workshop on organisational open systems design. They also visited other plants applying new forms of work organisation and management systems. The plant manager and his management group also met consultants who had worked in plants of a company which is reported to use open socio-technical systems designs.

The internal organisation development manager of the company and an external socio-technical consultant, working together with the plant manager and his team, developed and implemented their own version of an open socio-technical system design for the new plant. The "design team" of managers continues as a planning team today.

In the open systems socio-technical model, the idea is to encourage responsibility at its lowest possible level. It is taken as a premise for such designs that people enjoy making decisions, taking on responsibility and reaching a level of task ownership. In this case ownership meant participating in work-related decision-making, setting goals and implementing them.

The work design process involved three steps. The first was to formulate the purpose, or "mission", of the plant. Simply stated, the plant mission was to produce high-quality integrated circuits (in other words the individual microprocessor circuit "chips") at a minimum economic and social cost. A key distinction here is that the goal was to produce good circuit chips, _not_ the silicon wafers upon which a number of identical circuits are simultaneously fabricated. The importance of this distinction cannot be over-emphasised. It would force each worker to think in terms of an end product, a _result_, instead of an isolated process or an interim step. The conventional semiconductor manufacturing facility has two separate processes: wafer fabrication ("wafer fab") and wafer sorting. Wafer fabrication refers to the manufacturing process itself, in which as many as 250 separate circuits are constructed on and in the silicon wafer. Wafer sorting, on the other hand, refers to testing the finished circuits on the wafer and identifying the faulty ones. This distinction has no meaning in the new plant. At every step of the process, workers think in terms of the number of good circuits, not the number of wafers they can move through the line. In effect each team is responsible for its own quality control.

The second step in the work design was to analyse the technical process involved in producing good circuits. To do this, "variance control analysis" was carried out to discover what factors could destroy or reduce the opportunity for each circuit chip to be considered "good". As a result, 120 significant variances were identified. For example, a wafer might not be perfectly flat, or it might not have the proper thickness of film deposits, or the circuit designs might not be perfectly aligned on the wafer. It was found that in the process of building the "gate structure" (the basic element in microprocessors, and part of the numerous transistors comprising them) alone, there are 15 different things that can go wrong.

Ways had to be developed for these 120 variances to be controlled. The variances could be grouped into 10 logical units, designated "unit operations" or state-change units because they constituted identifiable changes in the circuits as they progressed from raw silicon to finished "chips". For instance, the photolithographic impressions of the circuit designs upon an oxide layer of a silicon wafer changes the state of the evolving circuits on that wafer, but the alignment of those circuit impressions on the wafer does not.

The third step involved designing a social organisation (including the division of labour, the design of jobs and the location of internal departmental boundaries) both to match the needs of people who would staff the organisation and to control the key technical variances described above. The social organisation was to guide the limits of authority, of standards for co-operation and of roles for co-ordination. It was based on a clearly articulated set of values and beliefs about people; in the plant start-up, the premise was accepted that all employees are trustworthy, that they can learn and grow, that they can accept responsibility for their product and that they can work together in harmony. The organisational structure described below was designed on the basis of these values and beliefs.

2.3.3 The design structure

The system is based on a philosophy of organising around a set of results rather than activities. In addition, the integration of people with technology and the involvement of those who build the product with identifying and solving potential problems and with setting goals, are central to the system.

Initially, 23 employees of the company went to Nampa to set up the plant, representing engineering, management, purchasing and materials control. Except for management and the training group, few had operating skill in manufacturing. They set about hiring and training local people who had no experience in semiconductor manufacturing. Since November 1979, about 100 manufacturing technicians have been taken on. There are currently a total of 146 employees in the plant.

Originally, when the plant was small and growing, and only one product was being built, the production process was organised into four work teams of manufacturing technicians, each team with its own manager. Some six to eight months after start-up, other products (similar in process, but differing in execution from the original product) were added to production; and the number of manufacturing technicians in each of the four teams was approaching 25. The management design team, working with the technicians and their team managers, took that opportunity further to subdivide the four teams into smaller units while retaining the most important "key variance control" within the team structure. The plant now has eight teams of technicians interacting among four work "families" (which correspond to the four original teams).

The plant manager and seven others currently comprise the management structure, which is flat, with two levels above the manufacturing technician. Decisions are made by consensus on the basis of a set of goals. There are still only four team managers, all of whom are part of the original start-up group. These managers focus on development of the eight team groups and on the problem-solving process. They are involved with helping technicians understand their roles.

In the new plant, employees are taught the manufacturing process from beginning to end, so that they know why they are there. Efficiency is not measured in wafers per hour, but rather in "die" or "chip" costs. In fact, the plant manager prefers to call the plant a "die generation" rather than a "wafer fab" facility.

The team of employees assigned to a given unit of production has control over all technical variances affecting the "dies", and which are associated with that unit. Such process monitoring is not new in itself. However, in the past in this industry it has not been carried out by the workers building the product, i.e. the people in the best position to solve problems connected with it. Process-monitoring data has traditionally been used by those higher up the organisational ladder, who seldom, if ever, returned feedback.

Plant start-up began with building the social part of the sociotechnical design. The key to this was that each work team became responsible for building an identifiable part of the product. The team became truly accountable, in terms of cost and quality, for the successful completion of a state change. To give a team this kind of accountability required a major simplification in organisation structure. A conventional "wafer fab" organisation (as described earlier) contains highly specialised operators for diffusion and photolithography who communicate not among themselves but through higher-level intermediaries, the engineers and supervisors. If the new plant were to be structured in this conventional way, every wafer would have to change hands 43 times. There would be no way to pinpoint who was responsible for building the product at any point, with the justifiable result that no one would care.

The socio-technical design permitted the structuring of the circuit-building process so that only eight transfers of product responsibility take place. A sense of ownership is achieved because each of the eight state changes that happens to the product is large and important enough to warrant this. Each of the four work team "families" is responsible for two state changes, and for the quality, quantity and cost of its output. Thus each team has a concrete, tangible goal to achieve. Because these team goals can be so easily identified, individual productivity ceases to be an issue for its own sake. However, as part of the team contribution, individual productivity is of paramount importance.

Central to this concept of team work groups is team decision-making. Decisions are not made by managers and handed down to the team to implement. Instead, each team member is fully involved in the decision-making process and understands the reasons behind the decision. The team's level of decision-making grows as the team's knowledge and expertise grow to meet the demands of the tasks it faces.

It has been said that this "decision by consensus" is much more time consuming than decision-making imposed by management. This may be true, but at the new plant it is also proving to be more effective. Even if everyone does not agree with a decision, each worker has had a chance to influence that decision and is prepared to support it. And more importantly, once a decision is reached, its implementation is very rapid. There is no need to communicate that decision three or four levels down the ladder, and there is no questioning of why the decision was made. At the same time, there is no blame attributed to a higher-level authority for an unpopular decision.

2.3.4 The technicians and their teams

The manufacturing technicians, as the operators are called, are organised into teams, each of which is responsible for building a segment of the product. All the technicians understand the entire process and perform a variety of functions aimed at carrying the product through to the next stage.

Acquiring enough information and skill to be an effective team member obviously requires a much broader level of training than what is usually given to a "fab operator". Every worker must be knowledgeable not just in a narrow task, but in the technical system as a whole, so that the consequences of individual actions can be understood. This is achieved by a two-week (80-hour) initial training and product orientation course given to all new employees (not just manufacturing technicians), and based on the concept that the entire plant has a hand in building good circuit "dies". Over the two years that the plant has operated so far, this training course has not been shortened in length or reduced in its content.

The manufacturing teams are deeply involved in analysing what can go wrong in the fabrication process and in solving such problems, so that quality is built in throughout the procedure. In fact, there is no quality control organisation but an entire plant focused on quality improvement.

The teams are responsible for many of the tasks of a supervisor in a traditional set-up: they hire, train and discipline their members.

In the first two months, there were over 4,000 applicants for entry-level positions, and the screening process was a considerable task. Once formed, the teams themselves began to help with the screening process as well as selection. Once a team makes the decision to employ someone, it is committed to making sure that the individual is an effective team member. This commitment was evidenced from the very beginning.

The teams manage their own hours. Goals are set for the week and the month, and the work hours needed to accomplish them are planned accordingly. Typically, the equipment is operated by each team for 18 hours out of every 24. In addition, the teams are involved in budgeting, forecasting and even salary decisions. Salaries are competitive in the area. The original salary system was a fixed salary scale based on development of skill and knowledge, and there is a qualification process. The team itself decides on increases for its members.

Figure 35 presents the tasks and responsibilities of team members, and of management. At regular goal-setting meetings, the team determines what it needs to do during the week to achieve the required quality, quantity and cost levels. Team members decide among themselves how many hours they will need to work that week, which variances they have to watch out for and what personnel problems they have to tackle. The team's authority extends to screening and hiring new members when it needs them, as well as managing individual performance of its members.

What is management's role in all this? Often the plant manager knows very little of what is going on in any given team; and this is not considered a necessity. The role of management is to provide the teams with enough information so that they can set reasonable, compatible goals. Because the team members have the information and skills to achieve those goals, they are trusted to make the decisions on <u>how</u> to achieve them.

Clearly, demands are often made upon the team from external sources, e.g. from organisations outside the company or even other teams. It is management's task to interpret and clarify these demands for the team. This is known as "boundary management", i.e. defining where the team's responsibilities begin and end, and what is expected of members both inside and outside the team group. Management tasks and responsibilities outside of those of the teams are shown in figure 35.

One essential ingredient for success was that plant management entered this process fully committed to, and supportive of, the design team idea. In spite of scepticism from some outsiders, the experiment so far appears to be working. We give some examples below.

At one time or another, every company in the industry has faced problems connected with the process of growing intermediate oxides on silicon wafers. Their susceptibility to poor visual quality can lead to low-quality yields during the subsequent photolithographic process. In addition, temperature shock occurring when the wafers are removed from the

Figure 35: Integrated circuit fabrication: manufacturing team and management responsibilities and tasks

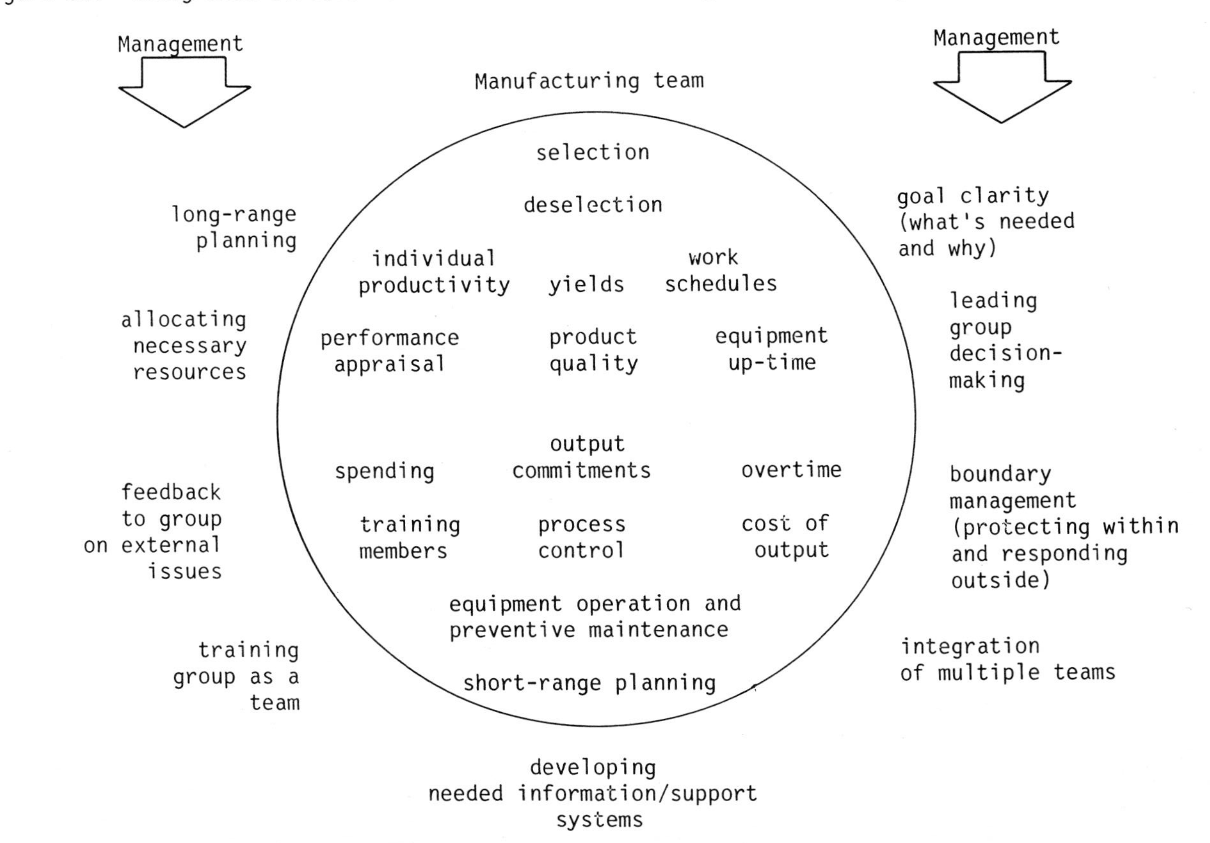

heated chamber of the diffusion furnaces can cause breakage levels higher than desired. At the new plant these problems were addressed totally by the teams responsible for this phase of the manufacturing process. The team members worked together to come up with a solution to a problem that had long plagued the company's Cupertino plant as well.

Personnel problems are solved in much the same way. There are no time clocks in the plant, but team members monitor each other's contributions. An individual who is not meeting his commitment to the group will first receive verbal feedback from co-workers. If the problem persists, the group will draw up an actual written contract with the individual to change his or her behaviour. The team is fully supportive of the member, often working with him or her to identify the problem and find a solution.

We cannot go into detail here about the preventive maintenance programmes, and even equipment changes, that the teams have made. However, we attempt to answer two key questions: Did the teams achieve the goals set for the new plant and, even more important, have these achievements been maintained after two or three years of operation?

2.3.5 Results

Good results seem to support this approach. Table 91 shows actual results. The plant has lost six operators and three managers in its two years of operation, a less than 6 per cent annual turnover. The typical rate for "Silicon Valley" fabrication plants for the same period was 55-60 per cent. Sometimes, as in mid-1979, when the industry was booming, turnover rates sometimes peaked at 100 per cent or more. Fabrication plants outside the "Valley" have been known to have turnover this high, although on the average it is generally slightly lower than these figures. Turnover rates at the remote plants of other companies are low, but not as low as at the Nampa plant. This low attrition rate is translated into higher performance and greater efficiency, as well as cost savings, given the time taken for an operator to become proficient.

The plant saw its costs go down after manufacturing began in the fourth quarter of 1979. Output was increased to the point where the plant was serving its customers and obtaining a satisfactory return on investment by early 1980. Since the start-up time was cut by half, the investment cost, which is closely related to how quickly a plant comes up to standard, was also reduced proportionally.

A dramatic increase in product quality yield has been achieved as compared with experience of the company in "Silicon Valley", and with reports of yields at other companies. Yields at the new plant exceed those of comparable facilities by more than 25 per cent. It thus performs well above the industry average as cited by independent research organisations. Success, however, is not only measured in yields; the team approach has reduced the amount of reworking, and therefore also the cost of labour.

2.4 The current situation: problems and prospects

The new plant is geared to high-volume "chips", and is currently set up for two basic processes: one for most logic products and microprocessors, and the other for memories.

The biggest problem in making the system work was within the management team. It was difficult to shed the traditional management roles where boundaries are clearly defined, and to make a commitment to a new philosophy. The plant management team had to solve the problem of how to interact with the outside world and to translate exterior demands (i.e. the company's Cupertino plant, and customers). Management's role, too, was to teach. According to the plant manager, "we are obliged to clarify what's expected and why, and we leave the freedom on how to do it up to the teams". "It took three months to a year before the managers felt comfortable", he continued. "It was particularly difficult at first to stand back and let the teams decide how they were going to solve problems. Nevertheless, when the results started coming in, everyone relaxed." Only one manager so far has been unable to adjust. He was hired as personnel manager and found the lack of restrictive rules, such as attendance requirements and set hours, difficult to contend with.

Recently, as the plant moved into its third year of operation, the following question was asked of members of the organisation: To what do they attribute their success and feel are necessary ingredients for continued improvements? The principles discussed below highlight critical factors.

First on the list was corporate support. Without the initial and continued support and willingness to risk doing something new with the design of a new plant from the President and Group Vice-President of Operation, the results might have been very different. This new manufacturing plant was the company's first domestic expansion effort, and it would be the home for future growth. This new method of organisation design (never before tried in the semiconductor industry) was no easy decision with a new plant and new equipment, 700 miles from the technology base, as well as new employees, new technologies and products soon to be delivered.

The second factor was the management commitment of the initial site members who made up the Nampa design team. They seemed to be convinced that there must be a better way of managing and operating a manufacturing facility. This fact helped them to overcome the many problems and unanswered questions that accompany new plant start-ups. Soon after these new managers were hired, they attended a week-long workshop on open systems design theory and team building. This workshop was attended by both those allocated to the new plant and key company personnel, including the President. This workshop helped the design team to begin to come together as a real management team with a vision of "something different".

Workshops played an important part in the overall development of the plant. Soon after the initial 40 or so individuals were gathered at the site, all members of the plant participated in a week-long workshop to explain the design technologies to be utilised and to encourage team building. This was an experimental workshop attended by manual workers as well as managers and professionals, and followed up by numerous workshops, team-building sessions and external innovative plant visits by members of the plant.

The next area mentioned was that of design technology. The methodologies of socio-technical systems design, open systems planning, general system theory and transition planning were all extremely useful in the overall plant organisation design.

Somehow, the geographical separation of 700 miles from corporate headquarters and other departments in the division was mentioned as helping the new plant pull together as a team internally and utilise the strength and expertise that existed in the plant.

Other areas mentioned as important to the success of the plant include the felt need or pressure because of the business requirements to design and operate a high-quality plant in terms of product and working life. This felt need was no accident, but grew from an important early decision to initiate and to emphasise statements of mission and philosophy. The mission of the organisation, to "build high-quality dies (electrically good integrated circuits) at minimum economic and social cost", distinguished the plant as one that manufactures quality products as opposed to one that merely "moves material". This then led easily to identifiable goals for the plant. The statement of the mission, and the objective focus of the design team on a statement of philosophy about organisational values and benefits, together made a powerful and visible keystone upon which the organisation could operate.

Also mentioned as essential to the success of the organisation was the importance of communication, clarity of expectations and openness about the goals of the organisation. This candour, along with open office space, facilitated the openness and trust that bonded the organisation together. It seemed that communication was a key element that began breaking down traditional barriers between departments and helped to foster trust throughout the organisation.

Instead of the syndrome that exists elsewhere in the semiconductor industry, where one individual stands out as taking all the risk and consequences of a decision, the system in the new plant emphasises the importance of consensus decision-making and good technical resources, where expertise is pooled together to help solve challenges the organisation may face. Consensus risk-taking, where key parts of the organisation which were knowledgeable were brought into taking the necessary risks, required a high-performance organisation.

"Ego death" was one of the key elements that most of the managers felt was an important factor which helped facilitate the growth and development of the organisation as a whole. This refers to a change from traditional management beliefs that decision-making, risk-taking and being recognised as the unique expert are the primary sources of management pride and satisfaction. When managers foster skill development, information flow, decision-making and risk-taking at lower levels in the organisation, they learn that devolution derives a team reward in which they share and from which their satisfaction is enhanced.

Some other areas mentioned as responsible for success were the importance of a good selection process, orientation and assimilation process which are managed by the teams with whom a potential organisation member would work. The teams are involved completely in the interview and selection process, and with the orientation and planned assimilation programme, along with their new members. Although a large available labour pool has offered selection ratios as high as one technician selected for every 40 applicants, the selection criteria have remained simple - good vision, an interest in other people and a willingness to try something new. Flexibility, adaptability, patience and the ability to deal with ambiguity are all important characteristics that the organisation has found to be essential.

Three areas where much more work will be needed in future expansions at the site (and which are felt to be necessary for high performance) are manufacturing, equipment and plant layout prior to construction of the total building, social preparation (to prevent culture shock), especially when moving from the San Francisco bay area, and integration and clarification of expectations with corporate and other external company members upon whom plant employees depend, and who depend upon and use the plant's products and resources.

The new plant operation has been a model for making some changes in the Cupertino manufacturing operation that have resulted in lower costs. The company is also in the process of implementing a socio-technical approach in several non-manufacturing departments in Cupertino. Since no two designs are alike and a redesign is more difficult than a design from start-up, the current projects bear little resemblance to the start-up described in this case study. It was emphasised by the company President that the Nampa design could not be imposed on other plants, simply because it had worked in that particular case. Ownership could only be created if people believed in what they were doing, and were able to take the initiative themselves. Nevertheless, since the Cupertino plant wanted to implement a redesign, the company supported that move. Since the design process has started, both the President and the Nampa plant manager agreed that the resulting organisation for the Cupertino plant will probably be quite different from the Nampa structure because of the differences in the needs of the technical systems, the people and the external environment.

Notes

[1] United States Department of Labor, Bureau of Labor Statistics: "Current Labor Statistics", in Monthly Labor Review, Nov. 1981, pp. 61-99.

[2] Howard Fullerton, Jr.: "The 1995 labor force: A first look", ibid., Dec. 1980, pp. 11-21.

[3] David North and Phillip Martin: "Immigration and naturalisation: A need for policy co-ordination", ibid., Oct. 1980, p. 47.

[4] Leon Bouvier: "Immigration at the crossroads", in American Demographics, Oct. 1981, p. 17.

[5] Joseph P. Goldberg: "The law and practice of collective bargaining", in Federal policies and worker status since the thirties, Industrial Relations Research Association Series (Wisconsin, Madison, 1976), pp. 13-42; also Benjamin Aaron: "Industrial relations law", in Lloyd Ulman (ed.): Challenges to collective bargaining, American Assembly (Englewood Cliffs, Prentice Hall, 1967), pp. 113-133.

[6] Thomas Kochan: Collective bargaining and industrial relations (Homewood, Richard D. Irwin, 1980), Ch. 3; and Theodore J. St. Antoine: "The role of law", in United States industrial relations, 1950-80: A critical assessment, Industrial Relations Research Association Series (Wisconsin, Madison, 1981), pp. 159-198.

[7] United States Department of Labor, Bureau of Labor Statistics, 1981.

[8] United States, National Labor Relations Board: Forty-Fourth and Forty-Fifth Annual Reports of the National Labor Relations Board (Washington, DC, 1979 and 1980).

[9] Daniel J.B. Mitchell: "The employment relationship in the 1980s", Working Paper No. 30 (Los Angeles, UCLA, Institute of Industrial Relations, 1981).

[10] Fred K. Foulkes: "Large non-unionised employers", in United States Industrial Relations, 1950-80, op. cit., pp. 129-158.

[11] Kochan, op. cit., p. 22.

[12] Doris B. McLaughlin and Christine E. Miller: The impact of labor unions on the rate and direction of technological innovation (Ann Arbor, University of Michigan and Iowa State University, Institute of Labor and Industrial Relations, 1979).

[13] United States Department of Labor, Bureau of Labor Statistics: Characteristics of major collective bargaining agreements, Bulletin 2065 (Washington, DC, United States Government Printing Office, 1978); Kevin Murphy: Technical change clauses in collective bargaining agreements (Washington, DC, American Federation of Labor-Congress of Industrial Organisations - AFL-CIO, Department of Professional Employees, 1981); Bureau of National Affairs: Basic patterns in union contracts (Washington, DC, 1979); AFL-CIO, Industrial Union Department: Comparative survey of major collective bargaining agreements, manufacturing and non-manufacturing (Washington, DC, 1979).

[14] Murphy, op. cit., p. 2.

[15] Bureau of National Affairs, op. cit.

[16] AFL-CIO, Industrial Union Department, op. cit., pp. 154-155.

[17] United Automobile Workers and Ford Motor Company: Letters of understanding covering agreements dated October 4 1979, pp. 38-41.

[18] International Union of Electrical Workers (IUE): Resolution Number 61, 19th Annual Constitutional Convention, Detroit, Michigan, Sep. 1980.

[19] The Machinist, Aug. 1981, p. 10; also William Winpisinger, International President, International Association of Machinists: Skilled manpower and the rebuilding of America, written testimony presented to the Subcommittee on Economic Stabilisation of the Committee on Banking, Finance and Urban Affairs, House of Representatives, Washington, DC, July 1981.

[20] The Machinist, op. cit.

[21] Communications Workers of America (CWA) and American Telephone and Telegraph National Joint Committee on Working Conditions and Service Quality Improvement: Statement of principles on QWL, Apr. 1981; also Ronnie J. Straw (CWA): The development of worker participation in the communications industry, presentation to International Conference on QWL and the 80s, Toronto, Canada, Aug.-Sep. 1981.

[22] For cumulative reports and discussion of this literature, see "Automation, productivity and industrial relations", in Annual Proceedings, Industrial Relations Research Association, 1954, pp. 114-150; "Crucial problems posed by automation", in Annual Proceedings, Industrial Relations Research Association, 1958, pp. 20-75; Morris Phillipson (ed.): Automation - Implications for the future (New York, Vintage, 1962); John T. Dunlop (ed.): Automation and technical change, American Assembly (Englewood Cliffs, Prentice Hall, 1962); United States Department of Labor, Bureau of Labor Statistics, Impact of automation, Bulletin 1287 (Washington, DC, United States Government Printing Office, 1960); Gerald Somers et al. (eds.): Adjusting to technological change, Industrial Relations Research Association (New York, Harper and Row, 1963); United States Senate Committee on Labor and Public Welfare: Exploring the dimensions of the manpower revolution, Vol. 1 of Selected readings in employment and manpower (Washington, DC, 1964); National Commission on Technology, Automation and Economic Progress: Technology and the American Economy, Vol. 1 (Washington, DC, United States Government Printing Office, 1966); Rudolph Oswald: Adjusting to automation, AFL-CIO Publication No. 144 (Washington, DC, AFL-CIO, 1969). For annotated bibliographies of this period, see Einar Hardin and Associates: Economic and social implications of automation, Vol. 2 (1957-60) and Vol. 3 (1961-65) (East Lansing, Michigan State University, Labor and Industrial Relations Center).

[23] Norbert Wiener: The human use of human beings (Boston, Houghton Mifflin, 1950), p. 189.

[24] Charles Silberman: "The real news about automation", in Fortune, Jan. 1965, pp. 125-126, quoted in George Tarborgh: The automation hysteria (Washington, DC, Machinery and Allied Products Institute and Council for Technological Advancement, 1965), pp. 31 and 96.

[25] Garth Magnum: "Manpower research and manpower policy", in Industrial Relations Research Association: A review of industrial relations research, Vol. 2 (Madison, Wisconsin, 1971), p. 81.

[26] United States Department of Health, Education and Welfare Task Force: Work in America (Cambridge, MIT Press, 1973).

[27] Harold L. Sheppard and Neal Q. Herrick: Where have all the robots gone?: Worker dissatisfaction in the 70's (New York, Free Press, 1972).

[28] Now the American Center for Quality of Work Life and no longer associated with the University of Michigan.

[29] National Commission on Productivity and Work Quality: Fourth Annual Report (Washington, DC, 1975), p. 36.

[30] National Center for Productivity and Quality of Working Life: Annual Report to the President and Congress (Washington, DC, 1976).

[31] United States General Accounting Office: Stronger federal effort needed to foster private sector productivity, Report to Congress, 18 Feb. 1981 (AFMD-81-29), p. 6.

[32] Charles G. Burck: "Working smarter", in Fortune, 15 June 1981, pp. 72-73.

[33] Wickham Skinner: "The impact of changing technology on the working environment", in Clark Kerr and Jerome M. Rosow (eds.): Work in America: The decade ahead (New York, Van Nostrand Reinhold/Work in America Institute Series, 1979), p. 229.

[34] Robotics Institute of America, quoted in Eli Lustgarten et al.: Robotics and its relationship to the automated factory, paper presented at Robotics Workshop, Office of Technology Assessment, United States Congress, Washington, DC, July 1981, p. 5.

[35] James Albus, National Bureau of Standards: Industrial robots and productivity improvement, paper presented at Robotics Workshop, op. cit., pp. 13-16; also "The speed-up in automation", Business Week, 3 Aug. 1981, p. 62.

[36] Dennis Wisnosky: "Using computers to improve manufacturing productivity", in Defense Management Journal, Apr. 1977, pp. 41-50; and United States Air Force: ICAM program prospectus, Air Force Materials Laboratory, Air Force Wright Aeronautical Laboratories, Air Force Systems Command, Wright-Patterson Air Force Base (Ohio, 1979).

[37] Ray Wild (ed.): Management and productivity readings (New York, Penguin, 1981), pp. 290-291, quoted in Daniel S. Appleton: "A strategy for manufacturing automation", in Datamation, Oct. 1977, pp. 64-70.

[38] Louis Davis and James Taylor: Design of jobs (Santa Monica, Goodyear, 2nd ed., 1979), p. x-xvii.

[39] Robert Boguslaw: The new utopians: A study of system design and social change (Englewood Cliffs, Prentice Hall, 1965), excerpt reprinted in Davis and Taylor, op. cit., p. 16.

[40] Reginald Newell, International Association of Machinists: Labor's view of automation and productivity, Society of Manufacturing Engineers, Technical Paper Series AD 75-766 (Michigan, Dearborn, 1975).

[41] Louis Davis and Albert Cherns (eds.): Quality of working life, Vol. 2 (New York, Free Press, 1975); also J. Richard Hackman and J. Lloyd Suttle: Improving life at work (Santa Monica, Goodyear, 1977); and Richard Walton: "Work innovations in the United States ", in Harvard Business Review, July-Aug. 1979, pp. 88-98.

[42] Honeywell Systems and Research Center: Human factors affecting ICAM implementation, Final Report, United States Air Force Materials Laboratory, Wright-Patterson Air Force Base (Ohio, 1981).

[43] National Research Council, Assembly of Engineering: Committee on Computer-Aided Manufacturing, Annual Report 1979 (Washington, DC, National Academy of Sciences, 1980).

[44] Westinghouse Research and Development Center: Report of Committee to Define Worker-Related Research for Adaptable-Programmable Assembly Systems, prepared for the National Science Foundation (Pittsburgh, 1980), pp. 37 and 53.

[45] United States General Accounting Office: Follow-up on use of numerically controlled equipment to improve defense plant productivity, Report to Congress, 17 Jan. 1979 (LCD-78-427), p. 26.

[46] MIT, Center for Policy Alternatives: Integrated Computer-Aided Manufacturing: Social and economic impacts (Cambridge, Mass., 1979), pp. 50-72.

[47] Donald Gerwin and Thomas Leung: "The organisational impacts of flexible manufacturing systems: Some initial findings", in Human Systems Management, Vol. 1, 1980, pp. 237-246.

[48] Donald Gerwin: The do's and dont's of computer-integrated manufacturing systems (Milwaukee, University of Wisconsin, School of Business Administration, 1981); also Donald Gerwin and Jean-Claude Tarondeau: Uncertainty and the innovation process for computer-integrated manufacturing systems: Four case studies (Milwaukee, University of Wisconsin, School of Business Administration); Trondheim, Norwegian Technical

University, Institute for Social Research in Industry; and Cerq, France, ESSEC, 1981).

[49] United States General Accounting Office: Follow-up on use of ... productivity, op. cit.

[50] United States General Accounting Office: Manufacturing technology - A changing challenge to improved productivity, Report to Congress, June 1976 (USG-AO), pp. 44-49.

[51] "The speed-up in automation", in Business Week, 3 Aug. 1981, pp. 58-67.

[52] David Noble: "Social choice in machine design", in Andrew Zimbalast (ed.): Case studies on the labor process (New York, Monthly Review Press, 1979), p. 39.

[53] Melvin Blumberg and Donald Gerwin: Coping with advanced manufacturing technology, paper presented at CORS-TIMS-ORSA Joint National Meeting, Toronto, Canada, May 1981 (Milwaukee, University of Wisconsin, School of Business Administration), p. 22.

[54] Personal communication; source confidential.

[55] Robert Blauner: Alienation and freedom: The factory worker and his industry (Chicago, University of Chicago Press, 1964), p. 167.

[56] R.J. Hazelhurst et al.: "A comparison of the skills of machinists on numerically controlled machines", in Ocupational Psychology, Vol. 43, Nos. 3 and 4, 1969, pp. 169-182.

[57] Joel Fadem: "Fitting computer-aided technology to workplace requirements: An example", in NC/CAM: The new industrial revolution, Proceedings of the 13th Annual Meeting and Technical Conference of the Numerical Control Society, Cincinnati, Mar. 1976, p. 9.

[58] ibid., p. 10.

[59] ibid.

[60] Melvin Blumberg and Antone Alber: The human element: Its impact on the productivity of advanced batch manufacturing systems (Milwaukee, University of Wisconsin, School of Business Administration, and Peoria, Bradley University, College of Business, 1981), pp. 1-20.

[61] ibid., p. 15.

[62] George England et al.: "Automation and the American automobile worker: Routes to humanised productivity", in Jan Forslin et al. (eds.): Automation and industrial workers: A fifteen-nation study, Vol. 1, Part 1 (Oxford, Pergamon Press, 1979), pp. 65-94.

[63] ibid., p. 88.

[64] Phillip and Betty Jacob: "Life on the automated", in The Wharton Magazine (Philadelphia), Fall 1976, p. 59.

[65] Irving Bluestone: "Comments to 'Life on the automated' by the UAW", ibid., p. 60.

[66] Personal communication, Delmar Landen, Director of Organization Development, General Motors Corporation, Aug. 1981; see also Thomas Hayes: "At GM's Buick unit, workers and bosses get ahead by getting along", in New York Times, 5 July 1981, pp. F4-5; and Irving Bluestone: "Quality of work life: Its status and future", in Proceedings of the 1980 Executive Conference on QWL, Apr. 1980, pp. 21-35.

[67] Further understandings developed between the major industry union, the UAW, and the corporation led to the eventual suspension of GM's active anti-union programme in the Southern and Sunbelt States.

[68] Delmar Landen: "Comments to 'Life on the automated' by GM", in Jacob, op. cit., p. 60.

[69] Carl Boyer, Director of Organization Development, and Lee Crawford, Plant Manager, Warren, Ohio Operations: Case study of Packard Electric, presentation to Tenth Annual Residential Course on improving the quality of working life and organisational effectiveness, UCLA, Institute of Industrial Relations, Center for Quality of Working Life, Sept. 1981.

[70] See note 67.

[71] Lawrence B. Evans: "Impact of the electronics revolution on industrial process control", in Science, Vol. 195, 18 Mar. 1977, quoted in The impact of chip technology on employment and the labour market (London, Metra Consulting Group Ltd., 1980), p. 258.

[72] For a summary, see United States General Accounting Office: "Three Mile Island - The most studied nuclear accident in history", in Report to Congress, 9 Sep. 1980, pp. 56-62.

[73] United States, President's Commission: Report on the accident at Three Mile Island - The need for change: The legacy of Three Mile Island (Washington, DC, 1979), p. 27.

[74] ibid., p. 8.

[75] ibid., p. 10.

[76] Ronald Eytchison: "Report of the technical assessment task force: Selection training, qualifications, and licensing of TMI reactor operating personnel", in Staff Reports to the President's Commission on the accident at Three Mile Island (Washington, DC, 1979), pp. 36-37.

[77] United States, President's Commission: Report on the accident at Three Mile Island, op. cit., p. 22.

[78] The operators had attempted to offset poor instrumentation design by adding their own explanatory labels and colour coding to switches and levers. This included "shape coding" in the form of tap handles from two popular brands of beer to control the motion of fuel rods inside the reactor. See United States General Accounting Office: "Three Mile Island ... history", op. cit., p. 28.

[79] United States, President's Commission: Report on the accident at Three Mile Island, op. cit., p. 23.

[80] ibid., p. 20.

[81] Richard Walton: "Establishing and maintaining high commitment work systems", in John Kimberly et al.: The organisational life cycle (San Francisco, Jossey-Bass, 1980), pp. 208-290; Edward Lawler: "The new plant revolution", in Organizational Dynamics, Winter 1978, pp. 3-12; Ernesto Poza and M. Lynne Markus: "Success story; the team approach to work restructuring", in Organizational Dynamics, Winter 1980, pp. 3-25; Carl Bramlette, Jr. et al.: "Designing for organisational effectiveness", in Atlanta Economic Review, Part 1, Sep.-Oct. 1977, pp. 35-41 and Part 2, Nov.-Dec. 1977, pp. 10-15; Robert Keidell: Quality of work life in the private sector: An overview and developmental perspective (Washington, DC, United States Office of Personnel Management, Office of Productivity Programs, 1980).

[82] Louis Davis: "Optimising organisation-plant design", in Organizational Dynamics, Autumn 1979, pp. 3-15; see also "Organisation design", in Gavriel Salvendy (ed.): Handbook of industrial engineering (New York, Wiley-Interscience, 1982).

[83] Davis: "Optimising organisation-plant design", op. cit., p. 9.

[84] ibid., p. 4.

[85] See Louis Davis and Charles Sullivan: "A labour-management contract and quality of working life", in Journal of Occupational Behaviour, Vol. 1, No. 1, 1980, pp. 29-41; Roy LaBerge: "The Shell chemical plant at Sarnia - An example of union-management collaboration", in Quality of Working Life - The Canadian scene, Spring 1979, pp. 1-4; Neil Reimer: "Oil, Chemical and Atomic Workers International Union and QWL - A union perspective", in Quality of Working Life - The Canadian scene, Winter 1979, pp. 5-7.

[86] Davis and Sullivan, op. cit., p. 34.

[87] Agreement between Shell Canada Ltd. (Sarnia chemical plant) and the Oil, Chemical and Atomic Workers International Union Local 9-848, Jan. 1978, p. 2.

[88] Davis: "Optimising organisation-plant design", op. cit., pp. 11-12.

[89] Personal communication; source confidential.

[90] ibid.

[91] ibid.

[92] ibid.

[93] Oliver Sheldon: "Taylor the creative leader: An analysis of Taylor's contribution to the problems of human welfare", in Bulletin of the Taylor Society: Critical essays on scientific management, Vol. 10, No. 1, Feb. 1925, p. 77.

[94] Quoted in above, pp. 76-77.

[95] Duncan Gallie: In search of the new working class: Automation and social integration within the capitalist enterprises, Cambridge Studies in Sociology No. 9 (Cambridge, Cambridge University Press, 1978).

[96] ibid., p. 35.

[97] L.E. Davis and J.C. Taylor: "Technology, organisation and job structure", in R. Dubin (ed.): Handbook of work, organisation and society (Chicago, Rand McNally, 1976).

[98] idem: "Design of jobs overview", in L.E. Davis and J.C. Taylor (eds.): Design of jobs, op. cit.

[99] T.G. Cummings: "Sociotechnical experimentation: A review of sixteen studies", in W.A. Pasmore and J.J. Sherwood (eds.): Sociotechnical systems: A source book (La Jolla, University Associates, 1978); W.A. Pasmore: "The comparative impacts of socio-technical system job-redesign, and survey-feedback interventions", ibid.; and S. Srivastva et al.: Job satisfaction and productivity (Cleveland, Department of Organizational Behaviour, Case Western Reserve University, 1975).

[100] L.E. Davis and C.S. Sullivan: "A labour-management contract and quality of working life", in Journal of Occupational Behaviour, Jan. 1980, pp. 29-42; F.E. Emery: "Designing socio-technical systems for 'greenfield' sites", ibid., pp. 19-28.

[101] E.J. Miller: "Socio-technical systems in weaving, 1953-70: A follow-up study", in Human Relations, Vol. 28, pp. 349-386.

CHAPTER XVII

INTRODUCING NEW EQUIPMENT AND CHANGING WORKING LIFE IN YUGOSLAVIA

(by Valentin Jež)

1. Introduction

The introduction of new equipment does not imply merely a change in the technical and technological components of production, but also has much broader implications. In particular, those factors affecting human beings inside and outside the plant where a change is taking place are often neglected. Some years ago it was believed that there was no technological choice and that all the benefits and drawbacks of a technological change had to be taken together (i.e. exclusive technological determinism). However, many experiences in changing the design of technological processes have weakened such beliefs.

Technology influences the working life of people, but these influences can be modified in advance of and during their work. Planning new equipment is the first stage at which people may become involved and the most important influence on future interactions between people and technology. However, many factors play an important role in reducing their power to modify this interaction.

This chapter, which describes and analyses the long-term introduction of new equipment in a steel plant in Yugoslavia, reveals some factors preventing the design of jobs with a view to improving QWL. During the period of implementation, many conditions changed. Profound changes occurred in the economic situation of the enterprise and in the country as a whole, as well as in the social and political system, which had a direct or indirect impact on the technological change and on the QWL of people within the plant concerned.

2. The national context

2.1 Socio-economic conditions

During the last three decades, the rate of industrial growth in Yugoslavia has been among the highest in Europe. Over the last ten years, the GNP of the country has been mainly determined by industrial production, whose share in GNP rose from 21.7 per cent in 1952 to 37.9 per cent in 1977. The share of the different republics (provinces) in national industrial production also showed important changes, as did their proportions of industrial production in the GNP of the republic concerned (table 92).

Substantial developments may also be seen in GNP per head over the past 25 years. A considerable increase was registered in all regions of Yugoslavia, the highest rates being in Slovenia (494 per cent), Vojvodina (452 per cent) and Croatia (428 per cent) and the lowest in Kosovo (243 per cent), Bosnia-Herzegovina (293 per cent) and Montenegro (369 per cent).

Table 92: Yugoslavia: proportion of industrial production in GNP and share in national industrial production by republic (province), 1952-77 (percentages)

Republic	1952		1977	
	Proportion of industrial production in GNP	Share in national industrial production	Proportion of industrial production in GNP	Share in national industrial production
Bosnia-Herzegovina	17.6	12.2	41.3	13.3
Montenegro	11.2	1.1	30.4	1.6
Croatia	23.4	28.7	35.3	24.3
Macedonia	18.0	4.3	40.8	6.1
Slovenia	33.9	23.9	45.5	19.8
Serbia	16.4	19.0	35.7	23.3
Kosovo	23.3	2.2	40.4	2.2
Vojvodina	21.7	8.5	33.2	9.4
Total		100.0		100.0

Source: Statistical Yearbook of the Socialist Federal Republic of Yugoslavia (SFRY) (Belgrade), for the years concerned.

These changes in GNP are due to the development partly of industry, as well as of other divisions of the economy. However, one may note the association with the growth of the population in various regions of the country. The highest rate of population growth exists in Kosovo, where it grew from 1.0 in 1957 to 27.6 in 1977 per 1,000 inhabitants. In all other regions a strong decrease in the rate of population growth was registered during that period, especially in Vojvodina where the rate was 21.6 in 1957 and only 3.7 in 1977 per thousand inhabitants.

According to the historical features of industrial development, the labour market in Yugoslavia varies considerably according to region. The proportion of population employed in agriculture for the country as a whole was 38.2 per cent in 1971, as against 67.2 per cent in 1948, and all republics registered decreases of between 25 and 40 per cent over the period. Figures show that Slovenia has the lowest and Kosovo the highest proportion of the population in agriculture (20.4 per cent and 51.5 per cent respectively in 1971). However, the highest migrations from primary to secondary economic sectors over the last 20 years took place in Montenegro and Bosnia-Herzegovina.

Almost parallel with the proportion of the population in agriculture are the figures for the education of the inhabitants in those regions.

The highest proportion of the population over 10 years of age with less than primary education is found in Kosovo and Bosnia-Herzegovina (76.9 per cent), and the lowest in Slovenia (37 per cent), according to 1971 census data.

Of the total workforce of Yugoslavia, unskilled and semi-skilled workers comprise 32 per cent and skilled workers about 43 per cent. Approximately 36 per cent of the workforce are under 30 years of age, 40 per cent between 30 and 45, and the rest between 46 and 60 years of age. Women represent about 36 per cent of the total, their activity rate ranging from 43.4 per cent in Slovenia to 20 per cent in Kosovo.

The high percentage of agriculture prevailing in the economy, the educational structure of the population and the rate of population growth, are the major reasons for the level of unemployment in some regions of Yugoslavia. This situation has resulted in continuous migration within the country and emigration to the more developed countries of Europe over the last two decades. Data on the number of people seeking work bear out this point.[1]

With regard to the educational system, education is compulsory between 7 and 15 years of age. After primary school, secondary schooling comprises various occupational schools, offering work-related training lasting from two to four years, and grammar schools which prepare students for higher education (colleges and universities). There is also a well-developed system of education for workers, the so-called "people's university", which offers all stages in the educational system to mature students.

As an illustration of the expansion of education at all levels over the past 30 years, the number of children at primary schools increased from 29,000 in 1951/52 to 323,700 in 1976/77 and at secondary schools from 53,400 to 202,400. Over the same period, the number of students at universities increased from 5,500 to 26,000. It is clear that this rise in educational levels has had far-reaching effects.

2.2 The industrial relations system

The industrial relations system in Yugoslavia is based on self-management, which represents not only an economic system within the economy but also the overall organisation of society. We do not intend to describe the whole system of self-management but merely to present those aspects of it which are relevant to industrial relations and to the case study under consideration.

The fundamentals of the industrial relations system, as set out in the Federal Constitution, are based on the power of the working class and of all working people, and on relations among people as free and equal producers. These relationships are the basis of various rights in society, among them the right of self-management. Within the self-management system every worker, on an equal footing with other working people, decides on his or her own work and on the conditions and results of that work, on his or her own common interests, and on the progress of social development and other social affairs.

Before the 1974 Constitution, the self-management system in organisations was more centralised. However, the degree of centralisation depended on the complexity of the organisation. In simpler working organisations there was one workers' council with its committees and board of management. This workers' council had the same power as that of more complex organisations. In complex organisations each unit had its own workers' council, but the central workers' council of the organisation had the supreme authority.

The members of unit workers' councils in complex organisations were elected directly by workers every two years. These councils had the power to decide on the disposal of incomes (i.e. division between investment funds, personal incomes, set differentials, etc.) on the basis of internal prices at different stages of the production process. In general, they were able to decide on matters affecting only the workers in the unit. Decisions affecting other units and the organisations as a whole were a matter for the central workers' council.

The members of central workers' councils (i.e. the council of the working organisation) were elected for four years by the workers' councils of all units within the working organisation. These councils established production plans, adopted development programmes, fixed prices, elected the management board and appointed the manager of the working organisation. The main functions of the management board were to implement the decisions and policies of the workers' council and to supervise all technical services. The manager was an ex officio member of the management board, whose function was to manage the working organisation in accordance with the decisions of the workers' councils, board of management and the law.

The central workers' councils also elected special committees as their executive bodies. The functions of these committees were to prepare documents and proposals relating to the social and economic life of the organisation, in order to make the decision-making process more effective.

Since the adoption of the 1974 Constitution, the organisation of self-management has been less centralised. In complex organisations, some units which fulfilled the underlying principles of autonomy were organised into so-called basic organisations of associated labour which, as the smallest socio-economic units in society, represent the basis of the whole economic system of the country. These units are grouped into larger ones known as working organisations, and these again into broader, composite organisations. Mutual rights, obligations and responsibilities stemming from the above-mentioned pooling of labour and resources are regulated by workers in basic organisations of associated labour through self-management agreements, which ensure their constitutional rights.

The basic organisations of associated labour are the organs whereby the workers directly realise their socio-economic and other rights and decide on other related questions through workers' assemblies, referenda and other forms of personal expression. Workers' assemblies are the supreme self-management body of the basic organisations of associated labour. They adopt the main policies of an organisation, decide on plans and consider suggestions of the workers' council and other executive bodies.

The workers' council is the executive body of a basic organisation of associated labour, consisting of workers' delegates in all areas of the work process, elected by workers in that organisation. The composition of the workers' council must correspond to the social composition of the workers belonging to the basic organisation. However, basic organisations with a small number of employees do not set up workers' councils, but workers make decisions exclusively through the assembly.

The workers' council of a working organisation or a composite organisation is made up of delegates of the workers in the basic organisations of associated labour directly elected by the method specified by the self-management agreement on association. Each basic organisation must be represented in the workers' council of the working organisation. The delegates work in accordance with guide-lines issued by workers or the workers' council of the basic organisation of associated labour which elected them, and to which they are responsible. The rights and obligations of delegates are laid down by the self-management agreements on association. Members of a workers' council or their committees are elected for a term not exceeding two years, and cannot serve for more than two consecutive terms.

Besides the basic rights guaranteed by the Constitution, industrial relations are regulated by the law on associated labour and other labour legislation. Also very important are the self-management agreements mentioned above which, along with social compacts, serve to regulate the complex socio-economic structure within the domain of labour.

2.3 Trade unions

The trade unions in Yugoslavia underwent many changes in organisation over the period 1955-80. During this time the number employed in the public sector increased considerably, swelling the trade union membership. At present each republic (and province) has its own trade union association, and all these together form the Confederation of Trade Unions of Yugoslavia.

It was originally the trade unions who suggested the system of self-management in 1950. Nowadays, their main functions are to -

- realise the constitutionally defined status of the working class;
- achieve self-management in all fields of work and life;
- educate workers in self-management and social participation;
- ensure democratic choice and election of delegates;
- ensure the broadest possible participation of workers in the exercise of power and management of social affairs;
- protect workers' rights;

- ensure their social welfare and standard of living; and develop and strengthen solidarity and the raising of class consciousness and responsibility among self-managing groups.

Trade unions also initiate self-management agreements and social compacts and take a direct part in their negotiation. They submit proposals to the managing bodies of self-managing organisations and communities, the assemblies of the socio-political communities, and other state and social agencies concerning questions relating to the economic and social position of workers. Moreover, trade unions are an important force in the system of workers' assemblies, and participate in it in various ways. Within the working organisations, the trade union comprises an integral part of self-management and plays a political role in order to safeguard the interests of its members. The self-managing bodies, on the other hand, are bound to consult the trade union organisation before taking any decision.

Each basic organisation of associated labour has, beside the self-management agreements and social compacts, its own internal rules regulating its structure as well as specifying workers' rights and duties. Social compacts are negotiated and concluded by the major social partners, whereas self-management agreements are of two kinds. The first is a result of negotiations between various basic organisations and is signed directly by their representatives and by the appropriate trade union organisation. The second is discussed in workers' assemblies and adopted by referenda. This kind of agreement regulates the relations within the basic organisation or working organisation. The organisational rules should be adopted by workers' assembly, too.

2.4 The development of automation

In Yugoslavia, the first technological steps towards automation were taken before the sixties. Among the most advanced enterprises, various elements for electronic control devices, as well as complete control systems, were produced and implemented in some phases of production processes within industries. Many technical institutes also began to study the possibilities of introducing automated control devices into factories at that time.

In 1955 the Federal Committee for Electronics, Telecommunications, Automation and Nuclear Technology was founded. The Yugoslav Seminar for Measuring and Regulatory Technology in Zagreb started its work the same year. Some five years later, the Federal Commission for Automation began to co-ordinate the production of automated equipment and its introduction into enterprises. Oil refineries, power stations and chemical industries were the first to use automatic equipment. Within the motor vehicle industry, a type of "Detroit automation" was installed. In other industries, too (paper, textiles, metallurgy, etc.), various systems of automated control were partially applied.

At the universities, scientific work on advanced technologies and their implementation in some branches of Yugoslav industries was carried out over a period of years. However, social sciences lagged far behind the technical aspects in understanding the social consequences of automation. In spite of this, some theoretical work has been carried out in this field.[2] The approaches taken drew mainly on the experience and theory of more developed countries. Automation and its social consequences were discussed as a stage towards technical and economic development, with the stress on the benefits rather than the disadvantages. The position of Yugoslavia on a linear scale of technological progress was described.

It is extremely difficult to make any comparison of the extent of automation in Yugoslavia in relation to other countries, as official statistics do not produce any comparable evidence. The only figures that exist are those on the structure of means of production according to their cost price in Yugoslav industry. On the basis of these and other data, especially those on labour force structure, some estimates on the level of automation in Yugoslavia were made.[3,4] According to them, the level of automation in the country seems to be approximately 10-15 years behind the most developed countries in Europe. However, these estimates are unreliable. Some exceptions are apparent, within industry, banking, insurance, hospitals, etc. In 1980, the first robots made in Yugoslavia were demonstrated.

Some authors expressed critical reservations concerning automation in market economy countries. According to them, the negative aspects of some social consequences were explained by the rigid application of scientific and technical knowledge in technological applications. The main criticism was that these constraints were due to production relations and to the profit motive governing economic behaviour, and therefore modifying the extent and methods of technological innovation. It was considered that technological development, and automation as a stage within it, formed the basis for developing a new kind of production relations, and was a precondition of a socialist society. It was hoped that such a society would be able to undertake the successful introduction of automation.[5]

In 1967 the first report on an empirical investigation in this field was reported.[6] Five years later, extensive research was designed on the impact of automation on industrial workers, and completed in 1978.[7] These studies, together with some new aspects emerging from a round table discussion held in 1979 at Novi Sad, led the present author to undertake a new study on the interaction between technology, organisation and self-management.[8] In the meantime, many theoretical discussions were held on the social consequences of automation, most of them dealing with its relations with self-management.

Overall, it is evident from empirical research and theoretical studies that -

- technology is not ideologically natural;

- along with technologies, underdeveloped countries import some elements that might conflict with their accepted ideologies;

- technology has a certain level of deterministic influence on the organisational structure of the plant;

- technology structures the workforce, not always in a positive sense;

- technology also has negative implications for workers' attitudes;

- there are some technological factors constraining workers' participation in decision-making;

- technology influences, directly or through its determinants of organisational structure, QWL and the quality of life in general, both in positive and negative directions.

Many articles on automation and its positive and negative social effects have appeared in the press in the last few years. However, there is still evidence that the average layman has a positive attitude to automation.

Within the official policies of federal and republic governments, there appears to be no critical view of advanced technology. However, the trade unions (the Confederation of Slovene Trade Unions) have shown considerable interest in this field, contributing a large proportion of financial support to research in automation and its social consequences during the last ten years. A resolution on technology was also accepted by the Congress of Slovene Trade Unions in 1980.

3. Case study - Introducing new technology in the steel mill of the Ravne steel works

3.1 The development of the enterprise

The geographical position, the layers of iron ore and extensive woods in the surroundings were probably the main reasons for iron smelting at Ravne even in Roman times. As early as 1620, two furnaces built not far from the present steel works were the forerunners of the future development of steel production. In 1870 steel casting started and, only ten years after, the first smelting furnace was built with a capacity of 5 tons, soon to be followed by a second, larger one. The annual rate of production grew to about 10,000 tons before the First World War, when stagnation set in. The solid finish, density, toughness and homogeneity, for which the Ravne steel works was renowned, was disappearing, with a reduction in high-quality steel production, and the plant was lagging behind in technological progress. By 1939 the rate of production had dropped to only 7,500 tons.

During the first years after the Liberation the Ravne steel works produced rolled, forged and cast steel of average quality. Soon after the Second World War, plants were enlarged and supplied with new equipment, modern technology was introduced and production started to grow. Again, geographical position and closeness to sources of energy directed the development of production; although the layers of iron ore in the vicinity had been exploited, the experience and knowledge of producing steel had accumulated through generations. The steel works again became the producer of high-quality constructional steel, tool steel and other special kinds of steel.

Technological change in the steel works began in the fifties and continued through different stages over the following decades. The process continues within the boundaries of the financial resources of the enterprise. From the technical point of view, the introduction of new equipment was very important. In 1950 the forging mill was erected, and two presses with 1,200 and 1,800 tons of pressure were installed. (Currently, many small presses and ten forging hammers produce about 26,000 tons of forged products per year). Continuous forging produces machine bars, tubes and various kinds of axles. Two years later the smelting plant and the foundry were modernised. First SM furnaces were built, and later modern electric-arc furnaces were implemented. The electrical slag remelting, as the first of its kind in the country, improved the production of high-quality steel. Two small and one large moulding machines replaced the traditional way of moulding, and four new machines were purchased some years later. Two sandslingers were also introduced into the production process. The new rolling mill with remote control started production in the early sixties. At the same time, diversification of final products was extended and modernised. Modern equipment and methods of quality control were introduced in plant laboratories.

The rate of production grew very fast (figure 36) and the quality of steel was improved. In 1950 the production schedule included 67 per cent of plain carbon steel grades, 12.5 per cent of quality and only 11.5 per cent of high-quality steel, but by 1978 the figures were 0.6 per cent of plain carbon, 16.3 per cent quality and 83.1 per cent high-quality steel.

Parallel to the production growth was the growth of employment (figure 36). Not only has the number of employees increased massively, but the educational structure of the labour force has greatly improved. For instance, in 1969 the proportion of skilled workers was 34.1 per cent, including 6.9 per cent with secondary, 0.7 per cent with college (two-year higher education) and 1.5 per cent with university education. By 1980, the proportion of skilled workers had grown to 46.3 per cent (8.6 per cent with secondary, 2.3 per cent with college and 1.8 per cent with university education). Some experts with higher degrees have also been employed in the last few years.

In 1950 the first workers' council was established, consisting of 69 members. Over 25 years, the number of workers participating in various kinds of self-management bodies reached 3,729. After 1974, when the new Constitution was adopted, the number of workers participating in self-management increased substantially.

Figure 36: Steel works: growth in steel production and in employment, 1945-80

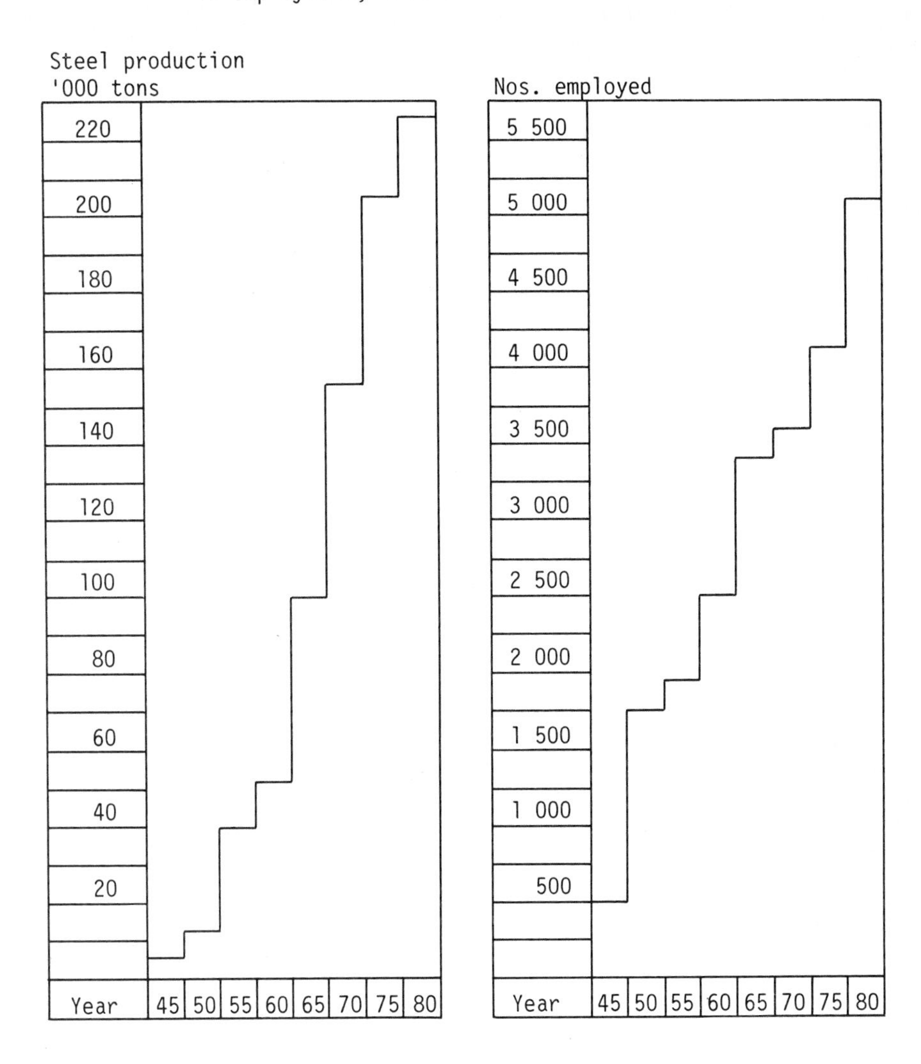

Parallel with the investments in new equipment, the community has been developing, and educational, health, social and leisure facilities have been established in the town as a result of the expansion of the steel works.

3.2 The old steel mill and the new decision

Our case study is that of the former manufacturing departments of the Ravne steel works. From 1975 it was one of the basic organisations of associated labour of the enterprise. It employs about 9 per cent of personnel of the working organisation and contributes 45 per cent of total output. As a basic organisation of associated labour it is relatively autonomous, with its own self-management bodies which control the socio-economic life within the system itself and its relations with the other 20 basic organisations, and with other working organisations. It has full control over investment, wages, prices and all allocations of income, as well as in setting internal policies. Like other basic organisations, it has genuine rights of merging or concluding business arrangements with other basic organisations in the working organisation or outside.

As an autonomous organisation, the unit has its own organisational objectives. However, they are incorporated into those of the working organisation and broader society. Since the unit was previously part of the centralised system of the Ravne steel works, the technological change described in this paper started in a different way from the way it would nowadays. Thus, the central planning system within the working organisation has had an impact on the change.

Over the last few years, the rolling mill unit made a considerable advance both from the technical and social point of view, as reflected in its scale of production, quality of products and QWL.

The technical equipment of the rolling mill was extremely antiquated when the change was initiated, having remained unchanged since shortly after the Second World War. The rolling stands and other equipment had been given as reparation for war damages, and brought from various countries, mainly from Romania. All machines were manually controlled and there was no continuous production process. The obsolete machines produced only 11,648 tons per annum of plain steel products in various shapes: round, square, flat and hexagonal bar steel, etc.

The ingots for rolling were small sized. After manual transportation, they were reheated in a pusher-type furnace. The ingots were then loaded onto a cart and transported to the rolling stand. The rollers in the stands were set manually for each shape, and fixed. First the ingots were rolled into billets, which were then cut by the hot cutter, and transported for reheating to the batch furnace. The reheated billets were rolled in the pass rolling mill: the pass roller workers inserted billets manually between the pass rollers with gripping-tons each time they passed through, until the billets reached the desired shape and size. The mills were driven by steam power. The work was physically exhausting and the working conditions extremely bad.

In the early fifties, the market demand for steel production world-wide, as well as in Yugoslavia, altered. Changes in the technology of producing steel resulted in many new and large-scale producers. Large plants with modern equipment and technology were built in Yugoslavia at that time. Their extensive production, appropriate geographical position, and nearness to sources of iron ore influenced the position of the Ravne steel works on the market. The plant was unable to maintain its competitiveness with the small amounts of plain carbon steel which could be produced with the out-of-date equipment. A demand for alloyed and high-alloyed steel, a type that was mainly imported, increased at the same time.

After considering these circumstances, and the economic, geographical and other possibilities of the plant, the management staff suggested new strategies in production and marketing to the workers' council.[9] The enlargement of production output through new investment and diversification of products were the general objectives approved for the coming period.

The bottleneck for large-scale production was the smelting shop. The old batch furnaces of small capacity, obsolete technology, physically strenuous manual work and extremely bad working conditions constituted obstacles to increasing the quantity of production. The lack of knowledge in producing alloyed steel also hindered the introduction of new products. There was a shortage of financial resources and technological experts to realise all the objectives over a short period. A decision was therefore taken to implement them step by step.

The change was made in the fifties, when the smelting shop was renewed. New furnaces were built and other changes made. However, because of the rapid increase in the production rate in the smelting shop, the other departments of the plant were not able to process all the steel produced. The processing capacities of the rolling mill, the forging shop and the foundry were small, and the equipment and technology too obsolete to improve them. For this reason, the plant was forced to sell ingots instead of higher processed products for some years.

The next step in realising the objectives set by the workers' council was investment in the rolling mill. The team which put this plan into operation consisted of several young engineers, who had recently completed their university degrees and who had helped to improve the production in the smelting shop, and some supervisors and foremen together with the most experienced workers from the existing rolling mill. During the discussion on the workers' council meeting, the following proposals of the team were analysed:

- production output of 50,000 tons of alloyed steel;
- production of round, square and flat steel bars of gauges between 8 and 100 mm;
- a broad assortment of steel.

The suggestions were accepted and the team was charged with elaborating a more detailed programme as a basis for investment. The workers' council also proposed that investment should first be made in those machines which were indispensable to producing the desired products. The decision was also made to instal a partly manual and partly remote-controlled mill in order to achieve greater flexibility of production.

At that time there was no R & D department in the plant. Since the above-mentioned team was not sufficiently experienced, they visited other plants and consulted experts there, before formulating the programme elements. The project for the rolling mill was prepared by a Swedish firm producing equipment for steel mills. Only the intermediate and light-section tracks were planned at this stage.

The workers did not participate in this planning stage, but were informed on the decisions taken by the central workers' council through different media. Nor were they informed about the type of new equipment or the possible social consequences of the new technology. However, they showed very little interest in the new project at that time, as the changes did not appear to affect their employment. On the other hand, they had expectations of amelioration of their working conditions, which were extremely poor, and hoped that their working life would also be improved in terms of physical and mental strain, better productivity and pay, etc. The difficulties that usually accompany every installation of new equipment were not realised by anyone within the plant.

3.3 Problems with the new equipment

The equipment was purchased from the same firm that designed the mill project. The first problems arose right at the beginning of installation. Six rolling stands had been foreseen in the project but the breadth of the hall permitted only five. Furthermore, the length of the building did not allow the optimal distance between the rolling mills. The need to change the dimensions of gauges arose, and many bottlenecks appeared because there were not enough experienced experts present at the planning and purchasing stage, and because the scarce financial resources prevented the purchase of continuous cooling equipment.

In addition, the instructions given by the equipment supplier were neither precise nor appropriate enough, having been taken from a similar rolling mill. The technological procedures referred exclusively to the rolling process, but the conditions are not always the same, and heating and controlled cooling are very important. Neither the experiences from former work in the old mill nor suggestions gained from visits to other plants were adequate. The furnaces of the new mill were different, and the input to rolling and dimensions of billets required another approach.

Personnel problems arose simultaneously with technological problems. The chief manager of the old mill, who was appointed as the head of the new mill, was unfamiliar with the new technology. The new personnel had much technological knowledge but they needed to acquire practical experience in a short time. The mill operators who came from the old mill were committed to innovation but unaccustomed to the speed of rolling; indeed, some of them refused to work at the new rolling mill after a few days. As a result, new operators had to be trained, although some of the old operators remained.

To rectify the situation described above, it was necessary for the unit to develop its own technological procedures and to restructure the team. Starting from the beginning, the restructured team began with the heating process, and tested each procedure in practice. The staff of experts, together with the workers from the old and new mills, were extremely committed at this stage of introducing the new technology and co-operated very closely. For example, in testing the production of rolling steel with 12 per cent of manganese, they stayed at work for two-and-a-half days continuously.

Because the intermediate and light-section rolling mills were placed close to each other, and therefore could not operate to full capacity, the original installations were changed. The lifting tables were improved and one automatic table was installed at the second rolling stand. At the light section track, by-passing loops were built. This increased the rate of production and the safety of workers at the same time.

In contrast to the troubles at the rolling mills, the smelting shop was able to produce high-quality steel almost from the beginning, and could melt all qualities of steel. The raw materials for the rolling mill have therefore never been a problem.

After all the problems had been rectified, the quantity and quality of production increased continuously, and after one year the production rate exceeded that planned. At the same time, selling prospects were extremely good, and mill production could be optimal, which would assure relatively low production costs and competitiveness on national and foreign markets.

Other installations were also improved later. An extensive reconstruction of the cleaning shop took place. The cleaning of billets had been carried out manually, with the result that the work was difficult, dangerous and dirty. Automated grinding machines were introduced step by step and the obsolete manually operated machines phased out. The inspection of billets, which was formerly visual, was reorganised and new equipment installed.

Among the other things, a new roller straightener and other equipment was installed in the dressing shop. Instead of the traditional method, continuous horizontal tunnel furnaces were built and new technological procedures for improving the quality of steel were implemented.

The next step in the process of introducing new equipment in the steel works was that of installing the blooming track (i.e. a rolling mill for making blooms from ingots). The need for this arose after 1963, when the rolling mill started to produce. A blooming track with reheating furnaces was designed on the basis of the experiences of other steel producers, while the producers of the equipment in Czechoslovakia carried out the detailed design. Soaking pits with automated regulation and forked crane loading and unloading were installed soon afterwards. Next, a reverse rolling mill (i.e. a series of rolls operating at different speeds for grinding and crushing) with a remote control system were installed. However, as became clear later, this rolling mill was not the best choice. It was too light to roll the high-alloyed steel, and especially for anti-friction bearing steel. Moreover, the electrical installation proved to be obsolete when the computer control came to be installed ten years later.

The above circumstances directed the adaptation of technology. It was essential to develop the appropriate technological procedures from the stage of reheating ingots in the soaking pit to that of processing it at the intermediate track. The only solution was to change the format of ingots to 1.3 and 2 tons according to the quality of steel.

The manning of the blooming mill was also inappropriate. Fewer operators were employed than the Czechoslovak constructors of the equipment had anticipated, because space in the workplace was limited. The operators received additional training courses for operating the blooming mill, as well as practising the rolling operations at similar mills in Czechoslovakia and Yugoslavia for three weeks. They later became very experienced and trained operators for other steel mills in the country.

The production at the blooming mill reached a high rate very quickly. Furthermore, the cobble (a form of wastage) decreased in a short time and remained within the boundaries of European standards. Only 0.9 per cent of cobble is normally present throughout the whole rolling mill, whereas in other steel mills it is usually between 1.1 and 1.15 per cent. (An exception to this rate was in 1976, when the cobble reached 1.8 per cent of a type of steel processed in the mill.) In the smelting shop, the SM furnace was removed in the meantime and two electric-arc furnaces, each of 40 tons' capacity, were installed. This investment, together with those in the rolling mill, allowed the production of the best qualities of steel. An additional soaking pit was built at the beginning of the rolling area.

Because of the high production rate (about 180 per cent of that planned), the working area in the rolling mill became constrained and caused great difficulties as to internal transport as well as processing itself. The intermediate and light-section rolling mills were already working at full capacity. Further, a much larger storage area was needed for such a broad assortment of products.

Bottlenecks persist in the cleaning shop, at the controlled cooling process, at the mills and at the end of the process flow. For these reasons the investment is being continued. A new continuous flow cooling process is being designed and a separate flow for grinding inspection with the most modern equipment is under preparation. A new light-section mill will be built as a joint investment by the enterprise and by the customers.

3.4 The present flow of production

The furnaces in the smelting shop are charged with scrap iron and cast alloyed steel scrap. These scraps are smelted in electric-arc furnaces to various ferro-alloys, and cast as ingots. From the smelting shop ingots are transported to the rolling mill and stored before further processing. In the soaking pits, which lie at the beginning of the rolling process flow, ingots are reheated to about 1300°C, the temperature suitable for rolling. Heating is controlled by two workers who take care that the ingots are at the required temperature when they are needed for further processing. When the temperature is suitable for the next stage, the incot-drawing crane transports them to the blooming track.

On the blooming track the ingots are shaped into billets. Two operators control the rolling on the roughing stand, from a control tower. One controls the rolling itself, the other the turning of the ingot casting. When the ingot is the appropriate size, it is transported to the crop shears where both ends are cropped and the ingot is finally transformed into billets. After the billets have been cooled, they are carefully checked (the checking determines the cleaning operation at the grinding machines), weighed and stored in a buffer stock. (Alternatively, some of them are packed and despatched.)

The next step in the process flow is the reheating of billets, after which they are transported to the rolling mills by roller table. The intermediate rolling mill has four rolling stands. The billet is rolled successively through them, each one reducing the thickness until the desired size is reached. The rolling process is controlled from the control tower by two operators, who monitor the speed and gauge of material running through the rollers, while four general operators work on the stands themselves. The correct temperature is achieved by control of the timing and speed of rolling, and by the use of water sprays to cool the steel.

After rolling, the bar passes to the crop shears where both ends are cropped and the bar cut into lengths, before being cooled. The cooling process is strictly controlled to achieve the appropriate quality of bars. Both cropping and cooling are remote controlled, with one operator for each operation.

The light-section rolling mill is similar to the intermediate one, but there are ten rolling stands. The main difference is in gauge of steel bars produced, and number and continuity of operations. The bars are cut to commercial dimensions or wound to form a uniform coil. The cut bars follow the same kind of process as that in the intermediate rolling mill described above. After cooling, the coils are temporarily stored in a buffer stock to await further processing. As with the intermediate rolling mill, two operators control the rolling operation from the control tower.

3.5 The introduction of computer control

The introduction of computer-controlled processing started in the Ravne steel works at the beginning of 1976. The experimental use of the first program was initiated in the smelting shop in May 1977.

From the smelting shop, the implementation of computer control was continued in the rolling mill. The first stage was to inform the supervisors about the characteristics of ingots coming from the smelting shop to be rolled, so that the rolling process could be prepared in advance. The quality and quantity of ingots is communicated some hours before the rolling process, and in this way a higher quantity and quality of rolling is ensured, and lower energy and material costs achieved. However, this application of the computer was merely a completion of the communication loop. Before it was applied, information about the rolling plans were usually communicated to the smelting shop.

The control of buffer stocks for billets and stocks of final products of the rolling mill represents the next stage of computerisation in the plant. Stocks of billets are held at the end of the blooming track, at the reheating furnaces, in the grinding area and in the dressing shop; and there are also stocks of final products. The computer provides data about the number of billets in each stock, according to series. This may consist of a list of all billets with regard to one order or a list of billets of one type of steel. Further, more detailed data are also available on demand. Thus, planning and other operations in the rolling mill and its sections are facilitated.

The cutting up of billets is an important operation in the rolling mill. During this process, bloom and billet scraps can account for a large proportion of production costs. Therefore, the implementation of computer control in these stages of the production process can result in considerable financial savings. Because of the difficulties arising from the relatively old electrical equipment at the roughing mill and at both rolling mills, this will be the next stage of computer implementation.

3.6 Task structure and organisational features

The workforce allocation at the new steel rolling mill was not the same as in the old one. The rolling process changed in that the near batch production before the change was replaced by continuous rolling, thus modifying the structure of tasks and organisation of work. Changes also occurred in the composition of the workforce. We shall briefly examine both types of organisation below.

3.6.1 The old mill

Within the old mill, the tasks were more or less well balanced. Workers performed more skilled, but physically more strenuous tasks.

The jobs in each shift consisted of two kinds of mill operators, a furnace operator, general operators, maintenance workers and a foreman.

(1) Mill and furnace operators. The first and second mill operators operated at the stands. Their tasks included setting the rollers to the appropriate gauges, handling the cooling system, inserting the ingots (or billets) between the rollers, visually controlling the heat of the piece rolled and the time of rolling, transporting the final products to the store area, inspecting the quality of products, etc. They also assisted in cleaning and maintenance work. The jobs of first and second mill operator were not fixed. Operators rotated freely from one mill to another upon mutual agreement. Everyone was competent to perform all rolling jobs.

The furnace operators loaded the pusher-type (or batch) furnace, inserted ingots into the furnace, controlled the heat according to the colour of the heated piece, controlled the level of reheating the ingot, drew it out of the furnace and put it on to a cart.

Operators were skilled, training being carried out over several years. Moreover, some special physical and mental qualities were required, which could only be developed over an extended period.

(2) General operators. These unskilled jobs were the first step in the career process for every worker. Cutting, transporting, cleaning, general lubrication, etc., were tasks within this category. The operator at the shears performed mainly cutting operations, and was responsible for the appropriate length of billets, besides helping with general machine cleaning at the time of maintenance. The tasks of the transporting operator included manual loading and unloading of ingots, transporting them to the furnace, assisting the furnace operator, cleaning ingots and general lubrication in the working area. There were also some other unskilled jobs in the old mill, such as labourers and apprentices, who helped with various jobs in the rolling process.

(3) Maintenance workers. These skilled jobs involved repair of all the machines and electrical equipment in the mill either manually or with the aid of simple tools driven by electrical power. Simple instruments were used for detection and checking of mechanical failures, and slightly more complicated equipment for electrical faults.

(4) Foremen. There were three foremen in the old mill, one for each shift. Their main role was supervision of workers, co-ordination of the technological process, inspection of produced goods, and co-ordination of maintenance and lubrication of equipment. Their daily tasks consisted of planning the product mix for each shift, assigning individual tasks, and supplying the workers with information needed for executing their duties and with the required resources. They also assisted the workers with the daily inspection of the quantity and quality of finished goods, and wrote reports to the plant manager on the fulfilment of daily plans.

The foremen were former mill operators who had advanced through seniority, experience and skill. Their knowledge of producing rolled steel bars of different sizes and shapes constituted a craft. In the case of an exceptional situation that the mill operators were not able to solve, it was the foremen who found the optimum solution. The information flow within the plant was extremely simple in the old mill. The plant manager drew up the plan for weekly production and set the technical standards for different products according to enterprise specifications and customer orders. Plans were drawn up individually for each shift. This data was then communicated to the foremen in the form of written schedules accompanied by oral explanations given at weekly meetings with the plant manager. At these meetings the plans were discussed, and corrected if necessary. Thus communication between workers and foremen was mainly oral.

The system of feedback information was the reverse of the information process described above. In cases of stoppages due to quality or supply of raw materials, equipment damage, etc., there were two information loops. In the case of minor interruptions, workers were to inform the foreman, who would find the appropriate solution. With serious stoppages, the plant manager was responsible for communicating with the maintenance or other departments of the enterprise.

The rolling mill was very dependent upon other parts of the working organisation at the time when the technological change started. Planning, staffing, directing, controlling and other managerial functions were centralised. All functions of goods exchange, policy-making, programming, budgeting, etc., were executed at the level of the working organisation.

After installation of the new equipment, there were other changes, not only because of the new division of labour but also as a consequence of the increase in productivity, change in quality of steel produced and in assortment of goods, increase in the number of employees and greater competitivity of the plant on the market, as well as external factors.

3.6.2 The new mill

The division of labour in the new mill is much more specialised than in the old one. This aspect of work organisation is built into the machines, and any conscious effort by organisational engineers is fruitless. In a steel mill, at this level of mechanisation/automation, the former functions of operators are split up into many tasks executed by different workers. The physical distance from the start to the finish of the rolling process also has an effect on social interactions on the mill. Moreover, the information process changed considerably because of the increase and diversification of production, division of labour and dispersion of workers in the mill area. Change has also been influenced by acceptance of (relatively traditional) organisation theories. The jobs in the new mill are outlined below.

(1) Crane operators. These workers are very specialised, each of them having his own area of materials handling. In spite of differences among the cranes, the functions of crane operators are almost all the same. They sit in cabins 10-15 m above the other workers, communicating

with them by visual signals, calling their attention by buzzers or, in exceptional situations, communicating by telephone. They remain for hours in this isolated position. Crane operators are trained, semi-skilled workers. Their training consists of attending a theoretical course and practical training on the job, supervised by an instructor.

(2) Rolling-mill operators. These are also specialised. There are four kinds of operators: blooming-mill operators, intermediate mill operators, light-section mill operators and operator assistants. The first three types sit in control towers, but the operator assistants stand at the work stands and react in cases where stoppages occur.

Mill operators monitor the production process from the control tower. They control the speed and gauges of rollers, cooling system, and other parameters, according to written information and feedback data gained from the instruments on the control panel. They react to visual information from the rolling area too.

(3) Furnace operators. Their task is to control the heating of ingots and the reheating of billets. The control of the process is made on the basis of information from various instruments and on the basis of the work schedule. The process of heating and reheating is controlled by regulating the fuel supply and the heating period.

Both rolling operators and furnace operators are skilled workers, previously trained on courses and on the job. However, their skills are narrower than those of the operators in the old mill, and less experience is needed on these jobs than was the case some years ago.

(4) Other jobs. Other jobs on the shopfloor in the rolling mill include cutting operations at the crop shears, cleaning at the grinding machines, checking of billets, dressing operations and other unskilled tasks.

(5) Maintenance workers. The maintenance of the new rolling mill is much more complicated than it was in the old one. The few maintenace workers within the rolling mill unit itself are responsible for maintaining the process in the sense of correcting smaller mechanical and electrical defects. The majority of maintenance workers for larger repairs are in a separate department of the working organisation.

(6) Foremen and supervisors. There is one supervisor for each shift in the rolling mill. In addition, foremen supervise parts of the rolling process, i.e. blooming mill, intermediate mill, light-section mill, dressing department, cleaning department, etc. Rolling-mill operators usually become foremen after many years of service and with additional training.

The control of the production process is executed through work schedules drawn up in advance by technical staff aided by computer. These are produced for each month, week and day, and every worker receives a daily schedule containing the assortment of daily production, production standards, etc. Additional information is communicated by supervisors and foremen. Because of the continuous process, the vulnerability of the plant is great. Each work station effectively depends on previous stages, and the knowledge and experience of each foreman and operator are essential, in addition to scheduling procedures. The organisational structure of the new mill remained essentially the same as in the old mill. Besides the previous hierarchical structure of functions, the supervisors and the managers' technical staff exist. There are also more fixed task assignments than before. In general, the system of work organisation within the rolling mill is more formalised and centralised than elsewhere in the plant. Some functions of the organisation are more tied to other departments of the working organisation (e.g. maintenance), whereas others are more autonomous.

3.7 Working conditions

The concept of working conditions as treated in the literature includes various factors. Here we deal with physical aspects of the workplace environment, the nature of work, level of physical exertion, social contacts between workers and, lastly, overall job satisfaction.

3.7.1 Physical aspects of the workplace environment

The physical aspects of the working environment depend not only upon the technological process, but also on other factors such as climatic conditions, quality and design of buildings, and working schedules. According to these intervening variables the physical aspects of the working environment have a greater or lesser effect on workers. Organisational factors, as for instance rest pauses and length of work, also play a role here. Thus the true influence of physical factors represents some kind of mix including technological, organisational and other aspects of working life.

As is known from the literature, at high levels of effective temperature the performance of workers usually drops. It is also clear that temperature is significantly related to accident rates in certain industries.

Lighting is usually regarded as important in improving the performance of workers in some industries, such as assembling, textiles, graphics, etc. Although it is often believed that bright lighting is important only for fine work, some experiments have shown that production in steel mills was raised by 10 per cent, when lighting was increased from 3.0 to 11.5 foot-candles.[10]

Exposure to low rates of noise (up to 80 dB) over a long period influences workers' vegetative nervous system and are a source of neuro-vegetative problems if the frequencies are high. When the intensity of noise continues for several hours a day, the results could be traumatic, even including injuries to the inner and middle ear. At a lower rate of noise, workers may react by feeling excitement, fatigue, boredom, etc.

In the present study, the physical aspects of the working environment were measured by engineers according to national standards. Effective temperature, lighting, noise and dust in the rolling mill are shown in table 93. It is obvious that dust and noise have the greatest influence on workers in this unit. The highest rate of noise is found at the machines in the dressing shop, followed by many rates which exceed the standards for the job, which are fixed at 90 dB for almost all those quoted in table 93 (the two exceptions are blooming-mill operators and assistants to dressing-shop operators). Only on three of the quoted jobs is the noise below acceptable standards (i.e. crane operators, shears operator and operator at the continuous heating furnace).

Table 93: Steel works: physical aspects of the working environment in the rolling mill

Job	Effective temperature (°C)	Lighting (lux)	Noise (dB)	Dust (particles/cm^3)
Soaking pit: crane operator	30.2	-	84	-
Blooming-mill operator	21.6	-	82	9 039
Shears operator	21.9	100	81	1 281
Transporter of billets	19.3	120	91	1 305
Grinder operator	16.4	-	92	2 339
Reheating billets: furnace operator	17.0	128	96	2 857
Reheating billets: assistant	27.0	22	95	2 537
Rolling-mill operator	26.1	20	96	16 994
Continuous heating: furnace operator	12.7	70	84	1 379
Dressing-shop operator	11.6	110	106	-
Dressing-shop assistant	14.7	110	85	-
Maintenance at rolling	18.5	150	95	4 640

- = data not available.

Dust was not measured in all the work areas in the rolling mill. Among those measured, however, it exceeds the standards on six of the jobs. The highest rate is found at the work areas of rolling-mill operators at the intermediate and light-section rolling mills. Blooming-mill operators also have a very high exposure to dust. Dust levels below national standards are found on the job of crane operator, shears operator in the blooming shop, transporter of billets to the cleaning shop, and in the dressing shop.

Lighting in the rolling mill corresponds more closely to acceptable standards than any other factor, being below standard only on two jobs, those of assistant to the billet-reheating operator and of rolling-mill operator.

Effective temperature is generally not high. However, it exceeds the acceptable rate in three jobs: crane operator, rolling-mill operators and assistant to billet-reheating operator.

Of all aspects measured, rolling-mill operators' jobs include the highest rate of dust, the lowest level of lighting, almost the highest noise (relating to standards),[11] and the third highest temperature. Taking into consideration the fact that there are ten rolling-mill operators on a shift, the proportion of workers exposed to poor working conditions is relatively high. Working conditions of the assistant to the billet-reheating operator are also very poor, all measurements exceeding the national standard.

The ratings in table 93 on the exposure of workers to adverse physical working conditions show that approximately 40 per cent suffer frequent to continuous exposure to poor lighting, almost 70 per cent to high temperature and about 35 per cent to continuous noise.

Comparing the working conditions in the present mill, as discussed above, with those before the change took place, we may conclude that the improvement was significant. In spite of the lack of objective data for comparison, such a conclusion is acceptable on the basis of interviews with those workers who worked in the mills before the change started. According to them, the new process and its environment (new buildings) are much more comfortable to work in. There is less noise and dust, the lighting is much better and the temperature is more acceptable, in spite of the higher rate of production.

Regarding other working conditions in the rolling mill which were rated but not measured, the highest was the exposure to draughts. These were frequent or continuous in about 70 per cent of work areas. Less than one-third of work areas were exposed to dirt, wetness and humidity.

Only 12 per cent of workers have a cramped work area, while some work in confined spaces only occasionally. This is especially true for workers operating the grinding machines and the rolling-mill operators in the control towers.

One of the most frequent characteristics concerning working conditions in the rolling mill is the high dependency of work processes on those of preceding phases, due to the continuous rolling process. This applies to 78 per cent of workers in the mill, among them being all operators and their assistants, transporters, etc. The exceptions are maintenance workers, supervisors and storekeepers.

The high exposure time of workers in the rolling mill is due to the length of time that the worker is tied to his work area. By comparing the relative figures, we find that this is a frequent feature of working conditions in the mill. About 81 per cent of workers in the mill are obliged to remain at their work area for the major part of the working day.

3.7.2 Nature of work

Besides environmental factors, the nature of the work in the rolling mill is tedious and all jobs are repetitive in nature. About 33 per cent of jobs have a work cycle shorter than one minute and an additional 49 per cent have a work cycle of less than five minutes; only the supervisors and maintenance workers have a longer work cylce. The work pace is predetermined by the process of rolling, and this affects about 68 per cent of jobs, especially rolling-mill operators in all divisions, shears operators, continuous furnace operators and some workers at the dressing shop.

3.7.3 Physical exertion

The level of physical exertion in the mill is low. Such physical operations as lifting, carrying, pulling and pushing heavy objects are rare, and exist only in a small number of jobs. Workers perform their work continuously sitting (or occasionally sitting) at 60 per cent of work areas, and continously standing at 33 per cent of work areas (mainly mill operators at the rolling stands). Many workers (43 per cent) are frequently or continually walking during their work. Only maintenance jobs demand body positions which may be uncomfortable or awkward for a worker.

3.7.4 Social contacts

Social contacts between workers are normally made possible by the technical work environment. Only in 20 per cent of all cases are other persons available only within signalling distance or via technical media of communication. Personal contacts with co-workers and supervisors required by the job are frequent or continual.

3.7.5 Job satisfaction

It is surprising how well workers adapted themselves to the new situation. In 1975, when the mill started to produce at full capacity, workers were interviewed about their jobs. Only 11 per cent declared that they were more satisfied before the change, while 21 per cent considered that work in the old mills was worse than in the new one. Very few of them (7 per cent) stated that there was a higher workload in the new mill but 26 per cent said that the work in the old mill had been physically more strenuous, and only 3 per cent considered that it had been mentally more strenuous. A negligible proportion of workers (7 per cent) felt that they had greater opportunities for promotion, while a similar percentage (5 per cent) found less chance to improve occupational skills within the new situation. However, approximately the same proportion (6 per cent) said that there had been a greater need to learn new skills, in spite of the fact that the new equipment had been installed some years before.

Many workers (20 per cent) were convinced that the new jobs yielded more independence, but no one felt any change in responsibility. Very few (3-7 per cent) experienced a decrease in variety and interest of work, and in the chance to develop new abilities. However, 8 per cent commented on the improved chances to use their knowledge, and the same percentage saw more possibilities of working out new and better ways of doing their jobs in the new situation. Almost nobody noticed any difference in opportunities to talk to co-workers, but 8 per cent of workers saw greater possibilities of helping each other.

3.8 Accidents and sickness absence

The rate of accidents in the rolling mill is always higher than in the enterprise as a whole. This is quite understandable if we recall the production rate and other circumstances of work in the mill. The other departments of the enterprise have a much more human job design, better working conditions and less risk in the work areas, the exceptions being the forging shop and the smelting shop. In addition, the structure of jobs in various departments influences the accident rate. Table 94 gives accident and sickness rates for the enterprise as a whole and for the rolling mill.

Table 94: Steel works: rate of accidents and sickness absence for the enterprise as a whole and for the rolling mill, 1975-80 (percentages)

Year	Enterprise			Rolling mill		
	Accident	Sickness < 30 days	Sickness ≥ 30 days	Accident	Sickness < 30 days	Sickness ≥ 30 days
1975	0.5	3.1	2.6	1.4	4.1	2.5
1976	0.4	3.2	2.0	1.0	4.1	2.9
1977	0.5	3.0	2.1	1.4	4.1	3.2
1978	0.5	3.9	2.2	0.8	4.5	2.5
1979	0.8	4.6	2.6	1.0	6.9	2.9
1980	0.6	4.0	2.7	1.4	4.2	3.8

We may see from table 94 that the accident rate in the rolling mill is always higher than that of the enterprise as a whole, except in 1979, when there was a significant trend in the other direction. The frequency of absence due to sickness is also significantly higher in the rolling mill than the enterprise as a whole, both for absences of under 30 days and for longer periods. However, the absence rate under 30 days shows a more significant difference.

As mentioned above, accidents are affected by working conditions. In our case study, there is no direct evidence of this relationship, but our data leads us to suggest certain correlations.

The most important factor relating to working conditions which affects accidents is temperature. Osborne and Vernon found that the rate of accidents is considerably higher when the temperature is below 18°C or above 24°C, and that the amount of time lost is greater for accidents occurring at such extremes of tempeature.[12] In our case it is obvious that many workers in the smelting shop and especially in the rolling mill are carrying out their jobs at the upper limit of, or above the temperature quoted. However, if we compare the temperature in the smelting shop with that in the rolling mill, out of accident and sickness rates only the data on short-term sickness rates confirm these findings.

It has also been shown that many accidents are due to poor lighting, and that intensity, diffusion and quality of light are factors correlated to accident rate. Our data, however, do not confirm these relationships.

Length of the work period, fatigue, and shift work may also be correlated to accidents. Workers in the rolling mill work a 42-hour week in three shifts: 6.00-14.00, 14.00-22.00 and 22.00-6.00, each with a half-hour meal break. The rolling mill operators have a rest of one hour every third hour, with two operators on duty and one off duty in the control tower. This kind of rotation is necessary because of the high stress which the operators experience during their work. This system of rest periods during working hours could be an important factor influencing the accident rate, as it enables workers to recover from fatigue and from other negative influences of working conditions.

3.9 Wages and fringe benefits[13]

The wage system is common to all basic organisations of associated labour in the steel works. In 1979 all workers concluded a self-management agreement upon the basis of dividing the funds for their personal income (wages) and common consumption. This agreement was adopted by referenda, and was signed by the president of the workers' council of the work organisation and by a delegate of the trade union organisation.

The share of enterprise net income allocated to wages is divided into individual wages by three main criteria: present work, previous work and innovative work. Each of these is determined separately, as follows:

(1) The current individual contribution of workers to the joint success of the basic organisation of associated labour is evaluated according to three aspects: the nature of the job performed, the effectiveness of an individual worker, and substitute payments (there are various substitutes for absence from work because of sickness, holidays, education and training, etc.). The nature of the job performed is evaluated by a points system specifically designed for the enterprise. The three groups of factors included are:

- complexity of job (skill, effort, responsibility);
- working conditions;
- working time (shift or weekend work, etc.).

The effectiveness of labour is measured individually for every worker in terms of quantity and quality of work produced, and economy.

(2) The previous work of every worker is evaluated on the basis of his or her labour at the time of a particular investment. It is the workers' right to share in the benefits of increased productivity achieved by accumulating the results of their labour over many years. Therefore the length of service in a basic organisation is taken into consideration.

(3) Innovative work is calaculated on the basis of increasing the income of the basic organisation due to an innovation or rationalisation made by a worker. The type of innovation, the worker's job and other factors are considered in evaluating suggestions.

After these aspects of labour have been evaluated for every worker, the sum allotted to wages is divided by the sum of all the points gained by workers in the basic organisation. In this way the value of a point is calculated, and each worker's wage is composed of the number of points as evaluated by the three criteria described above.

This sytem of wage determination assures a fair remuneration and workers are usually satisfied in this respect. It also motivates workers towards higher production, better quality and economy in their work, as well as encouraging many innovative suggestions. Eight innovative suggestions were approved in the rolling mill and one in the smelting shop in 1979 (out of 67 for the enterprise as a whole), increasing the income of the enterprise by about 25 million Dinars (Din).

The share of net income allotted to common consumption of workers in a basic organisation of associated labour is further divided into the following parts:

- investment in housing; workers can obtain a long-term loan at low interest for individual houses or apartments; the basic organisation also buys apartments and rents them to workers;
- investment in holiday and recreational centres for workers;
- subsidies for workers and their families for holidays;
- subsidies for works canteens;
- solidarity contributions;
- rewards and other grants.

3.10 Counteracting the negative effects of technological change on self-management

As described in section 3.6 above, the division of labour and organisational features in the rolling mill underwent considerable changes which had an impact on the roles of workers and on interactions between them.

The physical separation of workers, the dispersion of their roles and their increased confinement to their work areas created some difficulties for workers in relation to their self-managing function. The previous system of plant management did not demand frequent higher levels of interaction among workers in executing their right to self-manage. There was no great need to gather in order to make decisions because their workers' councils delegates fulfilled this purpose. The assembly was called only very rarely.

After the adoption of the new Constitution, the new system of self-management resulted in higher demands on the workers' assembly. The right to self-manage the basic organisation of associated labour had to be executed mainly through this body. However, since the new organisation, together with technological demands, did not allow for the adequate functioning of communications, changes were necessary to improve workers' participation in self-management. Workers had to be informed on matters to be discussed in the workers' assembly, and had to have enough time and opportunities to discuss these topics in advance. Many subjects needed to be discussed within smaller groups, especially those which concerned only one or two parts of the unit or a few workers within the production process.

Similar experiences in other basic organisations of associated labour and working organisations led to the idea of organising self-managing working groups. This idea, initiated in another steel works some years before, had proved effective in solving many of the problems in self-management which had resulted from size of the workforce, largeness of the plant, difficulties in communication and technological hindrances, and had subsequently spread to almost all organisations with a large number of employees.

Twelve self-managing working groups were formed in the rolling-mill unit according to three criteria. First, since about 67 per cent of workers were working in four shifts, the system of self-managing groups was adapted to shift work. The second criterion was a technological one; groups corresponded to the flow of production. The third criterion was the possibility of calculating production costs for the part of the production process in which the group was organised.

Of these 12 groups, four were set up in the blooming mill and four in the intermediate and light-section rolling mill, one for each shift. Two groups were formed in the dressing shop, one in the maintenance shop and one for information-processing workers.

Self-managing working groups discuss proposals for self-management acts and for development programmes, balance sheets, suggestions for division of income, achievement of plans, and problems of work and self-management. Meetings are held regularly each month, or on an ad hoc basis. An ad hoc meeting may be held on the suggestion of at least five members of each group, the president of the assembly, workers' council, trade union organisation or the manager of the basic organisation of associated labour.

The meeting is called by the leader of the group, who is elected by workers in the groups for two years. The leader's duties are -

- to call the meeting at a time when the production process is disturbed as little as possible;
- to manage the meeting;
- to inform the members of the self-managing working group of work and self-management problems and to obtain responses to their questions;
- to inform workers about achievement of the plans of their unit, of their basic organisation and of the working organisation;
- to collect workers' suggestions and opinions on various topics relating to their work and self-management, to communicate them to the self-management bodies, commissions or manager, and to inform the workers of acceptable solutions.

The workers of a self-managing working group have the right to recall the leader of a group who is not executing his or her duties.

A self-managing working group is also involved in decision-making. It decides on first drafts of acts or suggestions before they are proposed to the workers' assembly. Decisions are made by majority. In cases where the various groups within the basic organisation do not agree on a topic, a special delegation of three members is nominated to negotiate with similar delegations of other self-managing working groups in order to find an acceptable solution.

Every self-managing working group proposes at least one delegate to the workers' council of the basic organisation.

This kind of organisational change in self-management functions very well. Workers are more active in expressing their rights and in making suggestions about their work and other aspects of their life within the basic organisation. The individual interests of workers, as well as of various groups, may be considered more effectively than before.

4. Discussion

4.1 The causes of change

Two groups of factors influenced the introduction of new equipment in the case studied. The first of these is internal and the second external in character. The internal group will be called enterprise-level factors because they are more or less characteristic of the plant where the change took place, and the external will be termed environmental factors, as characteristic of the more general organisational environment. The components of these two groups are as follows:

(a) enterprise-level factors which can influence the implementation of new equipment or initiate technological change:

  (i) economic efficiency (energy costs, supply and costs of raw materials, competivity of the firm on the market, capacity and other technological characteristics);

  (ii) labour force (number of employees, employment structure, knowledge, employment, skill, psycho-physiological and social characteristics such as health, abilities, motivation, values, etc.);

  (iii) working conditions (physical work environment, work performance characteristics, work content, social aspects);

  (iv) production relations (characteristics related to the processes of communication, decision-making and control);

(b) environmental factors:

  (i) productive forces (potential capacity of the means of production, natural resources, knowledge, culture, etc.);

  (ii) production relations (socio-economic relations within the sphere of production derived from the ownership of the means of production).

Not all, but some of these enterprise-level and environmental factors influenced the introduction of new equipment in the case studied. They not only provided the initial impluse but also modified its course and intensity.

The decrease in the economic efficiency of the Ravne steel works had been caused by factors described in section 3.1. To summarise, the fact that the layers of iron ore in the neighbourhood of the plant had been exploited necessitated supply of iron from more distant parts of the country. In addition, the obsolete technology and equipment permitted neither a commercial rate of steel production nor high-quality products.

Other enterprise-level factors also pointed to technological change. The most influential were the following:

- closeness to sources of large amounts of electrical energy and scarcity of other kinds of energy, which suggested a new smelting technology and permitted production of high-quality steel;

- enlargement of production made possible by an available labour force from villages and towns near the plant and from other parts of the country;

- availability of specialists in producing steel from universities and technical schools;

- new values relating to work, which opposed the working conditions in the old mill as dangerous to workers, and rendered them unacceptable;

- strong pressure from workers towards technical innovations and economic progress;

- the Directions or the Establishment and Work of Workers' Councils, issued in 1949, which was reflected in enthusiasm on the part of the workers, who felt that the plant was under their ownership and started to behave accordingly.

On the other side, environmental factors were appropriate in the above-described situation within the organisation to stimulate technological change. We review some of them below.

In the fifties there was a great scarcity of steel in Yugoslavia. In 1979, with only 0.5 per cent of world steel production, the country was still on the lower end of the scale. At the start of the technological change in the Ravne steel works, steel was an extremely important product since the country was undergoing reconstruction after the Second World War. An enormous demand for steel (especially high-quality steel, which was mainly imported) directed many large investments in steel plants throughout the country. Owing to the international situation, Yugoslavia had to produce its own high-quality steel as far as possible in order to reduce its imports. As table 95 shows, the production of steel grew very fast over that period.

Table 95: Yugoslavia: growth of steel production, 1952-79

Type of steel	1952	1959	1969	1974	1977	1978	1979
Steel ('000 tons)	442	1 299	2 220	2 836	3 184	3 451	3 505
Index	100	294	502	642	720	781	793
Rolled steel ('000 tons)	293	861	1 570	2 235	3 329	4 142	4 090
Index	100	294	536	763	1 136	1 414	1 396

Source: Statistical Annuals of the Socialist Federal Republic of Yugoslavia, 1972-80.

Despite the rapid growth of steel production in Yugoslavia steel consumption overtook production during recent years, with the result that the import of crude steel and iron increased by about 35 per cent in five years. The development of other industries, especially machine manufacturing, as well as consumption in other areas, increased the shortfall in steel. Investments in this type of industry represented a large part of the financial resources of each enterprise, as well as an important percentage of national income. Therefore the investment policies at enterprise level, as at national level, have been oriented towards developing production in those kinds of steel which represent the largest part of steel import costs.

The division of steel production among various producers in Yugoslavia had been agreed upon many years before. This specialisation lost its importance within the country when greater decentralisation in the economy and development of republics was introduced. With regard to the costs of imports, it was still important to maintain an overall policy along with the autonomy of each republic or province. Some specialisation therefore took place within republics, but with co-ordination on the federal level.

The new approach did not lead so much to a higher level of steel production as to restructuring according to type of products and quality of steel. Some of the steel mills oriented their production towards standard hot rolled bar steel, some towards special sections of hot rolled bar steel and still others towards rolled wire, for instance. Those mills with the highest level of experience specialised in high-alloyed special steels, and those with less experience in plain carbon and low-alloyed steels. However, the type of specialisation was chosen not only on the grounds of experience and tradition. The existing equipment, the geographical position of a mill which determined the proximity of markets, the costs of production, etc., were also factors in the decision to restructure steel production.

In Slovenia, where this case study took place, steel production represented 60 per cent of the Yugoslav national total before the beginning of the Second World War. In 1946, the Slovene steel manufacturers produced 142,000 tons of crude steel (about 70 per cent of Yugoslav steel production). After that year, new plants were built and up-to-date equipment and processing technology were applied. According to the plans mentioned above, the orientation of production towards high-quality and fine sorts of structural, tool and special steels, and final steel products was chosen. Energy, resources and technological and labour force conditions were the deciding factors in selecting this programme.

Rapid industrialisation and deagrarianisation of the country gave an impetus to many farmers to leave their unproductive farms and seek employment in various kinds of industries. At the same time, the better living conditions and greater possibilities for employment, as well as other factors, stimulated many skilled and unskilled workers from other republics and provinces to move to the north of Yugoslavia to search for better jobs (see section 2.1).

4.2 The consequences

From the above description of the long-term process of introduction of new equipment in the rolling mill, we are able to detect the enterprise-level and environmental factors which influenced and modified the extent and the quality of the change. The scarcity of financial resources on the one hand, and lack of know-how on the part of existing personnel at the beginning of the change on the other, led to a choice of equipment which was not optimum at the time of implementation.

The workers were extremely committed to change, as was obvious from their endeavours at the time of installing the equipment. Many of them realised that the new technology would not offer them adequate job content, but they also knew that working conditions would be improved to the benefit of all of them, and that the modernised technology would imply better prospects for the plant as a whole. The consciousness of workers and professional staff was not however developed in the sense that they could understand the negative social consequences of advanced technology; these emerged only during and after the introduction of new equipment. From this case and others it becomes obvious that the implementation of advanced technology results in the restructuring of some levels of the organisation as well as the workforce within the plant.

The implementation of new equipment had an impact on the division of labour within the organisation. In our case there was no great change in this respect, except in the rolling mill. The jobs in the old rolling mill were more complex and skilled labour was imperative. The jobs in the new rolling mill are specialised, and instead of long training and experience they demand de-skilled work after a training of some weeks. Instead of interesting, varied and fulfilling monotonous, tedious work which destroyed the initiative of individual workers. However, there were also some positive consequences. The increase in the production rate, together with the new division of labour in the rolling mill after the changes, required more workers. In this way the introduction of new equipment opened up new possibilities of employment.

In general, advanced technology requires more highly skilled workers at the upper levels, and more sophisticated and better trained professional staff. At the same time it causes deskilling of workers at lower levels of organisation.

The participation of workers in introducing the new machines was partially a reflection of the above-mentioned factors but was also a result of increased participation in decision-making through self-management. This system of production relations within the organisation, as a reflection of the general system throughout the country, opened up possibilities for distribution of income produced by investment, with the aim of improving the lives of workers. It permitted workers themselves to make decisions on dividing the income among wages, common and general consumption according to their interests, or for further investment in new equipment. As a result of these decisions, schools, libraries, a centre for arts and culture, housing, etc., have been established.

The self-management in the organisation was poorly developed at the time of the change and had no positive effects on selecting the equipment best adapted to human needs. The level of technological development in the country was low, so that there were no possibilities for learning from the experiences of other plants. This state of underdevelopment was reflected in many mistakes and abuses.

The main negative effects of the new technology are reflected in the physical aspects of the working environment. Working conditions have been partly improved but still remain inadequate. Although it is true that they are to some extent a consequence of the specific conditions of steel processing, many of the negative aspects could have been avoided if other equipment and a different approach to the change had been chosen. These experiences are a lesson for workers in making further decisions on implementation of new equipment in the mill, which will be important in current investments.

The above factors and their positive and negative influences changed over time. The developing enterprise-level and environment context in which the change took place resulted in improvement of the process. However, some mistakes and deficiencies remained.

The introduction of new equipment in the rolling mill rendered possible the implementation of computer control. There are still no reflections of the social consequences, but this could be a result of the method of introduction. The participation of workers in introducing new computer-aided systems of data communication could also be one of the factors influencing such a result.

The changes in technology resulted in an important increase in the rate of production. The quantity and the quality of production, accompanied by a broader assortment of products, contributed to the increased income of the basic organisation and of its employees, as well as those funds which serve to cover the social needs of workers.

All these factors allowed the organisation to make improvements in the working life of workers through more sophisticated advanced technology (computer control). The first steps in this implementation, as described above, reflect a broader and more humanistic understanding of the technological consequences. With the application of this kind of technology, the jobs should be more acceptable to human beings in the long term. The introduction of computer control in the Ravne steel works should lead not to greater constraints on workers but to give them the choice of using a new, additional tool. It is hoped that in the future these kinds of improvements will lead to restructuring of jobs in the sense of their enrichment, and that they may also form a nucleus for restructuring the organisation as a whole.

Notes

[1] Figures from 1977, for instance, illustrate this point clearly. In the more developed regions of Slovenia and Croatia, the number of workers seeking jobs was extremely low in comparison with other republics and provinces: 1.4 per cent and 4.3 per cent of active population respectively. However, in Vojvodina this percentage was 9.2, in Serbia 9.3 and in Bosnia-Herzegovina 7.5, while the less developed regions were as follows: Montenegro 10.9 per cent, Kosovo 14.0 per cent and Macedonia 14.1 per cent.

[2] See Ilija Perič: Automatizacija, njena suština i društveno-ekonomski značaj [Automation, its essence and socio-economic importance]

(Belgrade, Savezni zavod za produktivnost rada, 1957); Dušan Čalić: Automatizacija u tehničkom i privrednom razvitku Jugoslavije [Automation in the technical and economic development of Yugoslavia] (Zagreb, Jugoslovenska akademija znanosti i umetnosti, 1962); and Rudi Supek: Automatizacija i radnička klasa [Automation and the working class] (Zagreb, Božidara Adžija, 1965).

[3] Valentin Jež: Tehnološki napredek in planiranje našega razvoja [Technological advance and planning our development] (Ljubljana, Universum, 1981).

[4] Marjan Manfredo: Materialno energetska vsebina družbene reprodukcije in mednarodna primerjava povojnega razvoja proizvajalnih sil v SFRJ [Material and energetic content of social reproduction and international comparison of development of productive forces in Yugoslavia after the Second World War] (Ljubljana, 1980; unpublished MA thesis).

[5] Miroslav Pečuljić: Prihodnost, ki se je začela [The starting future] (Ljubljana, Komunist, 1969).

[6] Josip Obradović: "Djelovanje tehničkog nivoa proizvodnje i učešća u organima upravljanja na stavove radnika prema radu" [The impact of the technical level of production and participation in management bodies on workers' attitudes to work], in Naše teme (Zagreb), 1967, No. 7.

[7] Jan Forslin et al.: Automation and industrial workers, Vol.I (Oxford, Pergamon Press, 1981). This study included 15 countries and was co-ordinated by the European Co-ordination Centre for Research and Documentation in Social Sciences, Vienna.

[8] Valentin Jež: Tehnologija, organizacija in samoupravljanje [Technology, organisation and self-management] (unpublished MA thesis; 1981).

[9] At that time, one workers' council acted as supreme self-managing body for the working organisation as a whole. See section 2.2.

[10] W.G. Darley and L.S. Ickis: "Lighting and seeing in the drafting room", in Illuminating Engineering, 1941, No. 36, pp. 1462-1487.

[11] The noise level at the rolling mill reaches 250 Hz as compared with 1,000 Hz at the grinding machines, with the result that noise at the rolling mill is less traumatic.

[12] E.G. Osborne and H.M. Vernon: Contributions to the study of accident causation, Industrial Patents Research Board, Report No. 19, (London, 1922); quoted by J.S. Gray: Psychology in industry (New York, McGraw Hill, 1952), p. 224.

[13] We use this terminology for the sake of clarity, even though it is inadequate to describe the notions expressed here.